NORTH EUROPEAN SYMPOSIUM FOR ARCHAEOLOGICAL TEXTILES X

edited by

*Eva Andersson Strand, Margarita Gleba,
Ulla Mannering, Cherine Munkholt and Maj Ringgaard*

Oxbow Books
Oxford and Philadelphia

ANCIENT TEXTILES SERIES VOL. 5

Published in the United Kingdom by
OXBOW BOOKS
10 Hythe Bridge Street, Oxford OX1 2EW

and in the United States by
OXBOW BOOKS
908 Darby Road, Havertown, PA 19083

© Oxbow Books and the individual authors 2010
Reprinted in paperback 2015

Paperback Edition: ISBN 978-1-78570-156-6

A CIP record of this book is available from the British Library

Library of Congress Cataloging-in-Publication Data

North European Symposium for Archaeological Textiles (10th : 2008 : Copenhagen, Denmark)
  North European Symposium for Archaeological Textiles X / edited by Eva B. Andersson Strand ... [et al.].
     p. cm. -- (Ancient textiles series ; 5)
  Contributions in English and German.
  Includes bibliographical references.
  ISBN 978-1-84217-370-1
  1. Textile fabrics--Analysis--Congresses. 2. Antiquities--Analysis--Congresses. 3. Textile fabrics--Europe, Northern--Congresses. I. Strand, Eva B. Andersson. II. Title.
  CC79.5.T48N67 2008
  930.1028--dc22
                                              2009039555

All rights reserved. No part of this book may be reproduced or transmitted in any form or by any means, electronic or mechanical including photocopying, recording or by any information storage and retrieval system, without permission from the publisher in writing.

For a complete list of Oxbow titles, please contact:

UNITED KINGDOM
Oxbow Books
Telephone (01865) 241249, Fax (01865) 794449
Email: oxbow@oxbowbooks.com
www.oxbowbooks.com

UNITED STATES OF AMERICA
Oxbow Books
Telephone (800) 791-9354, Fax (610) 853-9146
Email: queries@casemateacademic.com
www.casemateacademic.com/oxbow

Oxbow Books is part of the Casemate Group

Front cover image by Marianne Bloch Hansen

Contents

Dedication	vii
The participants of NESAT X	viii
Editors' Preface	xi
List of contributors	xiii

Lise Bender Jørgensen
A Brief History of the North European Symposium for Archaeological Textiles (NESAT) — xvii

1. Eva Andersson Strand — 1
   Experimental Textile Archaeology

2. Eva I. Andersson — 4
   The Perfect Picture – A Comparison between Two Preserved Tunics and 13th-century Art

3. Tereza Belanová Štolcová and Karina Grömer — 9
   Loom-Weights, Spindles and Textiles – Textile Production in Central Europe from the Bronze Age to the Iron Age

4. Sophie Bergerbrant — 21
   Differences in the Elaboration of Dress in Northern Europe during the Middle Bronze Age

5. Lena Bjerregaard, Ute Henniges and Antje Potthast — 26
   Avoiding Nasty Surprises: Decision-Making based on Analytical Data

6. Milena Bravermanová — 31
   Archaeological Textiles from Prague Castle, Czech Republic

7. Maria Cybulska, Tomasz Florczak and Jerzy Maik — 36
   Virtual Reconstruction of Archaeological Textiles

8. Camilla Luise Dahl — 41
   The Use of Terminology in Medieval Scandinavian Costume History: An Approach to Source-based Terminology Methodology

9. Anna Drążkowska — 52
   Haberdashery Elements made of Metal Thread: Conservation Problems

10. Andrea Fischer — 57
    Current Examinations of Organic Remains using Variable Pressure Scanning Electron Microscopy [VP-SEM]

11. Karin Margarita Frei — 63
    Textiles, Wool, Sheep, Soil and Strontium – Studying their Paths: a Pilot Project

12. Ruth Gilbert — 65
    Not so much Cinderella as the Sleeping Beauty: Neglected Evidence of Forgotten Skill

13. Annelies Goldmann und Eva-Maria Pfarr — 69
    Die Rekonstruktion des Vaaler Bändchens – ein archäologisches Kammgewebe aus Dithmarschen: Gemeinschaftsarbeit der Wollgruppe des Museumsdorfes Düppel, Deutschland

14. Kordula Gostenčnik — 73
    The Magdalensberg Textile Tools: a Preliminary Assessment

15. Dawid Grupa — 91
    Silk Ribbons from Post-Medieval Graves in Poland

16. Malgorzata Grupa — 95
    Silks from Kwidzyn Cathedral, Poland

17. Sunniva Wilberg Halvorsen — 97
    Norwegian Peat Bog Textiles: Tegle and Helgeland Revisited

18. Susanna Harris — 104
    Smooth and Cool, or Warm and Soft: Investigating the Properties of Cloth in Prehistory

19. Anne Hedeager Krag — 113
    Oriental Influences in the Danish Viking Age: Kaftan and Belt with Pouch

20. Viktoria Holmqvist — 117
    A Study of Two Medieval Silk Girdles: Eric of Pomerania's Belt and the Dune Belt

21. Linda Hurcombe — 129
    Nettle and Bast Fibre Textiles from Stone Tool Wear Traces? The Implications of Wear Traces on Archaeological Late Mesolithic and Neolithic Micro-Denticulate Tools

22. Katrin Kania — 140
    Construction and Sewing Technique in Secular Medieval Garments

23. Lise Ræder Knudsen — 150
    Tiny Weaving Tablets, Rectangular Weaving Tablets

24. Annika Larsson — 157
    Warriors' Clothing in the *Rigsþula*

25. Christina Margariti, Dinah Eastop, Georgianna Moraitou and Paul Wyeth — 162
    Potentials and Limitations of the Application of FTIR Microscopy to the Characterisation of Textiles excavated in Greece

26. Susan Möller-Wiering — 167
    Evidence of War and Worship: Textiles in Roman Iron Age Weapon Deposits

27. Britt Nowak-Böck — 174
    Bewahren und Erfassen – Anmerkungen zum Umgang mit mineralisierten Strukturen auf Metallen in der Denkmalpflege

28. Ruth Iren Øien — 181
    Medieval Textiles from Trondheim: An Analysis of Function

29. Judit Pásztókai-Szeőke — 187
    Curry-Comb or Toothed Weft-Beater? The Serrated Iron Tools from the Roman Province of Pannonia

30. Elvyra Pečeliūnaitė-Bazienė — 189
    Textiles from the 3rd–12th Century AD Cremation Graves found in Lithuania

31. Silja Penna-Haverinen — 195
    Patterned Tablet-Woven Band – In Search of the 11th Century Textile Professional

32. Riina Rammo — 201
    Social and Economic Aspects of Textile Consumption in Medieval Tartu, Estonia

33. Antoinette Rast-Eicher — 208
Garments for a Queen

34. Éva Richter — 211
Our Threads to the Past: Plaited Motifs as Predecessors of Woven Binding Structures

35. Virginija Rimkutė — 217
The Neolithic Mats of the Eastern Baltic Littoral

36. Maj G. Ringgaard and Annemette Bruselius Scharff — 221
The Impact of Dyes and Natural Pigmentation of Wool on the Preservation of Archaeological Textiles

37. Elisabeth Ann Stone — 225
Wear on Magdalenian Bone Tools: A New Methodology for Studying Evidence of Fiber Industries

38. Amica Sundström — 233
A Bronze Age Plaited Starting Border

39. Synnøve Thingnæs — 235
Textile Craftsmanship in the Norwegian Mirgation Period

40. Klaus Tidow — 240
Textilfunde aus Ausgrabungen in Heidelberg

41. Kristýna Urbanová and Helena Březinová — 242
Textile Remains on a Roman Bronze Vessel from Řepov (Czech Republic)

42. Ina Vanden Berghe, Beatrice Devia, Margarita Gleba and Ulla Mannering — 247
Dyes: to be or not to be. An Investigation of Early Iron Age Dyes in Danish Peat Bog Textiles

43. Marianne Vedeler — 252
Dressing the Dead: Customs of Burial Costume in Rural Norway

44. André Verhecken — 257
The Moment of Inertia: a Parameter for the Functional Classification of Worldwide Spindle-Whorls from all Periods

45. Elizabeth Wincott Heckett — 271
Elite and Military Scandinavian Dress as Portrayed in the Lewis Chess Pieces

46. Anna Zanchi — 276
Headwear, Footwear and Belts in the Íslendingasögur and Íslendingaþættir

47. Irita Žeiere — 285
The Use of Horsehair in Female Headdresses of the 12th–13th Century AD Latvia

48. Hanna Zimmerman — 288
Two Early Medieval Caps from the Dwelling Mounds Rasquert and Leens in Groningen Province, the Netherlands

49. Elena S. Zubkova, Olga V. Orfinskaya and Kirill A. Mikhailov — 291
Studies of the Textiles from the 2006 Excavation in Pskov

Colour Plates — 299

Dedication

Lise Bender Jørgensen is one of the founders of the North European Symposium for Archaeological Textiles (NESAT), which today is one of the most important conference series for archaeological textile research (see her article in this volume). Trained in prehistoric archaeology at the University of Copenhagen (Mag. art. 1976), she was awarded the Danish higher doctorate title, Dr. Phil. in 1993. After first being employed at Langelands Museum, and later as a research fellow at the University of Copenhagen and the Royal Danish Academy of Arts, School of Conservation in Denmark, Lise Bender Jørgensen moved to the University of Gothenburg's Department of Archaeology in Sweden in 1993. In 1996, she became Associate Professor and in 2002 Professor at the Norwegian University of Science & Technology, Institute of Archaeology in Trondheim, Norway. From 2002 to 2009 she was also Visiting Professor at the University College Borås, The Swedish School of Textiles, and in 2006 Visiting Professor at the Danish National Research Foundation's Centre for Textile Research, at the University of Copenhagen.

With her impressive and long list of publications, Lise Bender Jørgensen is one of the European textile researchers who have had the greatest impact on the development of archaeological textile research in recent decades. Her comprehensive publications on textiles from graves from Denmark, Sweden, Norway and Schleswig-Holstein (1986) and the follow up on the North European grave finds (1992) have provided the European textile research field with an invaluable tool that will stand as a milestone for many decades. Likewise, her work on textiles from Mons Claudianus and Abu Sha'ar, Egypt has not only shed new light on Roman textile production, but also initiated the development of the new and important methodology of the visual evaluation of textiles.

Throughout her professional career Lise Bender Jørgensen has always worked in close connection with textile craftspeople and has always been in strong favour of incorporating craft knowledge in archaeological research. Lise Bender Jørgensen has promoted and developed textile research throughout a lifetime. She is a source of inspiration and a role model to us all, always encouraging students and young scholars to continue their work within the field of textile research.

We dedicate the 10th NESAT jubilee publication to Lise Bender Jørgensen as a small token of appreciation and wish her many fruitful years of research to the benefit of all archaeological communities.

Ulla Mannering, Eva Andersson Strand and Margarita Gleba

The participants of NESAT X

1.	John Peter Wild	29.	Antje Potthast	57.	Eva Andersson Strand	85.	Britt Nowak-Böck
2.	Vibeke Ervø	30.	Annika Larsson	58.	Ave Matsin	86.	Maria Cybulska
3.	Annemette Bruselius Scharff	31.	Katherine Larson	59.	Riina Rammo	87.	Eva Richter
4.	Felicity Wild	32.	Kirsten Toftegård	60.	Saskia Thijsse	88.	Anne Elisabeth Stone
5.	Sascha Mauel	33.	Lise Ræder Knudsen	61.	Frances Pritchard	89.	Karin Margarita Frei
6.	Judit Pásztókai-Szeőke	34.	Eva Jordan-Fahrbach	62.	Anna Nørgaard	90.	Elizabeth Wincott Heckett
7.	Eva-Maria Pfarr	35.	Andrè Verhecken	63.	Carol Christiansen	91.	Andrea Fischer
8.	Annelies Goldmann	36.	Irita Zeiere	64.	Ulla Isabel Zagal-Mach	92.	Ina Vanden Berghe
9.	Kirill Mikhailov	37.	Susanna Harris	65.	Nahum Ben Yehuda	93.	Kristýna Urbanová (Poppová)
10.	Hanelle Köngäs	38.	Hanna Zimmermann	66.	Katrin Kania	94.	Synnøve Thingnæs
11.	Antoinette Rast-Eicher	39.	Gabriele Zink	67.	Camilla Louise Dahl	95.	Sophie Desrosiers
12.	Klaus Tidow	40.	Chris Verhecken-Lammens	68.	Hanna Wilkinson	96.	Lidden Boisen
13.	Sunniva Halvorsen	41.	Karina Grömer	69.	Eryk Rawicz-Lipinski	97.	Nicole Reifarth
14.	Irene Skals	42.	Kathrine Vestergaard	70.	Lise Bender Jørgensen	98.	Ulla Mannering
15.	Julia Galliker	43.	Elizabeth Völling	71.	Sylvia Crumbach	99.	Amica Sundström
16.	Silja Penna-Haverinen	44.	Tereza Belanová (Štolcová)	72.	Heini Kirjavainen	100.	Marianne Vedeler
17.	Virginija Rimkute	45.	Barbara Klessig	73.	Dawid Grupa	101.	Ingalena Hyrkäs
18.	Elvyra Pečeliunaite-Baziene	46.	Christina Margariti	74.	Dietlind Hachmeister	102.	Lena Hammarlund
19.	Sandra Comis	47.	Johanna Banck-Burgess	75.	Chris Wenzel	103.	Ann-Dorothee Schlüter
20.	Viktoria Holmqvist	48.	Martin Ciszuk	76.	Andreas Franzkowiak	104.	Linda Mårtensson (Olofsson)
21.	Astrid Geimer	49.	Ida Demant	77.	Bettina von Stockfleth	105.	Annelies Hayes Andersen
22.	Eva I. Andersson	50.	Linda Hurcombe	78.	Ruth Iren Øien	106.	Stella Steengaard
23.	Elinor Sydberg	51.	Piia Lempiäinen	79.	Milena Bravermanová	107.	Dorota Rawicz-Lipinska
24.	Malgorzata Grupa	52.	Susan Möller-Wiering	80.	Sue Harrington	108.	Joy Boutrup
25.	Margit Petersen	53.	Gudrun Böttcher	81.	Elizabeth Peacock	109.	Maj Ringgaard
26.	Anne Hedeager Krag	54.	Ruth Gilbert	82.	Kordula Gostenčnik	110.	Margarita Gleba
27.	Sophie Bergerbrant	55.	Karen Hanne Stærmose Nielsen	83.	Helena Březinová	111.	Frederik Nihlén
28.	Claudia Gross	56.	Julian Subbert	84.	Jerzy Maik		

Editors' Preface

The North European Symposium for Archaeological Textiles (NESAT) was founded in 1981 to promote the study and publication of textiles from archaeological sites in Northern Europe. It is a forum for archaeologists, historians, conservators and craftspeople with expertise and interest in the many aspects of North European textile history from prehistoric to recent times. The conferences have been held every three years in various North European cities and all proceedings have been published by the organizers. A full account of the history of NESAT is presented by Bender Jørgensen in this volume.

The *Tenth Jubilee Symposium* took place in Copenhagen on the 14th–17th May 2008, organised by the Danish National Research Foundation's Centre for Textile Research (CTR) in cooperation with the Department of Conservation at the National Museum of Denmark and Land of Legend, the Lejre Experimental Centre. The organising committee consisted of Marie-Louise Nosch (Director, CTR), Ulla Mannering (Research Programme Manager, CTR), Margarita Gleba (Research Programme Manager, CTR), Eva Andersson Strand (Research Programme Manager, CTR), Maj Ringgaard (Conservator, National Museum/CTR), Jesper Stub Johnsen (Director of the Conservation Department, National Museum), Irene Skals (Conservator, National Museum) and Marianne Rasmussen (Head of Research, Lejre Experimental Centre).

As the textile archaeology community has grown substantially since the beginning of NESAT we felt that it was important to open NESAT to a wider group of researchers. While participation in the previous symposia were based on initial attendance or invitation by the organisers, NESAT X was based on open call and the papers presented were selected by a scientific committee, consisting of the members of the NESAT steering board and members of the organising committee.

During NESAT X, 73 researchers presented 41 papers and 20 posters, with more than 130 researchers from 23 European countries and the USA attending the symposium and about 20 listeners tuning in via the live webcast, making this the largest NESAT ever.

In addition to the sponsorship of the three organising institutions, the conference was generously supported by the following: *Dronning Margrethe og Prins Henriks Fond*, *Nordisk Kulturfond*, *Knud Højgårds Fond*, Copenhagen City Council and The Danish Research Council for the Humanities. The publication of this volume was supported by a grant from *Knud Højgårds Fond*. The volume includes 49 contributions by 65 authors, organised in alphabetical order of the authors' last names, and an introductory article on the history of the NESAT by Lise Bender Jørgensen. We would like to thank the following people for their help with the preparation of this volume: Bettina von Stockfleth for her invaluable help with the editing of the articles written in German; Birgit Lyngbye Petersen for her indefatigable work with all the illustrations; Marianne Bloch Hansen for the graphic work; Vibe Maria Martens and Louise Malmer Rasmussen for their assistance with the editing.

Almost 30 years have passed since the first Symposium and fortunately, today no one can claim anymore that the field of textile research is neglected or unimportant. New methods are constantly being developed and thus challenging our understanding and interpretations of textiles, textile production and ancient society in general. Collaboration among various scholarly disciplines has become essential and makes textile research one of the most dynamic fields of archaeological research. Likewise, the economic and cultural impact of textiles and textile manufacture on past societies is becoming increasingly accepted in the wider scholarly community. It is of utmost importance to be open to new theoretical and methodological approaches and we hope that in the future NESAT will continue to provide the forum for discussion and scholarly exchange.

Editors
Eva Andersson Strand, Margarita Gleba, Ulla Mannering, Cherine Munkholt and Maj Ringgaard

List of contributors

Eva Andersson Strand
The Danish National Research Foundation's
Centre for Textile Research
University of Copenhagen
102 Njalsgade
2300 Copenhagen S
Denmark
evaandersson@hum.ku.dk

Eva I. Andersson
Institutionen för Historiska Studier
Göteborgs Universitet
Box 200
405 30 Göteborg
Sweden
eva.andersson@history.gu.se

Tereza Belanová Štolcová
Archaeological Institute
Slovak Academy of Sciences
Akademická 2
SK-94921 Nitra
Slovakia
tereza.stolcova@gmail.com

Lise Bender Jørgensen
Norwegian University of Science & Technology
Institute of Archaeology and Religion
Dragvoll
7491 Trondheim
Norway
lise.bender@vm.ntnu.no

Sophie Bergerbrant
Department of Archaeology and Classical Studies
Stockholm University
106 91 Stockholm
Sweden
sophie.bergerbrant@ark.su.se

Lena Bjerregaard
Ethnologisches Museum
Arnimallee 27
D-14195 Berlin – Dahlem
Germany
l.bjerregaard@smb.spk-berlin.de

Milena Bravermanová
Prague Castle Management
Art Collection Department
Prague Castle
CZ-11908 Prague 1
Czech Republic
Milena.Bravermanova@hrad.cz

Helena Březinová
Institute of Archaeology
Academy of the Sciences of the Czech Republic
Letenská 4
118 01 Prague 1
Czech Republic
brezinova@arup.cas.cz

Annemette Bruselius Scharff
School of Conservation
Esplanaden 34
DK-1263 Copenhagen K
Denmark
abs@kons.dk

Maria Cybulska
Institute of Architecture of Textiles
Faculty of Material Technology and Designing of Textiles
Technical University of Lodz
Żeromskiego 116
PL-90924 Lódź
Poland
cybulska@wipos.p.lodz.pl

Camilla Luise Dahl
The Medieval Centre, Department of Clothing and Textiles
Ved Hamborgskoven 2–4
Sundby L.
4800 Nyk. F.
camillaluise.dahl@yahoo.com

Anna Drążkowska
Instytut Archeologii UMK
Szosa Bydgoska 44/48
87–100 Torun
Poland
annadr9@wp.pl

Dinah Eastop
The Textile Conservation Centre,
University of Southampton
Winchester Campus
Park Avenue
Winchester
Hampshire SO23 8DL
UK
dde@soton.ac.uk

Andrea Fischer
Staatliche Akademie der Bildenden Künste Stuttgart
Am Weißenhof 1
70191 Stuttgart
Germany
a.fischer@abk-stuttgart.de

Tomasz Florczak
Institute of Architecture of Textiles
Faculty of Material Technology and Designing of Textiles
Technical University of Lodz
Żeromskiego 116
PL-90924 Lódź
Poland

Karin Margarita Frei
The Danish National Research Foundation's
Centre for Textile Research
University of Copenhagen
102 Njalsgade
2300 Copenhagen S
Denmark
kmfrei@hum.ku.dk

List of contributors

RUTH GILBERT
31 Grange Cottages, Marsden
Huddersfield HD7 6AJ
UK
plainweave@freeuk.co.uk

MARGARITA GLEBA
The Danish National Research Foundation's
Centre for Textile Research
University of Copenhagen
102 Njalsgade
2300 Copenhagen S
Denmark
mgleba@hum.ku.dk

ANNELIES GOLDMANN
Suarezstr. 27
14057 Berlin
Germany
goldmann@dueppel.de

KORDULA GOSTENČNIK
Archäologische Park Magdalensberg
Liniengasse 9/3
A-1060 Wien
Austria
kgosten@gmail.com

KARINA GRÖMER
Museum of Natural History Vienna
Department for Prehistory
Burgring 7
A-1010 Vienna
Austria
karina.groemer@nhm-wien.ac.at

DAWID GRUPA
Instytut Archeologii UMK
Szosa Bydgoska 44/48
87–100 Torun
Poland
d.m.grupa@gmail.com

MALGORZATA GRUPA
Instytut Archeologii UMK
Szosa Bydgoska 44/48
87–100 Torun
Poland
m.grupa@wp.pl

SUNNIVA WILBERG HALVORSEN
Prestegårdsaleen 6
3490 Klokkarstua
Norway
sunnivawh@yahoo.no

SUSANNA HARRIS
Institute of Archaeology
University College London
31–34 Gordon Square
London WC1H 0PY
UK
susannaharris@hotmail.com

ANNE HEDEAGER KRAG
Centre for Medieval Studies
University of Southern Denmark
Knoldene 1. sal
Campusvej 55
DK-5230 Odense M
Denmark
Hedeager@stofanet.dk

UTE HENNIGES
University of Natural Resources and Applied Life Sciences
Department of Chemistry
Muthgasse 18
A-1190 Vienna
Austria
ute.henniges@boku.ac.at

VIKTORIA HOLMQVIST
Jordhyttegatan 7
41473 Göteborg
Sweden
lanam_fecit@hotmail.com

LINDA HURCOMBE
Dept of Archaeology
Exeter University
Exeter, EX4 4QE
UK
L.M.Hurcombe@exeter.ac.uk

KATRIN KANIA
Buckenhofer Weg 54
DE-91058 Erlangen
Germany
katrin.kania@pallia.net

LISE RÆDER KNUDSEN
Conservation Centre Vejle.
Maribovej 10
7100 Vejle
Denmark
lrk@konsv.dk

ANNIKA LARSSON
Uppsala University
Department of Archaeology
St. Erikstorg 5
S-75310 Uppsala
Sweden
Annika.Larsson@arkeologi.uu.se

JERZY MAIK
Institute for Archaeology and Ethnology
Polish Academy of Sciences
Tylna 1
PL-90364 Łódź
Poland
archeo@cbmm.lodz.pl

ULLA MANNERING
The Danish National Research Foundation's
Centre for Textile Research
University of Copenhagen
102 Njalsgade
2300 Copenhagen S
Denmark
manner@hum.ku.dk

CHRISTINA MARGARITI
Directorate of Conservation of Ancient and Modern Monuments
Hellenic Ministry of Culture
81 Peireos Avenue
10553 Athens
Greece
chmargariti@culture.gr

List of contributors

Kirill A. Mikhailov
Institute for the History of Material Culture
Russian Academy of Sciences
Dvortzovaja nab., 18
191186 St. Petersburg
Russia
mikhailov_kirill@mail.ru

Georgianna Moraitou
Directorate of Conservation of Ancient and Modern Monuments
Hellenic Ministry of Culture
81 Peireos Street
10553 Athens
Greece
Georgia.moraitou@dsa.culture.gr

Susan Möller-Wiering
Archäologie und Textil
Moldeniter Weg 60
D-24837 Schleswig
Germany
smw@archaeologie-und-textil.de

Britt Nowak-Böck
Bayerisches Landesamt für Denkmalpflege
Restaurierung Archäologie
Schloss Seehof
D-96117 Memmelsdorf
Germany
Britt.Nowak-Boeck@blfd.bayern.de

Ruth Iren Øien
Håkon Herdebreisvei 4
7046 Trondheim
Norway
ruthiren@gmail.com

Olga V. Orfinskaya
The Russian Research Institute for Cultural and Natural Heritage
Kosmonavtov 2
129366 Moscow
Russia
heritage@mtu-net.ru

Judit Pásztókai-Szeőke
Juharfa utca 13. VII/20
9400 Sopron
Hungary
judit@hum.ku.dk
mamrad@freemail.hu

Elvyra Pečeliūnaitė-Bazienė
Gegliškių g.18
LT-14193 Vilnius
Lithuania
elvyrap@hotmail.com

Silja Penna-Haverinen
Turku University
School of cultural studies
Department of Archaeology
Yo-kylä 49 A 6
20540 Turku
Finland
silja.pennahaverinen@gmail.com

Eva-Maria Pfarr
Museumsdorf Düppel
Clauerstr. 11
D-14163 Berlin
Germany

Antje Potthast
University of Natural Resources and Applied Life Sciences
Department of Chemistry
Muthgasse 18
A-1190 Vienna
Austria
antje.potthast@boku.ac.at

Riina Rammo
University of Tartu
Chair of Archaeology, Lossi 3
50090 Tartu
Estonia
riina.rammo@ut.ee

Antoinette Rast-Eicher
Archeotex
Kirchweg 58
CH-8755 Ennenda
Switzerland
archeotex@hluewin.ch

Éva Richter
Harmat u. 77
1104 Budapest
Hungary
richtervica@gmail.com

Virginija Rimkutė
Department of Archaeology
Faculty of History
Vilnius University
Universiteto 7
LT-01513 Vilnius
Lithuania
v.rimkute@gmail.com

Maj G. Ringgaard
The National Museum of Denmark
I.C. Modewegsvej, Brede
DK-2800 Kgs. Lyngby
Denmark
maj.ringgaard@natmus.dk

Elisabeth Ann Stone
Department of Anthropology
MSC01-1040, Anthropology 1
University of New Mexico
Albuquerque, NM 87131
USA
elisabethastone@gmail.com

Amica Sundström
Spireabågen 69
16559 Hässelby
Sweden
amicasundstrom@spray.se

Synnøve Thingnæs
Majorstueveien 33b
N-0367 Oslo
Norway
synnovething@gmail.com

Klaus Tidow
Zur Ziegelei 18
D-24598 Boostedt
Germany
doerteklaus@online.de

Kristýna Urbanová
National Museum
Václavské náměstí 68
115 79 Praha 1
Czech Republic
urbanova-kristyna@post.cz

Ina Vanden Berghe
Royal Institute for Cultural heritage KIK/IRPA, Belgium
Jubelpark 1
B-1000 Brussels
Belgium
ina.vandenberghe@kikirpa.be

Marianne Vedeler
Department of Archaeology
Museum of Cultural History
University of Oslo
P.O. Box 6762 St Olavs plass
NO-0130 Oslo
Norway
marianne.vedeler@khm.uio.no

André Verhecken
Ed. Arsenstraat 47
B-2510 Mortsel
Belgium
andre.verhecken@village.uunet.be

Elizabeth Wincott Heckett
Department of Archaeology
University College Cork, Ballymore House
Cobh./Co. Cork
Ireland
heckette@indiso.ie

Paul Wyeth
The Textile Conservation Centre,
University of Southampton
Winchester Campus
Park Avenue
Winchester
Hampshire SO23 8DL
UK
P.Wyeth@soton.ac.uk

Anna Zanchi
via El Alamein 8
20157 Milano
Italy
orcassassina@hotmail.com

Irita Žeiere
National history museum of Latvia
Vecpilsetas iela 7
LV-1050 Riga
Latvia
irita.zeiere@apollo.lv

Hanna Zimmerman
Aldringaweg 20
NL-9892 PG Feerwerd
Netherlands
hannazim@planet.nl

Elena S. Zubkova
The State muzeum preserve of Pskov
Nekrasova st. 7
180000 Pskov
Russia
zuba.pskov@gmail.com

A Brief History of the North European Symposium for Archaeological Textiles (NESAT)

by Lise Bender Jørgensen

The first NESAT was held at the *Textilmuseum* Neumünster, Germany in 1981, organised by Klaus Tidow and myself. The real beginnings go a little further back in time, to my first meetings with Klaus some years earlier, and were founded on the loneliness and isolation that so many scholars of archaeological textiles have experienced[1]. To both of us, it was simply wonderful to be able to discuss really interesting issues – such as the significance of whether yarn was twisted z or s, or that of displacement in Iron Age twills – with another person who immediately understood, had ideas and opinions, and could contribute to the discussion. In 1979, after a long evening of such discussions accompanied by a great deal of beer (Klaus' wife and daughters had given up on us and gone to bed), we decided that we wanted to meet the other scholars of archaeological textiles, and started to make a list. The result was the *Textilsymposium* Neumünster, later to be known as NESAT 1 (photograph below).

The 23 participants of the first NESAT represented Germany, the Netherlands, Britain, the three Scandinavian countries and Poland (Table 2). Most participants were in the early or middle stages of their career. The most senior ones were Marta Hoffmann from Norway and Hans-Jürgen Hundt from Germany, both recently retired. We did not invite our very old colleagues, such as Agnes Geijer and Margrethe Hald who both were octogenarians; nor Karl Schlabow, founder of the host institution, who had turned 90 shortly before the Symposium. He only attended the opening. We were keen to establish an open, inclusive forum for the study of archaeological textiles, and expressly wanted to focus on a younger generation. Participants included archaeologists and other scholars from the humanities, textile engineers, conservators and craftspeople; this mixture seems to have been a happy one, and has remained a hallmark of the NESAT.

Participants of NESAT 1 (Photo: © Egon Vogt).

Numbers of Participating Countries and Delegates

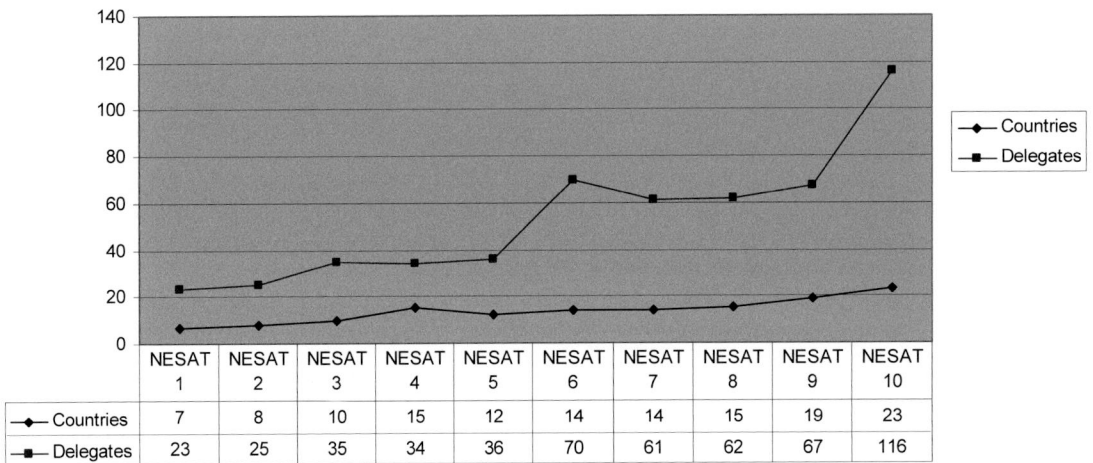

	NESAT 1	NESAT 2	NESAT 3	NESAT 4	NESAT 5	NESAT 6	NESAT 7	NESAT 8	NESAT 9	NESAT 10
Countries	7	8	10	15	12	14	14	15	19	23
Delegates	23	25	35	34	36	70	61	62	67	116

Table 1.

The Growth of NESAT (Table 1)

The second NESAT, held in Bergen, Norway in 1984, was organised by archaeologist Bente Magnus together with textile conservators Inger Raknes Pedersen and Aud Bergli, all of the Historical Museum, University of Bergen. It had 25 participants, and added France to the original 7 countries. The third, held in York, England in 1987, raised the number of participants by ten, and happily welcomed Spain to Northern Europe. NESAT 3 was organised by Penelope Walton, Textile Research in York, and John Peter Wild of the University of Manchester. At the fourth NESAT, in Copenhagen 1990, the lifting of the Iron Curtain was reflected. In addition to our faithful Polish member, Jerzy Maik, we were delighted to include participants from what was then called the USSR, the DDR, Czechoslovakia, and from what soon after became Latvia. Northern Europe was no longer lopsided. With the appearance of Switzerland, Central Europe had begun to fill in too. NESAT 4 was organised by Elisabeth Munksgaard of the National Museum of Denmark and myself, then at the Royal Danish Academy of Arts, School of Conservation in Copenhagen (Table 3). In 1993, NESAT returned to its place of origin, the *Textilmuseum* Neumünster. Organisers of this fifth NESAT were Gisela Jaacks of the *Museum für Hamburgische Geschichte* and Klaus Tidow; the number of participants was limited by the size of the auditorium in the *Textilmuseum*, but nonetheless included several fresh faces and our first delegate from Belgium. Heidemarie Farke, who three years earlier had represented the DDR, had now moved to Schleswig, and the two Germanies had merged into one. The sixth NESAT was held in Borås, Sweden in 1996, organised by Christina Rinaldo of the Swedish School of Textiles, the University College Borås, and myself, then at Göteborg University. This NESAT saw a large increase in the number of participants, as the venue in a teaching institution offered the opportunity for students and others who did not contribute actively to the symposium to attend. Some of them even braved the cool Swedish spring in their eagerness to participate and camped out in a tent. Proposals of papers greatly exceeded the time available; this was remedied by including a poster session, held at the Textile Museum in Borås.

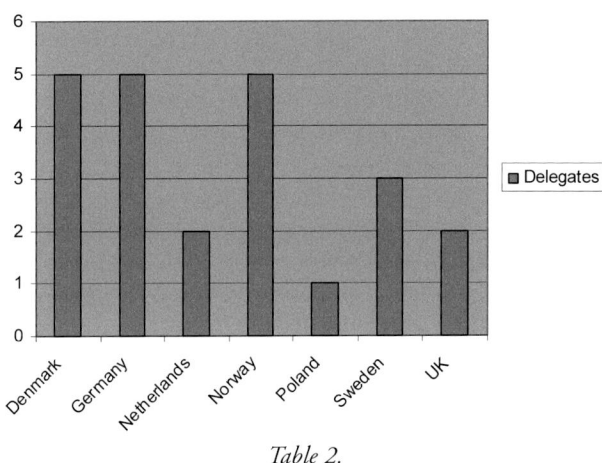

Table 2.

In 1999, the seventh NESAT was held in Edinburgh at the National Museums of Scotland, organised by Thea Gabra-Sanders and by Frances Pritchard of The Whitworth Art Gallery, University of Manchester. It saw a stabilisation in the number of participants, though still finding room for new scholars, including the first delegate from the USA and several textile scholars from the host country. The eighth NESAT took place in Łódź, Poland, in 2002, and was organised by Jerzy Maik of the Institute for Archaeology and Ethnology, the Polish Academy of Sciences. This conference also had slightly over 60 participants representing 15 countries, a large number of Polish participants emphasizing Eastern Europe's emergence from the shadows left by the 2nd World War (Table 4). At the ninth NESAT, held in Braunwald, Switzerland, in 2005 and organised by Antoinette Rast-Eicher, Archeotex, and Renate Windler, *Kantonsarchäologie Zürich*, the number of countries represented increased to 19. The

A Brief History of the North European Symposium for Archaeological Textiles (NESAT)   xix

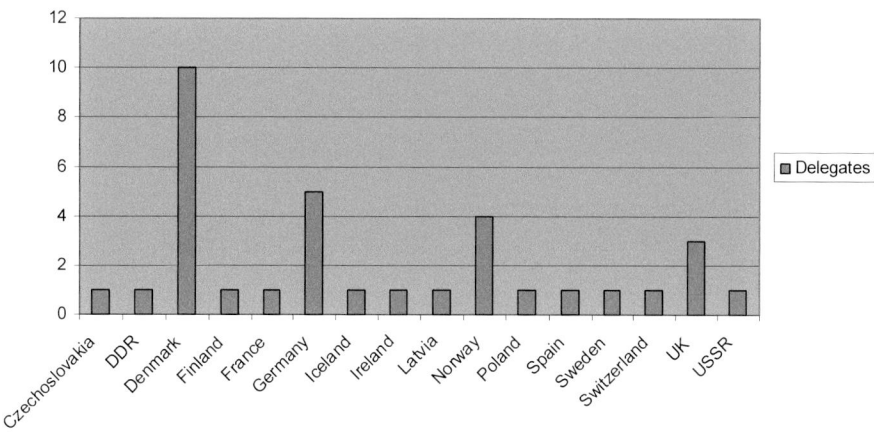

Table 3.

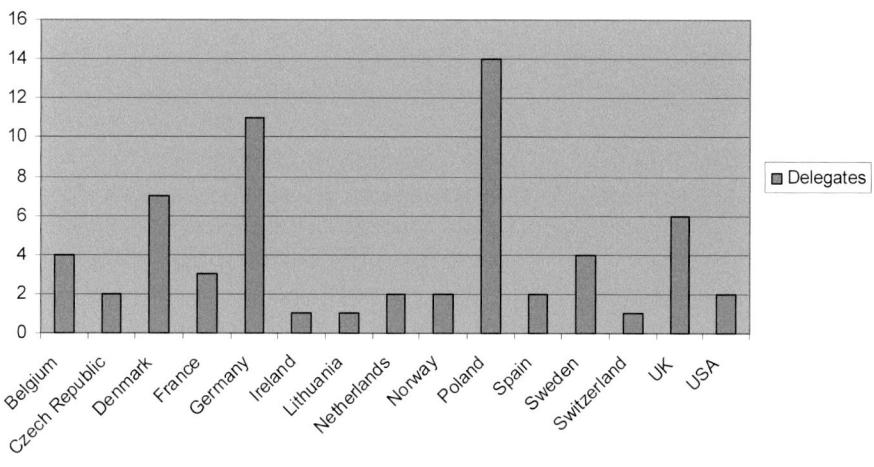

Table 4.

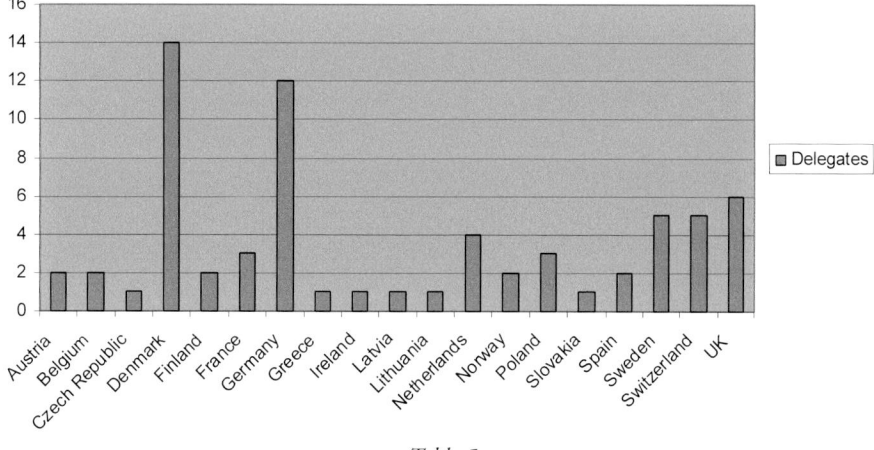

Table 5.

advent of textile scholars from Austria and Slovakia signified an enhancement of Central Europe; NESAT further saw its first delegate from Greece (Table 5). At the tenth NESAT, organized by Margarita Gleba, Ulla Mannering and Maj Ringgaard held at the Danish National Foundation's Centre for Textile Research, University of Copenhagen, Denmark,

NESAT 1-10

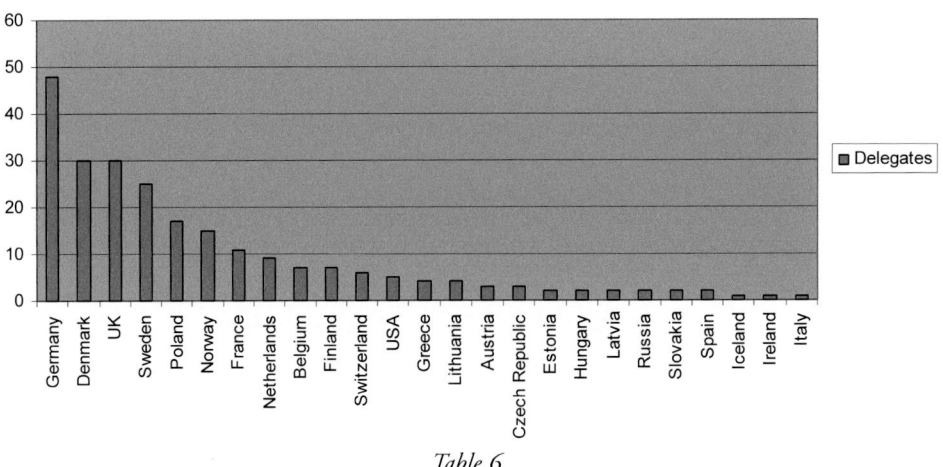

Table 6.

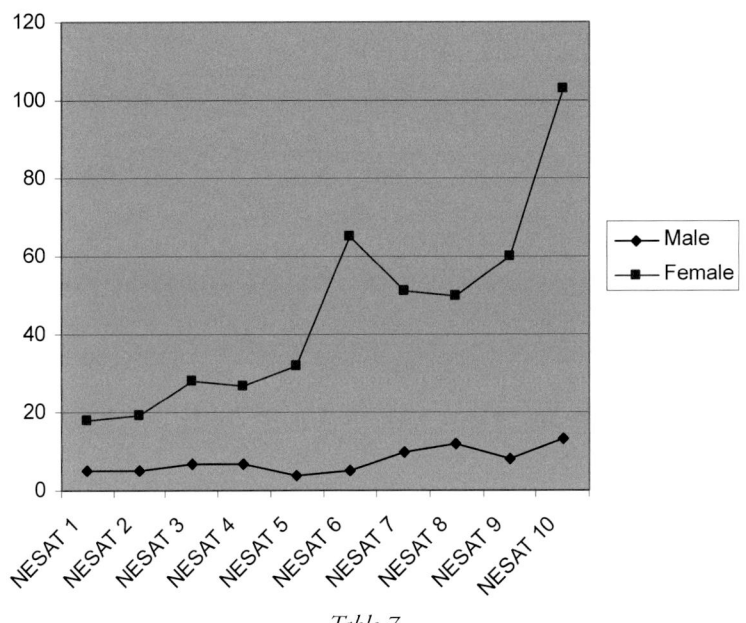

Table 7.

everything has increased: the number of countries to 23, that of participants to well over a hundred. Italy now included, few parts of Europe remain unrepresented.

By now, 239 individual scholars representing around 25 different countries have participated in at least one NESAT[2] (Table 6). When adding them all together, it is conspicuous that the majority of participants come from countries represented at the first NESAT: Germany, UK, Denmark, Sweden, Poland and Norway; only France has squeezed in before the Netherlands. As the six first-named countries all have hosted NESAT meetings, this is no great surprise. The great majority of delegates have always been women; male textile scholars constituting between 7 and 22% (Table 7). No less than 9 of the 23 participants of NESAT 1 in 1981 also took part in the tenth in 2008, although only 5 scholars have managed to attend all the conferences, meriting the title of NESAT stalwarts: Klaus Tidow, Karen-Hanne Nielsen, Jerzy Maik, Frances Pritchard and John Peter Wild.

Organisation

The mere fact that ten NESATs now have been held plainly demonstrates that it fills an important need. Our particular mixture of representatives from the Arts, Crafts and Sciences seems not only to be a happy one, but also a potent one. The fact that NESAT never has been turned into a formal organisation, but has remained an informal one, is, perhaps, another of our strengths. NESAT is best defined as a network and a meeting-place for a wide range of people. Participants engage in all kinds of textile studies: investigations of new finds, reassessments of older ones, conservation methods, identification and interpretation of fibres, weaves or dyes,

Volume	Papers	Volume	Papers
NESAT 1	23	NESAT 6	33
NESAT 2	21	NESAT 7	24
NESAT 3	25	NESAT 8	27
NESAT 4	25	NESAT 9	40
NESAT 5	25	Total	243

Table 8.

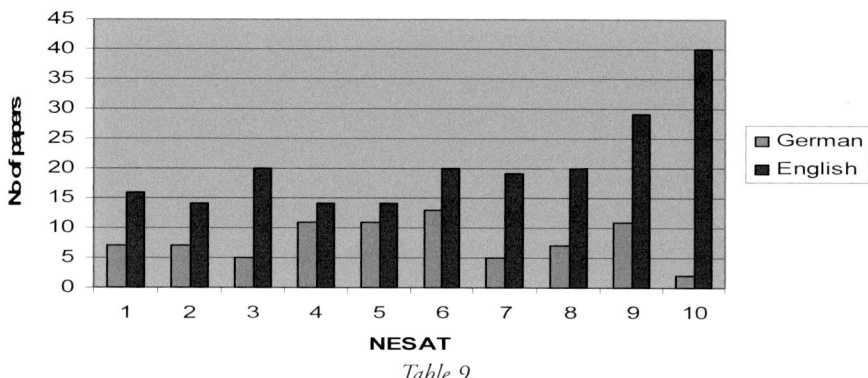

Table 9.

studies of tools, techniques and technologies, of dress and items of clothing, on how to reconstruct ancient textiles and garments, but also in more intangible subjects such as theoretical and ethical matters. The only common denominators have been 'Textiles in Archaeology' and, in principle, 'Northern Europe', although that was soon amended to 'Europe'.

The decision to meet at three-year intervals was taken at the end of the first NESAT. Up to now, NESAT has been run by an informal board. This was set up at the end of the second NESAT, to lend support and legitimacy to the organisers of the next symposium and support the one just held that faced the task of getting the proceedings published. The board has slowly grown from four members to six; legitimacy has been obtained by putting proposals for replacements and additions to a business meeting at the end of each symposium.[3] No one has ever been appointed chairman; meetings are usually chaired by the most senior board member. The membership, in principle, consists of everyone who has been registered as a participant in a NESAT meeting. Until the tenth NESAT, attendance (and, as follows, membership) has been by invitation, the decision of whom to invite that of the organiser(s) assisted by the board. A rule of thumb has been that new members ought to make a presentation at their first NESAT. NESAT X was by invitation as well as being announced in open call.

As yet, the main prerequisite for the continuation of NESAT has been the willingness of one or two individuals, with more or less institutional backing, to take on the task of organising the next meeting and publishing its proceedings. The continued growth of interest in NESAT makes this an increasingly daunting task. It may also one day necessitate the formalisation of membership and board.

Publications

Our success in getting the conference proceedings published certainly has had a great impact on the significance of NESAT. Printed in widely different formats, produced by eight different publishers, the NESAT logo nonetheless allows the nine volumes to emerge as a series that is recognised, acknowledged, and in demand. Lists of contents are to be found on the Internet; and queries of how to get hold of NESAT volumes regularly turn up, particularly the early ones that long have been out of print. The first nine NESAT proceedings contain no less than 243 papers and form a considerable part of the texts – *i.e.* documentation and discussions on the subject of archaeological textiles – that are available to us as essential tools and raw materials of scholarly work (Table 8). Seventy seven of them, *i.e.* 32% are in German, the rest in English. The two official languages of NESAT have afforded advantages as well as disadvantages. Several NESAT members do only understand one of them. This *e.g.* applies to many of the first generation of NESAT members from Eastern Europe, for whom German is the main or only foreign language. Communication across the frontiers of the Cold War has been of fundamental importance. Native speakers among the NESAT board members have helped by revising the sometimes halting German or English of manuscripts by authors not writing in their mother tongue. By the tenth NESAT, the use of German has almost been abandoned, emphasizing English as the *lingua franca* of current scholarly exchange (Table 9).

By the tenth NESAT, the nine volumes of previous proceedings may easily be taken for granted. The irregular dates of printing and varying formats and publishers are, however, tell-tale evidence that this is by no means the case. It has rarely been an easy task to get the symposia reports published. Although contributors throughout have submitted their manuscripts quite promptly, raising money for printing has always been a difficult hurdle. In the case of the second NESAT, it proved impossible to find funding in the host country, and it took a joint Norwegian-Danish effort to get it published. This proved a crucial point, as the second symposium report constituted NESAT as a series.

In the same way as political changes in Europe have been reflected in the history of NESAT, the proceedings bear evidence of developments in publishing technology. The first two NESAT reports were produced on typewriters; the first to be typeset was the third volume. By the time of the fourth, personal computers had begun to be available, and four or five authors were proudly able to submit their manuscript in electronic format. The rest was laboriously typed into the computer by the editor. At the sixth NESAT, this was reversed: only four or five authors did not submit on disc. Happily, by then scanning technology was sufficiently advanced to make it possible to convert the typewritten manuscripts into electronic format. The first illustrations submitted electronically appeared at the same time; nowadays it appears unfeasible to submit anything for publishing in other formats. Similarly, until about the mid-1990s, snail-mail was the main form of communication between authors and editors. Mail between Eastern and Western Europe was unreliable, at best very slow. To send requests for corrections, different illustrations, queries or proofs to authors from Poland, the USSR, DDR or Czechoslovakia was not worth the effort, as nobody could tell how long it would take or whether letters reached the recipient at all. The advent of e-mail has changed all that in a way that appears almost miraculous to someone who has had to cope with the situation before.

Fora for textile research

NESAT is by no means the only forum for scholars of ancient textiles, nor the first one. The Centre International d'Études des Textiles Anciens (CIÉTA) was already founded in 1954, had well over 300 members when the first NESAT was held, and now lists 525 members from 34 countries. Why didn't we just join them, instead of starting our own? The main reason for that was that we felt that the research material and methodological approaches applied to archaeological textiles of Northern Europe did not fit very well into the world of CIÉTA that very much focused on the history of silks, of composite weaves and patterns, and perceived ancient textiles as Art rather than Artefacts. The continued growth of both the CIÉTA and NESAT indicates that our feeling was right. Many scholars are members of both. Since then, other associations and symposia have been established, in Europe as well as in North America, demonstrating a need for a variety of fora for textile research. Several of these may be seen as related to or even spin-offs of NESAT, like the Early Textile Study Group in Britain, the Dyes in History and Archaeology, or the *PURPUREAE VESTES*, all of which have been initiated or run by, and certainly participated in by NESAT members. Conferences, research projects, journals, and exhibitions focusing on archaeological textiles are multiplying these years, attracting increasing interest from scholars, funding bodies and the general public. Textile studies are very much alive and kicking. I confidently look forward to many happy returns of NESAT, and to an exciting future for further textile research.

Notes

1 My first visit to the Textilmuseum Neumünster was in 1975, while still a student of prehistoric archaeology at the University of Copenhagen. Textile engineer Klaus Tidow, assistant curator at the Textilmuseum, was put in charge of the visitor from Denmark by the then Director, Rudolf Ullemeyer. Klaus took me to lunch at the Rathauskeller; that was when our discussion took off, and I never managed to finish the delicious meal. Thanks are due to the Landsdommer V. Gieses Legat for funding my journey to Neumünster and inadvertently igniting the spark that turned into NESAT.

2 Since 1981, some countries have changed their names, split, or merged. In addition, several NESAT members have moved around. In overall statistics, countries are entered according to their current name and/or format.

For the purposes of this paper, individuals have been catalogued as delegate of the country they represented at their first NESAT. The only exception is Margarita Gleba, who appeared as a delegate of the USA on the list of participants of NESAT 9, but of Denmark in the published report. A national of Lithuania, she is entered as such.

3 The first board consisted of Klaus Tidow and myself (organisers of NESAT 1), Bente Magnus (organiser of NESAT 2) and John Peter Wild (co-organiser of NESAT 3). Magnus was replaced by Frances Pritchard in 1990; Tidow by Johanna Banck-Burgess in 1996. Wild retired in 1999 and was replaced by Antoinette Rast-Eicher and Jerzy Maik. In 2005, Ulla Mannering was elected as a sixth member.

1 Experimental Textile Archaeology

by Eva Andersson Strand

Experimental archaeology is a method that can provide important insights into archaeological research. This is done primarily by practical tests performed on the basis of questions related to archaeological data.

Experimental archaeology has been seen as having developed within the positivistic research tradition, which was current in processual archaeology in the 1960s and 1970s (Brattli and Johnsen 1989, 49; Olsen 1997, 53, 59–62). According to this tradition, archaeological material had to be interpreted via procedures similar to those of the natural sciences with the aim of reaching objective knowledge without subjective influences. This way of arguing has been criticized in the post-processual research tradition of the 1980s. The criticism was directed towards the idea of experimental archaeology as a method of conducting objective studies. One of the arguments today is that designs for experiments are influenced by subjective values, which will affect their outcome. However, this critique was aimed at archaeology in general, implying that archaeological interpretations can hardly avoid the influence of cultural, economical and political ideology (Trigger 1993, 454–457). At the same time, and perhaps as a consequence of this criticism, various scholars began to question what constituted experimental archaeology (*e.g.* Coles 1983, 79–81; Johansson 1983, 81–83; Malina 1983, 69–78), and the subject of defining and developing experimental archaeology was debated (*e.g.* Johansson 1987, 2–4). Several attempts at defining the concept were made; some of them proposed to divide experimental archaeology into different topics such as experiments within archaeology and experimental archaeology (Olausson 1987, 7). Today, experimental archaeology is understood as a very wide term and may be seen as an umbrella under which several activities are undertaken. The three most important, in my opinion, are:

- Ethnographic Studies
- Experience Archaeology
- Experimental Archaeology (this is also what makes the definition complicated as experimental archaeology can be seen as both the umbrella and a defined method).

The three approaches are also sometimes used in combination and it can be difficult to separate them from one another. All three approaches are extremely valuable for textile research, and by using and applying the information and/or results to archaeological material, we can fill important lacuna in not only textile archaeology, but also archaeology in general.

Using experimental archaeology as a method in textile research has a long tradition but it is still important to develop and systematize the concept and definition of 'experimental textile archaeology' and to discuss its limitations and possibilities.

Ethnographic Studies

As archaeologists working with prehistoric textiles and textile techniques, we have the advantage that many of the techniques that were used during prehistoric times are still in use today. Through ethnographic sources, we often have knowledge of tool functions and different processes, such as fibre preparation, spinning and weaving. Experimental archaeology is closely connected to the use of ethnographic parallels in archaeology. Both experimental archaeology and ethno-archaeology are seen as having being developed within the same research tradition, which is related to the use of analogies in archaeology. As some handicrafts known from prehistory continue to be practised today, valuable information is readily available and much has already been recorded. For example, in the 1950s, the Norwegian researcher Marta Hoffman began her work on the warp-weighted loom by focusing on the living traditions of its use in the Nordic countries (Hoffmann 1974). Ethnographic knowledge of textile production and tools, like that of the warp-weighted loom (*e.g.* Crowfoot 1931; Sylwan 1941, 109–125), has played an important role for the understanding of ancient spinning and weaving, and in attempts to revitalize ancient textile technology.

The use of ethnographic parallels has been criticized in that the ethnographic records represent situations far removed in both time and space from the ancient context under examination (Coles 1979, 39). Regrettably, this criticism has not been discussed specifically within textile research and ought to be considered in this field as well.

One of the challenges one has to be aware of is that craftspeople today do not always work with the same type of tools or techniques, or even the same type of fibre material as in the past. For example, the weavers whom Martha Hoffmann worked together with used very heavy loom weights, compared to loom weights from, for instance, the Bronze Age Mediterranean. It is also important to remember that different regions have their own traditions; for example, on Crete spinners today spin on a suspended low whorl spindle, while on Iceland spinners often spin on a high whorl spindle.

Experience Archaeology

This approach has a reputation of not being a scientific method since it cannot be used for specific archaeological interpretations; nevertheless, it is a highly important approach. Textile technology should be taught in all basic courses in archaeology in the same way students learn about flint napping or bronze casting. It is necessary for all archaeologists to acquire 'experience' of different techniques and types of textiles. It is our responsibility to teach the student why this knowledge is important, and how it can be used. It is not enough to allow them to merely attempt spinning on a spindle and weaving on a warp-weighted loom for a couple of hours. This knowledge and experience is strictly necessary and important to work within all in other archaeological fields.

Experimental Archaeology

An important component in the methods of experimental archaeology is the testing of function and efficiency of textile tools and equipment (Peacock 2001). As such, experimental archaeology forms a link between textiles and textile tools and contributes to a better understanding of textile production and its complexity. The results from experiments form an important basis for the interpretation of the function of different tools and for the evaluation of the types of textiles that have been produced at different sites and regions. These results can also help visualise textiles in places, where none have been preserved.

In the *Textile and Tools – Texts and Contexts research programme* at the Danish National Research Foundation's Centre for Textile Research, we have conducted several experiments and used experimental archaeology as an important method.

One of our missions in the research programme has been to develop experimental textile archaeology (http://ctr.hum.ku.dk/research/tools_and_textiles_/). The reason for developing our methods within experimental archaeology is that we want the results to be reliable, clear, and easy to relate to. It was therefore essential from the beginning to make guidelines for our experiments. These guidelines are as follows:

- *The primary parameter to be investigated is the function of tools*
- *The raw materials are selected according to our knowledge of Bronze Age fibres and work processes*
- *The tools are reconstructed as precise copies of archaeological artefacts*
- *Each test is to be performed by at least two skilled craftspeople*
- *Every new test should be preceded by some practice time*
- *All processes must be documented*
- *All products must be analysed by external experts*

Experimental tool testing combined with knowledge of fibres and tools in the period under investigation can help us understand how and for what purpose specific textile tools may have been used. It is essential that tests are performed by several skilled craftspeople, otherwise it will not be possible to evaluate if the end product is affected by the tool or by the craftsperson.

Whatever the primary parameter to be investigated, thorough control and evaluation of every step of an experiment is essential. The raw materials should be selected according to the knowledge of the given period and area, and the tools tested should be reconstructed as precise copies of archaeological artefacts.

By testing textile tools we can obtain valuable information about tool function, their qualities and limitations, and the amounts of time consumed in the various production stages. For example, systematic spinning experiments with suspended spindles have demonstrated that it is primarily the fibre quality, and the weight and the diameter of the spindle whorl that affect the finished product, *i.e.* the spun yarn.

The tests also demonstrated that variations within a specific tool type determine variations in the final textile product (Andersson 2003; Mårtensson 2007; Andersson Strand and Nosch forthcoming), and the variations in time consumption connected to the use of the tools (Andersson *et al.* 2008). Experiments have also demonstrated that not only the weight, but also the thickness of a loom weight plays an important role in the weaving process, and hence that the choice of loom weights influences the fabric (Mårtensson *et al.* 2009; Andersson Strand and Nosch forthcoming). By recording the weight and maximum thickness of loom weights, and combining this data with the results of experimental weaving, it is possible to suggest the kind of textiles that the tools could have produced with a given type of yarn. In an archaeological context where textiles do not survive, the range of tool parameters may be used to infer the range of cloth and/or thread that could have been produced with these tools.

Combining the results from experimental testing with contextual analyses can further help to answer questions which are difficult to address by studying the tools alone. Experimental tool testing can, for instance, be used to investigate if the tools and their combinations at the site could have been used to produce yarn or fabrics corresponding to the surviving textiles. It also provides an insight into the variation in the production of yarns and fabrics at a given site, and allows for economic and social interpretations of whether the same kinds of textiles were produced in households and in workshops, thus approaching the concepts of skill and specialisation.

Conclusions

Experimental archaeology is important. Yet, the definitions of what is included in this term can be confusing. Ethnographic studies, experience archaeology and experiments are all very important methods and should of course be included in textile research. Yet, how, and which methods have been used in the interpretation should be clearly stated, as basing the results on ethnographic studies, experience archaeology and/or experimental archaeology can make a difference to the outcome. Therefore it is important to continue the discussion of the definition of experimental textile archaeology, how it can be used, its possibilities but also its limitations.

Bibliography

Andersson, E. (2003) *Tools for Textile Production from Birka and Hedeby*. Birka Studies, Vol. 8. Stockholm.

Andersson E. B., Mårtensson, L., Nosch, M-L., and Rahmstorf, L. (2008) New Research on Bronze Age Textile Production. *Bulletin of the Institute of Classical Studies of the University of London*. London.

Andersson Strand, E., and Nosch, M-L. (forthcoming) *Tools, Textiles and Contexts. Investigations of Textile Production in the Bronze Age Eastern Mediterranean*. Oxford, Oxbow Books.

Brattli, T., and Johnsen, H. (1989) Noen kritiske kommentarer til den eksperimentelle arkeologien. In E. Backman and C. Fredriksson (eds), *Experimentell Arkeologi, Kontaktstencil XXXIII*, 49–52. Umeå.

Coles, J. (1979) *Experimental Archaeology*. London, Academic Press.

Coles, J. M. (1983) Comments on Archaeology and Experiment. *Norwegian Archaeological Review* 16:2, 79–81.

Crowfoot, G. M. (1931) *Methods of Hand Spinning in Egypt and the Sudan*. Bankfield Museum Notes. Second Series No. 12. Halifax.

Hoffman, M. (1974) *The warp-weighted loom*. Oslo.

Johansson, T. (1983) Comments on Archaeology and Experiment Technical Processes of the Past. *Norwegian Archaeological Review* 16:2, 81–83.

Johansson, T., ed. (1987) Experimentell arkeologi. *Forntida teknik*, nr 15. Sveg.

Malina, J. (1983) Archaeology and experiment. *Norwegian Archaeological Review* 16:2, 69–78.

Mårtensson, L. (2007) Textilteknologiska studier av sländspinning – träsländan från Hjortspring. Lejre experimental report HAF 05/07. http://www.english.lejre-center.dk/THE-SPINNING-STICK.610.0.html

Mårtensson, L., Nosch, M-L., and Andersson Strand, E. (2009) Understanding a loom weight. *Oxford Journal of Archaeology*, 28:4.

Olausson, D. (1987) Experiment på gott och ont. *Forntida teknik. Experimentell arkeologi*, nr 15, 5–13. Sveg.

Olsen, B. (1997) *Fra ting til tekst. Teoretiske perspektiv i arkeologisk forskning*. Oslo.

Peacock, E. E. (2001) The contribution of experimental archaeology to the research of ancient textiles. In P. Walton Rogers, L. Bender Jørgensen and A. Rast-Eicher (eds), *The Roman textile industry and its influence. A Birthday Tribute to John Peter Wild*, 181–192. Oxford, Oxbow Books.

Sylwan, V. (1941) *Woollen Textiles of the Lou-lan People*. The Sino-Swedish expedition. Publication 15. Stockholm.

Trigger, B. G. (1993) *Arkeologins idéhistoria*. Stockholm.

2 The Perfect Picture – A Comparison between Two Preserved Tunics and 13th-century Art

by Eva I. Andersson

It has been said that medieval art cannot be used as a source for understanding medieval dress since religious and other symbolic considerations have too much influence on the way clothing is depicted (Vedeler 2007, 245–246). Even those who use art as a source of medieval dress acknowledge that it is a problematic source – it rarely shows enough detail to determine how the garments were constructed, and there is also an element of idealization in art: people are depicted as more beautiful and in more luxurious clothing than was generally worn. Yet a further argument is that symbolic and aesthetic considerations as well as tradition may be what dictated for example, the artist's use of colour rather than a wish to capture reality (Newton 1980, 102; Jaacks 1998). However, regarding the use of colour, I found in my study of clothing in Swedish and Norwegian medieval documents that, there is a strong correspondence between the colours of clothes mentioned in these documents and colours used for clothes in medieval art, especially when people of higher social status are depicted, *i.e.* the same group which dominates in the written sources (Andersson 2006, 191–193). Further research on colours and materials in other European documents may provide more insight into the relation between actual garments and representations in art in this respect.

This paper however, focuses on the correspondence between preserved garments and medieval art, showing that, with an intimate knowledge of the conventions and themes of medieval art and a source critical approach, art *can* be used as a source when studying medieval clothing. The basis for the discussion is the reconstruction of two 13th-century women's tunics: the St. Clare tunic and the St. Elisabeth of Thuringia tunic, both from the decades before 1250 and preserved as relics. This paper therefore also argues for using reconstructions as a scientific method.

The Tunics of St. Clare and St. Elizabeth of Thuringia

Preserved medieval garments are rare, and women's clothes are even more rare than men's. Thus, to have two female tunics from roughly the same period is unusual indeed. The tunic of St. Clare, which is kept in her native town of Assisi, Italy has been dated, through the connection with the saint, to before AD 1253, when she died. It is mentioned already in an inventory of AD 1348 and there is nothing to contradict it as having belonged to St. Clare (Fig. 2.1). The tunic is made from apparently undyed wool in four-shaft twill. The cut is of the standard medieval type with two straight panels widened with gores. Unlike the earlier Kragelund, Moselund and the later Herjolfsnæs finds, it has no central gores. This, on the other hand, is a characteristic it shares with St. Elisabeth's tunic, and by no means the only one (Grönwoldt 1979, 407–410). Two features that stand out when looking at figures 2.1 and 2.2 are the shape of the back and the placement and shape of the sleeves. The armscyes on the back are very straight; in fact, the tunics start widening, with the help of a gore, right at the shoulder. This is not otherwise a common feature in medieval garments, except for the roughly contemporary tunic of St. Elisabeth of Thuringia (Fig. 2.2).

St. Elisabeth's tunic, which is now kept in the Catholic parish church of St. Martin in Oberwalluf in Hesse, Germany, is dated to the first third of the 13th century and like St. Clare's tunic is made of undyed brownish wool (Schorta 2007, 120). As may be seen from figure 2.2, it is cut in much the

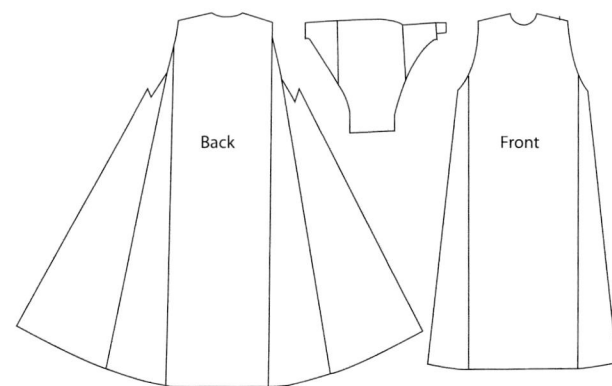

Fig. 2.1. Diagram of St. Clare's tunic (After Grönwoldt 1979).

same way, the main difference being the cut of the sleeves: St. Elisabeth's sleeve resembles a modern sleeve pattern to a great extent, except for the placement of the seams. When made up, however, both sleeves have a similar shape. The back of the tunic also shows an almost identical widening from the shoulder on the back, as St. Clare's tunic, while the armscyes on the front, again like St. Clare's tunic, are cut deep.

If the patterns of St. Elisabeth's and St. Clare's tunics are compared with, for example, the Herjolfsnæs tunics, which are mostly dated to the second half of the 14th century, the differences in the cut of sleeves and armscyes are not that great. In the Herjolfsnæs tunics, the armscyes are scooped out both in the front and the back and have a more rounded shape in general. This means that the back pieces do not cover the shoulder joint as they do on St. Elisabeth's and St. Clare's tunics. Before making my first reconstruction some years ago, I was not, however, aware that these little differences in pattern cuts were what made the tunics appear as they do in 13th century illuminations – in fact, I was unaware that these kinds of details were visible in paintings from as early as the High Middle Ages. The differences in cut over the shoulders between the two 13th-century tunics and those of the 14th century from Herjolfsnæs are small, but highly significant.

This little difference is exactly what creates the special look seen in contemporary manuscripts.

Regrettably, no comparison can be made with male 13th-century tunics, since the other extant examples, the so-called Rønbjerg tunic, which actually belongs to the Herjolfsnæs finds, and the Söderköping tunic, cannot be securely tied to a specific gender, and both lack sleeves or upper parts of the tunic. However, considering the similarity of male and female tunics in medieval art and how often garments were bequeathed from men to women, it is likely that the male tunics were constructed in much the same way, with the exception that if the Rønbjerg and Söderköping tunics *were* male, both appear to have centre gores, which then could be a gender-specific trait (Nockert 1992; Østergård 2003, 140; Andersson 2006, 276–280).

Leaving the extant garments for a while, I now turn to medieval art, in this case manuscript illuminations, to investigate armscyes, sleeves and back shape.

Clothing in 13th-Century Art

There are hundreds of preserved 13th-century manuscripts with illuminations; those shown in this article are chosen for clarity and, it must be admitted, for aesthetic reasons – the Maciejowski Bible, a French manuscript from *c*. AD 1250 (Cockerell 1927) is, in my opinion, one of the most beautifully illuminated manuscripts in existence. In this manuscript, which illustrates the Old Testament up to the story of David, men and women wearing contemporary 13th-century clothing are seen, as was the custom in art for most of the Middle Ages. It was not until the 15th century that artists introduced foreign, 'exotic' or 'ancient' elements into their paintings to represent other times and lands.

The illumination, which depicts the Benjaminites taking wives, portrays women in festive clothing, dancing to music. They wear loose, belted gowns, which sweep the floor (Fig. 2.3).

The rounded lines that emphasize the shoulder and the joint between arm and body on this and other images in the manuscript may appear to be only an artistic convention, or a sign of a naive painting style intended more as a specific message rather than as a realistic depiction. While the preoccupation with symbolic messages certainly is true of most medieval art, no conflict between symbolism and realism is necessary. I would argue that the artist attempted to the best of his ability to depict humans in a realistic way, and the rounded lines of the shoulders in the front were not only an artistic convention, but also a consequence of how garments were cut in this period. Since medieval art differs so much from modern art, the attempts to depict clothing realistically may not, however, be readily apparent to the modern observer – this requires not only knowledge of both medieval art and its conventions and techniques, and of preserved garments, but also experience of cutting and sewing historic garments.

Two typical features of tunics seen in 13th and early 14th century manuscripts are the shape of the sleeves and the shape of the armscyes. The sleeves are baggy at the upper arm and tighter at the lower, and the armscyes appear to be deeply

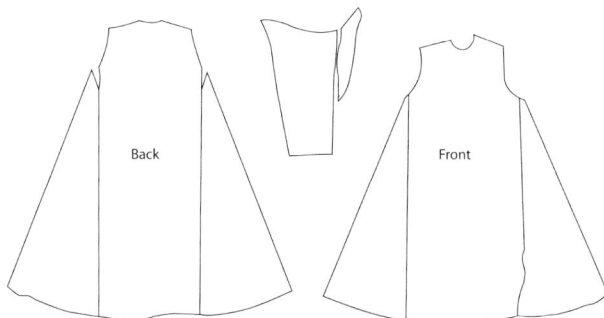

Fig. 2.2. Diagram of St. Elisabeth of Thuringia's tunic (After Grönwoldt 1979).

Fig. 2.3. Maciejowski Bible, French, c. 1250 (Pierpont Morgan Library MS M.638; After Cockerell 1927). Reproduced as a colour plate on page 299.

scooped out at the front, and not at all, or very little, at the back. In the illumination from the Maciejowski Bible, a woman wearing a tunic where the sleeves are only partly sewn to the armscye and hanging down her back can be seen. This is a not uncommon detail in both male and female dress in the manuscript. The absence of sleeves in the front allows a good view of the shape of the armscye (Fig. 2.4).

Looking at the shoulder of the woman without sleeves, one can see that the direction of the armscye when it reaches the shoulders indicates that at the back of the shoulder it is wider than it is at the front, and that it falls somewhat outside the shoulder joint. It may be a part of the hanging sleeve, since seams are not shown in the painting, but it may also be a depiction of the cut we find in the preserved tunics of St. Clare and St. Elisabeth, where the back piece widens directly from the shoulder.

The shape of the armscye is of course easiest to see on a tunic without sleeves, that is a sleeveless surcoat. These are, however, not seen on women before the end of the 13th century. In later manuscripts, as in the German manuscript *Grosse Heidelberger Liederhandschrift*, from the first decades of the 14th century (Walther and Siebert 2001, IX–XV), it is clear that the armscye is deeper and more rounded in the front than in the back, and that the back of the surcoat covers the back of the shoulder (Fig. 2.5).

It is of course by no means certain that the sleeveless surcoats had the same construction as the tunics. Looking at the general development of the sleeveless surcoat in medieval

Fig. 2.4. Maciejowski Bible, French, c. 1250 (Pierpont Morgan Library MS M.638; After Cockerell 1927). Reproduced as a colour plate on page 299.

Fig. 2.6. Der Jungfrauenspiegel c. 1200; Missal from Rheims 1285–97 (No. 15326. Rheinisches Landesmuseum, Bonn. Bildarchiv der Kunst und Architektur, Philipps-Universität Marburg; Lat.Q.v.1.78. National Library of Russia).

Fig. 2.5. Grosse Heidelberger Liederhandschrift, c. 1300–1325 (Cod. Pal. germ. 848, Ruprecht-Karls-Universität Heidelberg). Reproduced as a colour plate on page 299.

art, it however appears to have started as a garment whose only difference from the tunic was the lack of sleeves; the deeply scooped side openings of the sideless surcoat are a much later development.

What tunics and sleeveless surcoats in 13th-century art have in common is that the armscyes, and thus also the sleeves, go *fairly* deep down. In the manuscripts from the 13th century, this may be seen very clearly (Fig. 2.6).

The funnel shape of the sleeve is obvious, and it can be seen in many manuscripts throughout the entire 13th century. When compared with the two extant tunics, as seen in figures 2.1 and 2.2, both have funnel shaped sleeves, but these are most obvious in St. Clare's tunic.

The Reconstructions and Art

The reconstructions of the tunics of St. Clare and St. Elisabeth of Thuringia (Fig. 2.7) were made according to the proportions of the original garments, as reproduced in the patterns from the catalogue of the exhibition, *Die Zeit der Staufer* from 1977 (Grönwoldt 1979). The exceptions are the sleeves, which in both cases had to be made wider to fit into the armscyes drawn in the diagrams. This is not a cutting or sewing mistake, but is also evident in the original diagrams. Since my daughters, for whom I made the tunics, are taller than the original wearers, *c.* 165 cm, the gowns were made slightly larger than the originals. This of course affected the fit of the sleeves so that they are a bit looser on the lower arm than the originals probably would have been. While it can be argued that a reconstruction always must be adjusted to the person for whom it is made, there is an important reason for constructing the garments according to the original proportions: when dealing with such subtle details as the shape of armscyes, it is vital to keep the original proportions to see the effect of the cut, compared to other ways of cutting. Adapting the garment to a body which is different from the original wearer's could distort the result and make the reconstruction useless for the purpose of comparison with contemporary art.

Thus, to make the comparison with art easier, the reconstructions were made for my daughters, and not, for instance, for me, since their body type, tall and slender, matches the elongated ideal body of the 13th century. The only deviation from the originals is the size of the neck holes, which are a little bigger than on the originals to make them easier to put on.

Fig. 2.9. Funnel shaped sleeve on the St. Clare tunic and in art (Photo: Eva I Andersson; Lat.Q.v.1.78., The National Library of Russia).

Fig. 2.7. Reconstructions: St. Elisabeth to the left, St. Clare to the right (Photo: Eva I. Andersson).

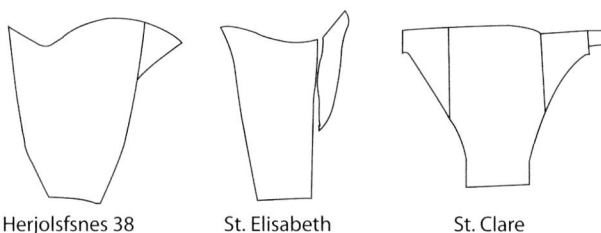

Fig. 2.8. Diagram of sleeves (After Nørlund 1924 and Grönwoldt 1979).

Fig. 2.10. Reconstructions from the back (Photo: Eva I. Andersson).

In this article, space does not permit me to examine more than the shape of the sleeves, the armscyes and the drape of the tunic, *i.e.* the same details that were in focus in the previous discussion on art and which are distinctive of the style of the 13th and early 14th centuries. The cut of the sleeves of the two 13th-century tunics and also the 14th-century Herjolfsnæs tunics are not that different. However, the sleeves from the 13th century are less shaped at the arm and at the shoulder, which gives the funnel shape typically seen in 13th-century art (Fig. 2.8).

The Herjolfsnæs and St. Elisabeth's sleeves are similar, although not exactly identical. St. Elisabeth's have an elongated funnel shape, while the Herjolfsnæs sleeve is more rounded. The St. Clare's sleeves have the most pronounced funnel shape and also represent the more typical 13th-century sleeves of the two (Fig. 2.9).

The back pieces of St. Clare's and St. Elisabeth's tunics are, as mentioned previously, angled out directly from the shoulders, with little or no, shaping of the armscye. What the reconstructions demonstrate is that this cut, which may seem random, does in fact serve a purpose. The way the fabric falls in the back, with folds that give an elongating visual effect for the body, and the way the back piece covers the back of the shoulder joint but not the front, are not accidental or unrelated to visual ideals shown in art (Fig.10). Exactly the same folds can, for example, be seen at the back of the 14th-century effigy of St. Julia in the Scaligeri Tombs, Verona (Contini 1965, 73).

Another feature of the St. Clare and St. Elisabeth gowns is that there is very little scooping out of the back of the neck, which explains why neck holes of tunics in art from the High Middle Ages go so high up in the back of the neck, as can, for example, be seen in figure 2.3.

Conclusions

This article has only touched upon a few features of 13th-century dress. It is hard to convey in photographs exactly how the tunics behave when worn in real life – another good argument for making reconstructions, since the experience of wearing and moving in the reconstructed tunics adds further to our knowledge of medieval dress. Making a reconstruction, affords the opportunity to observe how fabric drapes and behaves, seeing that even small details in cut can affect the general look of a garment. While the general similarity between the preserved garments and contemporary art may be easy to see merely by looking at them, the way 13th-century art depicts even specific details in the construction of women's tunics is not as clear. Therefore, it is useful to reconstruct preserved, or partly preserved, garments. The position of art as a source of medieval dress is strengthened by the results of this experiment, since it has now been proven that the same typical features can be found both in art and in extant garments from the same period; indicating that, with the application of source criticism, art certainly can be used as a source of medieval dress.

Acknowledgements

I wish to thank the following institutions for granting me the right to reproduce examples of medieval art:

Bildarchiv der Kunst und Architektur, Philipps-Universität Marburg, Germany

The National Library of Russia; St. Petersburg, Russia

Ruprecht-Karls-Universität Heidelberg, Germany

www.medievaltymes.com which provided the scans from the 1927 publication of the Maciejowski Bible.

Bibliography

Andersson, E. I. (2006) *Kläderna och människan i medeltidens Sverige och Norge*. Avhandlingar från Historiska Institutionen i Göteborg, Göteborg University.

Contini, M. (1965) *Fashion, from ancient Egypt to the present day*. London, Paul Hamlyn Ltd.

Cockerell, S. C. (1927) *A book of Old Testament illustrations of the middle of the thirteenth century sent by Cardinal Bernard Maciejowski to Shah Abbas the Great, king of Persia, now in the Pierpont Morgan Library at New York*. Cambridge.

Grönwoldt, R. (1979) Miszellen zur Textilkunst der Stauferzeit. In *Die Zeit der Staufer. Geschichte – Kunst – Kultur. Katalog der Ausstellung Stuttgart 1977. Band V, Supplement*. Stuttgart, Württembergisches Landesmuseum.

Jaacks, G. (1998) Mittelalterliche Bilder als Quelle. In L. Bender Jørgensen and C. Rinaldo (eds), *Textiles in European Archaeology. Report from the 6th NESAT symposium, 7–11th May 1996 in Borås*. GOTARC Series A, Vol.1, 243–251. Göteborg, Department of Archaeology, University of Göteborg.

Newton, S. M. (1980) *Fashion in the Age of the Black Prince. A Study of the Years 1340–1365*. Woodbridge, Boydell & Brewer.

Nockert, M. (1992) "Unam tunicam halwskipftan" Ett sensationellt dräktfynd i Söderköping. In *St. Ragnhilds Gilles årsbok 1992*. Söderköping.

Nørlund, P. (1924) Buried Norsemen at Herjolfsnes. *Meddelelser om Grønland*, 67. København.

Østergård, E. (2003) *Som syet til jorden. Tekstilfund fra det norrøne Grønland*. Århus universitetsforlag.

Vedeler, M. (2007) *Klær og formspråk i norsk middelalder*. Oslo.

Schorta, R. (2007) Bussgewand der heiligen Elisabeth. In D. Blume and M. Werner (eds), *Elisabeth von Thüringen. Eine Europäische Heilige, Katalog*. Michel Imhof Verlag.

Walther, I. F., and Siebert, G. (2001) *Codex Manesse. Die Miniaturen der Grossen Heidelberger Liederhandschrift*. Insel Verlag.

3 Loom-weights, Spindles and Textiles – Textile Production in Central Europe from the Bronze Age to the Iron Age

by Tereza Belanová Štolcová and Karina Grömer

Introduction

This paper surveys the evidence of prehistoric textile production based on textiles and tools from archaeological contexts within Central Europe (the west Carpathian and the circumalpine region), focusing on the territory of Austria and Slovakia and with reference to adjacent areas such as the eastern part of Czech Republic – Moravia. The Bronze (2300/2200–800/750 BC, BzA–BzD, HaA–HaB) and Iron Ages (Hallstatt Period: 800/750–450 BC, HaC–HaD and La Tène Period 450–15 BC, LtA–LtD) are the primary focus, but to understand changes in textile technology better, finds from preceding periods – Early and Middle/Late Neolithic (6000/5000–4000/3900 BC) and Late Neolithic/Eneolithic (4000/3900–2300/2200 BC)[1] – are included in the discussion.

Textile work is very complex – a great deal of resources and tools are necessary for the various stages of production (Fig. 3.1). When speaking of raw material, the areas of cultivation for flax and dyestuffs or pastures for sheep are very important. The organization of such resources in our region have been researched by means of landscape archaeology, partly by Jiří Waldhauser for the La Tène settlement Radovesice in Bohemia (Waldhauser 1993) and Raimund Karl for the La Tène Period settlement Göttlesbrunn in Lower Austria (Karl

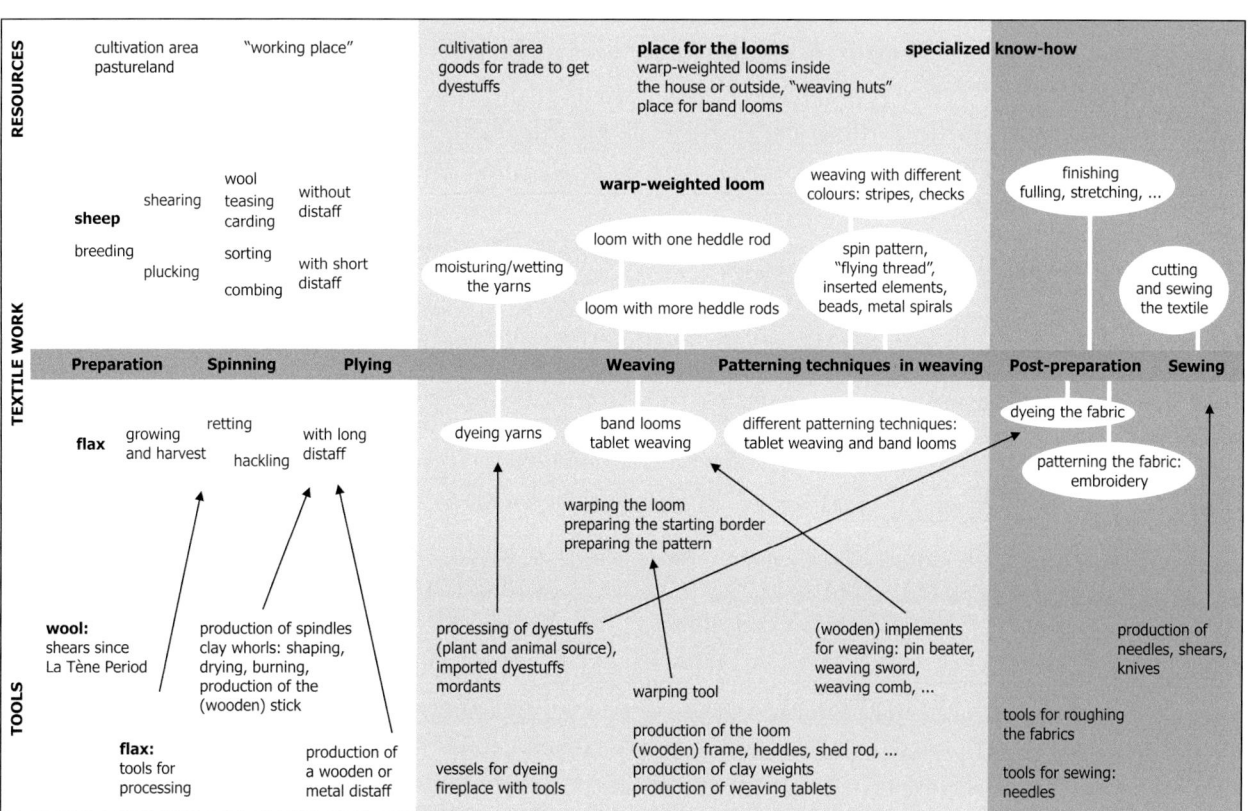

Fig. 3.1. Scheme of textile techniques, resources and tools (© K. Grömer).

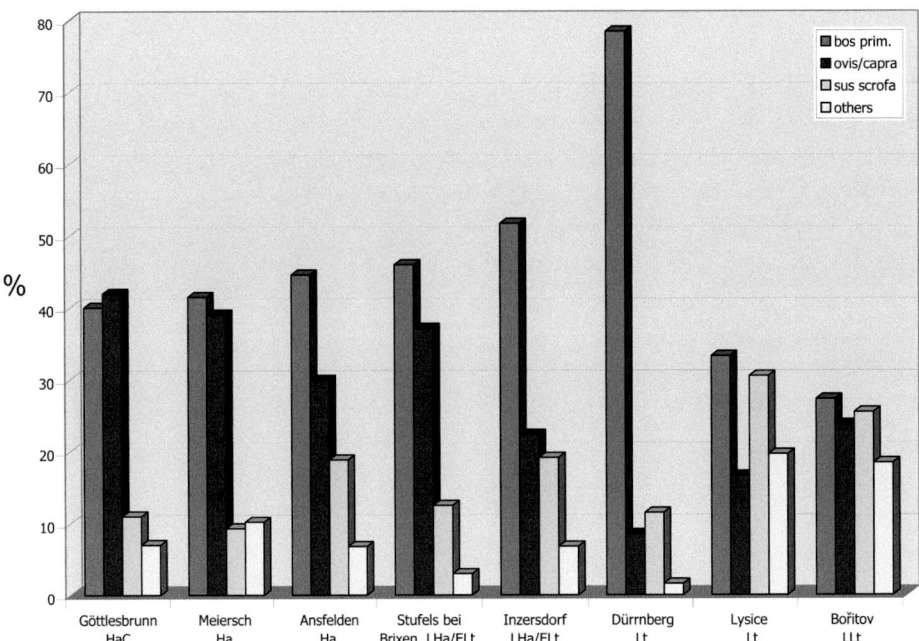

Fig. 3.2. The distribution of animal bones on the Iron Age sites in Austria and Moravia (after Pucher 1998; 1999; 2004; Peške 2003; Trebsche et al. 2007; Karl 1996). Ha – Hallstatt Period, Lt – La Tène Period, LHa – Late Hallstatt Period, ELt/LLt – Early/Late La Tène Period.

1996, 156–158, 216–218). Textile work requires some space, and thus there must have been special places for looms and space for the processing and storage of flax and other fibres in settlements. Another important aspect of textile production is specialised *know-how*, which was needed from the Bronze Age and significantly from the Iron Age onwards, when particular methods of patterns and weave types appear. Also, trade was necessary to obtain imported dyestuffs. For example, a red dyestuff containing kermesic acid was detected in the textiles from the Hallstatt salt mines, which probably would have been imported from the Mediterranean area (Hofmann de Keijzer *et al.* 2005, 65).

For Bronze and Iron Age textile production, a great deal of information about tools from archaeological excavations is available. Among the clay tools which predominantly survive, are spindle-whorls, loom-weights, as well as a few weaving tablets. Some bone and metal material, like needles, shears or weaving swords are preserved, too. A short overview of textile techniques is presented in the following sections, with a focus on preparation, spinning and weaving.

Raw material and its preparation

In the Neolithic and Early Bronze Age around the Alps, textiles were made of bast fibre (Rast-Eicher 2005, 118–119). From the Middle Bronze Age until the Hallstatt Period, wool dominates in our region, at Bronze Age sites like Hallstatt-Tuschwerk and Mitterberg (Grömer 2007). Textiles dated to the Early Iron Age salt mines in Hallstatt are 100% made in wool. Sometimes, other fibres like horsehair (tail), used as weft for bands could be identified (Grömer 2005a, fig. 6). Textiles from graves of the same period in Austria are also usually wool fabrics and represent the same weave types as the textiles from the salt mines (Grömer forthcoming). The Early La Tène salt mine in Dürrnberg provides the first hint of the increased use of flax, as about 25% of the fabrics are linen (Stöllner 2005, fig. 9). In the Middle and Late La Tène Period, flax is the main fibre for textiles recovered in Austrian, Slovak and Czech graves (Belanová forthcoming; Grömer forthcoming).

What can we recognise, when the archaeozoological material is examined (Fig. 3.2)? Some well-analysed Hallstatt and La Tène animal bone material from settlements is known from Austria and Moravia. Hallstatt Period settlements have usually about 30–40% sheep/goat, with sheep predominating (*e.g.* Göttlesbrunn: Pucher 2004; Maieresch and Stufels: Pucher 1998). Based on the slaughtering age of these animals, there is a high quantity of older males and females, which indicates wool production. When spindle-whorls and loom-weights are found at the same site, archaeologists perceive a connection between them. Usually, no textiles have survived at these sites.

At the recently excavated and analysed fortified settlement of Ansfelden (Trebsche *et al.* 2007) dated to Hallstatt period, tools for spinning and weaving were found in the huts. Of the bones, 29.6% were sheep/goat, if identifiable, most of them sheep. It is remarkable that on this site, where extensive archaeobotanical analysis took place, no flax could be detected. This evidence is a good correlation to textile finds. During the Hallstatt D Period, we know of some graves (Berg/Attergau or the graveyard in Hallstatt), where about 95% of textiles were made of wool. Additionally, Ansfelden has a connection with the Hallstatt site, because it is a fortified settlement and a trading place as interpreted by archaeologists. It was situated on the Traun River, which was a waterway from Hallstatt northwards into the river Danube.

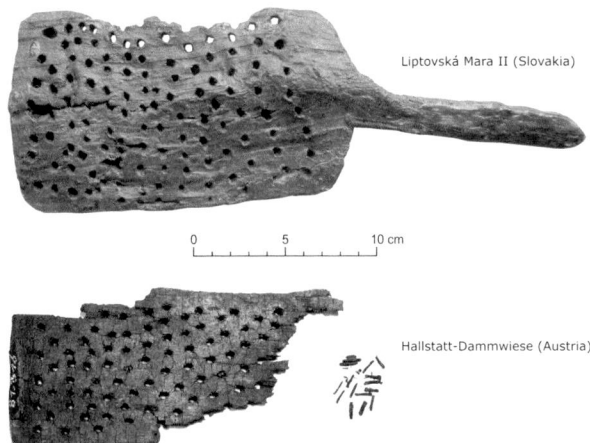

Fig. 3.3. Two similar La Tène Period tools used for raw material preparation (Liptovská Mara II: © Archaeological Institute of SAS, Nitra; Hallstatt-Dammwiese: © Natural History Museum Vienna).

In the La Tène Period, when the number of linen textiles increases, the evidence of sheep/goat fibres decreases. At Moravian La Tène settlements like *e.g.* Bořitov, 23.8% of all animal bones found were sheep/goat bones (Peške 2003, Tab. 1), whilst in Lysice the corresponding figure was 16.6% (Kratochvíl 2003, Tab. 2).

A special case is the settlement of Dürrnberg-Ramsautal, a large production and trading centre. The number of flax textiles recovered from the Dürrnberg salt mines is high, and the archaeozoological evidence from the settlement indicates that sheep/goat is represented by only 8.5%. The analysis of the slaughtering age proves that a great many young animals, rather than older ones were slaughtered. Therefore, sheep/goat was especially meant for meat production; wool production was not so dominant (Pucher 1999, 50–53). Additionally, cultivated flax/hemp was identified from Dürrnberg-Ramsautal (Gewerbesiedlung), dated to the LtA-LtB (Swidrak und Schmidl 2002). According to Stöllner, many tools for textile production were found on the Dürrnberg settlements, but it is not known if all of the textiles found in the salt mines were produced at the settlement of Ramsautal itself, or if some pieces were brought to Dürrnberg from the hinterland as trading goods, or merely for exchange of goods in the salt trade.

What kind of evidence of raw material preparation do we have from our region? Not many tools have survived for the processing of wool and flax fibres. The process of preparation usually requires only hands or some wooden tools, which do not survive well. The exception is a wooden tool from the Middle La Tène layers of the settlement at Liptovská Mara II (Pieta 1999, fig. 9:6) and the La Tène find from Hallstatt-Dammwiese.[2] These could have been used for the processing of flax as hackles or for carding wool (Fig. 3.3).

In the Hallstatt Period (for example from the Hallstatt salt mines) we can find evidence in the textiles for yarn, where the fibres show different kinds of preparation. On the one hand, there is a case of a fluffy yarn with fibres sticking out in all directions (teased wool), on the other hand, there is a very smooth yarn with fibres lying parallel (combed yarn).[3] In the Hallstatt Period, this kind of quality was necessary for making clear and visible spinning patterns.

It is interesting, that there are some changes at the beginning of the Hallstatt Period: a better preparation of wool (combing) occurs, and at the same time short bronze distaffs appear. Perhaps there is a connection between the use of a short distaff and the use of well prepared combed yarn, which is stored as a combed top (*Kammzug*) on the distaff. Additionally, the Hallstatt Period iconography of female spinners shows the technique of spinning with a short distaff, *e.g.* the famous urn of Sopron from Western Hungary and the *tintinnabulum* of Bologna from Northern Italy (Eibner 1986). On both images women spin standing up and they use the suspended spindle (both are possibly low-whorl dropspindles). Such a short bronze distaff was found in a grave in Frög from the Hallstatt Period and later in Unterradlberg and Mautern from the Roman Period (Grömer 2003), both in Austria.

Spinning

Spinning in Central European prehistory was usually done with wooden spindles and clay whorls, which are frequently found in prehistoric graves and settlements (Fig. 3.4). There are usually smaller spherical whorls in the Early and Middle Neolithic (5500–3900 BC). For the Late Neolithic/Eneolithic Jevišovice Culture and Chamer Culture, extremely large whorls are common in various shapes (spherical, discoid, bell-shaped, conical or biconical) and weights between 50 and 120 g. They were frequently found in settlements, sometimes together with loom-weights, *e.g.* Pulgarn or Krems-Hundssteig (Grömer 2006, fig. 5). In Slovakia, spindle-whorls seem to be less frequent in the Early Neolithic, but later, during the Middle Neolithic they appear in higher numbers, together with loom-weights. At the settlement of Blatné dated to the Želiezovce Group, over 300 spindle-whorls were found throughout the site (Pavúk 1994, 123). In Budmerice (Pavúk 2003, 455–456), the settlement dated to the Lengyel III–IV phases (Late Neolithic–Early Eneolithic), seven loom-weights were found in a concentration, possibly as remains of a warp-weighted loom.[4]

There is an interesting gap between the Late Neolithic Period and the Urnfield Culture (BzD-HaB). From the Early Bronze Age up to the end of BzD (Middle Bronze Age), whorls made of clay are almost completely missing in archaeological material from Austria. Few isolated pieces were found, such as two Early Bronze Age whorls from Jetzelsdorf (*Fundberichte aus Österreich* 43, 2004, 786). As for Slovakia, the state of research of the Early and Middle Bronze Age settlements does not allow us to consider the quantity of spindle-whorls. They occur within every settlement, but no significant amounts have been recorded as yet.

Whorls from the Urnfield Culture (*e.g.* Gars-Thunau site) and the Hallstatt Period (*e.g.* Hallstatt graveyard) are very frequent and smaller in shape. Conical and biconical forms are still in use, but also various new shapes with concave tops appear. Spindle-whorls from the Urnfield and Hallstatt

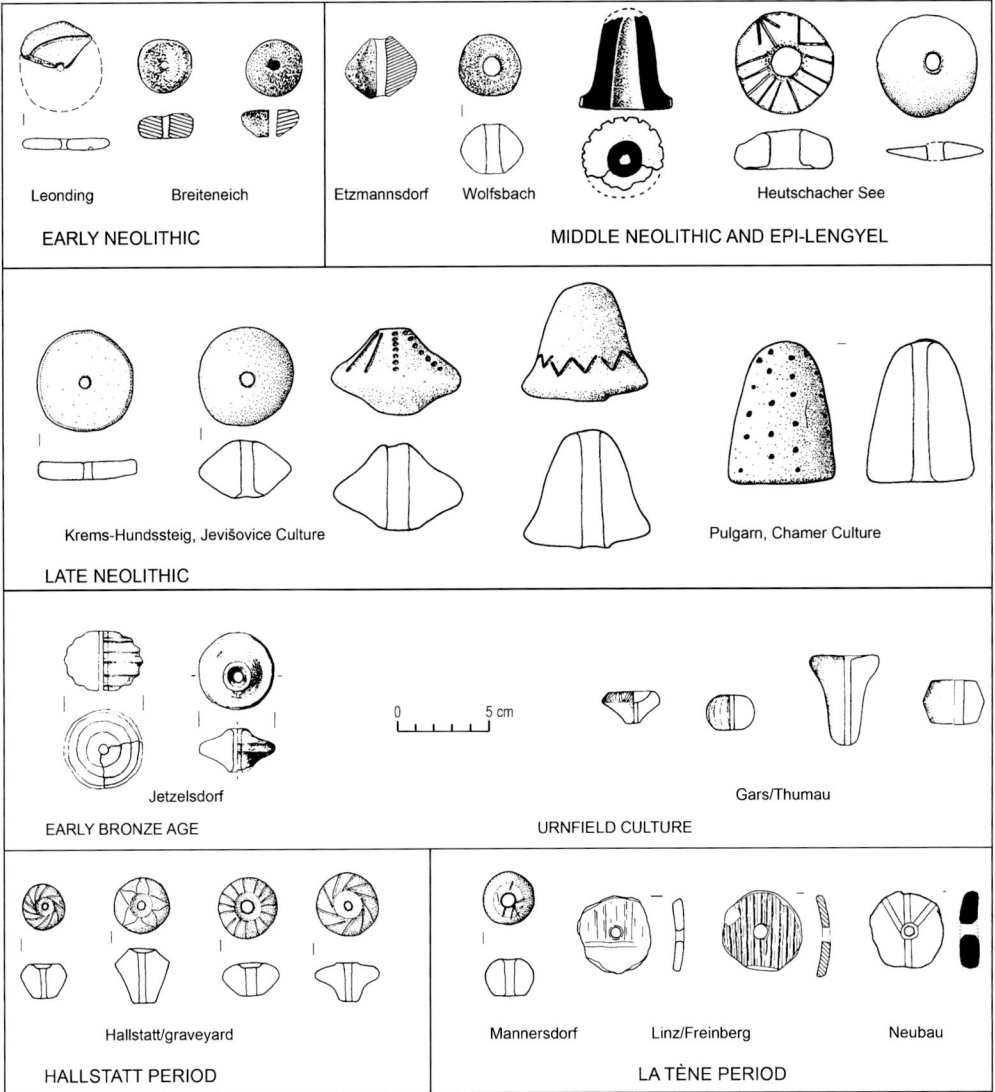

Fig. 3.4. Types of spindle-whorls from the Neolithic to the La Tène Period in Austria (© K. Grömer).

Periods are elaborately decorated. In contrast, from the Middle La Tène Period (LtB2-LtC1) onwards there are whorls made of potsherds, *e.g.* from Linz-Freinberg, Neubau (Grömer 2004) or the oppidum at Plavecké Podhradie – Pohanská (Paulík 1973).

How can the lack of clay spindle-whorls from the Early and Middle Bronze Age be explained, in contrast to their frequent use in the Late Neolithic/Eneolithic and an increasing number of them again in the Urnfield Culture? Is this due to the lack of research or is there another reason? It is possible that during this period spindle-whorls were made of organic materials such as wood. During the Early and Middle Bronze Age, many wooden spindles and whorls from lakeside settlements in Northern Italy have been recovered (Bazzanella *et al.* 2003). They are similar to the clay Neolithic ones in shape (usually flat discoid) and size. On the other hand, it is possible, that in the change of raw material used for the whorls, a shift in technology may be observed. In the Late Neolithic/Eneolithic and Early Bronze Age, linen textiles occur (*e.g.* Italian sites, in Franzhausen in Austria), usually made of plied yarn, sometimes very thin, 0.5–0.7 mm in diameter. Large whorls known from the Late Neolithic/Eneolithic fit very well with the production of flax yarn. Textile culture changed at the turn of the Early to the Middle Bronze Age in the north-east, south and south-east Alps. The Stone Age textiles made of plant fibre were replaced by wool ones: Mitterberg woollen textiles from Hallstatt-Tuschwerk dated to 1500–1100 BC (Grömer 2007, 158–166), and the fragment from Northern Italian Castione Marchesi dated to the 16th century BC (Bazanella *et al.* 2003, 200[5]). Wool textiles are usually thicker and made of single yarn (1–2 mm thick). Regrettably, due to the lack of spindle-whorls this important change cannot be observed in the tools. What is known is that, textiles become thinner and finer from the Urnfield Culture onwards and especially in the Hallstatt Period.[6]

When clay whorls appear again in the Urnfield Culture, they are very small and finely made, as in the Hallstatt Period. The size, weight and shape of clay whorls vary little in the Iron Age.

A good example from the Hallstatt Period (HaC2–HaD1) is an exceptional assemblage of spindle-whorls from the hill fort of Molpír near Smolenice, West Slovakia (Belanová 2007,

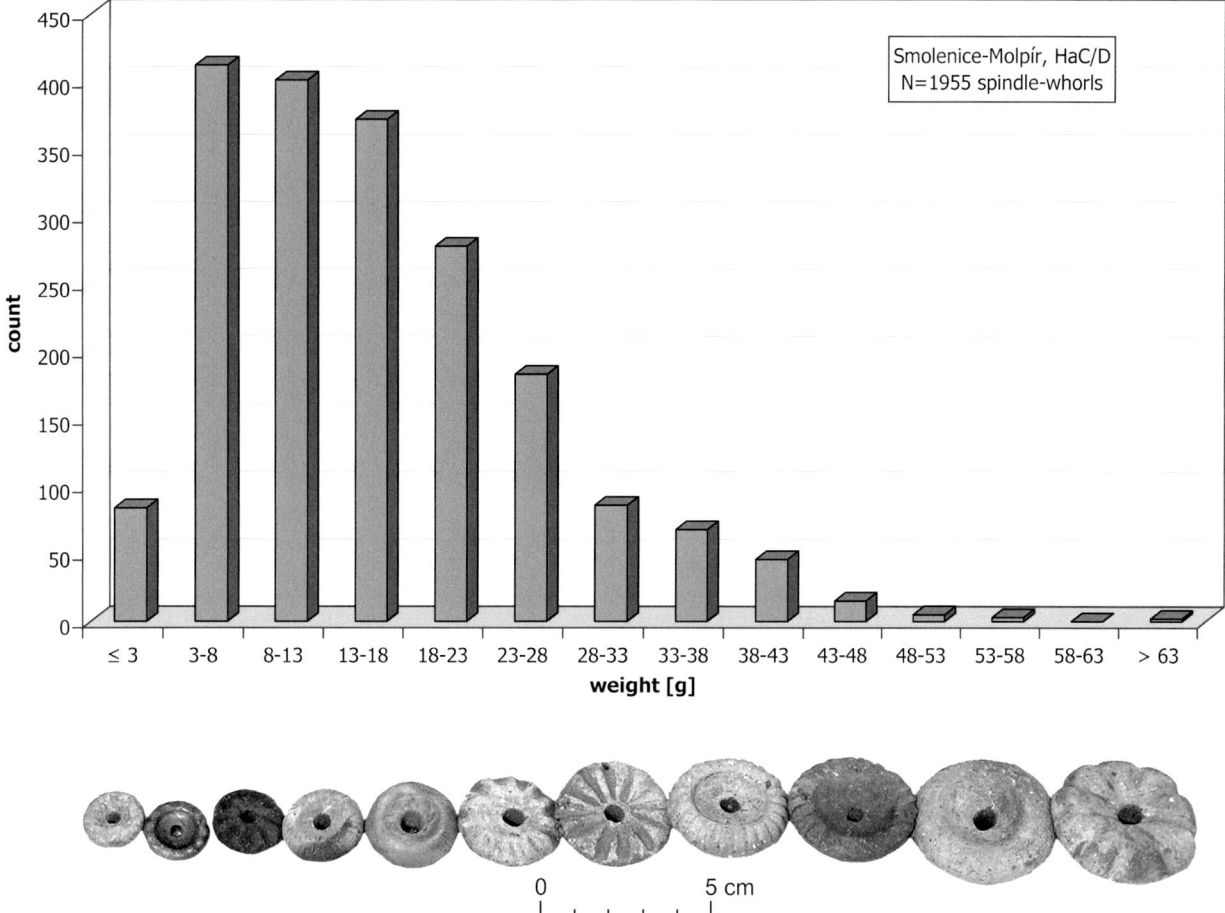

Fig. 3.5. Smolenice-Molpír: distribution histogram of spindle-whorls by weight and selection of spindle-whorls (© T. Belanová).

41–43). The site is strategically situated on the eastern slope of the Little Carpathians, within the north-east Alpine Hallstatt region, where many cultural influences intermingled during the Late Bronze Age and Hallstatt Period. Beside the vast number of pottery and small finds, the site is well known for its over 2200 spindle-whorls and almost 200 loom-weights.[7] All spindle-whorls were weighed in order to obtain general statistics (Fig. 3.5). Out of 2243 spindle-whorls, 1955 complete pieces were measured. The average weight of these spindle-whorls is 15.8 g and the vast majority of them range between 6 to 26 g. A certain amount of very tiny and light spindle-whorls (1–3 g) was found within this assemblage. They have the same shape, material and decoration as the heavier ones and therefore they were included in the statistics. Alongside these clay spindle-whorls, 22 made of bone were found here. The average weight of the measurable pieces is 9.2 g.

As is known from experiments, the size and weight of spindle-whorls have an influence on the produced yarn (*cf.* Andersson 2003b; Mårtensson *et al*; Grömer 2006; 2005b). The very fine qualities of 0.1–0.2 mm thin wool threads from the Early Iron Age can only be produced with very small and light whorls as those found in the Hallstatt Culture. The majority of the Hallstatt and La Tène spindle-whorls weigh 10–20 g each. This corresponds well to yarn qualities which are known from the textiles, of 0.3–0.6 mm diameter.[8] To ply two threads into one, it is advantageous to use the power of a slightly heavier whorl (40 g and more), because it runs well and long after one twist.

In contrast to the carefully produced and elaborately decorated Hallstatt period spindle-whorls, those of the Middle and Late La Tène appear more functional, with a spherical shape, as those from Mannersdorf (Grömer 2004). From the La Tène Period, there is evidence for the reuse of broken pottery. Potsherds were ground into a circular shape with a hole in the centre and were used as whorls (Fig. 3.4). Examples can be found at Linz-Freinberg (Grömer 2004), Plavecké Podhradie – Pohanská (Paulík 1973), Liptovská Mara[9] or Nitra-Šindolka (Březinová 2000, *e.g.* Taf. 94:8, 9, 10).

The visual appearance of the potsherd-whorls is merely a symptom of the times: they are efficient and functional, but not beautiful, because they are produced en masse, from recycled broken pottery. Finely decorated Hallstatt Period spindle-whorls and loom-weights demonstrate the creativity and pride of the owners and producers who worked with them, which is not the case for the La Tène Period tools. These symptoms are also visible in the textiles themselves. While the creativity of the Hallstatt Period weavers (examples from the Hallstatt salt mine) is represented by different weave

Fig. 3.6. Selection of band- and tablet-woven textiles from prehistoric Austria. Mitterberg: Bronze Age; Hallstatt-Ostgruppe: Hallstatt Period (© Natural History Museum, Vienna; after Grömer 2007). Reproduced as a colour plate on page 300.

types, patterns and colours, a reduced repertoire is seen in the Early La Tène Dürrnberg textiles (Stöllner 2005). The pattern is even more evident in the Middle and Late La Tène textiles. They are still of high quality, but appear standardised even in thread count and yarn diameter and only tabby can be recorded (Belanová 2005; Belanová forthcoming; Grömer forthcoming). The preference for z-spun yarn during the La Tène Period fits well in terms of the standardised textiles, with well organised and planned production with maximum output. It appears that serial workshop production was common (Andersson 2003a, fig. 1). Therefore the ergonomically most efficient method for spinning was used – a free hanging spindle by a right-handed person.

The first hint for such standardised mass production was already noted by Katharina von Kurzynski for the Dürrnberg textiles with the predominance of fabric types which can be produced quickly, easily and *en masse* (von Kurzynski 1996, 33–36). Dürrnberg is important, not only for its salt mining industry, which was responsible for its economy, but also for its slaughterhouses and production workshops for pottery, glass, metal, as well as trade.

Different weaving techniques

Apart from the usual finds of loom-weights, such as remains of warp-weighted looms, other weaving tools have been recovered from Austrian and Slovak sites, as for example, wooden weaving swords from the Late La Tène settlement Hallstatt-Dammwiese.[10] A very important source for the reconstruction of various weaving techniques is the Bronze and Iron Age textiles themselves. Their technical details give us a hint of the tools used for production.

Band looms

From Neolithic times, 10–15 cm wide bands from Swiss lake dwellings are known. According to Annemarie Feldtkellner's reconstructions (2003), they could have been produced with paired crescent-shaped loom-weights. Such pieces also occurred at Melk-Spielberg in Austria (Grömer 2006, fig. 18).

From Bronze Age Mitterberg and the Bronze and Iron Age Hallstatt salt mines (Fig. 3.6), narrow wool repp bands (1–1.5 cm wide), and also broader twill and tabby bands (about 6–17 cm wide) from Hallstatt are known. Iron Age narrow repp bands are sometimes patterned by using different dyed yarns for the warp. Check, stripes and complex weaves may be observed. The patterning is different on broader bands from Hallstatt and Dürrnberg (Klose 1926). We know of a checked pattern made by floating warp threads.

Narrow repp bands could have been produced separately or as starting borders. It is hard to say from the band itself what kind of loom was used to make it. For repp or tabby bands it could be a rigid heddle, or just one heddle rod placed on a bundle of warp threads with a fixed natural shed (*e.g.* fixed with a cord). For weaving twill, only three or more sticks are used as heddles and/or some sticks for fixing the pattern on the warp (*Trennstäbe*). If more than 5 cm wide bands are woven, it is good to fix the warp threads in their position with a frame or to fix the warp on sticks at both ends.

Tablet weaving

Narrow tablet-woven borders with a width of 1–3 cm are known from the Hallstatt and Dürrnberg salt mines. All of them were produced using four-holed tablets. Some of the tablet borders from Austria are elaborately patterned, made in various techniques involving the movement of the tablets (Grömer 2005c). Thus far, there is no evidence of tablet-woven bands from the territory of Slovakia. Outside our region, famous finds of weaving tablets, such as the one from Abri Altmühltal, Germany are known (Grote 1994). Triangular tablets with holes in the corners (Fig. 3.7) are recorded from the Hallstatt Period settlement of Smolenice-Molpír (Dušek 1995, Taf. 67:18) and from the Roman Period in Linz-Altstadt, where it was found in a house next to a warp-weighted loom (Karnitsch 1962, Taf. 30).

Warp-weighted loom

For the reconstruction of the prehistoric art of weaving in the circumalpine area, our sources of knowledge are multifaceted. Apart from one possible illustration of a horizontal ground loom at Rabensburg (Franz 1927; Barber 1991, fig. 8.1), there are only depictions of warp-weighted looms in Central Europe, *e.g.* from Val Camonica, Bologna and Sopron (Fig. 3.8).

Nevertheless, textiles with starting borders indicate the use of warp-weighted looms, with one or more heddles. Warping of a loom is shown on the Bologna depiction. Here two women are working together and they are possibly producing warp with a repp (?) starting border (*cf.* Barber 1991, 116, fig. 3.32). Such repp starting borders of 1–1.5 cm width can be found frequently on the Iron Age textiles from the Hallstatt salt mines.

In the Neolithic, Bronze and Iron Age periods, there are textiles woven on a simple warp-weighted loom with a shed rod and one heddle rod. The weaves are tabby, basket weave, half-basket weave and repp. Neolithic and Bronze Age textiles are usually tabby, whilst basket weave appears first in the Urnfield Culture (Vösendorf). During the Early Iron Age (Hallstatt

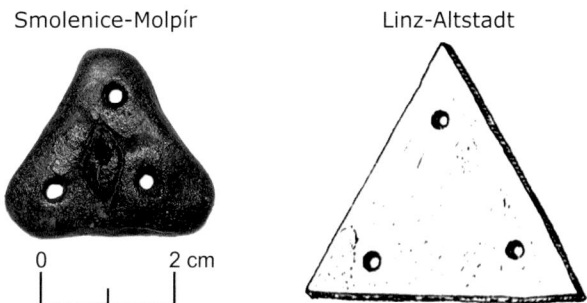

Fig. 3.7. A possible clay weaving tablet from Smolenice-Molpír, Slovakia (© West-Slovakian Museum in Trnava) and a bone weaving tablet from Linz-Altstadt, Austria (Karnitsch 1962, Taf. 30:10).

Fig. 3.8. Sopron, Western Hungary, tumulus 27: 3D scan of the spinning and weaving scene (Scan: T. Böhm, M. Singer, Archäologie-Service 2008, © Natural History Museum, Vienna).

Period), tabby is replaced by many twill variants, but in the La Tène Period tabby is the most common weave again.

The female weaver on the *tintinnabulum* of Bologna produces a weave with two rows of loom-weights, a shed rod and one heddle rod. Maybe she is weaving tabby or its variants. The loom-weights are depicted as dots.

A loom with more than one heddle rod for weaving twill and its variants is characteristic for the Hallstatt Period, as illustrated on the famous urn of Sopron (Fig. 3.8). The diagonal crossings in the woven part could be interpreted as a depiction of the weave, possibly a diagonal twill. In Central Europe, twill first appears at the Bronze Age salt mines of the Hallstatt site, *e.g.* Grüner Werk, *c.* 1500 BC and Christian-Tusch-Werk (Grömer 2007). From Malanser, Liechtenstein we know of a 2/2 twill dated to the end of the 14th century BC, BzD (Bazanella *et al.* 2003, 273). In the Hallstatt Period, 2/2 twill is the most common weave type. More complicated weaves like herringbone or lozenge twill demonstrate the creativity of the Hallstatt weavers. Twill occurs seldom during the La Tène Period, although some do occur at the beginning of the La Tène, usually 2/2 twills, or sometimes 2/1 twill, as at the Dürrnberg salt mine (Stöllner 2005, fig. 6).

In our region we know of loom-weights from Neolithic times onwards. The common Late Neolithic/Eneolithic and Early Bronze Age types are spherical or cylindrical ones. In Austria, the oldest in situ find of a warp-weighted loom is known from the Jevišovice Culture in Krems-Hundssteig, (Grömer 2006). The shape of the Late Bronze Age and Hallstatt Period weights is usually pyramidal, spherical, or discoid, but flat oval weights with a hole on top are known too (Fig. 3.9). Many of the loom-weights in the whole Eastern Hallstatt Culture bear various marks on the top, usually in the form of crosses, or dots. They can be functional as signs of the producer or owner, or they could be useful during weaving as markings for the patterning (*e.g.* in the flying thread technique?).

Apart from the usual truncated pyramidal loom-weights, some flat and heart-shaped ones were found on the hilltop settlement of Molpír near Smolenice, Slovakia, as well as few atypical shapes or decorated pieces. Two small pieces

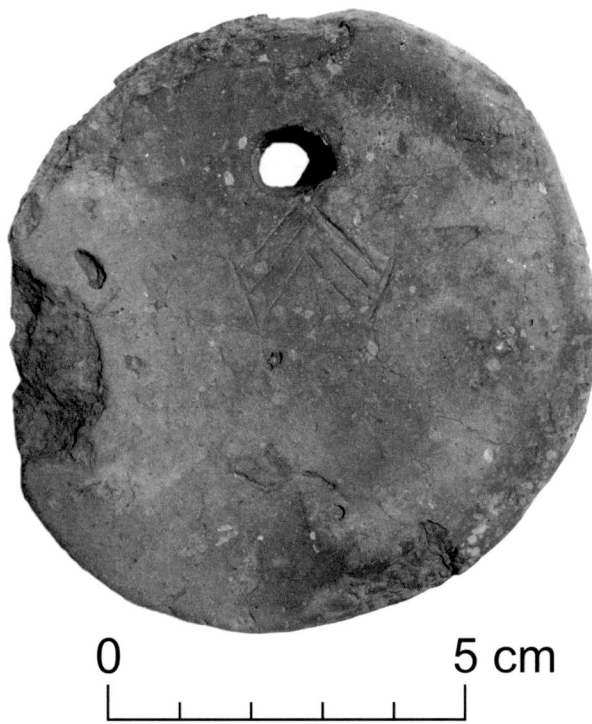

Fig. 3.9. Hallstatt-Salzbergtal: flat disc-shaped loom-weight (© Natural History Museum, Vienna).

from House 17 were covered by unusual patterns: they bear geometric, zoomorphic and anthropomorphic motives and are considered to be idols meant for ritual use (Stegmann-Rajtár 1998, 278–282).

In comparison with the Hallstatt period, loom-weights are not so frequent in the La Tène Period. Beside the state of research, the reason for this could be the introduction of other (new) weaving techniques, *e.g.* the use of the two-beam loom (Stöllner 2005, 173). The Hallstatt Period depiction from Rabensburg possibly shows a horizontal ground loom. Large textiles known in the Iron Age could have been produced on such a loom, too.

How does the shape of the loom-weights influence the final fabrics? Late Neolithic and Bronze Age weights are very large and cylindrical. From the Urnfield Culture and Hallstatt Period, weights become narrow and pyramidal, but do not change in weight (Fig. 3.10). A certain amount of weight per thread is necessary to stretch the warp on the loom. Thus loom-weights do not need much space – and it is thus possible to create a higher thread count with the same weight. This is seen in the known textiles: in the Middle Bronze Age we have lower thread counts, than later in the Hallstatt Period.

The presence of warp-weighted looms in prehistoric settlements is testified by rows of loom-weights *in situ*. From the recorded area, most of them date to the Hallstatt Period. Preliminary analyses have shown that, in the Hallstatt Period, three standard loom sizes were in use. In Stillfried (Eibner 1974), there are 60–90 cm wide looms meant for weaving a narrow fabric. Wider looms of 120–160 cm are known from Michelstetten (Lauermann 2000, 19–20) and Smolenice-Molpír (Dušek and Dušek 1995, 49). It is a loom for the standard fabric where weaving can still be handled by one person. Finally, there are very large 3–4 m wide looms for weaving a special cloth which required the cooperative effort of at least two people. For example, when weaving on the 4 m wide loom from Nové Košariská, about 84 kg of loom-weights had to be moved. In the East Hallstatt region, looms over 3 m wide were found not only at Kleinklein (Dobiat, 1990), but recent excavations have brought to light new finds from lowland settlements, *e.g.* Hafnerbach (Fig. 3.11), (Preinfalk 2003, fig. 12), Freundorf (Blesl and Kalser 2005, 88) and Nové Košariská (Čambal and Gregor 2005, 36–43; Belanová *et al.* 2007). It is noteworthy that the large loom from Kleinklein is not an exceptional find. The presence of such wide looms was not connected only with hilltop settlements as earlier supposed, because they often appear at the lowland settlements mentioned above.

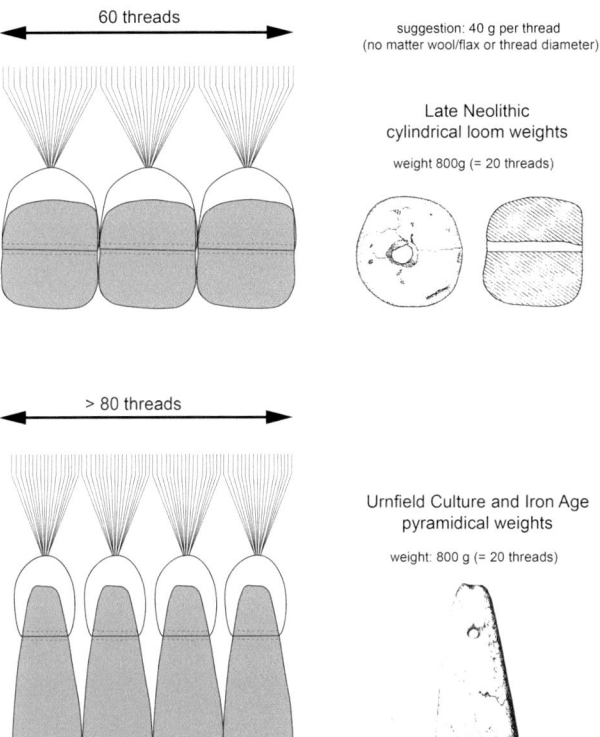

Fig. 3.10. Comparison between the thread spacing of a fabric produced on a Late Neolithic and Urnfield Culture/Iron Age warp-weighted loom (© K. Grömer).

Conclusions

The aim of our paper was to present a general overview of textile production in Austria and Slovakia during the Bronze and Iron Ages. The emphasis was placed on the connection between the most representative and important selection of tools and textiles.

Bronze Age textile craft in Central Europe is presumed

Fig. 3.11. Hafnerbach, Austria: in situ find of 4 m wide loom with over 50 loom-weights, Hallstatt Period (after Preinfalk 2003, fig. 12, © Bundesdenkmalamt, Austria).

to be household production. It is organised on a household level and is not full-time work. From the mass occurrence of textile tools and highly developed textile art, it appears that, in the Hallstatt Period, specialised textile production was attached to the site. In the La Tène Period, textiles were still of a high quality, but they were much more standardised and manifest no striking expressions of creativity involving a wide range of different types, qualities and patterns as in the preceding period. This serial mass production was connected to workshop production for trading purposes and is well represented by the Dürrnberg site.

Apart from the Iron Age salt mines in Austria, textiles do not survive well in the area in focus. Likewise, some of the textile tools were made of perishable organic materials and therefore the absence of evidence does not always have to signify their absence, as *e.g.* for spindle-whorls made of wood during the Early and Middle Bronze Age. It is also important to follow the occurrence of loom-weights within certain periods, because their shape, size and weight can provide more information about the character of textile production. In the future it will be important to obtain more detailed data on as many *in situ* finds of loom-weights as possible, because then we would be able to say more about the quality of textiles and how they were produced. These *in situ* finds fill a lacuna, especially in places, where no textiles survive in the archaeological record.

In conclusion, it is important to note that textiles and tools should not be analysed separately, because together they reveal a much more comprehensive picture of textile production in our prehistory.

Notes

1. Middle Neolithic in Austria – *i.e.* Late Neolithic on the territory of Slovakia; period of Late Neolithic in Austria (*Spät- und Endneolithikum*) comprises Eneolithic in Slovakia.
2. Personal communication from Hans Reschreiter, Museum of Natural History, Vienna.
3. The textiles from the Hallstatt salt mine were analysed by Michael Ryder. Sometimes, wools from "flat" yarns and "fluffy" yarns belong to the same quality group.
4. Personal communication from Juraj Pavúk, Archaeological Institute of the Slovak Academy of Sciences, Nitra.
5. Fibre analysis by A. Rast-Eicher.
6. *E.g.* Urnfield Culture: Vösendorf (Grömer 2007, 142); Staré Město in Moravia (Hrubý 1968–1969), Jánosháza in Hungary (Gondár and Jankovits 2001); Hallstatt Period: Hallstatt, Berg/Attergau (Grömer forthcoming).
7. After the excavations in the1960s and 1970s (Dušek and Dušek 1984; 1995), the site was re-examined by the group of researchers under the guidance of Susanne Stegmann-Rajtár from the Archaeological Institute of SAS, Nitra (*e.g.* Slovak Academy of Sciences 1998, 2001, 2005). Analysis of the material is still in process, being prepared for a new publication. A recent revision of textile tools in May–June 2008 arrived at the final number of 2243 spindle-whorls and 198 loom-weights.
8. Experimentally tested by Grömer in 2004–2005. The wool used was "Tiroler Bergschaf".
9. Personal communication from Karol Pieta, Archaeological Institute of the Slovak Academy of Sciences, Nitra.
10. Personal communication from Hans Reschreiter, Natural History Museum, Vienna.

Bibliography

Andersson, E. (2003a) Textile Production in Scandinavia during the Viking Age. In L. B. Jørgensen, J. Banck-Burgess and A. Rast-Eicher (eds), *Textilien aus Archäologie und Geschichte. Festschrift für Klaus Tidow*, 46–62. Neumünster, Wachholz Verlag.

Andersson, E. (2003b) *Tools for Textile Production: from Birka and Hedeby*. Birka Studies 8. Stockholm

Barber, E. J. W. (1991) *Prehistoric Textiles. The Development of Cloth in the Neolithic and Bronze Ages with Special Reference to the Aegean*. Princeton, Princeton University Press.

Bazzanella, M., Maye, A., Moser, L., and Rast-Eicher, A. (2003) *Textiles – Intrecci e tessuti dalla preistoria europea*. Catalogo della mostra tenutasi a Riva del Garda dal 24 maggio al 19 ottobre 2003. Trento, Esperia.

Belanová, T. (2005) The State of Research of La Tène Textiles from Slovakia and Moravia. In P. Bichler, K. Grömer, R. Hofmann-de Keijzer, A. Kern and H. Reschreiter (eds), *Hallstatt Textiles – Technical Analysis, Scientific Investigation and Experiment on Iron Age Textiles*, 175–189. Oxford, BAR S1351.

Belanová, T. (2007) Archaeological Textile Finds from Slovakia and Moravia Revisited. In A. Rast-Eicher and R. Windler (eds), *Archäologische Textilfunde – Archaeological Textiles. Report from the 9th NESAT Symposium, 18–21 May 2005 in Braunwald*, 41–48. Ennenda, ArcheoTex.

Belanová, T. *et al.* (2007) Die Weberin von Nové Košariská – Die Webstuhlbefunde in der Siedlung von Nové Košariská im Vergleich mit ähnlichen Fundplätzen des östlichen Hallstattkulturkreises. In M. Blečić, M. Črešnar, B. Hänsel, A. Hellmuth, E. Kaiser and C. Metzner-Nebelsick (eds), *Scripta praehistorica in honorem Biba Teržan*, 419–434. Situla 44, Dissertationes Musei Nationalis Sloveniae, Ljubljana, Narodni Muzej Slovenije.

Belanová, T. (Forthcoming) Iron Age Textile Finds from Slovakia and Moravia. In M. Gleba and U. Mannering (eds), *Textiles in Context*. Oxford, Oxbow Books.

Blesl, Ch. and Kalser, K. (2005) Die Hallstattzeitliche Siedlung von Freundorf. In *Zeitschienen vom Tullnerfeld ins Traisental. Fundberichte aus Österreich Materialhefte*, Reihe A, Sonderheft 2, 86–89.

Březinová, G. (2000) *Nitra-Šindolka. Siedlung aus der Latènezeit. Katalog*. Archaeologica Slovaca Monographiae VIII. Veda, Bratislava-Nitra.

Čambal, R., and Gregor, M. (2005) *Dunajská Lužná v Praveku*. Obec Dunajská Lužná.

Dobiat, C. (1990) *Der Burgstallkogel bei Kleinklein I. Die Ausgrabungen der Jahre 1982 und 1984*. Marburger Studien zur Vor-und Frühgeschichte 13. Marburg.

Dušek, M., and Dušek, S. (1984) *Smolenice-Molpír. Befestigter Fürstensitz der Hallstattzeit I*. Nitra.

Dušek, M., and Dušek, S. (1995) *Smolenice-Molpír. Befestigter Fürstensitz der Hallstattzeit II*. Nitra.

Eibner, A. (1974) Zum Befund einer hallstattzeitlichen Webgrube aus Stillfried. *Forschungen in Stillfried* 1, 76–84.

Eibner, A. (1986) Die Frau mit der Spindel. Zum Aussagewert einer archäologischen Quelle. *Hallstatt-Kolloquium Veszprém 1984. Mitteilungen des Archäologischen Instituts der Ungarischen Akademie der Wissenschaften*, Beih. 3, 39–48. Budapest.

Feldtkeller, A. (2003) Nierenförmige Webgewichte – wie funktionieren sie? *ATN* 37, 16–18.

Franz, L. (1927) Eine niederösterreichische Urnenzeichnung. *IPEK Jahrbuch für Prähistorische und Ethnographische Kunst* 3, 96–99.

Gondár, E. and Jankovits, K. (2001) Die Untersuchung der Textilreste an einem Bronzearmband aus der Spätbronzezeit. *Ösrégészeti Levelek* 3, 37–40.

Grömer, K. (2003) Ein Spinnrocken aus einem spätantiken Grab von Mautern/Favianis. In *Berichte zu den Ausgrabungen des Vereines ASINOE im Projektjahr 2003. Fundberichte aus Österreich* 42, 465–469.

Grömer, K. (2004) Aussagemöglichkeiten zur Tätigkeit des Spinnens aufgrund archäologischer Funde und Experimente. *Archaeologica Austriaca* 88, 169–182.

Grömer, K. (2005a) The Textiles from the Prehistoric Salt-mines at Hallstatt. In P. Bichler, K. Grömer, R. Hofmann-de Keijzer, A. Kern and H. Reschreiter (eds), *Hallstatt Textiles – Technical Analysis, Scientific Investigation and Experiments on Iron Age Textiles*, 17–40. BAR S1351. Oxford.

Grömer, K. (2005b) Efficiency and Technique – Experiments with Original Spindle Whorls. In P. Bichler, K. Grömer, R. Hofmann-de Keijzer, A. Kern and H. Reschreiter (eds), *Hallstatt Textiles – Technical Analysis, Scientific Investigation and Experiments on Iron Age Textiles*, 107–116. BAR S1351. Oxford.

Grömer, K. (2005c) Tablet-woven Ribbons from the Prehistoric Salt-mines at Hallstatt, Austria – results of some experiments. In P. Bichler, K. Grömer, R. Hofmann-de Keijzer, A. Kern and H. Reschreiter (eds), *Hallstatt Textiles – Technical Analysis, Scientific Investigation and Experiments on Iron Age Textiles*, 89–90. BAR S1351. Oxford.

Grömer, K. (2006) Vom Spinnen und Weben, Flechten und Zwirnen. Hinweise zur neolithischen Textiltechnik an österreichischen Fundstellen. In *Festschrift für Elisabeth Ruttkay, Archäologie Österreichs* 17/2, 177–192. Wien.

Grömer, K. (2007) *Bronzezeitliche Gewebefunde aus Hallstatt – Ihr Kontext in der Textilkunde Mitteleuropas und die Entwicklung der Textiltechnologie zur Eisenzeit*. Unpublished dissertation, University of Vienna.

Grömer, K. (Forthcoming) Textile Craft in Iron Age Austria. In M. Gleba and U. Mannering (eds), *Textiles in Context*. Oxford, Oxbow Books.

Grote, K. (1994) *Die Abris im südlichen Leinebergland bei Göttingen. Archäologische Befunde zum Leben unter Felsschutzdächern in urgeschichtlicher Zeit*. Veröffentlichungen der urgeschichtlichen Sammlungen des Landesmuseums zu Hannover 43, Teil I,2. Isensee Verlag, Oldenburg.

Hofmann de Keijzer, R. *et al.* (2005) Dyestuff and Element Analysis on Textiles from the Prehistoric Salt-mines of Hallstatt. In P. Bichler, K. Grömer, R. Hofmann-de Keijzer, A. Kern and H. Reschreiter (eds.), *Hallstatt Textiles – Technical Analysis, Scientific Investigation and Experiments on Iron Age Textiles*, 55–72. BAR S1351. Oxford.

Hrubý, V. (1968–69) Nález tkaniny z mladší doby bronzové ve Starém Městě. (Der Fund eines Gewebefragmentes aus der jüngeren Bronzezeit in Staré Město). *Časopis Moravského Muzea – Acta Musei Moraviae* LIII/LIV, 51–58.

Karl, R. (1996) *Latènezeitliche Siedlungen in Niederösterreich. Untersuchungen zu Fundtypen, Keramikchronologie, Bautypen, Siedlungstypen und Besiedlungsstrukturen im latènezeitlichen Niederösterreich*. Historica Austria, Band 2 and 3. Wien, Österreichischer Archäologie Bund.

Karnitsch, P. (1962) *Die Linzer Altstadt in römischer und vorgeschichtlicher Zeit*. Linzer Archäologische Forschungen 1, Linz.

Klose, O. (1926) Ein buntes Gewebe aus dem prähistorischen Salzbergwerke auf dem Dürrnberge bei Hallein. *Mitteilungen der anthropologischen Gesellschaft in Wien* 56, 346–350.

Kratochvíl, Z. (2003) Osteologické nálezy z laténských sídlišť v Bořitově-Býkovicích a v Lysicích. Osteologische Funde aus latènezeitlichen Siedlungen in Bořitov-Býkovice und Lysice. In M. Čižmář, *Laténské sídliště v Bořitově. Latènezeitliche Siedlung in Bořitov*, 144–146. Pravěk Supplementum 10, Brno.

Kurzynski, K. v. (1996) *"… und ihre Hosen nennen sie bracas". Textilfunde und Textiltechnologie der Hallstatt- und Latènezeit und ihr Kontext*. Internationale Archäologie 22, Espelkamp.

Lauermann, E. (2000) Archäologische Forschungen in Michelstetten, NÖ. *Archäologie Österreichs* 11:1, 5–30.

Mårtensson, L., Andersson, E., Nosch, M-L., and Batzer, A. (2006) Technical Report. Experimental Archaeology Part 1, Tools and Textiles – Texts and Contexts Research Programme. The Danish National Research Foundation's Centre for Textile Research (CTR), University Copenhagen. Unpublished report.

Paulík, J. (1973) Hlinené praslený na keltskom opide Pohanská v Plaveckom Podhradí. *Zborník Slovenského Národného Múzea LXVII – História* 13, 31–60.

Pavúk, J. (1994) *Štúrovo. Ein Siedlungsplatz der Kultur mit Linearkeramik und der Želiezovce-Gruppe*. AÚ SAV, Nitra.

Pavúk, J. (2003) Hausgrundrisse der Lengyel-Kultur in der Slowakei. In J. Eckert, U. Eisenhauer and A. Zimmermann (eds), *Archäologische Perspektiven – Analysen und Interpretationen im Wandel (Festschrift für Jens Lüning zum 65. Geburtstag)*, 455–470. Rahden/Westf, Verlag Marie Leidorf GmbH.

Peške, L. (2003) Osteologické nálezy z laténského sídliště v Bořitově. In M. Čižmář, *Laténské sídliště v Bořitově. Latènezeitliche Siedlung in Bořitov*, 138–143. Pravěk Supplementum 10. Brno.

Pieta, K. (1999) Der archäologische park und die Feuchtschichtengrabung in Liptovská Mara. In E. Jerem and I. Poroszlai (eds), *Archaeology of the Bronze and Iron Age: Experimental Archaeology, Environmental Archaeology, Archaeological Parks. Proceedings of the International Archaeological Conference, Százhalombatta, 3–7 October, 1996*, 357–366. Archaeolingua, Budapest.

Preinfalk, F. (2003) KG Hafnerbach. In Chr. Farka, Die Abteilung für Bodendenkmale des Bundesdenkmalamtes. *Fundberichte aus Österreich* 42, 15–17.

Pucher, E. (1998) Der Knochenabfall einer späthallstatt-/frühlatènezeitlichen Siedlung bei Inzersdorf ob der Traisen (Niederösterreich). In P. Ramsl, *Inzersdorf-Walpersdorf. Studien zur späthallstatt-/latènezeitlichen Besiedlung im Traisental, Niederösterreich*. Fundberichte aus Österreich, Materialhefte A6, 56–67. Wien.

Pucher, E. (1999) *Archäozoologische Untersuchungen am Tierknochenmaterial der keltischen Gewerbesiedlung im Ramsautal auf dem Dürrnberg*. Dürrnberg-Forschungen 2. Rahden/Westf, Verlag Marie Leidorf GmbH.

Pucher, E. (2004) Hallstattzeitliche Tierknochen aus Göttlesbrunn, p.B. Bruck an der Leitha, Niederösterreich. In M. Griebl (ed.), *Die Siedlung der Hallstattkultur von Göttlesbrunn, Niederösterreich*. Mitteilungen der Prähistorischen Kommission 54, 309–328. Wien, Österreichische Akademie der Wissenschaften.

Rast-Eicher, A. (2005) Bast before Wool: the First Textiles. In P. Bichler, K. Grömer, R. Hofmann-de Keijzer, A. Kern and H. Reschreiter (eds), *Hallstatt Textiles – Technical Analysis, Scientific Investigation and Experiments on Iron Age Textiles*, 117–131. Archaeopress – BAR S1351. Oxford.

Ræder Knudsen, L. (2002) La tessitura a tavoletta nella tomba 89. In P. von Eles (ed.), *Guerriero e sacerdote. Autorità e comunità nell'età der ferro a Verucchio. La Tomba der Trono*, 220–234. Florence.

Ryder, M. (2001) The fibres in textile remains from the Iron Age salt-mines at Hallstatt, Austria. *Annalen des Naturhistorischen Museums Wien* 102A, 223–244.

Stegmann-Rajtár, S. (1998) Spinnen und Weben in Smolenice-Molpír. Ein Beitrag zum wirtschaftlichen und religiös-kultischen Leben der Bewöhner des hallstattzeitlichen "Fürstensitzes". *Slovenská archeológia* XLVI:2, 263–287.

Stegmann-Rajtár, S. (2001) Kultúrne vzťahy haštatského hradiska Molpír pri Smoleniciach na príklade hlinených predmetov kultového charakteru (Kulturelle Beziehungen des Hallstattzeitlichen Burgwalls Molpír bei Smolenice am Beispiel der Tongegenstände mit Kultcharakter). *Pravěk Nová Řada* 10/2000, 457–471.

Stegmann-Rajtár, S. (2005) Smolenice-Molpír. In H. Beck, D. Geuenich and H. Steuer (eds), *Reallexikon der Germanischen Altertumskunde* 29, 146–156. Berlin. New York.

Stöllner, T. (2005) More than Old Rags – Textiles from the Iron Age Salt-mine at the Dürrnberg. In P. Bichler, K. Grömer, R. Hofmann-de Keijzer, A. Kern and H. Reschreiter (eds), *Hallstatt Textiles – Technical Analysis, Scientific Investigation and Experiments on Iron Age Textiles*, 161–174. BAR S1351. Oxford.

Swidrak, I. and Schmidl, A. (2002) Pflanzengroßreste aus der latènezeitlichen Gewerbesiedlung im Ramsautal am Dürrnberg. In C. Dobiat *et al.* (eds), *Dürrnberg und Manching. Wirtschaftsarchäologie im ostkeltischen Raum. Akten des Internationalen Kolloquiums in Hallein/Bad Dürrnberg 7–11 Oktober 1998*, 147–156. Kolloquien zur Vor- und Frühgeschichte 7. Bonn.

Trebsche, P., *et al.* (2007) Untersuchungen zur Wirtschaftsstruktur eines hallstattzeitlichen Marktorts in Ansfelden (Oberösterreich). *Archäologie Österreichs* 18/1, 31–47.

Waldhauser, J. *et al.* (1993) *Die hallstatt- und latènezeitliche Siedlung mit Gräberfeld bei Radovesice in Böhmen*. Archeologický výzkum v severních Čechách 21/2, Praha.

Wild, J. P. (1988) *Textiles in Archaeology*. Shire Archaeology 56. Aylesbury, Shire Publications Ltd.

4 Differences in the Elaboration of Dress in Northern Europe during the Middle Bronze Age

by Sophie Bergerbrant

Introduction

This study uses textile evidence and artefacts to shed light on the elaboration of male and female dress in two contemporary Bronze Age cultures: the Nordic Bronze Age and the Lüneburg Culture. The two cultures appear in the archaeological record from *c.* 1500 BC and the focus of this article is the material from Montelius Periods II (1500–1300 BC) and III (1300–1100 BC) (for details about chronology and geography see Bergerbrant 2007). Even though weaving and related techniques are similar in the two cultures, there are significant differences in male and female appearance, particularly regarding the emphasis placed on the way an outfit was adorned. By discussing this, we obtain knowledge of the amount of time spent creating the clothing and which gender invested the most in achieving his/her 'look'.[1]

Inga Hägg (1996a) states that textile and clothing are important markers of cultural identity. She indicates that the making of costume is a complex and time-consuming task, a craft that is handed down from generation to generation. It is therefore important to study the amount of time and detail used in different areas, and specifically to identify whose clothing required the most effort.

Background to Bronze Age Textiles

According to Lise Bender Jørgensen, several geographical groups can be identified among Bronze Age textiles. The author argues that they correspond with well known cultural groups. The coarse wool tabby in Scandinavia and the North European Lowland forms one group. The textiles of these two cultural areas therefore belong to the same weaving tradition, *i.e.* wool tabbies with single and plied yarn (Bender Jørgensen 1992, 116–117). There are only a few examples of textiles made of different types of spun yarn or in other materials.

Elizabeth Barber has claimed that due to the coarseness of the cloth found in Scandinavia during the Middle Bronze Age, the weaving technique must be new here. However, she also asserts that the embroidery that adorns the cloth, which she views as elegant and highly skilful, must reflect millennia of knowledge of needlework of different kinds (Barber 1991, 176–177). All Middle Bronze Age cloth from southern Scandinavia may be categorised as 'coarse fabric', but, according to Broholm and Hald (1940, 110) it is not the product of beginners, *i.e.* the technology utilized during Period II had to have had a long tradition behind it. Thus, it is clear that there are divergent opinions amongst these writers on the length and the skill of the weaving tradition in this region. Karen Hanne Stærmose Nielsen (1989, 48) argues that cutting and stitching of fabric is a very unusual activity among societies with limited weaving technology. There is evidence of cutting and stitching in the Nordic material as seen on the blouses from the Danish oak-log coffins (Broholm and Hald 1940), a clothing making technique that is also assumed to be used in the Lüneburg Culture (Laux 1996), and this would indicate that the weaving tradition had an antecedent as assumed by Broholm and Hald.

Textiles in the Nordic Bronze Age

Despite the alleged coarseness of the Bronze Age textiles in the region, there were a number of different ways to make more elaborate textiles. By combining s- and z-spun yarn, different optical effects on the surface of the fabric could be obtained (Demant 2000, 355). The way the different s- and z- spun yarns are combined in the weave therefore creates different patterning effects, even when only the tabby weaving technique was used. This can be seen, for example, in the wool belt from Borum Eshøj, Denmark in which an optical illusion of a zigzag was created by mixing s- and z-spun yarn in the warp (Broholm and Hald 1935, 278–280; Barber 1991, 197). Spun yarns can be combined into two-ply yarns. Plying can create different types of yarns, such as Sz or Zs (Bender Jørgensen 1992, 13). These types of plied yarns are generally found in the Middle Bronze Age in southern Scandinavia on *e.g.* corded skirts or other types of strings, but are not used in the weaving of cloth (Bender Jørgensen 1986, catalogue; Ehlers 1998, catalogue).

It has also been shown that the textiles have different qualities. For example, the Borum Eshøj textiles are woven with less refinement than the textiles found in Trindhøj,

Denmark (Kristiansen 1979, 189). The treatment of the textiles might have become more sophisticated through time. In the Period III outfit from the grave in Skrydstrup, Denmark there is evidence that ten different yarns had been used, yet the weaving technique is the same as in other textiles from the period (Nielsen 1980, 12; Bender Jørgensen, Munksgaard and Stærmose Nielsen 1984, 39, 43).

After the cloth itself was finished, there were many ways of creating differences in the costume. For example, there are cases where the cloth had piled stitches added, making it resemble fur. The pile technique is mainly found on caps, but is also seen on the cloak from Trindhøj, and on textile fragments from the Melhøj grave in Denmark (Broholm and Hald 1948, 70; Nielsen 1988, 21; Stærmose Nielsen 1989, 36). Most examples of pile stitches, especially from early in the period, are from male burials.

As mentioned earlier, there is evidence of cutting and stitching in the material. This can be seen, for example, in the blouses from several female burials (Borum Eshøj grave C, Skrydstrup and Egtved in Denmark), as well as in the wrap-around garments from the male graves in Muldbjerg and Trindhøj in Denmark (Broholm and Hald 1940; Hald 1974, 63, 69–71). It is noteworthy, then, that this textile technique was used in both male and female costumes.

There is evidence of clothing with embroidery from a large part of northern Europe, from the Middle Bronze Age. The earliest evidence of embroidery found within the Nordic region is from a probable blouse, recovered from a Period II burial at Flintbek, North Germany (Ehlers 1998, 162–165, 222–225). However, it contains a deceased young woman aged 15–16, who, based on the metal objects that accompanied her, most likely came from the Ilmenau area of the Lüneburg Heath in Germany (Zich 1992, 186; Bergerbrant 2005, 165–166). Embroidery on female blouses also exists in Period III burials both in the Nordic region (Skrydstrup and Melhøj) and in Lower Saxony, Germany (Heiligenthal). There is only one known example of embroidery on male costume and that is from a bog find from Emmer-Erfscheidenveen, Holland dating to Period II or III (Bender Jørgensen, Munksgaard and Stærmose Nielsen 1989, 39, 43; Ehlers 1998, 166–170; Comis 2003, 193–197, van der Plicht *et al.* 2004; Fendel 2006, 35).

Colour is another way of both making more elaborate textiles and creating difference in the costume. The colour of the cloth for the Middle Bronze Age textiles has been subject to debate. According to some scholars, the now brown textiles could have become that way from spending millennia in a wet environment (Hedeager Madsen 1988, 249). However, microscopic examinations have shown that the wool was brown from the beginning. There are also a few instances where white wool was used in the Middle Bronze Age, for example, in the white belt from the Skrydstrup grave and a very light textile (probably a blanket, coat or shawl) from the Trindhøj grave (Stærmose Nielsen 1989, 57; Ryder 1990, 137–140). As far as may be ascertained, there are no archaeological traces from the Scandinavian Middle Bronze Age indicating that dyed yarn was used. The earliest known examples of dyed yarn in Scandinavia date to the Early Iron Age and, prior to this, it is most likely that only natural pigment was used to create patterning (Bender Jørgensen and Walton 1986, 186; Vanden Berghe *et al.* this volume). Bronze Age people probably created patterns by using different shades of yarn. This can be seen in the use of a belt of a lighter shade in the earlier mentioned Skrydstrup grave and the possibly darker yarn used for the embroidery on the textiles from Emmer-Erfscheidenveen (van der Sanden 1996, 124; Comis 2003, 193–197).

A hairnet made of horsehair was found in the Skydstrup burial, and beneath the woman's left cheek was a sprang cap with cords. Corded skirts are probably the most well known item of clothing from the Nordic Bronze Age. Corded skirts and the different kinds of hairnets are examples of accessories created by textile techniques that are different from that used to create larger items of woven fabric (Broholm and Hald 1939; 1948, 13–20; Hald 1974, 71–81).

Based on indirect evidence, the artefacts, we can deduce that leather objects were probably used as a part of many costumes. For example, the existence of a double button indicates the presence of a leather belt or strap. This claim can be substantiated as all cases of double buttons in the well-known oak-log coffins are related to leather belts or straps. This is equally valid for other examples of well preserved material such as those from Hvidegård, Lyngby-Tårbæk and Jægersborg, Gentofte (Anér and Kersten 1973), both in north Zealand, Denmark. In regions where some of the weapons were worn differently, *e.g.* tied to the leg, there are few or no belt hooks or double buttons (Bergerbrant 2007, 75–80).

Textiles in the Lüneburg Culture

Textile fragments from the Bronze Age have only been analysed from a small number of graves in Lower Saxony (Ehlers 1998, catalogue). There are, however, a number of unanalysed textile fragments from, for example, Wardböhmen, Lower Saxony (Bergerbrant and Malmius 1996).

One grave with textile remains in Lower Saxony is grave 2 in Heiligenthal, mound 7, Lüneburg where a large quantity of bronze artefacts was also found. Textile fragments were found in connection with all the bronze objects. All determinable textile fragments were in tabby and s/s-spun. According to Ehlers, the deceased woman had worn a short-sleeved blouse with a decorative edge at the neck opening, probably a belt and/or a skirt with jewellery attached, socks of some kind, and a possible cloak that covered the body (Ehlers 1998, 166–170; Fendel 2006, 32–35). Another female grave, this time from Quelkhorn, Verden (grave C), Germany, also had textile fragments that can contribute to our understanding of the clothing from this region. There were wool textile fragments of tabby of s/z-spun yarn. In contrast to Ehlers, Inga Hägg interprets these fragments as remnants of a long-sleeved blouse rather than a covering cloak. This is based on the fact that the textile fragments were found under the arm rather than over it (Hägg 1996b). No traces of any kind of skirt were found.

According to these interpretations, the clothing in these burials resembles the Nordic clothing in many ways, albeit with some important differences. If Ehlers is correct in her assumption that the textile fragments above the different types of rings in the Heiligenthal burials are the remains of a cloak and not part of a long skirt, then the skirt must have been shorter than the Scandinavian long skirts. In contrast to southern Scandinavia, it was common to wear ankle rings in this region, and the presence of ankle rings might suggest a preference for shorter skirts since these otherwise would be hidden by a long skirt. Thus, it could possibly be a string skirt, even though we do not have any clear evidence for its use in this area during the Middle Bronze Age. There is evidence of bronze tubes (Ehlers 1998, catalogue; Bergerbrant 2007, 82–84), but these were apparently associated with headgear, not string skirts as in southern Scandinavia.

Similarly, the head ought to have had some kind of textile headcovering, onto which the bronze sheet (diadem) had been sewn. However, this headcovering was apparently not preserved. It seems unlikely that these headcoverings were made of some kind of netting, as seen in the southern Scandinavian female graves. Other graves also show that some women in the Lüneburg Heath had headdresses that were heavily embellished with bronze objects (Laux 1996; Ehlers 1998, 196).

Although male graves with textile fragments occur, *e.g.* at Quelkhorn or Verden (Ehlers 1996), there is insufficient evidence for a serious discussion of male clothing. This is because the textile fragments are often found in association with a dagger and provide no other information about the context, or have not been analysed, such as grave IV in mound 13 in Schafstallberge, Wardböhmen, and Celle in Germany (Bergerbrant and Malmius 2006). Nevertheless, Inga Hägg (1996b, 233) argues that it is likely that middle and south Jutland and the northern German coastal area had a shared clothing tradition in the Bronze Age, as they belonged to the same cultural sphere in the Late Neolithic. Until all unanalysed textile fragments are analysed and brought into the discussion about male clothing in the Lüneburg area, we can only assume that they were similar to the pieces of clothing in the Nordic Bronze Age.

As for the females, many were heavily equipped with bronze objects, which would have made a distinctive visual impact. As indicated by the burial evidence, it would have been a heavy burden to the wearer. It seems likely that it was the norm in this area to sew bronze objects, such as studs, onto the clothing, which seldom can be documented in south Scandinavia. That the studs might have been sewn onto different kinds of clothing, such as headgear, cloaks or belts (Piesker 1958; Bergerbrant 2007, 83–84) is shown by the thread found in one of the studs in the burial at Heiligenthal (Fendel 2006, 33). This indicates that the bronze objects were a permanent part of the clothing in this region. Thus, this apparent taste for more ostentatious decoration also led to a more elaborate treatment of textiles than observed in southern Scandinavia, as indicated by the objects found in the graves. It seems likely, however, that the other kinds of elaboration of the cloth as those found in Scandinavia, such as the use of different yarns and the creation of different optical effects by combining s- and z-spun yarn, also existed in this region.

At present, however, the lack of textile analyses makes this hypothesis difficult to prove.

Discussion

Male Costume

There is little remaining evidence of male clothing from the Lüneburg Culture, which makes a comparison between the two cultures problematic. In the Nordic Bronze Age, many males were buried with a number of bronze objects, whereas male burials in the Lüneburg Culture possessed few bronze items. It is therefore clearly indicated that their costumes differed in some ways. In the Nordic region, men wore caps of different kinds, and many burials contain bronze or wooden double buttons, which indicate the use of leather belts or straps. Their cloaks were fastened with a fibula. In contrast, in the Lüneburg Culture, the cloak was fastened with a pin and there are no indications of the use of leather belts or straps. There is also evidence indicating that the headgear was different as seen in the finds of *Lockenrings*. Comparing the male costumes from the two cultures, it seems that more work must have gone into creating the Nordic costumes. In the Lüneburg Culture however, due to the lack of finds, it is impossible to say how much effort was put into creating elaborate cloth (*e.g.* with pile or embroidery). However, even if we assume that an equal amount of effort was invested in the cloth, the evidence of *e.g.* leather straps in the Nordic area reveals that the men there probably had a more complex and elaborate costume, comprising both textile belts and leather straps (for further details of the Nordic and the Lüneburg costume, see Bergerbrant 2007, chapter 4).

Female Costume

Where women are concerned, the picture is somewhat different. Here, again, there is a lack of evidence of textiles with different weaving techniques, *i.e.* displaying the creation of optical illusions within the fabric, from the Lüneburg Heath. This is however, probably due to the lack of finds, rather than real lack of technique in the region, as there is evidence of other kinds of elaborate textile techniques. For example, there are indications that embroidery was used earlier in this area than in the Nordic region (Bergerbrant 2008). However, it seems that the corded skirt techniques were not employed by the inhabitants of the Lüneburg Culture. Nevertheless, there are remains of cords in bronze tubes from this region, even if these come from headgear rather than skirts. We can therefore assume that cording technologies were known and used in the Lüneburg Culture even if it was used in a different way. In the Lüneburg area, there is clearly evidence of objects sewn onto the female costume, including examples of clothing embellished with over 200 studs, *e.g.* grave II mound 1 in Schaftstallberg. In comparison, there are only a few examples of artefacts with permanently sewn on decorations from the Nordic Middle Bronze Age. Evidently, there were different traditions concerning female dress, and more time and effort was put into creating the female costume in the Lüneburg area than in southern Scandinavia. This can be seen most clearly through the quantity of bronze objects sewn onto the female clothing (for details of the Nordic and the Lüneburg costume see Bergerbrant 2007, chapter 4).

Conclusions

There seems to be a difference in the effort that was invested in creating costumes for the different genders. Based on the burial evidence, males from Scandinavia were generally accompanied by more bronze objects than females and at least some of their clothing required a significant investment of labour, such as the pile on the cloak from Mulbjerg. In contrast, the men from the Lüneburg Heath had fewer bronze objects than the women in that region. The burials indicate that the women of Lower Saxony possessed both more bronze objects and more elaborate items of clothing. It seems that in the Nordic region, more effort was placed on the male appearance, whereas in Lower Saxony, the opposite occurred, and here, the females were the subject of display.

It also seems that some of the new textile impulses from central and south-eastern Europe reached the Lüneburg Heath first, and spread to Scandinavia from there. This can best be seen in the arrival of the technique of embroidery, which may be attributed to long distance marriage, for which there is evidence from Period IB, and the later evidence from Period II of intermarriage between Scandinavia and the Lüneburg Culture. In addition, there are bronze artefacts from the Lüneburg women that relate both to central Europe and Scandinavia, indicating contact and influences with both these regions (Bergerbrant 2007, 87–91). This might explain why certain textile innovations first appear in this area. An examination of all the unanalysed textile fragments in the Niedersächsisches Landesmuseum in Hannover, Germany may help us to understand these textile innovations and their spread even better.

Many textile innovations occurred in Central Europe during the Middle Bronze Age, such as the introduction of twill and the use of dye to colour yarn (Grömer 2007, 305). There is no evidence for these innovations and the use of colour in the North European material until the Early Iron Age (Bender Jørgensen and Walton 1986, 186; Vanden Berghe *et al.* in this volume). It seems that changes in textile production in these two regions occurred at about the same time, but several centuries later than in Central Europe.

It seems likely that both cultures had knowledge of textiles beyond the basics, which would indicate that inhabitants of the region had been weaving and making costumes for a while at this time. The degree of effort put into dress by the genders seems to vary between the two cultures, with the males of Scandinavia and the females from the Lüneburg Heath displaying evidence for the most elaborate costumes. There are also indications that the Lüneburg culture had gained access to new technology first, and that it spread from here to southern Scandinavia through intermarriage.

Note

1. In using terms such as cloth, clothing and costume, I follow Sørensen's (1991; 1997) definitions. Outfit as defined in this article consists of a number of items or pieces of clothing and accompanying artefacts.

Bibliography

Aner, E., and Kersten, K. (1973) *Die Funde der älteren Bronzezeit des nordischen Kreises in Dänemark, Schleswig-Holstein und Niedersachsen.* Vol. 1. Neumünster, Karl Wachholz Verlag.

Barber, E. J. W. (1991) *Prehistoric Textiles: The Development of Cloth in the Neolithic and Bronze Age with Special Reference to the Aegean.* Princeton, Princeton University Press.

Bender Jørgensen, L. (1986) *Forhistoriske Textiler i Skandinavien.* Nordiske Fortidsminder Ser B 9. København, Det Kongelige Nordiske Oldskriftsselskab.

Bender Jørgensen, L. (1992) *North European Textiles until AD 1000.* Aarhus, Aarhus University Press.

Bender Jørgensen, L., Munksgaard, E., and Stærmose Nielsen, K.-H. (1984) Melhøj-fundet. En hidtil upåagtet parallel til Skydstrupfundet. *Aarbøger* 1982, 19–57.

Bender Jørgensen, L., and Walton, P. (1986) Dyes and Fleece Types in Prehistoric Textiles from Scandinavia and Germany. *Journal of Danish Archaeology* 5, 177–188.

Bergerbrant, S. (2005) Fremde Frau eller i lånade fjädrar? Interaktion mellan Sydskandinavien och norra Europa under period I och II. In J. Golhahn (ed.), *Mellan sten till järn Rapport från det 9:e Nordiska bronsålderssymposiet, Göteborg oktober 2003–10–09/12,* 229–240. Gotarc Series C Arkeologiska Skrifter no 59, Gothenburg.

Bergerbrant, S. (2007) *Bronze Age Identities: Costume, Conflict and Contact in Northern Europe 1600–1300 BC.* Lindome, Bricoleur Press.

Bergerbrant, S. (2008) Weaving identity – cultural belonging and cultural change, 1600–1100 BC in southern Scandinavia and northern Germany. *Lund Archaeological Review* 13–14, 5–17.

Bergerbrant, S., and Malmius, A. (2006) A hidden treasure – Bronze Age textile remains from Lower Saxony. *ATN* 43, 2–5.

Broholm, H. C., and Hald, M. (1935). Danske Bronzealders Dragter. *Nordiske Fortidsminder* II:5 and 6, 1–347.

Broholm, H. C., and Hald, M. (1939) Skrydstrupfundet. *Nordiske Fortidsminder* III:2, 1–116.

Broholm, H. C., and Hald, M. (1940) *Costumes of the Bronze Age in Denmark.* Copenhagen, Arnold Busck.

Broholm, H. C., and Hald, M. (1948) *Bronze Age Fashion.* Copenhagen.

Comis, S. Y. (2003) Prehistoric Garments from the Netherlands. In L. Bender Jørgensen, J. Banck-Burgess and A. Rast-Eicher (eds), *Textilen aus Archäologie und Geschichte. Festschrift Klaus Tidow,* 193–204. Neumünster, Wachholz Verlag.

Demant, I. (2000) Die Textilfragmente von Skovgårde. In P. Ethelberg (ed.), *Skovgårde Eine Bestattungsplatz mit reichen Frauengräber des 3. Jhs n.Chr. auf Seeland,* 348–361. Nordiske Fortidsminder serie B 19. Copenhagen.

Ehlers, S. K. (1996) Die Textilien aus dem Schwertgrab bei Quelkhorn, Ldkr. Verden. *Die Kunde N.F.* 47, 237–241.

Ehlers, S. (1998) *Bronzezeitliche Textilen aus Schleswig-Holstein. Eine Technische Analyse und Funktionsbestimmung.* Dissertation zur Erlandung des Doktorsgrad der Philosophischen Fakultät der Christian-Albrects-Universität zu Kiel.

Fendel, H. (2006) *Eine bronzezeitliche Frauenbestattung aus Heiligenthal*. Hamburger Beiträge zur Archäologie Werkstatterreihe. Berlin, LIT Verlag.

Grömer, K. (2007) *Bronzezeitliche Gewebefunde aus Hallstatt – Ihr Kontext in der Textilkunde Mitteleuropas und die Entwicklung der Textiltechnologie zur Eisenzeit*. Unpublished PhD dissertation, Ur- und Frühgeschichte, University of Vienna.

Hald, M. (1974) *Oldtidsdragter*. Copenhagen, Nationalmuseet.

Hägg, I. (1996a) Textil und Tracht als Zeugnis von Bevölkerungsverschiebung. *Archäologische Informationen* 19:1–2, 135–147.

Hägg, I. (1996b) Mikrostratigraphische Analyse von Geweberesten auf Armspiralen aus einem Grabhügel bei Quelkhorn, Lkdr. Verden. *Die Kunde N. F.* 47, 223–236.

Hedeager Madsen, A. (1988) The wool Material in the Archaeological Textile Finds. In L. Bender Jørgensen, B. Magnus and E. Munksgaard (eds), *Archaeological Textiles. Report from the 2nd NESAT Symposium 1984*, 247–250. Copenhagen.

Kristiansen, K. (1979) The Consumption of Wealth in Bronze Age Denmark. A Study in the Dynamics of Economic Processes in Tribal Societies. In K. Kristiansen and C. Paludan-Müller (eds), *New Directions in Scandinavian Archaeology*, 158–190. Studies in Scandinavian Prehistory and Early History Volume. Copenhagen, National Museum of Denmark.

Laux, F. (1996) Tracht und Schmuck der Frauen und Männer. In G. Wegner (ed.), *Leben – Glauben – Sterben vor 3000 Jahren Bronzezeit in Niedersachsen. Eine niedersächsische Ausstellung zur Bronzezeit-Kampagne des Europarates*, 95–116. Oldenburg, Niedersächsisches Landesmuseum Hannover, Isenee Verlag.

Nielsen, K.-H. (1980). Ti Slag Garn. *Skalk* 5, 12–15.

Nielsen, K.-H. (1988) Melhøj – an Unheeded Parallel to Skrydstrup. In L. Bender Jørgensen, B. Magnus and E. Munksgaard (eds), *Archaeological Textiles. Report from the 2nd NESAT Symposium 1.–4 v 1984*, 7–25. Copenhagen.

Piesker, H. (1958) *Untersuchungen zur älteren Lüneburgischen Bronzezeit* (Veröffentlichung des Nordwestdeutschen Verbandes für Altertums forschung und der Urgeschichtlichen Sammlungen des Landesmuseums Hannover). Lüneburg.

Ryder, M. L. (1990) Danish Bronze Age Wools. *Journal of Danish Archaeology* 7 (1988), 136–143.

Sørensen, M. L. S. (1991) The Construction of Gender through Appearance. In D. Wade and N. D. Willows (eds), *The Archaeology of Gender. Proceedings of the 22nd Annual Charcmool Conference*, 121–129. Calgary.

Sørensen, M. L. S. (1997) Reading Dress: the Construction of Social Categories and Identities in Bronze Age Europe. *Journal of European Archaeology* 5:1, 31–49.

Stærmose Nielsen, K.-H. (1989) Bronzealdersdragterne som blev en messe værd. *Fynske Minder 1989*, 31–66.

van der Plicht, J., van der Sanden, W., Aerts, A. T., and Streurman, H. J. (2004) Dating bog bodies by means of ^{14}C-AMS. *Journal of Archaeological Science* 31, 471–491.

van der Sanden, W. (1996) *Udødeliggjorte i mosen. Historierne om de nordvesteuropæiske moselig*. Amsterdam, Batavian Lion International.

Zich, B. (1992) Eine Frauenbestattung der Ilmenau – Kultur aus Flintbek, zur Frage von Handels- und Personenkontakten in der älteren Bronzezeit. Symposium St. Jyndevad. *Archäologie in Schleswig/ Arkæologi i Slesvig* 2, 185–191.

5 Avoiding Nasty Surprises: Decision-Making based on Analytical Data

by Lena Bjerregaard, Ute Henniges and Antje Potthast

Introduction

Every time a museum object has to be treated or is exhibited externally or internally, the conservator has to state the condition of the object. Often this is a very personal and perhaps value-laden emotional statement such as: old feathers should not travel! Often it produces confrontations with curators or museum directors, who may have political, academic or other professional reasons why they want exactly this or that object to go on loan, whereas the conservator only looks at the ability of an object to survive an exhibition or transportation.

Estimating the state of conservation from a conservator's point of view is based on the appearance of the textile, the handle or tactility, the brittleness, the colour/bleaching – all subjective statements that will often change with experience, or differ from conservator to conservator. Sometimes, it can also be impossible to see that a textile has really deteriorated. For instance, a historic textile, cotton or linen, might seem to be in good condition to look at; and so is its tactility. Aqueous treatment seems to be feasible and performed without further tests. The textile can survive the washing wonderfully but as it lies drying – flat of course – the edges suddenly begin to go brown and hard, the textile is literally 'burnt' and partly disappears. There is no way to halt this rapid degradation.

Obviously, an easy analysis that could prevent such events occurring appears highly desirable. We also need information on the condition of an object to argue for, and decide on, or against planned activities dealing with conservation, restoration or exhibitions. Ideally, an accepted standard, with numbers that could be referred to, when evaluating an object's state of preservation would considerably improve decision-making in a restoration and exhibition context. The analysis using this standard should, of course, be low cost, with low material requirements, and fast results. The method presented in this paper represents a step towards this ambitious goal.

Two important parameters for evaluating cellulosic fabrics are molecular weight and state of oxidation. Cellulose is a natural polymer, consisting of linear molecules of sugar units. The amount of single units joined together defines to a large extent the strength of the molecule as a whole. Changes on the molecular level can be detected earlier than on a macroscopic level using strength tests. Oxidation on the other hand indicates sensitivity towards external influences and ability to form inter and intra molecular hydrogen bonding: the more oxidized functionalities that exist, the easier the cellulose backbone might be attacked and broken and the less opportunity there will be for hydrogen bonding (Strlič and Kolar 2005).

For measuring the condition of the conservation of cellulose textiles via molecular weight and state of oxidation, we have investigated the application possibilities of fluorescence labelling of oxidized cellulose functionalities followed by gel permeation chromatography and multi-angle laser light scattering (GPC-MALLS) to determine the molecular weight distribution of cellulose chains. This technique has been applied successfully in the field of paper conservation and fulfils the need for precise analysis with low material consumption (Henniges *et al.* 2006).

Experimental

Labelling

Carbazole-9-Carbonyl-Oxy-Amine (CCOA) labelling of carbonyl groups was performed as described earlier (Röhrling *et al.* 2002a; 2002b; Potthast *et al.* 2003).

In preliminary analysis, it was assured that auto-fluorescence of the aged and dyed samples did not interfere with the fluorescence of CCOA label. Historic cotton was chosen as reference material, as it reflects the age of the material, but has not undergone any processing or dying steps.

General Analytics

Gel permeation chromatography (GPC) measurements used the following components: online degasser, Dionex DG-2410; Kontron 420 pump, pulse damper; auto sampler, HP 1100; column oven, Gynkotek STH 585; fluorescence detector TSP FL2000; multiple-angle laser light scattering

(MALLS) detector, Wyatt Dawn DSP with argon ion laser (λ0 = 488 nm); refractive index (RI) detector, Shodex RI-71; Data evaluation was performed with standard Chromeleon, Astra and GRAMS/32 software.

GPC Method

The following parameters were used in the GPC measurements: flow, 1.00 mL min^{-1}; columns, four PL gel mixedA LS, 20μm, 7.5 × 300 mm; fluorescence detection, λ_{ex} = 290 nm, λ_{em} =340 nm; injection volume, 100 μL; run time, 45 min. N,N-dimethylacetamide/ lithium chloride (0.9% w/v), filtered through a 0.02 μm filter, was used as the mobile phase.

Textile Sample

Three different sets of tests were made to cover different angles of questioning. First, undyed archaeological cotton textiles from Peru dating between AD 1000–1550 have been analyzed to discover details of the conservation of pure, but processed cotton material. The second step extended analysis on dyed or painted archaeological Peruvian cotton textiles in comparison with untreated pieces of the same textiles (AD 800–1550). Paints for textiles most often consist of inorganic minerals that are fixed with a binding media to the textile. Most minerals lead to the degradation of the textile – and some of the binding media, too (Timar-Balazsy and Eastop 1998). In the third part, the question of the influence of aqueous washing treatments was also addressed. Therefore, the condition of washed samples was compared to the unwashed reference. The age of these samples is between 3000 and 200 years.

Results and Discussion

It is very difficult to assess the condition of a textile object from a cursory glance. Textiles might look very well preserved, or badly deteriorated (Fig. 5.1). Micro-analytics helps to determine the state of an object in a more quantitative manner. Another question that has to be addressed when assessing and rating the condition of an object is the homogeneity of the textile itself, *i.e.* if there are any tensions to be expected within the object itself due to different degradation conditions. As textiles may usually have quite large dimensions, slightly different conditions can also be expected. Nevertheless, the extent to which these differences occur can vary a lot.

Undyed Cotton

Molecular weight (Mw) is a very important parameter in estimating the condition of a polymer. The longer the average chain, the more stress it may still support before a fibre made out of it will break. The first set of tests on undyed cotton indicated that the Mw of the textile fibres were on average half the Mw of a sample of unprocessed cotton of comparable age and provenance. This indicates stress and degradation due to processing and use of cloth that has to be considered as being normal. There was, however, no evidence of the condition having deteriorated further only due to age, the samples being between 800 and 450 years old (Fig. 5.2).

One very interesting observation, however, was that textiles Inka_on1 and _on2 (two sub-samples of the same textile) were in much worse condition than all the other textiles. It would not have been possible to ascertain this without these analyses, as the textile in question did not look any worse than the other textiles.

Next to molecular weight, oxidation is a very important aspect in the rating of the condition of a textile. Oxidation causes cellulose to be more susceptible to degradation caused by external influences, especially light. The amount of oxidation can be followed by measuring carbonyl group content using fluorescence labelling. When comparing the theoretical number of reducing end groups, which are an inherent feature of the cellulose molecule, to actually measured carbonyl groups, the extent of oxidation in contrast to new carbonyl groups formed after chain rupture, *e.g.* due to hydrolysis, can be differentiated. The difference between the theoretical values (graphically shown in all figures as 'REG') and those determined by fluorescence labelling consequently originates from oxidative processes.

In the following two examples, VA_60274 and VA_

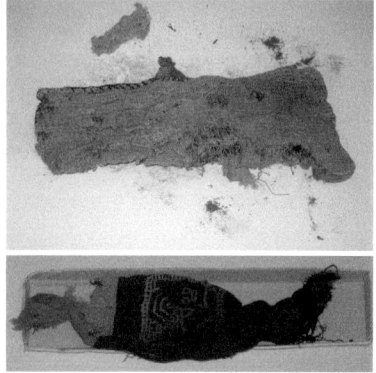

Fig. 5.1. Two archaeological textiles. Top: Inka_on6 looks quite deteriorated, but according to measurement, it is in an acceptable condition, pieces that contribute to the poor impression are merely dirt, but no fibre loss. Bottom: Inka_on1 and on2. The material looks well preserved, but has extremely low Mw (Photo by the authors).

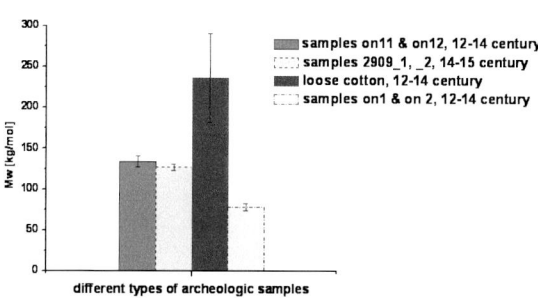

Fig. 5.2. Different types of archaeological samples have been compared to one another. It shows that all textiles have lower Mw than loose cotton. A sample of different age has been included to underline the strong degradation that occurred in samples Inka_on1 and _on2 that is not dependent on age.

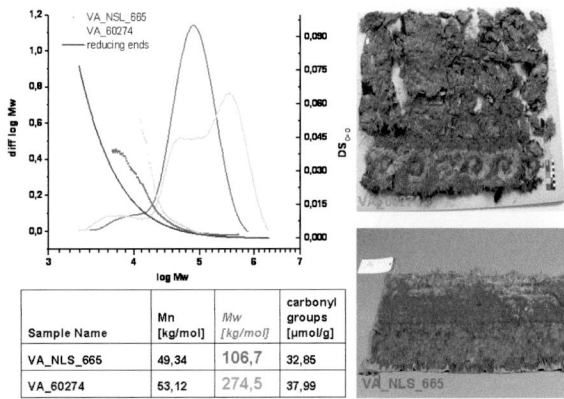
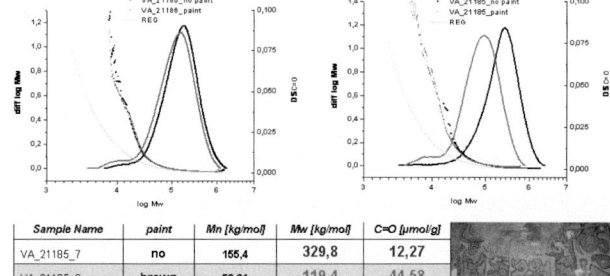

Fig. 5.3. Molecular weight distribution of two different samples, VA_60274 (above) and VA_NLS_665 (below). Pictures of the two objects and an overview of result are added. VA_60274 (Peru; bei Lima) reproduced with kind permission of the Staatliche Museen zu Berlin, Preußischer Kulturbesitz, Ethnologisches Museum.

Fig. 5.4. Comparison between painted and unpainted areas in one object illustrated by two examples. Left side: 21185, 7 undyed fibres, 8 painted fibres. Right side: 21186, 9 undyed fibres, 10 painted fibres. Below: overview of results.

NLS_665, the tested material is a cotton foundation of a feathered textile. Both textiles are woven of undyed cotton. The samples look very different: VA_60274 is brittle and discoloured, whereas VA_NLS_665 seems to be in really good condition. Nevertheless, when it comes to oxidation, both samples are very similar, their oxidation only occurs in the low molecular weight region, and, compared to other textiles of this age, is not very much pronounced. Looking at Mw, VA_60274 yields a more than twice higher average value (Fig. 5.3). At first glance this might be misleading, as it suggests that, despite its brittle appearance, VA_60274 seems to be in a better condition. After GPC analysis though, molecular weight distribution (MWD) is also obtained and the situation looks extremely different. While sample VA_NLS_665 has suffered from a relatively moderate degradation, leading only to a small low molecular weight peak, the molecular weight distribution of sample VA_60274 indicates heavy degradation even though the average molecular weight is higher than for VA_NLS_665. An explanation might be that the starting material had a much higher molecular weight and then degraded. There is not only the low molecular weight shoulder similar to sample VA_NLS_665, but a second far bigger peak where the bulk amount of cellulose chains is expected to be. This example underlines the importance of MWD as additional information. Merely by looking at the average molecular weight, sample VA_60274 might have been rated even better than VA_NLS_665, although the opposite is true. In addition, this example also demonstrates that degradation within one sample might be quite different. Fibres within a single sample do not age homogeneously, and the type and quality of materials within one object varies, a fact often not detectable to the naked eye.

Painted Textiles

To study the influence of paints on Mw and carbonyl group content, two fragments of the same textile, VA_21185 and VA_21186, have been chosen for investigation. It is a plain weave of undyed cotton yarn that has later been painted.

The differences between an undyed and an identical, but painted yarn can be very large, as shown in Figure 5.4. The type of paint on object VA_21185 has obviously caused heavy degradation of the sample, and more chain scission.

Compared to that, the Mw of sample VA_21186 appears a great deal more homogenous, even though slight differences, which are not considered to be significant in this context, occur between the undyed and the painted areas of the sample. Both sub-samples suffered to almost the same extent from oxidation, but there is a difference in molecular weight. This difference is assigned to a homogenous degradation that also affected high molecular weight regions leading to a decrease in Mw (Fig. 5.4).

The observations from molecular weight distribution and $DS_{C=O}$ plot are also reflected and summed up in Table 5.1.

Dyed Textiles

The object chosen for comparing dyed textiles is a *quipu* – a mnemonic device that the ancient Peruvians used for storing information. It consists of a number of plied strings tied to a main string. The strings are mostly in natural brown shades (the cotton in Peru grew in at least five different colours) and with a few dyed strings – mostly blue shades dyed with indigo.

From this example, we acquire an idea that dyes may be more or less aggressive towards the textile. As example in Figure 5.5 shows, it may also be possible that a dye even exerts some kind of protection against the degradation of the textile. An undyed sub-sample is compared to two dyed sub-samples (blue and brown). The undyed sub-sample is considered to reflect the natural degradation of the sample. Nevertheless, the best preserved sub-sample in this example is sub-sample 2 (dyed with indigo). It shows least oxidation and less decrease in molecular weight in comparison to the two other sub-samples, including the undyed sample. In contrast to the brown dye, which induced a higher degree of degradation compared to the undyed sample, the indigo

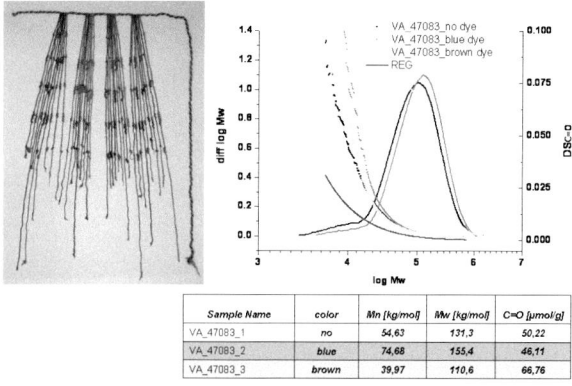

Fig. 5.5. Left: Quipu from Peru, sample VA_47083. Right: Three different sampling positions: sub sample 1 (no dye), sub sample 2 (blue dye), and sub sample 3 (brown dye). Below: Overview of results.

Sample Name	color	Mn [kg/mol]	Mw [kg/mol]	C=O [µmol/g]
VA_47083_1	no	54,63	131,3	50,22
VA_47083_2	blue	74,68	155,4	46,11
VA_47083_3	brown	39,97	110,6	66,76

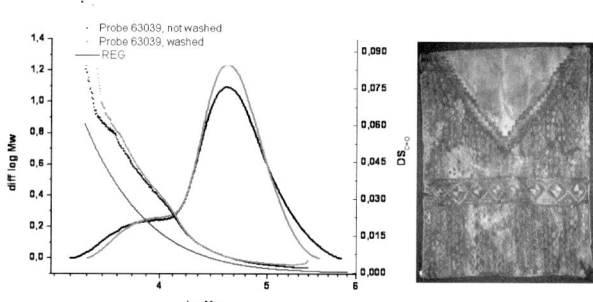

Fig. 5.6. Inca textile from the 15th century washed with non-ionic detergent. Left: molecular weight distribution of an Inca textile before and after washing. No significant changes in the molecular weight distribution and the degree of oxidation are observed. Right: Picture of the Inca textile (VA_62696 Peru; Pachacamac), reproduced with kind permission of the Staatliche Museen zu Berlin, Preußischer Kulturbesitz, Ethnologisches Museum.

	No dye [Mw in kg/mol]	Dye/Paint [Mw in kg/mol]	Ratio
VA_21185	329.8	119.4	2.8
VA_21186	219.1	175.3	1.3
VA_47083	131.3	155.4	0.9
VA_47083	131.3	110.6	1.2
VA_5811	257.6	100.6	2.6
VA_5811	257.6	307.1	0.8
VA_5811	257.6	102.9	2.5
VA_5811	257.6	168.4	1.5

Table 5.1. Overview of ratios between dyed/painted and undyed textiles. The higher the ratio, the more degradation has to be assumed; ratios below 1 indicate no degradation compared to undyed material.

protected the cellulose against both, hydrolysis (higher Mw) and also oxidation (less carbonyl groups). Obviously, the indigo protected the cellulose from light induced ageing. The chemical nature of the dye determines its role during the ageing of cellulose. The dye may act as both, promoter of hydrolysis and oxidation or, as in the case of indigo, inhibitor.

Nevertheless, it is difficult to compare different samples and determine their degree of degradation. In this context, the ratio between the best preserved sample (in this case, VA_47083_2) and the worst sample (VA_47083_3) might serve as a means to make values more comparable (Table 5.1).

Washed Textiles

The influence of aqueous cleaning on Mw and carbonyl group content was studied on fragments of several textiles. The oldest piece is Egyptian linen dating from *c.*1000 BC. Another textile is of Inca origin from the 15th century AD. The most recent textile was produced in the 19th century. The unwashed reference was in each case compared to a washed sample. The washing was performed with a non-ionic detergent.

No significant influence of washing on the cellulose molecule and carbonyl group content was found in the older textile samples. Even the molecular weight distribution remained largely unchanged by the wetting and drying procedure (Fig. 5.6).

However, washing is not always a safe procedure. On more recent textiles, *i.e.* those from the 19th century, the washing procedure caused a significant loss of molecular weight. This loss is reflected in a change of molecular weight distribution

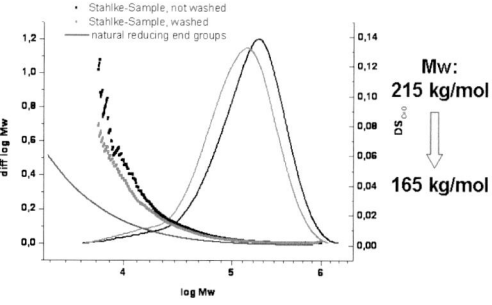

Fig. 5.7. Molecular weight distribution of a cotton textile from the 19th century washed with a non-ionic detergent. The loss of molecular weight is clearly illustrated by a shift of the distribution to the right side. Analytical data of Mw are added. (Sample courtesy of the Museum European Culture, Berlin.)

(Fig. 5.7). The phenomenon of more recent textiles being more sensitive towards aqueous treatments is currently under further investigation.

Conclusions

Fluorescence labelling followed by GPC-MALLS was successfully applied on archaeological cellulosic textiles. With a low amount of sample material that can be covered by single fibres, this technique yields detailed information on the degree of oxidation and hydrolytic degradation of cellulose. Additionally, molecular weight distribution helps to further interpret the results.

The true condition of an object may be different from its appearance. With the help of analytical results the influence of dyeing, painting, cotton processing and washing of a historic textile can be ascertained. It is also a very valuable tool to identify heterogeneity within one single object.

Coloured and dyed areas of a textile, but also samples of processed cotton are generally in worse condition than uncoloured areas or unprocessed material that, in contrast, are generally in excellent condition. When differences occur within one object, intra-object tensions may be present due to differing mechanical strength – caused by the paint or the adhesive used. This may be an important hint for further restoration treatments, especially for aqueous ones. Keeping in mind that also carbonyl group content can be very dissimilar, immersion in water will most probably lead to uncontrollable behaviour: some areas might be more hydrophilic than others. The whole object might disintegrate during treatment.

However, also the opposite behaviour could be demonstrated for indigo dye, which had a protecting effect on cellulose, as in sub-sample VA_47083 containing blue dye. When an object, like sample VA_47083, is quite homogenous in its intra-object condition, but analysis of oxidized groups indicates that there has been a great deal of oxidation, care should be taken when it is exhibited. Oxidized sites in the cellulose are considered to be hot spots for further degradation. Especially light and alkaline treatments may harm this object, lead to severe degradation and loss of mechanical strength.

The influence of washing has not been fully understood yet. Especially recent textiles seem to suffer from aqueous immersions. One possibility might be that processing parameters have been changed in the course of industrialization, exerting more stress on the fibres and leading to easier accessibility for aggressive influences. However, unless more experience has been gained on this topic, this explanation remains in the realm of speculation.

Further research into threshold values for certain treatment options or limitations for exhibitions and loan traffic are needed to define certain condition categories more accurately. The concept of calculating ratios between an undyed reference from the same object and dyed or painted segments may improve comparability between different objects. Furthermore, such ratio numbers can be the basis for decision-making and defining threshold values. In the examples investigated for this paper, moderately degraded samples have ratios between 1.2 and 1.5, while heavily degraded samples will have ratios around 2.5. This concept should be further elaborated and more thoroughly investigated to ascertain if it will hold true for a broader range of samples and degradation patterns. Additionally, it should be kept in mind that these ratios are an easy to handle parameter. However, they simplify a great deal and neglect valuable information obtained from molecular weight distribution.

Acknowledgement

The authors would like to thank Dr. Sonja Schiehser for practical assistance.

Bibliography

Henniges, U., Prohaska, T., Banik, G., and Potthast, A. (2006) A fluorescence labeling approach to assess the deterioration state of aged papers. *Cellulose* 13, 421–428.

Potthast, A., Röhrling, J., Rosenau, T., Borgards, A., Sixta, H., and Kosma, P. (2003) A Novel Method for the Determination of Carbonyl Groups in Cellulosics by Fluorescence Labeling. 3. Monitoring Oxidation Processes. *Biomacromolecules* 4, 743–749.

Röhrling, J., Potthast, A., Rosenau, T., Lange, T., Ebner, G., Sixta, H., and Kosma, P. (2002a) A Novel Method for the Determination of Carbonyl Groups in Cellulosics by Fluorescence Labeling. 1. Method Development. *Biomacromolecules* 3, 959–968.

Röhrling, J., Potthast, A., Rosenau, T., Lange, T., Borgards, A., Sixta, H., and Kosma, P. (2002b) A Novel Method for the Determination of Carbonyl Groups in Cellulosics by Fluorescence Labeling. 2. Validation and Applications. *Biomacromolecules* 3, 969–975.

Strlič, M., and Kolar J. (2005) *Ageing and stabilization of paper.* Ljubljana, National and University Library.

Timar-Balazsy, A., and Eastop, D. (1998) *Chemical Principles of Textile Conservation.* Oxford, Butterworth/Heinemann.

6 Archaeological Textiles from Prague Castle, Czech Republic

by Milena Bravermanová

Several unique garments originating in the graves in the Cathedral of St. Vitus in Prague were conserved between 2005 and 2007 in the conservation workshop of Prague Castle, Czech Republic.

The first garment is a mitre preserved in fragments, which was found in the grave of the bishop of Prague, Bernard († 1240) in the cathedral choir. The bishops of Prague were buried primarily at Prague Castle, in the church consecrated to St. Vitus. The bishops' remains were moved from the earlier Basilica to the new Cathedral in 1374. The graves were opened in 1928 and the remains were re-consecrated. Despite their fragmentary state, the remains of crosiers and shears, which emphasized the bishop's role as a shepherd, are unique. The other relics are fragments of clothing – mitres, chasubles, gloves, cingulums, stoles and stockings. Shoes are mostly made of leather (Bravermanová 2004).

The mitre from the grave of Bishop Bernard was conserved in 2005–2006 by Angelica Sliwka. The costly mitre (the so-called *auriphrygiata*), which is preserved only in a fragmentary state, is made of patterned silk fabric (lampas) woven somewhere in the Near East or Spain. The pattern of the fabric is not readable but it is evident that there were medallions bordered with a woven motif of small pearls. The mitre was probably white in the past (today it is brown) and was very richly embroidered, above all with metal threads (gilded silver with silk core), and in some parts with coloured threads.

Although the embroidery is damaged and the identification of the characters figuring on it is not certain, it can be assumed that the main motif is the figure of the suffering Christ. On the left side is the Virgin Mary, the figure on

Fig. 6.1. The graves of the Prague bishops in the Cathedral St. Vitus (© Photo: Jan Gloc, Picture Library of Prague Castle).

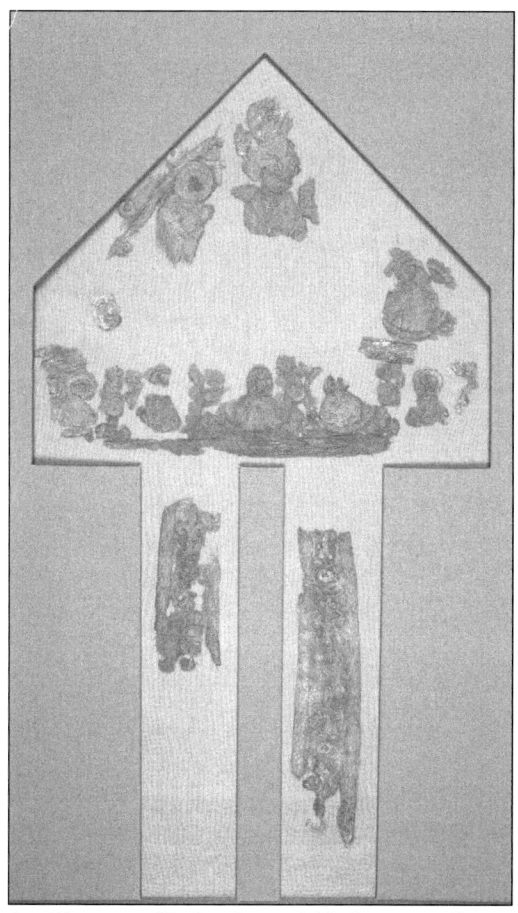

Fig. 6.2. The mitre of Bishop Bernard (©Photo: Jan Gloc, Picture Library of Prague Castle).

Fig. 6.3. The Royal Crypt, the present state (© Photo: Jan Gloc, Picture Library of Prague Castle).

the right side is missing (probably John the Baptist?). In the corners below these figures were probably St. Paul (with a book as an attribute) and St. Peter (this figure is now missing, but it was documented in the past). In the lower strip (*circulo*) there were five figures. Some details attest that they could have been Czech saints – St. Ludmila (with a veil as an attribute), St. Vitus (this figure is only preserved in fragments), St. Adalbert (a dalmatic, the rest of the mitre and the crosier), St. Wenceslas (a duke's cloak fastened together on one side) and St. Procopius (with the tonsure of an abbot). If these are in fact the Czech patron saints, then the mitre could have been embroidered in Bohemia, probably at St. George's convent. Although only a few historical embroideries from this time are preserved in the Czech Republic, the possibility of Czech provenance is documented by historical sources (Friedrich 1904, 171–172, n. 167).

The fanons of the mitre and the space among the figures are trimmed by an embroidered pattern of plants and the motif of a crescent.

In spite of the fact that Bishop Bernard's mitre is only preserved in a fragmentary state, it is a unique artefact. Such old mitres are rarely found. A few extant examples are those in the cathedral in Anagni, Italy; in the church of St. Zeno in Verona, Italy; in the cathedrals in Halberstadt and Bamberk, Germany; in the church of St. Emmeramus in Regensburg and in the museum of Nuremberg in Germany. Moreover, if the Czech patron saints were really depicted on Bishop Bernard's mitre, it is further proof of the increasing significance of local patron saints. However, it is logical that they appeared on a bishop of Prague's clothing as intercessors in a hierarchic configuration of Christ with the Virgin Mary, the Apostles, and the patron saints (Bravermanová, 2005a).

The other newly conserved textiles were discovered in the Royal Crypt (Fig. 6.3). It was built by Charles IV in *c.* AD 1350 directly below the chancel of the newly built cathedral. The first ruler to be buried there was the emperor himself in 1378, then his four wives and sons John of Görlitz and Wenceslaus IV. The remains of the kings Ladislaus the Posthumous and George from Podebrady were interred there in the 15th century. The remains of members of the Habsburg dynasty – Ferdinand I, his wife Anna of Jagiello and granddaughter Eleonora were also placed there.

On the initiative of Emperor Rudolf II, the construction of the new crypt connected with the mausoleum was commenced in front of the entrance to the chancel of the cathedral. The remains of Czech rulers and their relatives were relocated there in 1580. The mausoleum was opened and examined in 1677, 1743, 1804 and 1855. When the construction of the St. Vitus cathedral was near completion in 1928, the tomb was abandoned. Its contents were then gradually taken up to the church. Packed and conserved remains were placed in the newly made sarcophagi in 1933, but without the funerary assemblages, especially the textiles, which became part of the collections of Prague Castle. The textiles were forgotten in the ensuing period. However, since the beginning of the 1990s, they have been given special

Fig. 6.4. *The funerary dress of one of the wives of Charles IV* (© Photo: Jan Gloc, Picture Library of the Prague Castle).

Fig. 6.5. *The funerary cloak of Jan of Görlitz* (© Photo: Jan Gloc, Picture Library of the Prague Castle).

attention. The fragments of garments and other fabrics have been placed in air-conditioned depositories, and they have been conserved, exhibited and published (Bravermanová 2005b).

Here follows a description of some of the newly conserved garments. The first is the funerary dress of one of the wives of Charles IV. The garment was conserved by Romana Kloudová in 2007 (Fig. 6.4). This dress was made of lampas silk. The pattern was woven with two wefts: one was *lancé*, the second, which is now nearly completely disintegrated, was *broché*. We know, however, that it was gold (probably gilded silver on the core). The trimmed pattern was created by rosettes, filled by a stylized Islamic inscription and palmettes. Rosettes and palmettes always surround different pairs of dogs and birds. The fabric is probably of Italian provenance from about the mid-14th century.

The dress was sewn in one big piece, with slits both in front and back. After the reconstruction, it was clear that the bodice was narrow and relatively tight; the skirt, towards the lower edge was enlarged by four gores in a trapezoid shape. Only two gores have been preserved. It seems the dress was sleeveless, possibly a female surcoat, or the sleeves could be lost. A pillow was made of the same fabric. This indicates that the outfit was made specifically for the funeral. According to the dating of the fabric, the clothing was made for one of the first three wives of Charles IV, not for the last. We suggest it was his third wife Anna of Svídnice, primarily because the garment belonged to a woman of average height, not small in stature as his second wife Anna of the Palatinate. Furthermore, it is possible to attribute another dress with ogival palmettes and pairs of birds and llamas, which can be dated to the first third of 14th century to the first wife, Blanche of Valois. Possible parallels to the funerary garment that probably belonged to Anna of Svídnice include one of the Herjolfsnæs dresses and a slightly earlier funerary robe of the Danish queen Margrethe the First (Geijer *et al.* 1994; Østergaard 2004; Bravermanová and Kloudová In press).

The other newly conserved garment is the funerary cloak of John of Görlitz, the last son of Charles IV and Elizabeth of Pomerania, who died in 1396 (Fig. 6.5). Romana Kloudová undertook the conservation in 2005. The garment was made from plain silk velvet. Despite evidence that monochromatic

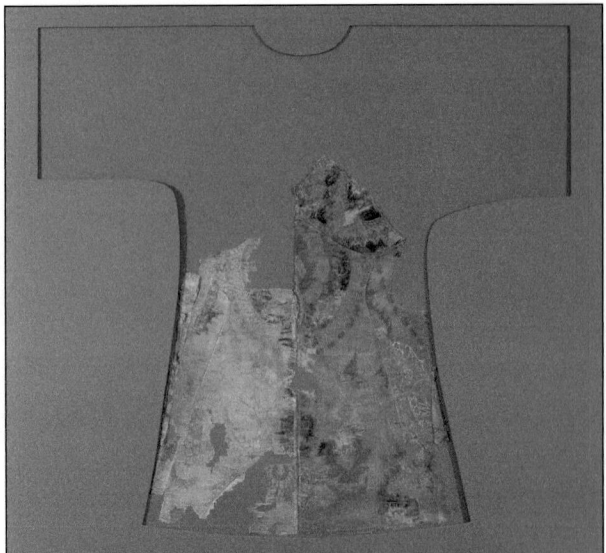

Fig. 6.6. The funerary dalmatic of Ladislaus the Posthumous (© Photo: Jan Gloc, Picture Library of the Prague Castle). Reproduced as a colour plate on page 301.

Fig. 6.7. The funerary shoes of Ladislaus the Posthumous (© Photo: Jan Gloc, Picture Library of the Prague Castle). Reproduced as a colour plate on page 301.

velvets were already produced in England, France and Germany at the time, we assume that the fabric from the end of 14th century was imported from one of the traditional areas of silk production – Italy or Spain. The garment was sewn from nine back and nine front long trapezoid-shaped pieces. The upstanding collar, probably fastened by some clasp in the past is preserved.

As John of Görlitz was not a Czech king, it can be assumed he was buried in contemporary secular clothing and not in the ceremonial robe – the dalmatic and semicircular cloak, which belonged to kings. The cut corresponds to the fashion worn at the end of the 14th century and the first half of the 15th century. Male outer wear of that time consisted either of short cloaks, or long cloaks with a complicated cut, fastened with belts and trimmed with collars and furs, termed *houppelande* (Boucher 1987, 195–196). These types of garments inspired the cut of the funerary cloak belonging to John of Görlitz. It was a hastily made robe as indicated by the fact that the cloak was left unlined and the careless work of the tailor's. We cannot establish its original colour with certainty, although it was probably black.

The newly conserved funerary garment of Jan of Görlitz, which was adapted into a three-dimensional shape, represents a unique object with few analogies except for the clothing from Herjolfsnæs or the cloak of patterned velvet, probably resewn, now deposited at the Historisches Museum in Bern (Flury-Lemberg 1988, 442–445, 462). However, this mantle is dated at the earliest to 1430. It is sleeveless and its interpretation has been questioned lately (Bravermanová 2006).

The final newly conserved garment from the Royal Crypt discussed here is the funerary dalmatic of King Ladislaus the Posthumous who died in 1457 (Fig. 6.6). The conservation was undertaken by Alžběta Vrabcová in 2006–2007. The king's dalmatic was found in many fragments and a large part of the garment is missing. It was discovered in the course of textile technical research that two fabric types with a different pattern were probably woven on the loom which was drawn in the same way, since the pairs are technically identical. A very similar design appears in both patterns, where the composition is identical and the differences are merely in the details. Trees, with pomegranate flowers, climb up in the background, leaning to the right and left, and tendrils with leaves and different flowers twine around the trees.

The original shape of the garment was not apparent at first. Remains of the plain silk lining were entangled in the seams on some fragments. Based on them, we managed to reconstruct all the lining. The imprints of a part of a pattern of the velvet were found here; they were of metal wefts. From its shape, it is clear that it was a lining of the dalmatic. Thus, it is evident that also Ladislaus the Posthumous, just as the Czech kings Charles IV and Wenceslaus IV before him, was buried in a ceremonial robe.

It was discovered that before the funeral, the dalmatic had been resewn from another garment, perhaps even from two. These garments were of a secular nature. The first garment probably reached the thighs. It was enlarged, probably not sewn at the sides, and was perhaps sleeveless and with deep armholes. Therefore, it may be a so-called tappert, *giornea* or a cloak in the Burgundian style. Selvages of the second (?) garment, the shape of which is not apparent today, were finished by tiny clipped arcs. It is not known if such clipped arcs were placed on sleeves, collar or around the lower circumference.

The pillow and the shoes (Fig. 6.7) that reached high above the ankles were reconstructed from remaining fragments of the velvet. The shoes have velvet soles, the side seams are on the outer side and a short slit is on the vamps (Bravermanová, In press).

All these newly conserved funerary garments of the Czech rulers and their relatives demonstrate the uniqueness of the collection of medieval clothing preserved at Prague Castle.

Bibliography

Boucher, F. (1987) *A History of Costume in the West*. London and New York, Thames & Hudson.

Bravermanová, M. (2004) Hroby pražských biskupů v katedrále sv. Víta na Pražském hradě. Předběžné sdělení. *Archaeologia historica* 29, 599–615.

Bravermanová, M (2005a) Fragmente der ausgenähten Mitra offenbar mit gestalten der böhmischen Schutzpatrone aus dem Grab Bischofs Bernard vor 1240 (Kirchenschatz bei St. Veit). In P. Sommer (ed.), *Der Heilige Prokop, Böhmen und Mitteleuropa*, Colloquia mediaevalia Pragensia, 4, 144–146. Praha.

Bravermanová, M. (2005b) Hroby knížat, Hroby králů, Hroby českých patronů, Hroby významných církevních činitelů. In K. Tomková (ed.), Castrum Pragense 7. *Pohřbívání na Pražském hradě a jeho předpolích*, díl I. 1, 47–140, Praha.

Bravermanová, M. (2006) Pohřební oděv Jana Zhořeleckého z královské hrobky v katedrále sv. Víta na Pražském hradě, *Archaeologia historica* 31, 403–412.

Bravermanová, M. (In press) *Pohřební výbava Ladislava Pohrobka z královské hrobky v katedrále sv. Víta*.

Bravermanová, M., and Kloudová, R. (In press) *Pohřební šaty jedné z žen Karla IV*.

Flury-Lemberg, M. (1988) *Textile Conservation*. Riggisberg, Abegg Stiftung.

Friedrich, G., ed. (1904) *CDB: Codex diplomaticus et epistolaris regni Bohemiae*. I. Praha.

Geijer, A., Franzén, A. M., and Nockert, M. (1994) *Margaretas gyllene kjortel i Uppsala domkyrka*. Stockholm.

Østergård, E. (2004) *Woven into the Earth. Textiles from Norse Greenland*. Aarhus.

7 Virtual Reconstruction of Archaeological Textiles

by Maria Cybulska, Tomasz Florczak and Jerzy Maik

Textiles as organic matter are relatively rarely found in archaeological excavations. In order to survive they would have had to stay in very favourable conditions, slowing down microbiological deterioration, one of the most important factors determining their durability.

Among many factors enabling the preservation of textiles are: the presence of tannins in the soil, proper temperature and humidity (for example, in the waterlogged strata of ancient towns), frozen earth (Greenland), or contact with metals which, due to their antibacterial properties, preserve textiles (*e.g.* in inhumation graves).

However, even under optimal conditions, textiles never survive in the state in which they were deposited in the soil. In the first place, they often survive in fragments only. Even the clothing found among the numerous offerings in bogs such as Thorsbjerg, Bernuthsfeld or Vehnemoor in northern Germany, which had favourable environmental conditions for the survival of textiles, can be deteriorated and fragmented. Their structure can be deformed, and the colour changed – they are usually found in different shades of brown.

The presence of metals has a dual effect on the state of archaeological textiles: on one hand, it results in preserving the clothes, on the other hand, supersaturating the textiles with metal particles results in the stiffness and in the change of colour to that of the metal oxide – green in the case of copper alloys, or rusty red in the presence of iron. Therefore, textiles from archaeological excavations are seldom the most interesting objects in archaeological exhibitions and, without specialist analysis, they cannot reveal information about ancient textile technologies even to archaeologists. That is why textile reconstruction is so essential and has been undertaken for so long. For instance, Karl Schlabow – founder, and director of *Textilmuseum Neumünster*, reconstructed two beautiful *prachtmanteln* – replicas of the items found in bog offerings in Thorsbjerg and Vehnemoor. However, this kind of reconstruction is difficult and time consuming (it took several months), and requires skills in weaving on ancient loom types (Schlabow 1976).

In 2003, at the Institute of Architecture of Textiles, Technical University of Łódź, Poland, the idea arose of applying new tools, using computer graphics methods to reconstruct textiles. This type of reconstruction requires the determination of all parameters constituting the final appearance of the textiles. The main factors are: the raw material, colour scheme and the structure of threads, fabric and the object as a whole. Despite the state in which the archaeological textiles were found, applying such analytical methods as SEM, IRS and ICP AES, HPLC, image analysis, and many others, it is possible to determine and identify the properties these textiles had in the past. Then, on the basis of the results obtained, it is possible to reconstruct them – not on the loom, which would be rather expensive, but virtually, using methods of 3D graphics.

The analysis of archaeological textiles

The analysis of archaeological textiles is difficult mainly due to the fibre degradation resulting in changes of physical, chemical and structural properties. From the point of view of the identification of textiles, the most important properties can be divided into three groups: the nature of fibres, the colour and the structure of threads and fabric. The methods of analysis of the raw material include: SEM (scanning electron microscopy), ICP (inductively coupled plasma), ATR/FTIR (attenuated total reflection using infrared Fourier transform spectroscopy), and AAS (atomic absorption spectroscopy) for the determination of the raw material and some other components of the object and archaeological environment, and TLC (thin layer chromatography) and HPLC (high performance liquid chromatography) for analysis of colour (Cybulska *et al.* 2008).

The main problem with the analysis of structural parameters of textiles is the quality of the analysed samples. Numerous computer methods for the identification of structural parameters of woven or knitted fabrics can be found in literature. However, they can only be applied in the case of high quality samples. They are useless in the case of samples that are partially destroyed, as archaeological textiles usually are. The other problem is that some fabric parameters, such as yarn crimp, cannot be determined using traditional methods due to unclear fabric cross-section and the brittleness of yarn fibres. Thus, it was necessary to develop the methods that could be applied in the case of such degraded textiles. The first stage of analysis is the re-construction of an image of the sample itself. It consists of the multiple applications of the proper filtering and non-linear image transformation to obtain the image of the recovered texture of fabric. It is also necessary to take into account the changes of the form of fabric due to shrinkage and other deformities caused by ageing in a difficult environment (Cybulska *et al.* 2005). Image analysis-based methods can be applied to a sample image processed in this way. Methods have been developed and implemented allowing the determination of the structure of yarn and fabric. They include such parameters of yarn as diameter, hairiness and twist (Cybulska 1999). For woven fabrics, the methods developed include, among others, the determination of the yarn crimp, yarn spacing, and cover factor (Cybulska and Pancer 2003). Information about these structural parameters of fabric coupled with the weave, colour and raw material identification provides us with sufficient data to reconstruct the fabric as it could originally have looked like.

The Application of 3D Graphics in the Simulation of the Appearance of Textiles

3D graphics methods are increasingly popular and widely applied not only in entertainment and advertising, but also in the field of science and technology. This also applies to textiles. Computer programs support engineers and designers of knitted and woven fabrics. The main advantage of these kinds of products is the ability to simulate the final product without costly and time-consuming trials. There are three main disadvantages of the existing and available software designed for textiles. The first is price: each specialised programme segment is quite costly. The second is the limitation of the software to strictly determined kinds of products based on traditional technology. The lack of elasticity and open programming codes make it unusable in the case of textiles produced using ancient technologies. The film, art and computer game industries are a good source of 'open' and flexible 3D graphics software. In the reconstruction of archaeological textiles, *Discreet 3ds max 7* software tools and the procedures developed with the use of *maxscript* programming language were applied (Cybulska and Florczak 2005; Cybulska and Florczak 2008).

The appearance of textiles is a very complex phenomenon. It is a combined effect of many different factors, such as raw material, yarn and fabric structure, finishing and many other parameters. This results in different product properties as the aerial mass or stiffness. These parameters determine some other factors, such as fabric texture or drape. On the other hand, the product's appearance is determined, among others, by light reflection and refraction, translucency and colour. All these parameters have to be taken into account in the reconstruction of textiles.

Modelling the textile product proceeds in several stages. First we model the basic element – fibre or yarn; then, on the basis of predetermined parameters of fibres, yarn and fabric, the textile's appearance is simulated. Procedures have been developed to create models of fibre characterised by different cross-sections, diameter and length.

Two different methods of thread modelling have been developed. The first method consists of giving the linear element the texture of properties determined by the yarn's structural properties. The presence of fibres is reflected by concavities and convexities on the cylindrical yarn surface. The second method consists of forming the yarn from previously created 3D models of fibres by wrapping, twisting or knotting them, according to the thread technology. The method allows setting some predetermined features such as yarn evenness or yarn hairiness for staple yarn. The simulation of the fabric proceeds in two stages. First, the fabric is formed from previously modelled linear elements. Then, elements are interlaced according to the previously chosen weave pattern with the relative position determined by the yarn spacing. The method allows the simulation of some fabric patterns obtained by different techniques (Cybulska and Florczak 2005).

Examples of Virtual Reconstruction of Archaeological Textiles

Woven fabric from Leśno, Pomerania

In grave 1 at Leśno, called the Princess's Grave due to the rich assemblage interred with a young woman, numerous fragments of wool fabric were found (Kanwiszerowa and Walenta 1985, 115–118). Most were probably the remains of clothing, as they were found just near the skeleton. However, several fragments were found in, under, and on the edges of a bronze vessel (Fig. 7.1). The arrangement gave the impression that the vessel was wrapped in the fabric, and was deposited together with burial goods for the woman for her journey to the other world. We cannot be sure if it was a piece of fabric or clothing (Maik 2005).

Fig. 7.1. The bronze vessel found in Leśno, Pomerania, in grave no. 1, dated to the 2nd century AD (Photo: Krzysztof Walenta).

Fig. 7.2. One of the surviving fragments of the fabric from the Princess's Grave, Leśno, Pomerania (Photo: M. Cybulska).

Fig. 7.3. The virtual reconstruction of the fabric from the Princess's Grave, Leśno, Pomerania (Graphics: © M. Cybulska and T. Florczak 2008). Reproduced as a colour plate on page 302.

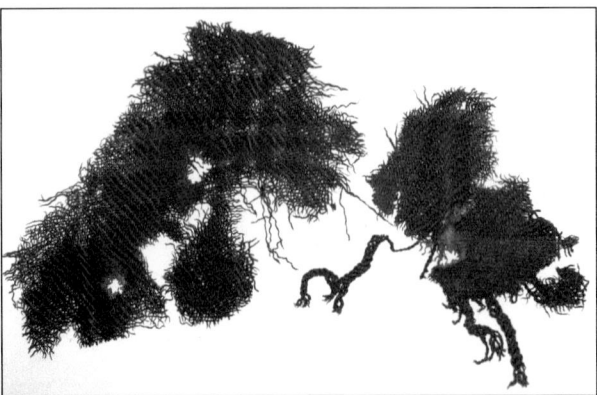

Fig. 7.4. Fragments of prachmantel found in the male grave 1 of mound 3 in Gronowo, Pomerania, dated to the B_2/C_1 phase of the Roman period. (Photo: M. Cybulska).

The image of one of the extant fragments of the fabric is presented in Figure 7.2. The fabric was initially classified as open-work fabric. However, microscopic and chemical analysis showed the residues of disintegrated fibres (probably linen) and the traces of crimp on the surviving wool threads. It allowed us to infer that the fabric was originally made from wool and linen, but linen fibres became completely putrified. The fabric was made in 2/2 twill weave. Using image analysis, it was possible to determine the yarn crimp and spacing, and the number of existing wool yarns, as well as the previously existing linen ones. The diameter of wool thread, the number of fibres in yarn cross-section, twist angle and thread spacing were also determined (diameter 0.52 mm, twist 29 twists per cm (z), thread spacing: 0.61 mm for warp and 0.69 mm for weft). On the basis of this information, it was possible to restore the structure of the whole fabric, including linen threads. The analysis of dyes undertaken by Penelope Walton-Rogers from York Archaeological Trust indicated the presence of indigotin, giving the blue colour to the wool (Walton 1993, 64). As no evidence of dyeing cellulose fibre has been found in this period, it was assumed the linen threads were in the natural colour of flax. The reconstruction of the fabric can be seen in Figure 7.3 (Cybulska and Florczak 2008).

Prachtmantel *from Gronowo, Pomerania*

In the male grave 1 of mound 3 at Gronowo, dated to the B_2/C_1 phase of the Roman period, numerous fragments of two fabrics and fringes were found next to a man's feet (Wołągiewicz 1973; 1974; 1979). They were all compacted into one ball with a metal spur inside. After cleaning the find, the fabrics were found to be remains of two *Prachtmanteln* with tablet-woven borders and fringes. One of the cloaks was monochrome, the second was checked polychrome with dark and light brown threads (Fig. 7.4). They were probably used to wrap the dead man's body (Maik 1979; Bender Jørgensen 1992, 246–247).

The analysis of dyes by Penelope Walton-Rogers was made for the polychrome fabric only. Tannin was found in the dark threads indicating a brown colour. There was no dye found in the light threads, so it was probably in the natural colour of wool (Walton 1993, 64). Further analysis allowed the determination of the textile's structure. The main fabric was made on a warp-weighted loom and is a 2/2 twill weave. The thread diameter was determined as 0.65 mm for the brown, and 0.55 mm for the light yarn. The weave pattern included 162 threads and was 12.5 cm wide. Threads in the tablet border had a smaller diameter, equal to 0.25 mm. Borders were made with 30 tablets, with a density of 13 tablets per cm, all from the light thread used as a warp. Fringes were in different colours, depending on their position in relation to the main fabric.

In the next stage, all elements were virtually reconstructed, including the main fabric, borders, corners and fringes (Fig. 7.5). These elements were used to reconstruct the whole cloak. Figure 7.6 presents the reconstructed object, and the way it was worn.

Silk Fabric from St. John's Church in Gdansk

Few years ago during a renovation of a floor in the Gothic church of St. John in Gdansk, built at the turn of the 14th century AD and modified later, a number of 18th century crypts were found directly under the floor. In some of them small residues of silk fabrics were found, probably fragments of the clothing of the dead buried there. One of them is a fragment of decorative openwork fabric (Fig. 7.7, on the left). Analysis has shown that it was made in a half-cross leno technique and decorated by ornamental strips woven with silk of a higher number of filaments. The analysis allowed the determination of some structural parameters of threads and fabric. Number of threads per cm equals 18 for warp and 15 for weft. Thread diameter in mm was equal to 0.19 for weft, 0.095 for basic warp and 0.37 for decorative yarn. The twist of threads per cm was determined as: weft 1.78, basic warp 3.18, decorative yarn 1.32 twists. The reconstruction of the weave pattern is shown in Figure 7.7 centre.

The shape and the size of the fragment did not indicate its function. For reconstruction purposes, it was assumed it could be a fragment of stocking, virtually reconstructed and presented in Figure 7.7 right.

Fig. 7.5. The virtual reconstruction of some details of the fabric structure – Gronowo' prachmantel (Graphics: © M. Cybulska and T. Florczak).

Fig. 7.6. The virtual reconstruction of the prachmantel from Gronowo (Graphics: © M. Cybulska and T. Florczak). Reproduced as a colour plate on page 302.

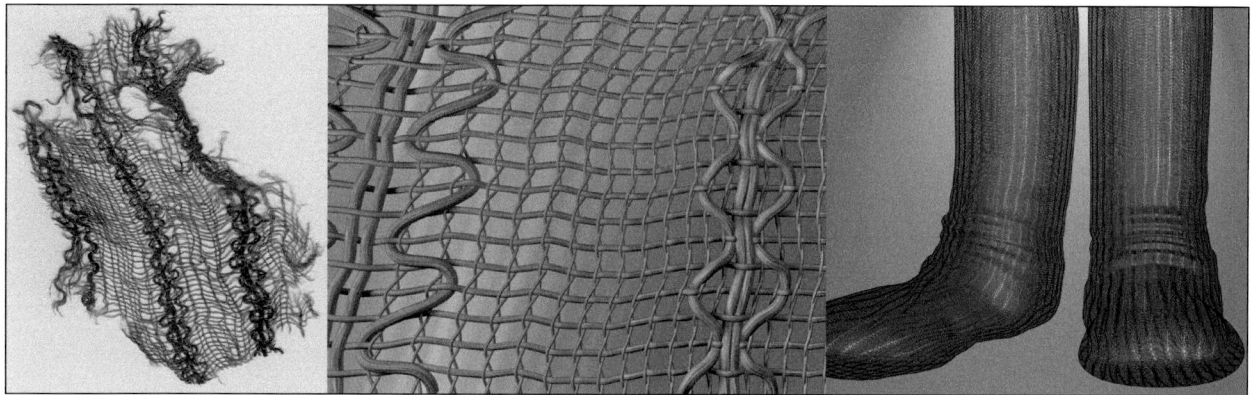

Fig. 7.7. Left: fragment of the silk fabric from St. John's Church in Gdansk (Photo: Maria Cybulska); Centre: the virtual reconstruction of the weave pattern; Right: an example of the virtual object made from the silk fabric (Graphics: © M. Cybulska and T. Florczak).

Conclusions

Archaeological textiles are not the most interesting of objects in museum exhibitions. They are preserved in fragments, colourless, and thus hard to appreciate and understand for the unprepared viewer. A reconstruction of these textiles can of course be made on the loom. However, this approach is not always possible, and sometimes even not reasonable. Reconstructed this way, the object needs proper storage, exhibiting and conservation, and is very costly. The main advantage of virtual reconstruction is that it can be easily accessible not only in the museum, but also on the Internet. This way it can be available not only for professionals interested in the subject, but also for the general public. Apart from the way it is presented, it can clearly show both the complete reconstructed object and some details of its structure.

The virtual reconstruction of archaeological textiles allows not only the preservation but also the restoration of the important testimony of art, history, technology and their evolution through the centuries. The educational role of the reconstruction is also very important. When studying any archaeological object in its current state, a chance to know its original structure and appearance is very valuable. Archaeological textiles constitute a very good example – colourless and shapeless remains can regain their form and beauty. Virtual reconstructions of archaeological textiles can also serve as a useful and stimulating supplement to museum collections.

Bibliography

Bender Jørgensen, L. (1992) *North European Textiles until AD 1000.* Aarhus, Aarhus University Press.

Cybulska, M. (1999) Assessing Yarn Structure with Image Analysis Methods. *Textile Research Journal* 69: 5, 369–373.

Cybulska, M., and Pancer, R. (2003) Analysis of Woven Fabrics Using Image Analysis. *Proceedings of VI International Conference ArchTex 2003*, 162–167.

Cybulska, M., and Florczak, T. (2005) Application of 3D graphics methods in textile designing. *Proceedings of VII International Conference ArchTex 2005*, 182–186.

Cybulska, M., Florczak, T. and Maik, J. (2005) Archaeological textiles – analysis, identification and reconstruction. *Proceedings of the 5th World Textile Conference AUTEX 2005*, 878–883.

Cybulska, M., *et al.* (In press) Methods of chemical analysis in identification of archaeological and historical textiles. *Fibres & Textiles in Eastern Europe.*.

Cybulska, M., and Florczak, T. (In press) Reconstruction of archaeological textiles using 3D graphics methods. *Fibres & Textiles in Eastern Europe.*

Kanwiszerowa, M. and Walenta, K. (1985) Grób książęcy nr 1 z Leśna na Pomorzu Wschodnim. *Prace i Materiały Muzeum Archeologicznego i Etnograficznego w Łodzi, seria archeologiczna*, 29, 101–127.

Maik, J. (1979) Tkaniny wykopaliskowe z cmentarzyska w Gronowie, woj. Koszalińskie. *Materiały Zachodniopomorskie* 22 (1976), 111–121.

Maik, J. (2005) Tkaniny z grobu książęcego w Leśnie. In H. Rząska and K. Walenta (eds), *Brusy i okolice w pradziejach na tle porównawczym*, 98–112. Brusy.

Schlabow, K. (1976) *Textilfunde der Eisenzeit in Norddeutschland.* Göttinger Schriften zur Vor- und Frühgeschichte, 15. Neumünster, Karl Wachholtz Verlag.

Walton, P. (1993) Wools and Dyes in Northern Europe in the Roman Iron Age. *Fasciculi Archaeologiae Historicae* 6, 61–68.

Wołągiewicz, R. (1973) Gronowo 1973 – Badania na cmentarzysku kurhanowym z okresu wpływów rzymskich. *Materiały Zachodniopomorskie* 19, 129–167.

Wołągiewicz, R. (1974) Gronowo 1974 – Badania na kurhanowym cmentarzysku kultury wielbarskiej. *Materiały Zachodniopomorskie* 20, 7–30.

Wołągiewicz, R. (1979) Cmentarzysko kurhanowe kultury wielbarskiej w Gronowie w świetle badań w latach 1973–1976. *Materiały Zachodniopomorskie* 22, 71–95.

8 The Use of Terminology in Medieval Scandinavian Costume History: An Approach to Source-based Terminology Methodology

by Camilla Luise Dahl

Introduction
Research into dress history, whether the approach is founded in history, art or archaeology, incorporates terminology in one way or another. Usually, terminology (the terms for items of clothing) is necessary to describe and characterize garments in a given context. Archaeologists may place less emphasis on terminology, as it is clearly of little consequence if an excavated dress from the Medieval Period would have been called a *warthækor*, a *kordhumbla* or a *kobe* in its own time. However, terms for dress acquire great significance when archaeologists, historians and art historians express their knowledge in writing. A clear and well-defined dress terminology is crucial when moving from the researcher's own sphere into another field.

In all regions of Europe, different traditions prevail as to how terminology is approached and used in medieval dress history. In France, there has been a long tradition of highly scholarly collections of glossaries and encyclopaedias of dress terms (de Laborde 1872; Gay 1887; Enlart 1916; Zangger de Saint-Gall 1945), whereas in German-speaking areas, most work on medieval costume has been based on visual evidence, or, to a limited extent, written tradition with little focus on terminology (Jaritz 1988, 7–8; Vavra 1988, 22–23), and even lesser focus on archaeological evidence (Kraft 2003, 77–79). Moreover, much of the medieval terminology use in dress literature is not based on the written sources themselves, but instead adopts dress terms from previous surveys. Thus, the apparent source-based terms may instead be dress history terms, (Vavra 1988, 21–23; Kühnel 1988, 11–19; von Wilckens 1988, 47), which can be even more problematic when applied to visual evidence or actual garments from the period (Blindheim 1980, 270, 278–280; Vavra 1988, 23–24; Jaacks 1998, 243–251; Netherton 2008, 151–152). Many of these popular European dress terms have often been identified in Victorian-era dress history without any reliable result. In recent years, scholars instead have sought to replace the misleading terminology applied to certain garments with more appropriate neutral terms or modern descriptive terms.[1] Consequently, various foci have influenced the traditions of methods of writing dress history in the various areas.

In a survey I previously conducted of terms for headwear used in modern Scandinavian dress literature, and in a selection of about 900 various documents such as laws, sumptuary legislation, literature, wills and inventories mentioning dress (Dahl 2007), a number of inconsistencies became evident. The survey showed examples of clear discrepancies between terminology used in modern dress literature and that used in medieval sources. In various works on Scandinavian medieval dress, the female headwear covered by the terms *vill* and *hviv* is said to be the commonest type of headwear in the Medieval Period (Nørlund 1941, 55; Andersson and Franzén 1975, 15–16; Hansen 1978, 64; Nockert 1996, 208;). Yet, among the approximately 900 various documents examined dating between *c.* AD 1200–1600, the term *hvif* could only be found mentioned twice (Dahl 2007, 80–82). Considering the quantity of examined written accounts, this result definitely does not support the common notion that these two types of headwear were the most common in the Medieval Period, especially as numerous other terms defining headwear appear in this corpus of texts (Dahl 2007, 127–176).

Traditional Praxis in Scandinavian Terminology Use
Traditionally, terminological studies have not been in much focus in Scandinavian dress research, and little has been added to the field since the work on medieval Scandinavian dress terms by Hjalmar Falk in 1918. As works on Scandinavian costume terminology are so rare, scholars have had to rely on the works from other areas in order to incorporate discussions of dress terminology. Consequently, Scandinavian dress history abounds with studies of French and English terms for dress, rather than Scandinavian terms. For instance,

nearly all modern works on Scandinavian medieval dress mention the garment term *cotehardie*, although this is hardly ever mentioned in medieval Scandinavian written sources, whereas the Scandinavian dress term *kapo*, which appears in numerous medieval documents, has not been mentioned in a single modern work on medieval dress until recently.[2] The use of dress terms in literature on medieval Scandinavian dress therefore displays a lack of consistency with the terms actually used in medieval Scandinavia.

Moreover, Scandinavian dress historians have primarily regarded terms for dress in a merely statistical manner, as a base for quantitative surveys. These surveys have concentrated on how many times a certain term appeared in a given material. However, the present survey demonstrates that it can definitely be more productive, in terms of typological definitions as well as grouping medieval terms, to look instead at the material in a more individual context.

Looking at Scandinavian costume literature at large, the terminology use by different scholars is based on various considerations and traditions. Whereas older dress literature often employs what might be defined as 'archaic' terminology, modern scholars have begun to develop a more well-defined terminological praxis. The first group – which I define as 'archaic terminology' – includes terms meant to simulate a historical vocabulary. In Scandinavian medieval dress history, terms such as *kåbe*, *kofte* and *kjortel* are used to provide a historical ambience, especially in writings on medieval Scandinavian dress. These terms are no longer part of modern dress terminology, though many of them were still in use well into the 17th, 18th, or 19th centuries. In modern dress literature, the terms are often used anachronistically, in the meaning they carried in these later periods, although this cannot be expected to be identical to the original meanings of the terms in medieval times.

Lately, a clearer use of terminology has developed. Especially in recent archaeological works, formerly confusing archaic terminology has been replaced by neutral or modern descriptive terminology, thereby avoiding previous complications in regard to historically accurate terminology (Østergård 2003; Mannering 2006; Vedeler 2007). Yet, within the field of terminology and dress history, a corpus of defining and using historical terminology needs to be developed.

Neutral terminology and modern descriptive terminology can be described as the use of generic and easily understood modern terms such as headwear, legwear, or covering. These can be combined with more specific modern terms like dress, shirt, hood and cloak, or with additional explanatory description. Neutral terminology is primarily used within a branch of dress history that is not based on written evidence. Archaeological studies use neutral terminology for describing and cataloguing actual items of clothing or depictions of garments on archaeological artefacts. Their purpose is not to link text-based dress terms to physical garments, but to give a precise description of clothing remains.[3]

Another way of incorporating terminology studies in works on medieval costume is to use what can be defined as 'source-based terminology'. This is preferable for studies based on general dress descriptions or on written evidence. Until now, this type of terminology has only been used to a limited extent in Scandinavian dress history. Source-based terminology uses the written accounts as the basis for the definition of terms for garments. The advantage of source-based terminology is the consistency achieved between terms used in contemporary medieval sources and in dress history. The danger of this approach, however, is that it is often based solely on the scholar's own understanding and interpretation of the typology of terms. Thus, a more clearly defined methodology is a prerequisite to the study of terminology and the use of source based terminology in dress literature.

Terminology Use in Contemporary Medieval Sources

Just as terminology systems differ from one researcher to another, so did terminology use differ from scribe to scribe in the Medieval Period. In medieval documents we find a broad range of terms, which at first glance appear to be used inconsistently. In reality, this is not strange, as medieval documents were written by a broad group of individuals. On closer examination, the seemingly inconsistent usage of terms shows a system: a terminology system. Some scribes have used overall generic terms for dress, others more precise names for specific garments.

In a modern setting, we can undoubtedly find a similar system in the use and definition of terms. If a random group of people had been asked to define a long dark brown trench coat, some would define it a coat, while others would perhaps be very specific and define it as a long dark brown trench coat and maybe even add details such as pockets, belt and plaid lining. This is also very much the case with medieval terminology usage. The difficulty for researchers is to discover which terms are used as precising definitions of specific terms *e.g.* trench coat, and which are overall generic terms *e.g.* just coat. Further descriptive elements can be added to these term groups, such as 'with a belt', 'dark brown', 'full' or 'with a train'. The difficulty, again, is then to clarify if and which of these are merely additional descriptions of a given object that could be applicable to any object, and which are additional characteristics specifically connected to a certain term.

There can be various reasons for medieval clerics and scribes to use different terms for garments, but it is clear that the term usage in contemporary documents is primarily dependent on the following conditions: the scribe's understanding and choice of language, the scribe's individual choice of terms and the scribe's knowledge of terminology. Thus, these conditions form and define the terminology usage in a scribe's text, which eventually has to be decoded by the modern scholar.

Language Programme

The linguistic factor (the scribe's understanding and choice of language), which influences the scribe's choice in terminology, can be defined as the language programme. The latter is based both on individual taste, as well as on external factors (*e.g.* language availability and language tradition) that constitute the terminology available to the scribe.

Language use in Scandinavian medieval sources traditionally varies between Nordic terms and Latin terms (Falk 1918; Geijer *et al.* 1994, 55; Andersson 2004) used in documents written in Nordic languages and Latin respectively. The language use is, however, more varied and complex than this. Terms of dress can in an otherwise Nordic text also be described by foreign loanwords, usually from French and German (Middle Low German), or in an otherwise Latin text by local Nordic terms (Falk 1918; Krogerus 1982).[4]

Latin, too, is not simply Latin. In Scandinavian written sources, both Classical and Medieval Latin terms appear in connection with terms for dress, each of which has its own interpretational frame. In my survey of Classical Latin and Medieval Latin terms for headwear, it became apparent that these two systems had significantly different typological features, which again is crucial in interpreting the garment in context (Dahl 2007). It was apparent that nearly all of the terms which appeared in Classical Latin were mainly overall generic terms for dress (Dahl 2007, 47–59). On the other hand, most of the Medieval Latin words for dress originated in loanwords from languages other than Nordic, mainly German and French. Furthermore, most of these were specified terms, relating to a specifically defined type of garment.

Basically, the language programme can be understood as the choices available to the scribe when describing a garment:

1) A garment can be described by a Nordic, Latin or to some extent a German term
2) A garment can be described by a loanword or a local term
3) A garment can be described by an overall generic term or by a more specifying term

This signifies that the medieval scribe had a choice of a broad range of different terms available to him, which could be combined in various ways. These terms were all interlinked, yet fundamentally very different. Some terms would be more obvious than others to the medieval scribe, for example, terms that were well known in Latin and Nordic languages or terms that were commonly used within a context or tradition known to the scribe. Some terms would be more or less synonymous and therefore mainly be based on the scribe's individual taste and choice, whilst others would be more difficult to translate directly into Latin from Nordic and *vice versa* and therefore prove challenging to the scribe's own understanding of language.

Individual factors, dependent on the scribe's choice and knowledge, are therefore obviously challenged by external non-individual factors dependent on more fixed elements of language. Scribes naturally depend on terms available in the given language and are assumed to use already existing terms for dress rather than to invent a whole new spectrum of dress terms. Other external factors that influence the use of terminology include developments in fashion and changes in terminology over a given period, for instance, when a term for a specific garment is replaced by a newer and more fashionable term, or when the same term is used over a length of time while the typology of the garments covered by the term changed a great deal during that given period.

An example of this is the Scandinavian term *kjortel* (tunic), which was in use from the early Medieval Period into the 17th century. The term did not change in that period, but garments known as tunics certainly did (Nørlund 1941; Geijer *et al.* 1994, 55, 58–60). Terms whose significance changed over time can be found alongside terms whose meaning was rigid and continued to be used unchanged throughout a long period. Everywhere in Europe, such terms covering garments of changeable typology can be found (Madou 1988, 78–80, 82–84). Some were so generic that they could refer to other objects than items of clothing, or cover a wide range of different items of clothing that would appear unconnected in terms of anatomical similarities; although the *essence* of a term's original typology would be shared (Chambers and Owen-Crocker 2008, 66).

Fluctuating Terms

Certain fluctuating factors of terminology use can be difficult to fully categorize in a more rigid system of definition. Yet, to place them in a term-based context, some of these factors and conditions can be described generally as term fluctuation due to changes in meaning of a given term over time and/or different meanings of a given term, which again constitutes a number of different results. These factors also had an impact on terminology use in contemporary documents. In Latin examples alone, it is clear how interchangeably the language programme was used and how it affected terminology use in various sources.

The fluctuating terms influenced by changes over time can, in short, be defined as 1) evolutative/progressive terms, 2) stagnated terms and 3) a combination of the two. These are based on three different lines of development: the first covers terms linked to fashionable dress, changeable in accordance to shifts in fashion and style; the second covers dress terms that were originally used for fashionable dress, but eventually were discarded from fashion, yet maintained in accordance with formal or formalized wear (*e.g.* official dress or ceremonial dress); and the third covers terms that are used within a closed context, independent and unaffected by developments in fashionable wear (*e.g.* clerical costume[5]). The three lines of development can be viewed as parallel developments. Clerical costume also used terms originating in fashionable terms, mainly from Roman dress. These were later formalized in official liturgical costume and, within that context, they developed according to their own logic, not influenced by the chronological development of fashionable secular dress (Table 8.1).

In dress literature, we often find clearly contradicting statements by scholars on the typology of given terms; contradictions that at first glance must lead us to assume that at least one of them must be in error (von Wilckens 1988, 54). The French scholar Boucher, for instance, defines the term *tabard* as an open short cloak that was later adopted into formal official dress of heralds in an archaic form of the previously fashionable short cloak (Boucher 1996, 197, 442). The German scholar Post instead translates the German term *tappert* deriving from the French *tabard* as a long, full gown

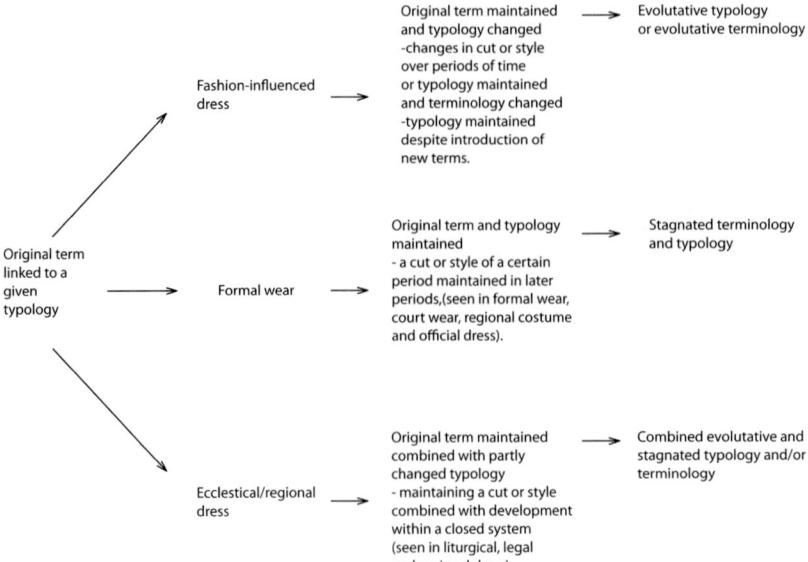

Table 8.1. *Three parallel lines of development in terminology use: evolutative/progressive, stagnated and combined evolutative-stagnated terminology use.*

(Post 1924, 42–44). The clear contradiction in defining the term has led scholars to take sides on the issue, either referring to Boucher's or Post's interpretation of the term (Wilckens 1988, 54). Post offers no sources for his definition, making Boucher's more comprehensive study far more reliable. The case is, however, not that simple. The two scholars use very different chronological and geographical frames in their study. Boucher, on one hand, defines the garment based on written evidence primarily from France in the 12th and 13th century. He also refers to an archaic use of the term in 15th-century France, where the term was preserved for herald's wear. His survey of 'European' dress is in this context therefore limited to France, and his description of *tabard* as a fashionable garment is restricted to high medieval dress. Post, on the other hand, writes about late 15th and early 16th-century fashionable German dress, and although Post himself offers no sources for his interpretation, several German written accounts do contain examples of *tappert* being defined as long gowns (*Ulm* 1411, 220, art. 434–435). Clearly, the term *tabard*, which in France was a term for fashionable wear in the 12th and 13th century, had been replaced by other fashions and new terms, yet continued in use as an archaic term. In Germany, the term instead was in use for fashionable wear throughout centuries, and the evolution of the typology continued, so that the original meaning of *tabard* as a short cloak (Carroll-Clark 2005, 84) gradually changed into a long gown-like garment. The example shows that even seemingly contradictory definitions can be technically correct and result from different chronological and geographical frames, which makes them not directly compatible. Moreover, the example shows how dress terms are employed differently throughout Europe.

Variable Terms for Similar Typology

As mentioned earlier, sources show examples of both specified terms and overall generic terms. Overall generic terms

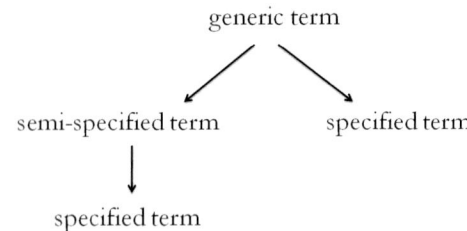

Table 8.2. *Relationship between generic terms, semi-specified and specified terms.*

naturally imply non-specific dress terms, which, although of a specific typology, do not specify for instance, the type of cloak, tunic, or dress, but rather merely name 'cloak', 'tunic' or 'dress'. Generic terms are not defined by a specific shape or cut of a garment, specific features, colours or material. They can, however, be followed by additional information, such as mention of colour, shape, or length. Specifying terms, on the other hand, are terms which imply a certain type of garment, a special cut or shape, a certain type of dress in a fashion sense, a specific look, material or other. Between these two there can even be found semi-specified terms, that can indicate a certain type of an overall term, yet they do not imply enough specified features to be a specific term. Thus, a *tunica* is an overall term, a *supertunica* a semi-specific term and *supertunica aperta* (open over-tunic) a specific term. A generic term can thereby cover a range of both semi-specified and specified terms, and a semi-specified term can cover one or more additional specified terms (Table 8.2).

In Scandinavian medieval documents, generic Latin terms are sometimes specified by an additional Nordic term. In these sources, we find mention of, for example, a *velum dictum strik* [a veil called *strik*] or a *cappa dictum tabærd* [a *cappa* called *tabard*]. My research has shown that Classical Latin words were in almost all cases generic terms, while

Classical Latin	Loanword/ Fashionable term	Latinised term/ Medieval Latin
supertunica ◦	*surcotte/surcot* ◦	*surcocio/surcocium* ◦
(*cappa*)*	*tabard* ▪	*tabardo/tabardum* ▪
(*cappa*)*	*gardecors* ▪	*gardecorso/ gardecorsum* ▪
(*cappa*)*	*corset/corsette* ▪	*corsetta/corsettum* ▪
(*pilleus/pilleum*)*	*bonette* ◦	*boneto/bonetta/ bonettum* ◦

Table 8.3. Examples of the terminological development between Classical Latin, loanwords and Middle Latin. Word in ( ) indicates a term which is not directly synonymous with the specified loan word. * indicates a generic term, ◦ indicates a semi-specified term and ▪ indicates a specified term.

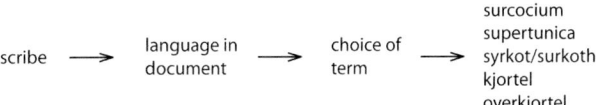

Table 8.4. Example of terms depending on language in document (Latin or Nordic) and choice of term (generic, semi-specified or specified term), which influences the final choice of local term, (loanword, Classical Latin term, or Middle Latin term). Depending on these choices, several terms would be available to the scribe merely to describe an over-tunic.

Scandinavian terms could be generic, semi- or specified terms (Dahl 2007).

In general, some terms were fashionable in the sense that they were directly influenced by a certain fashion in a given period of time, and then disappeared as the dress item went out of fashion, whereas other terms continued in use, even when fashions changed. In general, the generic terms were independent of fashion and could be used over a long period of time for different types of garments with a simple common denominator as regards typology. The many loanwords from French and German were, in contrast to the Classical Latin terms, more fashion-influenced terms. Usually, they appear for a limited period of time, in which a garment covered by a certain term was in fashion. Loanwords can be either semi-specific or specified terms, but rarely overall generic terms.

Sometimes, a specific term which first appears as a loanword can be found translated into Latin – or rather Latinised in Low Latin. An example of this is the Classical Latin term *cappa*, covering a broad range of cloaks, short as well as long, mantle-like garments and full gowns. Sometime during the Medieval Period, a variant of *cappa* became fashionable in France under the name *gardecors*. In 13th and 14th century Scandinavian written accounts, both the Classical term *cappa*, the loan word *gardecors* (usually almost unrecognizably transformed into *warthækors*, *vartekors* etc. in Scandinavian texts), as well as the Latinising version of the loanword, *gardecorsum*, appear (Dahl 2005a, 39–41, 51, 56; 2005b, 11–13, 16). In these cases, the Middle Latin term and the French term are synonymous, whereas the Classical Latin and the Middle Latin are not. This can even be seen in cases where a synonymous term can be found in both Classical Latin and as a foreign loan word. An example is, for instance, *supertunica* in Classical Latin, which appears in French as *surcot*, which again can be found in the Latinised term *surcocium*. In this case, the Classical Latin, the loan word, and the Middle Latin/Latinized term are all equivalent (Table 8.3). Examples of such equivalent terms available to a local scribe can be found all over Europe.

There are several examples of variable terms being used for garments of similar typology within Europe. In most cases, this is simply due to the scribe's individual choice in terminology, but occasionally, it is based on a tradition in a given area. For instance, certain terms are used in one part of Europe but not in another, or a specific term is outdated or replaced by others in one area of Europe but remains in another. It is not always possible to find sufficient material to arrive at conclusions on regional differences within Scandinavia, although some differences in terminology use can be found between East and West Scandinavia.

With the broad range of Classical Latin, Medieval Latin, German and French loanwords as well as local Scandinavian terms available to a Scandinavian scribe, there would have been many possibilities to choose from. To this can be added generic terms versus semi-specified or specified terms, all of which the scribe would choose from and use according to a given tradition in his work, area or other (Table 8.4).

Unravelling Medieval Terminology

Even though medieval scribes' use of terminology at first glance seems inconsistent and diffuse when treated as an overall statistic or collective material, it is evident from the individual documents that there is consistency and logic in each individual scribe's definition of garments, as well as in their choice of terms. In general, the scribes consistently use the same overall group of terms throughout a document. Thus, a scribe who includes semi-specified and specified terms for dress, will define nearly all garments with semi-specified and specified terms throughout the document and not include generic terms, whereas a scribe using generic terms, will mostly have chosen generic terms only to define garments and not include semi-specified and specified terms as well.

Similarly, additional information is rarely found in documents with specified terms, whereas additional information on typological features and local terms abound in documents with generic terms. Thus, if both *cappa* and *cappa manicata* are mentioned in the same document, then the scribe has attempted to separate two terms and types of *cappa* with additional information as well as specification (Dahl 2005a, 40–44; 2005b, 13–15). Other terms, such as *supertunica*, *surcocium* and *overkjortel* are never found in the same document, possibly because these terms could not be used to indicate differences, but were instead synonymous.

Furthermore, the example shows that generic, semi-specified and specified terms are interchangeable. This means that in documents where, for instance, *cappa* and *cappa manicata* are mentioned together, *cappa* is no longer a generic term, but a semi-specified term, as it is used as an indication that it was a different type of cappa than the *cappa manicata*. A similar situation arises with *tunica*, *supertunica* and *supertunica aperta*. In this case tunica, which is normally a generic term, becomes a semi-specified term with the appearance of the two others, and *supertunica* becomes a specified term as is apparent by the appearance of a *supertunica aperta* in the same document. In contrast, in documents where only *tunicas* or *cappas* appear, the terms remain generic.

Common Misconceptions – defining Equivalent Terms

In medieval documents, some objects can be referred to by double definition, meaning that an object can be defined with more than one term. This is, for instance, the case when a scribe writing in Latin chooses to refer to a garment with both a Latin and Scandinavian term (*e.g.* "unum toga dictum ærmekåbe" – a *toga* called *ærmekåbe*), or in some cases, where a generic Latin term is followed by a specific Latin term (*e.g.* "unum Cappa dictum gardecorsum" – a *cappa* called *gardecors*). In studies of dress incorporating source-based terminology, these doubly referenced objects have proven particularly problematic as they are often mistaken for simply equivalent terms directly translatable from Latin to Danish, Swedish or Norwegian.

They show, however, a rather complex system of describing one term by adding another. By adding a term from the native language, the scribe can specify the type of overall garment meant, or merely describe the garment in question. It must be noted in that respect, that a Scandinavian scribe writing a will for any nobleman, would have been presented with a description of the donations, including garments in the local language, which he would then translate into Latin terms. This is something that is often overlooked by modern historians who deal with wills written in Latin. They are seen as merely Latin terms for Latin garments rather than as a system of Latin terms translating a Scandinavian dress mode. In other words, a Scandinavian scribe may, in a document written in Latin, have added both the local term for a garment as well as the nearly equivalent Latin term, when the local term was not directly translatable into Latin or when the scribe was unsure which Latin term to use for the given garment.

Historians also tend to neglect these particular reasons for the scribe to add Scandinavian terms to a Latin text and the Latin and Scandinavian terms respectively are usually seen as simple synonyms, which leads to incorrect conclusions. In modern translations and definitions of medieval terms for dress, the conclusions made by treating seemingly equivalent terms as synonyms may not at first sight appear misleading, or strike the scholar as an incompatible set of terms, especially as medieval terms appear strange and unfamiliar to the modern eye.

Yet, if a similar analogy were to be used in regard to modern costume terms, the misleading results that can occur by such a method of definition will be apparent. If a 1920s text mentions 'underwear called garter belt', we could conclude that garter belt is a type of underwear, but not that the word underwear always means a garter belt, nor is it possible to apply the meaning of the example to other time frames, such as documents from 19th century probates, and finally conclude what percentage of men wore garter belts simply because the material mentions 'underwear'. Clearly, in this case 'underwear' is an overall generic term, whereas 'garter belt' is a specified one and therefore they are not synonymous terms. Again, a text may mention 'underwear called *lingerie*', in this case both terms are generic terms, but do not fully cover the same typology. And placing these two terms as fully synonymous in the above-mentioned example would lead to an equally incorrect conclusion.

Thus, treating two medieval terms as equivalent, and taking them out of a context in regard to time, gender and place, can lead to a serious misinterpretation of these terms. For instance, if a Latin text mentioned a *velum dictum glissing* then it was to be understood as a '*velum* called *glissing*'. With *velum* as an overall term and *glissing*, a specified one, it can thus be stated that a *glissing* is a sort of veil, but not that *velum* and *glissing* is always the same or that a veil is always a *glissing*. In this particular case, however, the two terms in Latin and Nordic clearly refer to the same object in question, and are therefore used synonymously in a specific context.

Related terms may only be related in typology in a particular context. Even when two terms, written in Latin and Nordic respectively, are both overall generic terms, each of them can still cover a broad range of meanings that are not directly compatible. For instance, a text can refer to a '*velum* dictum *strik*' but the *velum* (veil) is an overall generic term covering all sorts of veils, whereas the Nordic term *strik* is an overall generic term as well, but not an overall generic term for all sorts of veils (Dahl 2007, 70–79).

The Austrian scholar Elizabeth Vavra, for instance, points out what she calls *Überdefinition* (over-definition), referring to the popular method in modern dress history, when a medieval term of dress is translated into a specific definite garment. The suggested detailed meaning of a dress term leaves no room for further interpretation, or for the possibility that a term can refer to a slightly different garment in another time or context (Vavra 1988, 44). Vavra indicates how this over-definition leads to a series of misinterpretations of medieval costume. Her concept of *Überdefinition* can also be applied to the neglecting of generic terms when translated as if they were specified terms, or to using sources as a general quantitative material rather than as a broad spectrum of information on garments and their terms in a given context. Each document has its own setting in time and space, context and use. Removed from their context, they instead create misleading results. The importance of differentiating terminology in its accordance to region and time has been previously addressed by scholars (Kühnel 1988, 13–14; Vavra 1988, 44), but it cannot be emphasised enough.

Gender, Time and Type of Term

The common misconception that a given term in Latin corresponds fully to an additional Scandinavian term and that these are not influenced by gender, time or region can lead to unusual results. The simplified definition of terms as equivalent or synonymous, regardless of gender indications in the individual document, the confusion of generic terms with specified ones, and out of chronological and spatial context, can lead to serious misinterpretations of a given garment term.

Thus, in a will from AD 1328, which deals with a woman's belongings, one garment is described as a *toga dictum ærmakapæ*. If the two terms are regarded as two words that mean exactly the same, it is possible to conclude that every *toga* mentioned in medieval documents equals *ærmakapæ* (Andersson 2004, 107), and thus a *toga* mentioned in a man's will from around AD 1400 can be understood as an *ærmakapæ*. We may even conclude that, according to written evidence, 25 % of the male garments mentioned in Swedish noblemen's wills around the year AD 1400 are *ærmakapæs*. Yet, when going through the wills I used for the survey mentioned at the beginning of this article (Dahl 2007), it became clear that the word *ærmakapæ* or the like, never appears in connection with male costume, although the Latin term *toga* does. Furthermore, the term *ærmakapæ* is not found in documents dating around AD 1400, although it is present about 50 to 100 years earlier, whereas the term *toga* can be found throughout the Medieval Period. Consequently, the simple equation between *toga* and *ærmakapæ* would translate a High Medieval woman's garment into Late Medieval menswear.[6]

The above example demonstrates how crucial it is to define terms on an equal level. This necessitates grouping terms according to whether a term is a generic or a specified term, as well as accepting that a term in one particular time or context cannot be transferred directly to a different time, gender and context.

A study of the Scandinavian term *kåbe* reveals that there are certain conditions connected with terms in Latin and in Scandinavian. It is, for instance, evident that while most of the Latin terms are non gender-specific, most Scandinavian terms are.[7] The gender distinction, however, is not fixed, but rather can change from gender-neutral to gender-specific from one period to another. This means that a term connected to men in one period, can be connected to both genders, or to women in another. The gender aspect of terms can function in two different ways: it can refer to the term itself being gender-specific (like modern blouse and shirt that are similar items of clothing but are gender-specific), or it refers to a garment with a certain typology that was primarily or exclusively worn by either men or women. It is also evident that none of the terms used is directly translatable between Scandinavian languages and Latin; instead, it shows an intriguing system of grouping garments according to level of specification, gender of the wearer, and the period in which they appear.

For instance, within a timeframe of AD 1280–1350, for which a great deal of documentation exists, various terms can be seen in connection with the Latin term *cappa*. Some of these terms appear all over Europe in this period, while others had long been outdated elsewhere, but remained in use in Scandinavia. Some terms, which are used gender-specifically in Scandinavia, were gender-neutral elsewhere.

And furthermore, some of the terms appear only to be connected to this particular time frame – not before or after. The Danish term *kåbe*, for instance, is exclusively female wear during this period, but appears as a male term around AD 1400, when it is, however, no longer related to the Latin term *cappa* but rather to *toga*. Although both terms appear over a long period of time, they may not be connected to each other over the same period. While related terms, for instance, a *cappa dictum ærmakapæ* [*cappa* called *ærmakapæ*) provide information on the typological definition of the term *ærmakapæ* (modern Danish *ærmekåbe*, modern Swedish *ärmkappa*), non-related terms in the same document can be equally informative. Non-related terms signify terms that appear in the same document alongside a given term and can therefore be defined as something other than that particular term. For instance, if *mantle*, *cappa* and *cappa* called *ærmakapæ* appear in the same text, then the three terms are meant to relate to three different items of dress (Dahl 2005a, 49–50).

The Use of Additional Terms

Individual sources provide a large group of additional terms – corresponding terms – which however do not appear to be the same. For instance, the term *cappa* can be found mentioned alongside terms such as *ærmekåbe*, *kåbe*, *kapo*, *kappe*, *tabard* and *gardekors* (modern harmonized spellings). On the other hand, a term such as *ærmekåbe* can be found in connection with Latin terms such as *mantel clos*, *cappa*, *cappa manicata* and *toga*. Yet, at the same time, a term like *cappa* is never seen in connection with *toga* or *mantel clos*, nor is the Nordic term *kåbe* ever seen in connection with *tabard* or *gardecors*, and neither of these two can be connected to *toga*. So how is this to be understood? The explanation is that some terms in Latin are not gender-related, whereas some Nordic terms are; furthermore, some of these terms are generic, whereas others are specific.[8]

In addition, some of these terms for garments correspond with yet other terms that do not correspond with the first. It should also be noted that these constellations of terms can be defined as corresponding in a given period or context, but can change in another time and context, so that two terms that corresponded earlier are no longer defined by a similar typology. This can be due to many circumstances, mainly changes in fashion, the shift in the typology of a term over time, or the introduction of new garments that could prompt the adoption of new terms to define their characteristics. For instance, a Nordic term *kapo*, which is defined by the Latin term *cappa* around AD 1300, can instead be referred to as *toga* or *gonum* in Latin just 100 years later, simply because the garments covered by the Nordic term *kapo* changed appearance and characteristics in such a way that they could no longer be covered by the same term as previously (Tables 8.5–8.6).

In short, the typological definitions as well as the overall terms indicate that in the period AD 1280–1350, when *cappa* refers to men's wear, it refers to cloaks, but when applied to women's wear, it designates gowns. Mostly, men's *cappae*

Date	Source	Related/corresponding terms	Non-related terms (different from)	Additional typological data
1287	*Capa* SD: I, p. 39		*Mantel*	Double, of scarlet cloth lined with cloth of Gent.
1287	*Capa* SD: I, p. 39		*Mantel*	Double, of moreto cloth lined with white cloth, for riding
1287	Capa SD: I, p. 39		*Mantel*	Of cloth of Gent, for riding
1292	*Kobe* (DD + Test, 38)		*Mantel*	Of red scarlet cloth
1293	*Mantella clausa* (inventory of Isabella of Scotland, Queen of Norway, DN XIX, 426)		*Mantel* *Cappa*	Blue cloth
1293	*Capa* (inventory of Isabella of Scotland, Queen of Norway, DN XIX, 426)		*Mantel* *Mantella clausa*	Of white murret cloth lined with fur
1293	*Mantulkapu* (will, SD II, 158)	*Mantel + cappa*	Skin (*skingr*) *Mantel* *Cappa*	Of bruneto cloth lined with fur
1293	*Kopu* (will, SD II, 157)		Skin (*skingr*) *Mantel* *Cappa*	Silver trimmings for a cappa
Late 13th–early 14th century	*Kapor* (romance, Hertig Fredrik av Normandie, verse 3058)		Mantel	Of silk, two cappae worn by a royal couple for travelling on horseback, appearing to another royal couple.
1307	*Capa manicata* (will, DD +Test, 59)	*Cappa + manicata*	*Cappa* *Mantel*	Blue cloth
1307	*Capa manicata* (will, DD +Test, 59)	*Cappa + manicata*	*Cappa* *Mantel*	Dark blue cloth
1307	*Capa* (will, DD +Test, 59)	*Cappa*	*Cappa manicata* *Mantel*	Black murret cloth with a matching tunic
1308	*Capam ... manicatam* (will, DD +Test, 62)	*Cappa manicata + cappa*	*Cappa* *Mantel*	
1313	*Capa* (will, SD III, 756)		*Mantel*	Red
1313	*Capo* (will, SD III, 755)		*Mantel*	Costly decoration (trimmings) for a cappa
1319	*Capa* (will, SD III, 403)		*Mantel*	White, for daily wear
1328	*Togam meam dictam ærmæ kapæ* (will, SD IV, 82)	*Toga + ærmekåbe/kappa*	*Cappa* *Mantel* *Toga*	
1334	*Mantellum ... sibi eciam I erma kapo.* (will, SD IV, 370)	*Mantel + ærmekåbe/kappa*		Blue, lined with miniver
1339	*Capa ... ærmækapæ* (will, Test, 80)	*Cappa + ærmekåbe*	*Cappa* *Mantel*	Red, costly cloth bequeathed to a noble Lady
1339	*Capa* (will, Test, 81)		*Ærmekåbe* *Mantel*	Bequeathed to a servant woman
c. 1340	*Ærmækopæ* (King Magnus of Sweden and Norway's Swedish National law, CIS: X, 57)			For riding/travelling (for the transport of a bride from her old home to her new)
1345	*Ærmækopo* (Sumptuary law of Telge, Sweden, SD V, 479)			For riding/travelling (for the transport of a bride from her old home to her new)
1345	*Epitogium dictum ærmækapu* (King Magnus of Sweden and Norway, general law, SD: V, 484)	*Epitoga + Ærmekåbe*/kappa		For riding/travelling (for the transport of a bride from her old home to her new)

Table 8.5. Examples of female cappa, kåbe/kappa and ærmekåbe/kappa from a selection of sources dating c. 1280–1350. The selection shows examples of ærmekåbe mentioned in relation to toga, epitoga, cappa and mantel, as well as relations between kåbe and mantel, cappa and toga. Examples also show non-related terms, for instance, mentioning both cappa, cappa manicata and mantel in a single document, showing that cappa manicata is different from mantel and cappa in that specific source.

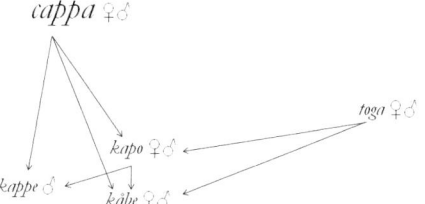

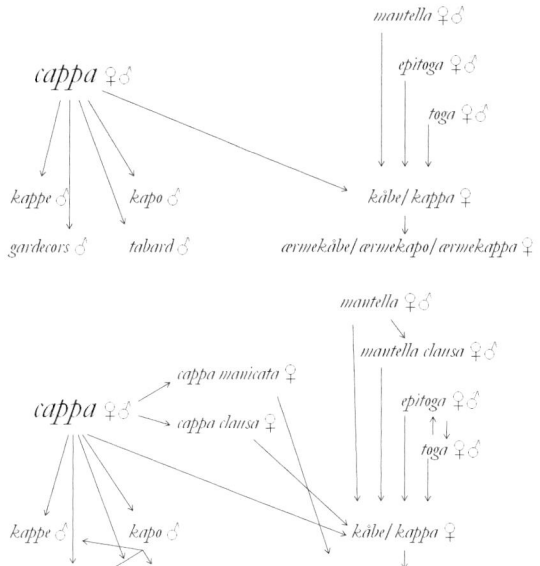

Table 8.6. Constellation of terms related to cappa *in the period c. 1280–1350. A simplified constellation (top) and a wider constellation including more specified terms (bottom). When a group of terms are placed in a constellation within a specific time frame, a certain pattern in terminology is revealed. In the period c. 1280–1350, it is noticeable that the generic Latin term* cappa *covers a range of male garments, such as the* kapo *which again could signify all the others, whereas* cappa *only covers a single female,* kåbe, *which again covers the specified term* ærmekåbe. *In contrast, the female* kåbe *can be covered by multiple generic, semi-specified and specified terms. The gender-neutral generic Latin term* toga *likewise covers the Scandinavian term* kåbe *in regard to female dress, but it was unclear from my survey, which Scandinavian term for male dress it covers, as it did not appear in the sources I examined. The term* mantella *is by definition a gender-neutral term but appears more often in connection with female dress in Scandinavia. The male* mantella *always refers to a mantle (general ceremonial). Only the female* mantella *can refer to both a mantle and a mantlelike gown. The Latin terms* cappa manicata *and* cappa clausa *are by definition gender-neutral, yet appear exclusively in regard to female dress in Scandinavia.*

Table 8.7. Constellation of terms related to cappa *and* kapo *in the period c. 1400–1480. The pattern in the constellation appears very different from the earlier period. There were no significant differences between the periods 1400–1440 and 1440–1480 in the use of terms, clearly suggesting that the terminology use was similar throughout the 15th century. Fewer terms were used to describe the groups of garments than in the earlier period of c. 1280–1350, and the overall pattern shows a shift toward a more gender-neutral terminology used in the later period. The Latin generic term* cappa *covers the Scandinavian terms* kappe, *kåbe and* kapo, *whereas the Latin term* toga *only covers* kåbe *and* kapo *but not* kappe. *In addition, the Scandinavian term* kapo *covers both* kappe *and* kåbe, *sometimes referring to one sometimes to the other.*

Conclusions

The study of individual additional typological definitions as well as the definition of connected terms can provide a wealth of information about garments within a specific group in a certain time and context. It is clear that a methodological approach to terminology may help us gather a great deal of information on the typology of certain garments in a given context. It is evident that these main conditions, which in short can be defined as language, scribe's choice and fashion development, are crucial to understanding the terminology used in the Medieval Period. In a merely quantitative or statistical study of dress terms, a broad range of information is neglected.

In surveying a number of documents mentioning for instance *cappa*, the number of *cappae* mentioned in the document group does not demonstrate how the term is used in a larger constellation of generic and specified terms, in correspondence with other terms and through means of additional information given to each single garment mentioned in a text. Clearly, as shown in this survey, the constellation of terms and their related terms infers information on the typology of garments within a given time frame, but these conclusions are not necessarily valid for another time frame. We cannot assume that the meaning of a term in one period is the same in another, that a garment that appears to be gender-neutral or gender-specific in one context, will have the same attributes in another, or that the relationship between Latin and vernacular garment terms remain the same in different periods. Instead, it is crucial that the dress historian analyse each document methodically, defining the terminology apparatus of language, term specification and context all of which form the terminology project within the given document.

In order to gain a deeper understanding of medieval dress, the description of which is so often hidden behind

were short and open, but even when they were longer and closed and therefore had the same typological features as the female version of the garment, the male garments would still be defined as cloaks, and the female as gowns.

In the following period from the late 14th to 15th century, the male fashions changed, and the Nordic term *kapo* was no longer limited to cloaks but could also describe gowns. In dress literature, the 13th century is generally described as the period of unisex fashion, yet in regard to terminology, terms of 13th century dress are highly gender-specific. Perhaps, dress that to a modern eye would appear gender-neutral was considered gender-specific in its own time or, perhaps, a period with less gender-specific traits in costume would specify differences in terms for dress and *vice versa* in periods of less gender specific terminology.

unfamiliar and strange terms, further investigation into many other aspects of medieval dress must be undertaken. As dress historians, we have probably all experienced falling short in determining typology of specific terms, either due to lack of sources, detailed description or systematic methodology in the process. An improved methodology and approach to interpreting terms in context will help to diminish some of the problems in the field. Avoiding the typical errors in translating and defining medieval dress terminology, as well as a better understanding of the conditions that form terminology use in a historical context, will strengthen the field. Hopefully, research in this direction will throw light on more dress terminology, its use and, eventually, further improve knowledge of medieval dress.

Acknowledgement

I extend my thanks to Adam Dutton for comments and suggestions on the manuscript.

Notes

1. This applies to the disputed term *cotehardie*, which has been mistaken for a specified term defining a tight-fitting dress. American dress historian Robin Netherton has introduced the more fitting modern descriptive terms 'gothic fitted dress' and 'gothic fitted gown' to replace the dubious terms that are now generally accepted to describe 14th-century tight-fitting fashions (Netherton 2005, 115–16; 2008, 143, 151–153).
2. Definitions of *kapo* and *kapu* in Icelandic are made by Andersen 2006, Straubhaar 2005 and Zanchi 2008, all of whom deal with dress in Icelandic sagas.
3. In recent works on prehistoric and medieval dress, neutral terminology has been used by Østergård 2003, modern descriptive terminology by Mannering 2006, and combined neutral and modern descriptive terminology by Vedeler 2007.
4. The language programme in Scandinavian texts is complex. Nordic languages include Danish, Norwegian, Swedish, Icelandic, Faeroese and, to some extent, Finnish. On a smaller scale, Scandinavian languages include Danish, Swedish, Norwegian and, to some extent, Icelandic and Faeroese (West Norwegian and Old Norse). German includes both High German and various forms of Low German. All these languages are used differently in the Scandinavian languages. In Finland, for instance, sources can be written in Finnish (from the early Modern Period), Swedish and Latin. Sources written in Low German mainly appear in Denmark and sources written in Latin are more common in Denmark and Sweden than in Norway, where the predominant written language is Norwegian. Icelandic sagas are written in West Norwegian or in *Norrøn* (Old Norse) (Andersen 2006, 6). There are also loanwords from other languages. In Norway, more English and French loanwords for dress can be found than in Denmark and Sweden (French words are loaned through England), whereas many German loanwords can be found in Denmark (Dahl 2007, 50–52, 57–58). The loanwords from Germany which appear in texts written in Danish or Latin, have nearly all originated in High German, but are loaned through Low German dialect, and sources written in German in Denmark are all written in Low German (Carlie 1925). Loanwords appear both as old loanwords that had been integrated into the Nordic languages and used for a long period of time, as well as new loanwords that had been introduced in the Nordic countries more recently.
5. And later on, to some extent, regional costumes that developed from an original fashionable item within a closed system that has a parallel line of development.
6. Several different interpretations of the term *ærmekåbe* have been given; English dress historian Newton suggests that the ærmekåbe was a "short cape…presumably long enough to cover the upper arms" (Newton 1980, 99). Swedish historian Andersson suggests that *cappa* and *mantel* are more or less identical and refers to sleeveless garments, cloaks, whereas the *ærmakapæ* was an upper garment or coat with sleeves mainly worn by men (Andersson 2004, 108; 2006, 99). Finally, Danish archaeologist Nørlund suggests that *ærmekaabe* was a predominately male garment, a long, full gown, identical with the French *houppelande* and German *tappert* (Nørlund 1941, 76–77).
7. This does not refer to the gender of the term itself (*Femininum, Masculinum, Neutrum*) as it does not affect the gender of the wearer.
8. More terms, such as *kordhumbla, corseto, kothardil cotehardi, skingr* and the male *toga*, can all be placed in various relationship in the constellation, but it is not possible to go in depth into the definitions and relations of these terms here.

Bibliography

Andersen, A. H. (2006) *Kapper som litterært motiv i Laxdæla Saga.* Doctoraal scriptie, wetenschappelijke variant. Master's thesis, Amsterdam University.

Andersson, A., and Franzén, A. M. (1975) *Birgittareliker inlånade till Historiska museets utställning "Birgitta och det Heliga landet" 30 november 1973–17 februari 1974.* Antikvariskt arkiv 59. Stockholm, Almqvist & Wiksell.

Andersson, E. I. (2004) Written Traces – Wills in 13th to 15th century Scandinavia. In J. Maik (ed.), *Priceless Invention of Humanity – Textiles, NESAT VIII*, 105–112. Acta Archaeologica Lodziensia: 50/1. Łodz.

Andersson, E. I. (2006) *Kläderna och människan i medeltidens Sverige och Norge.* Avhandlingar från Historiska institutionen i Göteborg 67.

Blindheim, C. (1980) Drakt. In *Kulturhistorisk leksion for nordisk middelalder*, bd. 3, 276–290. Malmø.

Boucher, F. (1996) *A History of Costume in the West.* London, Thames & Hudson.

Carlie, J. (1925) *Studium über die mittelniederdeutsche Urkundensprache der dänischen Königskanzlei von 1330–1430.* Nebst einer Übersicht über die Kanzleiverhältnisse, Lund.

Carroll-Clark, S. M. (2005) Bad Habits: Clothing and Textile References in the Register of Eudes Rigaud, Archbishop of Rouen. In R. Netherton and G. R. Owen-Crocker (eds), *Medieval Clothing and Textiles,* vol. I, 81–103. Woodbridge, Boydell & Brewer.

Chambers, M., and Owen-Crocker, G. R. (2008) From Head to Hand to Arm: The Lexicological History of "Cuff". In R. Netherton and G. R. Owen-Crocker (eds), *Medieval Clothing and Textiles,* vol. 4, 55–67. Woodbridge, Boydell & Brewer.

Dahl, C. L. (2005a) Cappa og kobe: forvirringen om dragten "kåbe" i middelalderen. In C. Oksen (ed.), *Middelalderdragter: Arbejdspapirer 2001–2005. Tekstilforskning på Middelaldercentret,* vol. 1, 39–74. Nykøbing, Middelaldercentret: Forsøgscenter for Historisk teknologi.

Dahl, C. L. (2005b) Fra cotehardie til kåbe: dragtbetegnelser i dragtlitteratur og i kilder til nordisk dragt i middelalderen. In C. Oksen (ed.), *Middelalderdragter: Arbejdspapirer 2001–2005. Tekstilforskning på Middelaldercentret*, vol. 1, 11–20. Nykøbing, Middelaldercentret: Forsøgscenter for Historisk teknologi.

Dahl, C. L. (2007) *Strige, Glissing, Skaut: Terminologiske undersøgelser af genstandsfeltet "hovedbeklædning" samt typologiske angivelser deraf i skriftlige kilder fra ca. 1200–1600, med særligt henblik på komparative studier af dragtforskningens terminologiprojekter og termanvendelsen i samtidige kilder.* MA thesis, Saxo Institute, History, Copenhagen University.

Enlart, C. (1916) *Manual d'archéologie francaise, depuis les temps mérovingiens jusqu`a la renaissance. Tome III: Le costume.* Paris. Edt. Auguste Picard.

Falk, H. (1918) *Altwestnordische Kleiderkunde. Mit besonderer Berücksichtigung der Terminologie.* Skrifter udgivet av videnskapsselskapet i Kristiania II. Kristiania/Oslo.

Gay, V. (1887) *Glossaire Archéologique du Moyen Age et de la Renaissance. Bd I–II.* Paris, Librairie de la Société Bibliographique.

Geijer, A., Franzén, A. M., and Nockert, M. (1985/1994) *Drottning Margaretas gyllene Kjortel i Uppsala domkyrka. The Golden Gown of Queen Margareta in Uppsala Cathedral.* Stockholm, Kungliga Vitterhets Historie och Antikvitets Akademien.

Hansen, H. H. (1978) *Politikens dragtleksikon.* Copenhagen, Politikens Forlag.

Jaacks, G. (1998) Mittelalterliche Bilder als Quelle. In L. Bender Jørgensen and C. Rinaldo (eds), *Textiles in European Archaeology. Report from the 6th NESAT Symposium.* Gotarc Series A, vol. 1, 243–251. Göteborg.

Jaritz, G. (1988) Realienkunde und Fragen vom Terminologie und Typologie. Probleme, Bemerkungen und Vorschläge am Beispiel Kleidung. In *Termnologie und Typologie mittelalterlicher Sachgüter: Das Beispiel der Kleidung. Veröffentlichungen des Institut für mittelalterliche Realienkunde Österreichs*, Nr. 10, 7–20. Wien, Österreichischen Akademie der Wissenschaften.

Kraft, K. (2003) Akademisches Puppenspielen? – Für eine Objektbasierte Bekleidungsforschung. *Waffen- und Kostümkunde* 45:1, 77–96.

Krogerus, G. (1982) *Bezeichnungen für Frauenkopfbedeckungen und Kopfschmuck im Mittelniederdeutschen.* Commentationes Humanarum Litterarum 72. Helsingfors, Societas Scientiarum Fennica.

Kühnel, H., ed. (1988) *Terminologie und Typologie mittelalterlicher Sachgüter: Das Beispiel der Kleidung. Veröffentlichungen. des Institut für mittelalterliche Realienkunde Österreichs,* Nr. 10. Wien, Österreichischen Akademie der Wissenschaften.

Kühnel, H., ed. (1992) *Bildwörterbuch der Kleidung und Rüstung vom alten Orient bis zum Mittelalter.* Berlin, Walter de Gruyter.

Laborde, M. L. de (1872) *Glossaire Francais du Moyan Age: A L'usage de l'archéologue et de l'amateur des arts précédé de l'inventaire des bijoux de Louis, Duc D'Anjou dressé vers 1360.* Paris, Adolphe Labitte Libraire.

Madou, M. (1988) Das mittelalterliche Kostüm in den Niederlanden. In H. Kühnel (ed.), *Terminologie und Typologie mittelalterlicher Sachgüter: Das Beispiel der Kleidung. Veröffentlichungen des Institut für mittelalterliche Realienkunde Österreichs,* Nr. 10, 77–91. Wien, Österreichischen Akademie der Wissenschaften.

Mannering, U. (2006) *Billeder af dragt. En analyse af påklædte figurer fra yngre jernalder i Skandinavien.* Ph.D. dissertation, Copenhagen University.

Netherton, R. (2005) The Tippet: Accessory after the Fact? In R. Netherton and G. R. Owen-Crocker (eds), *Medieval Clothing and Textiles,* vol. 1, 115–132. Woodbridge, Boydell & Brewer.

Netherton, R. (2008) The View from Herjolfsnes: Greenland's Translation of the European Fitted Fashion. In R. Netherton and G. R. Owen-Crocker (eds), *Medieval Clothing and Textiles,* vol. 4, 143–171. Woodbridge, Boydell & Brewer.

Newton, S. M. (1980) *Fashion in the Age of the Black Prince: A Study of the Years 1340 – 1365.* Woodbridge, Boydell Press.

Nockert, M. (1985/1998) *Bockstensmannen och hans dräkt. The Bocksten Man and his Costume.* Halmstad/Varberg, Varberg Museum.

Nockert, M. (1996) Textiler och dräkt 1350–1450. In *Margrete I. Nordens Frue og Husbond*, 200–210. Copenhagen, The National Museum.

Nørlund, P. (1941) *Klædedragt i oldtid og middelalder. Nordisk Kultur bd. XVb. Dragt.* Copenhagen, J. H. Schultz Forlag.

Østergård, E. (2003) *Som syet til jorden: Tekstilfund fra det norrøne Grønland.* Aarhus, Aarhus Universitetsforlag.

Post, P. (1924) Herkunft und Wesen der Schaube. In *Waffen- und Kostümkunde,* bd. 1 der Neuen Folge, 1923–1925. Berlin.

Straubhaar, S. B. (2005) Wrapped in a Blue Mantle: Fashions for Icelandic Slayers. In R. Netherton and G. R. Owen-Crocker (eds), *Medieval Clothing and Textiles,* vol. I, 53–65. Woodbridge, Boydell & Brewer.

Vavra, E. (1988) Kritische Bemerkungen zur Kostümliteratur. In H. Kühnel (ed.), *Terminologie und Typologie Mittelalterlicher Sachgüter: Das Beispiel der Kleidung. Veröffentlichungen des Instituts für Mittelalterliche Realienkunde Österreichs* Nr. 10, 21–45. Wien, Verlag der Österreichischen Akademie der Wissenschaften.

Vedeler, M. (2007) *Klær og formspråk i norsk middelalder.* Ph. D. dissertation, archaeology. Det humanistiske fakultet, Oslo University.

Wilckens, L. von (1988) Terminologie und Typologie spätmittelaterlicer Kleidung. Hinweise und Erläuterungen. In H. Kühnel (ed.), *Terminologie und Typologie Mittelalterlicher Sachgüter: Das Beispiel der Kleidung. Veröffentlichungen des Instituts für Mittelalterliche Realienkunde Österreichs*: Nr. 10, 47–57. Wien, Verlag der Österreichischen Akademie der Wissenschaften.

Zanchi, A. (2008) 'Melius Abundare Quam Deficere': Scarlet Clothing in *Laxdæla Saga* and *Njáls Saga.* In R. Netherton and G. R. Owen-Crocker (eds), *Medieval Clothing and Textiles,* vol. 4, 21–37. Woodbridge, Boydell & Brewer.

Zangger de Saint-Gall, K. (1945) *Contribution à la terminologie des tissus en ancien francais, attestés dans textes francais, provencaux, italiens, espagnols, allemands et latins. Thèse présentée à la première section de la Faculté de Philosophie de l'Université de Zurich pour l'obtention du grade de docteur.* Bienne, Art graphiques Schüler S.A.

9 Haberdashery Elements made of Metal Thread: Conservation Problems

by Anna Drążkowska

Haberdashery elements, the focus of this study, were excavated in several churches in Poland. The objects were made of metal thread as a whole, or merely decorated with silver plated or gold plated copper wire. They constitute the decorative element of grave clothing from the 17th and 18th centuries (Drążkowska 2004, 167; 2005a, 98; 2008). They were sewn both on garments made especially for funerals and on ordinary clothes worn during life. Laces accentuated dress fashion, and concealed seams joining two textiles together. Most often, the laces were stitched onto the front in single, double or triple rows. They ran from the neckline to the lower edge, and decorated sleeves and the bottom fringes of dresses and trousers. Children's bonnets, shoes and men's *calpacks* (caps) were hemmed with them (Grupa 2005a; 2005b, 41; Drążkowska 2008). Haberdasher's buttons were stitched onto the front and formed an element of fastening or served as mere decoration (Fig. 9.1). Silk cords with metal thread were sewn on *żupans* and outer garments along all cut open edges and also along the collar, around the neckline and the cuffs.

State of Preservation
The state of preservation of haberdashery elements is varied and depends on several factors, age and environment being of essential importance. Haberdashery and the clothing onto which it was stitched were exposed to chemical compounds formed during the body's decomposition, and the activity of mildew, fungi, bacteria and insects (Drążkowska 2005a, 98; 2005b, 57; 2007, 132). The raw material of the metal threads, their chemical composition as well as the production method, all have a great influence on the scale and type of damage. The elements of haberdashery with metal threads subjected to conservation have diverse structures such as laces, braids, open-work ribbons and buttons of metal threads, all with complex constructions. Their inside structure consists of a silk core, around which a flat narrow metal strip is twisted (Figs 9.1–9.2). Some laces, however, are decorated with narrow strips of tin-plated hammered wire (Fig. 9.3).

Laces are in the poorest condition; their delicate open-work construction being considerably compromised. In most cases, they had been sewn onto clothes that were found in crypts. The silk threads have been weakened and contaminated by corrosion products, and now they are strained or torn in places (Fig. 9.4).

A metal thread that constituted the element of haberdashery as a whole or only decorated it, was seriously damaged in places (see Figure 9.5). It has lost its lustre and is corroded in places due to the effect of chemical substances emanating from the decomposition process of the human body. Pinholes, only observable under the microscope, occurred on the metal strip of tin and weakened its condition considerably. In places the deterioration was so advanced that the strip had split, uncovering the inner core thread. In some places, the delicate core is also in the process of disintegrating.

All the haberdashery objects were coated with dust. Moreover, they were dirty with sand, coffin stuffing, and organic compounds, all of which formed a hard and dense crust in places.

Cleaning
The conservation of each of the haberdashery elements of metal threads was undertaken separately. Each time a seam was unpicked, the place on which they had been was marked in order to enable them to be resewn. The seams were unpicked after the fabrics of the garment decorated with haberdashery had been made flexible. This was done since starting these procedures before giving flexibility to the fibres, could increase their deterioration.

Cleaning haberdashery made of metal thread was very laborious. The difficulties were due to the complex structure of the threads, a strip of tin foil twisted around a core of silk thread, and also due to the complicated construction of each element of haberdashery. The coupling of metal and fibres made it difficult to find a cleaning agent, which would clean the metal efficiently, whilst at the same time keeping the silk safe from deterioration. Work was impeded in most cases by the unfavourable state of preservation of the metal strips, which were markedly corroded, split and very dirty. The silk fibres were over-dried, torn in places and this additionally complicated the

Fig. 9.1. A detail of a fragment of lace from a child's bonnet, 17th–18th century. (National Museum in Szczecin, Poland; Photo: A. Drążkowska).

Fig. 9.2. Another detail of a fragment of lace from a child's bonnet, 17th–18th century. (National Museum in Szczecin, Poland; Photo: A. Drążkowska).

Fig. 9.3. A further detail of a fragment of lace from a child's bonnet, 17th–18th century. (National Museum in Szczecin, Poland; Photo: A. Drążkowska).

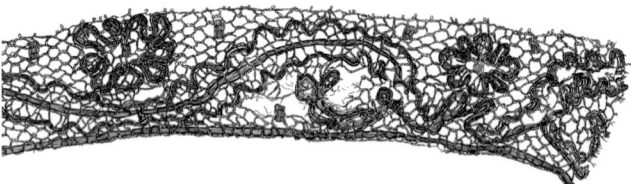

Fig. 9.4. A fragment of lace from an 18th century dress (Cathedral Museum, Lublin, Poland; Drawing by Wiesława Matuszewska-Kola).

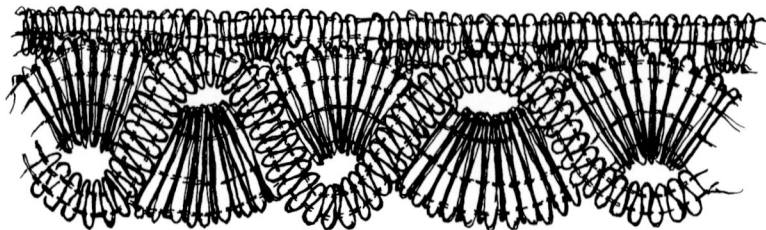

Fig. 9.5. A fragment of lace from a garment belonging to Bishop Michał de la Mars, 18th century. (Cathedral Museum, Lublin, Poland; Drawing by Wiesława Matuszewska-Kola).

cleaning. Therefore determining the sequence of operations and the cleansing methods was very important.

Dry-cleaning of metal laces, buttons, braids and strings was abandoned in order not to exacerbate the danger of their destruction. Besides, removing contamination by dry cleaning would only have enabled the removal of the surface layers. The space between silk fibres and metal threads could not be cleaned properly using this method, as it was impossible to remove particles of sand and organic remains. The treatment would only have been able to remove the dust, which was considered insufficient cleaning. As a consequence, the removal of contamination by a wet method was chosen. However, it was decided to increase the flexibility of the silk fibres before beginning the procedure. First, efforts were made to moisten fibres delicately and make them more flexible simultaneously. Spreading the garment fragments, flattening crumpled places and unstitching haberdashery were also avoided until the silk had recovered some flexibility.

Tests with various chemical agents were carried out before starting the procedure. Efforts focused on finding the substance, which would meet expectations of the desired results. The best results were achieved by moistening the over-dried fibres with 5% (or) 10% aqueous solution of PEG-200 or PEG-300 (polyethylene glycol). By using these substances, it was possible to make the fibres slightly more flexible. Moreover, they made the contaminations flaky, and as a result, it became easier to remove tough clots from the fabric surface without causing the destruction of the fibres.

When the moistening of fibres was completed, unstitching and cleaning haberdashery elements began. Metal laces, braids, buttons and strings were placed in a water bath with Pretepon G.[1] This substance additionally softened the fibres and facilitated the saponification of impurities, helping to pull them out from the nooks of the weaves. Preparatory needles, pencil-brushes and very soft brushes were applied to clean the haberdashery, which was spread on glass panels to avoid further damage. Layers were removed carefully, in order not to damage threads, which were already weakened. This manipulation was particularly laborious in places where the silk core appeared from under the corroded metal strip. Cleaning was a lengthy and repetitive process. Regrettably, this only enabled the removal of surface impurities. Corrosion layers resisted removal in the vast majority of cases. Corrosion products exfoliated and peeled off together with metal wrap fragments. Therefore, repeated strict cleaning of the metal strips was avoided in order to halt further damage. The cleaning treatment was performed under the microscope, which was helpful in defining the type of impurity and localizing it. Moreover, the microscope allowed the evaluation of the state of preservation of the haberdashery, thus enabling the treatment to be stopped the moment its continuation would have endangered the object's condition.

Numerous tests with different substances were undertaken while cleaning, searching for the optimal methods and agents, which would facilitate or improve the cleaning process. Among others, pads soaked with 3% sodium potassium tartrate were used in efforts to remove dirt from metal threads. These tests were only carried out on well-preserved objects. The action ought to have been repeated in order to remove most of the impurities, but there was danger that the tartrate could merely weaken the silk fibres. Therefore, this method was not used on most of the antique haberdashery, but solely limited to the tests.

Tests with ultrasonic rinsing were also carried out in the course of searching for new cleaning methods. Laces and buttons were immersed in water in the ultrasonic machine and subjected to minimal vibration effects. The condition of the metal strips and silk threads was checked after each single bath to determine the appropriate length of treatment. Laces and buttons were subjected to ultrasonic action from 5 to 8 minutes. The ultrasonic rinsing afforded very favourable results, as it accelerated the removal of dirt layers. Only haberdashery in a good state of preservation, having no broken metal threads and not many corrosion layers is suitable for ultrasonic treatment.

All instruments used while cleaning were rinsed with great care using only distilled water to avoid contamination.

Disinfection

The preliminary disinfection of the materials was carried out immediately after they were removed from the graves, when they were still stitched to the clothing. Historic objects were placed in a gas-chamber and subjected to ethylene oxide action. The disinfecting treatment was repeated for a second time, after cleaning the haberdashery. A chemical solution of 5% p-chloro-m-cresol (*parachlorometacresol*) in methanol (PCMC) (Strzelczyk and Karbowska-Berent 2004, 208) was applied by brush. Then, the haberdashery elements were placed in bags and left for two weeks. The disinfection treatment was repeated depending on the specific character of the environment, from which the fabrics were excavated (tomb-crypts or graves under floor) and also to avoid any remaining impurities from providing a medium for the growth of microorganisms.

Impregnation

Due to the poor condition of the metal thread, all the haberdashery elements were submitted to impregnation to strengthen the structure of the threads.

The mixture used for impregnation had to fulfil the following requirements: it should increase the objects' resistance to mechanical factors; protect them from external factors; make fibres flexible; penetrate the weave's spaces efficiently; not affect the objects' colour; not stiffen fibres; be resistant to temperature and humidity changes (non-hygroscopic) and the ageing processes; be easy to remove and thus be reversible. To strengthen and protect haberdashery, the same impregnant as for the textiles was used. It consisted of the following substances: Paraloid B72, polyethylene glycol 300 (PEG 300) in methanol and toluene.

The impregnant was brushed onto the object with a brush to cover the entire surface precisely, and carefully make the substance penetrate the spaces between fibres. After completion of the brushing, the textile was placed in a tightly closed plastic bag. Then the object was left to dry very slowly. Conducting the solvent's evaporating process very slowly and under strict control was essential, as a rapid solvent evaporation could cause the migration of the impregnant from inside the structures of the braids, laces and buttons onto their surface. In such a case, the strengthening substance would not serve its purpose and the haberdashery elements would still remain susceptible to mechanical factors.

After drying, the artefacts were straightened and repaired where necessary. During the garment reconstruction, the haberdashery was resewn onto its original places. Impregnation prevented, or sometimes only decreased the peeling off of a corroded metal strip, reinforced it and glued it together.

Conclusions

Conservation procedures strengthened the metal laces. The impregnant used enclosed the fragile fibres and increased their flexibility, improving their resistance to mechanical action. This reinforced the silk core of the metal threads. Regrettably, it was impossible to improve the state of the metal strip. The impregnant glued it in parts, but numerous fragments crumbled and fell off. The cleaning process was not always efficient, either. In case of very dirty laces, where corrosion layers adhered tightly to the metal, the effort of cleansing them completely was unsuccessful. The applied methods did not always result in fully satisfactory effects, therefore studies and work on their improvement will be continued.

Note

1   This is constituted of 9-Octadecen-1-ol, hydrogen sulfate, sodium salt, (Z)-, mixed with sodium hexadecyl sulfate.

Bibliography

Drążkowska, A. (2004) 17th–18th century clothing from children's graves discovered in the church at Kostrzyn, on the Oder, Poland. In J. Maik (ed.), *Priceless invention of humanity – textiles*. Acta Archaeologica Lodzieniesia 50:1, 167–169. Łódź.

Drążkowska, A, (2005a) Prace konserwatorskie przeprowadzone na jedwabnym wamsie wydobytym w kościele św. Mikołaja w Toruniu. In *Materiały do dziejów kultury i sztuki Bydgoszczy i regionuzeszyt* 10, 97–103. Bydgoszcz.

Drążkowska, A. (2005b) Konserwacja XVII i XVIII-wiecznej dziecięcej odzieży grobowej. *Ochrona Zabytków*, 1/2005, 55–67.

Drążkowska, A. (2006) Sprawozdanie z przebiegu prac konserwatorskich przeprowadzonych na kamizelce wydobytej w kościele p. w. św. Michała Archanioła w Brzozie. In *Lubuskie Materiały Konserwatorskie*, 20–23. Zielona Góra.

Drążkowska, A. (2007) *Odzież dziecięca w Polsce w XVII i XVIII wieku*. Toruń.

Drążkowska, A. (2008) *Odzież grobowa w Rzeczypospolitej w XVII i XVIII wieku*. Toruń.

Grupa, M. (2005a) *Ubiór mieszczan i szlachty z XVI–XVIIII wieku z kościoła p. w. Wniebowzięcia Najświętszej Marii Panny w Toruniu*. Toruń.

Grupa, M. (2005b) Kolekcja ubiorów pochodzących z badań archeologicznych w kościołach Torunia. In *Non omnis moriar. Zwyczaje pogrzebowe w XVII i XVIII-wiecznym Toruniu*, 40–53. Toruń.

Hryszko, H. (1994) Problemy związane z rekonstrukcją strojów pochodzących z sarkofagów książąt Pomorza Zachodniego. In A. Sieradzkiej and K. Turskiej (eds), *Ubiory w Polsce. Materiały III Sesji Klubu Kostiumologii i Tkaniny Artystycznej przy Oddziale Warszawskim Stowarzyszenia Historyków Sztuki, Warszawa 1992*, 141–152. Warszawa.

Strzelczyk, A. B., and Karbowska-Berent, J. (2004) *Drobnoustroje i owady niszczące zabytki i ich zwalczanie*. Toruń.

10 Current Examinations of Organic Remains using Variable Pressure Scanning Electron Microscopy [VP-SEM]

by Andrea Fischer

Introduction

The conservation and examination of grave goods recovered *in situ* is a huge challenge for the conservator of archaeological artefacts. In the first instance, it is not the metal artefacts which need the conservator's sole attention. Instead the emphasis is on the organic remains preserved in close proximity to the corroded metal (Höhne 1964; Edwards 1989; Fischer 1997). These are traces of clothing, blankets, belts or small bags, which are often very fragile and hardly visible to the naked eye. Investigations in the field of archaeological textile research have proved the great importance of organic material for archaeological investigations (*e.g.* Bartel and Knöchlein, 1993; Banck-Burgess 1999; Rast Eicher 2002; Rast Eicher and Burzler 2002; Ebhardt-Beinhorn 2003). The aim of the examination of organic remains in their context is to understand the original function and use of textiles, leather and other organic materials, as well as the nature of the materials and their manufacturing. The precondition of maintaining and utilizing all preserved information is a systematic and methodical approach in the documentation and examination of the archaeological finds. Soil blocks are excavated layer by layer in the conservation laboratory, using a binocular microscope. Several 1:1 plans are made, using drawing, photography and a written report, in order to record the position of all finds, along with the extent and nature of any preserved organic remains. Summarising all recorded information makes it possible to reconstruct the details of the burial.

In the Objects Conservation Programme of the State Academy of Fine Arts and Design in Stuttgart, the conservation process is accompanied by examination using a variable pressure scanning electron microscope (VP-SEM). This adds further information about the nature of the organic traces at a microscopic level, making use of the large depth of focus of the images and the possibilities of material analysis. At the same time, it is possible to have a closer look at different degrees of degradation and the effects of the conservation methods used for metal artefacts with associated organic remains, in relation to their state of preservation. Specific applications of the microscope were tested on *in situ* rescued burial artefacts from the early medieval burial field Lauchheim in Baden-Württemberg, Germany. These scientific investigations were closely aligned with the practical projects of the students.

Research to Date

Corroding metals such as iron or copper protect organic materials and preserve recognisable microstructures. If well preserved, the diagnostic features can be analysed by using transmitted light microscopy. However, if the processes of mineralisation and the micro-organism activity alter the material, identification can be difficult. Identification of the mineralised structures has greatly improved with the introduction of SEM in the 1970s. Considerable advantages arise from the higher magnification, the larger depth of focus and higher resolution of this method of investigation. Carol Keepax, a British conservator, investigated mineralised wood in 1975. She defined different types of mineralisation, mainly a coating and a replacement of the organic matrix. Figure 10.1 shows a negative cast – an impression of the outer surface of the fibre while organic structures have disappeared – and a positive cast – the replica of the fibre itself – of keratin fibres found on an iron artefact (Fig. 10.1). In the following years,

Fig. 10.1. (1) negative cast, (2) positive cast of keratin fibres (Image: © A. Fischer).

work was carried out to identify textile fibres, wood, leather, bone, horn and ivory (Janaway 1985, 29–34; Watson 1988, 69–73). Gillard *et al.* (1994) and Chen (1998) studied the process of mineralisation by simulating the reaction between metal solutions and textile fibres. However, the mechanism of fibre decay and mineralisation remains poorly understood.

In a conventional SEM, the examinations are conducted under high vacuum, which requires the preparation of samples removed from their context. Organic materials, such as non-conductive samples, have to be coated with a layer of gold or carbon. However, often the organic remains only exist in traces attached to the metal artefacts, or in the form of discolouration of the surrounding soil. Taking samples is equivalent to the destruction of the remains. Therefore, many interesting contexts cannot be fully interpreted by using conventional SEM.

Fig. 10.2. Cross section of mineralised horn (Image: © A. Fischer).

Ongoing Development

Since 2005, the Stuttgart Laboratory for Archaeometry and Conservation Science provides the opportunity for examinations using a VP-SEM. In contrast to a conventional SEM, it offers the possibility of working under different pressures. A small amount of remaining gas in the sample chamber prevents charging, which makes it possible to observe non-conductive samples without the need of a conductive coating. Furthermore, the sample chamber can easily accommodate small archaeological finds such as fibulae, fittings or amulet discs up to sizes of approx. 20 × 20 × 20 cm. These advantages mean that the analysis of samples without prior preparation is possible and traces of organic remains can be examined. During the conservation treatment, it is possible to visualise details and the changes the treatment might cause. Even small damp samples including soil blocks can be examined, as it is possible to induct water as remaining gas into the chamber and to control the temperature of the sample stage.

Examining Single Objects

For the first examinations using the VP-SEM, metal artefacts with adhering organic remains were selected after having removed them from their context and left to dry. It was found that the mounting in the sample chamber is quite easy. The equipment of the sample chamber with two cameras is essential for a safe and smooth alignment under the electron gun. To achieve an optimal performance, the interesting areas were studied and marked using light microscopy before being locked into the chamber. Several parameters such as the acceleration voltage, the chamber pressure, the detector settings or the working distance can be changed during the examination, leading to a significant effect on the quality of the image. The condition of the organic remains is of crucial importance for the parameter settings. If the organic substance is well preserved, the use of a high acceleration voltage can generate charge. With an increasing degree of mineralisation, this problem only occurs to a smaller extent.

The results of the examinations show that, even at low magnification, it is possible to visualize details of the weave structure and traces of use, which are often not so well defined using light microscopy. Moving the find is easy, and allows the analysis of both thread systems, different layers of textiles or the studying of a cross section. It is possible to study surface details which cannot be visualized using a conventional SEM, because the material is too fragile and brittle to be sampled. Already existing cross sections of mineralised organic material can be used for identification. Figure 10.2 shows a cross section of mineralised horn, which can be recognised by fine, rippled layers. The horn is the evidence of a handle, attached to the corroded iron blade of a knife. It was quite easy to investigate the knife *in situ* and to distinguish between horn, bone and wood.

Many organic remains associated with the metal artefacts from the Lauchheim cemetery were in such poor condition, that it was difficult to distinguish between corrosion products and organic remains under reflected light. Often, the only evidence of the remains of a textile is just a cross point of two threads or a negative cast of a fibre on a corroded metal surface.

The identification of mineralised leather is still a problem. It can easily be identified if hair follicles survive (Cameron 1991), but often, the surface has been removed by abrasion. Sometimes it can be observed as a powdery reddish-brown layer and easily mistaken for iron corrosion products. Figure 10.3 shows details of a leather surface found on a pair of iron scissors. By using VP-SEM, it is possible to establish a reference for making progress in the studies of leather identification.

Examining Waterlogged Material

No scientific studies have been carried out as yet in order to examine waterlogged organic remains *in situ* using SEM. To study hydrated samples, a Peltier cooling stage and a water control system can be fitted to the VP-SEM. In this application, water vapour is used as the gas to discharge the sample surface. The temperature of the Peltier element can be controlled from -30 to +35 °C.

Fig. 10.3. Surface of mineralised leather with hair follicles (Image: © A. Fischer).

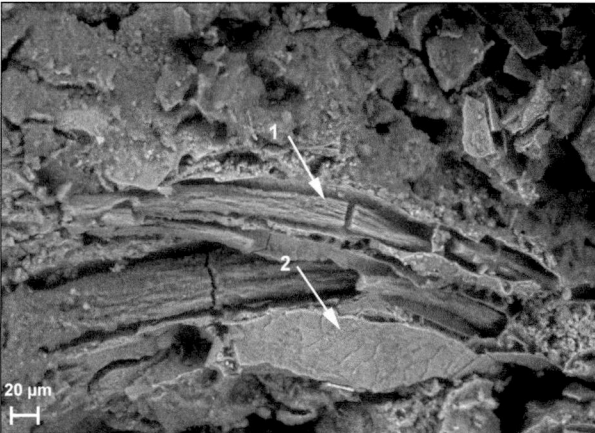

Fig. 10.5. (1) highly degraded fibre, the cuticle is lost (2) negative cast of fibres (Image: © A. Fischer).

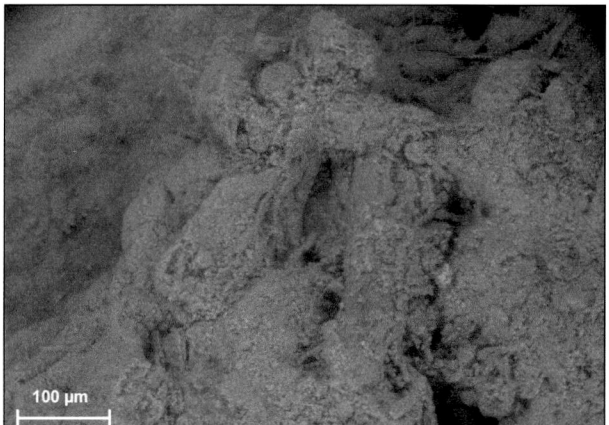

Fig. 10.4. Degraded textile remains examined in waterlogged condition (Image: © A. Fischer).

Fig. 10.6. Cross section of a chain, made of two layers of leather and a third layer of organic material (Image: © A. Fischer).

Compared to the application mentioned above, the complexity of coordinating all pivotal parameters increases. Preliminary investigations were carried out to explore the influence of humidity and temperature on the image quality and the preservation of the samples. Using a specimen holder made from copper with a diameter of 5 cm, new cotton fibres were embedded in clay. The rate of dehydration was observed by measuring the weight of the sample. Best results in the quality of the images were achieved by locking in frozen samples through the use of the back scattered electron detector. The working parameters were: a chamber pressure of 500 Pa, a temperature of the cooling stage near freezing point, a relative humidity just below 100 % and an acceleration voltage of 20 kV. However, the drying of the sample surface was obvious and could not be avoided by changing the parameters or by covering a part of the specimen holder with metal foil. Observing the drying process, it turned out that a working time of less than one hour was adequate for the examination of the original samples.

For the examination, small sections of waterlogged organic remains were lifted and placed on the above mentioned sample holder. Figure 10.4 shows degraded textile remains, preserved under an amulet disc, perhaps used as a bag or wrapping. The detail of Figure 10.5 indicates the state of preservation: the keratin fibre is highly degraded and the cuticle is lost. At the same time the scale structure is preserved as a negative cast.

Figure 10.6 shows another sample examined in waterlogged conditions: a cross section of a multilayered chain hanging from a belt. The chain was made of two layers of leather, with a characteristic dense, non-oriented fibrous structure and a third layer of another organic material. Unfortunately, this layer was too degraded to be identified.

The current examinations indicate that the imaging of waterlogged material *in situ* is possible, and provide an insight into structures which have not been investigated earlier. Even the identification of highly degraded material is possible, if characteristic features like scale structures are preserved. It would be expedient to accomplish a systematic study of fibre degradation. Furthermore, it is also possible to visualise dynamic processes like freeze drying (Wiesner and Krekel, in press) or the moisture expansion of fibres (Haller 2008). However, it is necessary to dissect a sample out of the block of soil because the cooling stage has a limited capacity as far as both weight and size are concerned. It is possible to avoid the risk of the sample drying due to a limited period of

Fig. 10.7. Textile fragment before cleaning (Image: © A. Fischer).

Fig. 10.9. Detail of a mineralised textile before treatment (Image: © A. Fischer).

Fig. 10.8. Detail after cleaning with an ethanol/water solution (50:50) (Image: © A. Fischer).

Fig. 10.10. After treatment in a 4% disodium EDTA solution, pH 5.5, 6 hours (Image: © A. Fischer).

exposure. Moreover, it is impossible to obtain an image in a magnification of less than 170×, which makes the examination more complex.

Examining Cleaning Treatments

Imaging the organic remains without prior dissection is a great advantage, particularly with regard to highly degraded samples, as each treatment can cause a loss of information. However, in the case of well-preserved organic remains, a preparation or cleaning treatment can increase the chance of observing characteristic features. Using VP-SEM, it is possible to record and demonstrate the process and any changes arising during the treatment. Figure 10.7 shows a textile fragment adhering to a bronze ring. Loose soil particles were removed with a dissection pin before taking the image. The spin direction and the thread count are visible in some areas, but adhering soil and fragments of fibres cover the fibre structure. Figure 10.8 shows the same detail after cleaning with an ethanol/water solution (50:50). Further adhering soil and fibre fragments are rinsed away and the weave structure is distinguishable, but the single threads appear still coated with soil and corrosion products.

To continue the cleaning treatment, it is possible to use sequestrants. In 1964, Höhne established the use of EDTA for the dissection of mineralised textiles (Höhne 1964, 213). EDTA is an excellent sequestrant for most di- and trivalent metals. Today sequestrants are applied only in particular cases, because mordants of dyes or inorganic colouring agents may be chelated precluding further analysis. Furthermore, the treatment itself poses a risk of damage to the mineralised organic remains, as they are also composed of metal compounds, depending on their state of preservation. However, as the chemical process is quite slow, it is easy to control. The complex forming process depends greatly on the pH and the temperature of the solution. Figure 10.9 shows a detail of a mineralised textile before treatment. Even though the spin direction of the thread is visible, the surface texture of the single fibres is covered with a corrosion layer. Figure 10.10 shows the same detail after a 6 hour treatment in a 4% disodium EDTA solution with a pH of 5.5. The corrosion layer dissolves and single fibres appear.

The VP-SEM is a useful tool for evaluating different cleaning methods at a microscopic level during the conservation treatment. To study the effect of the essential parameters within the complex forming process and the mode of

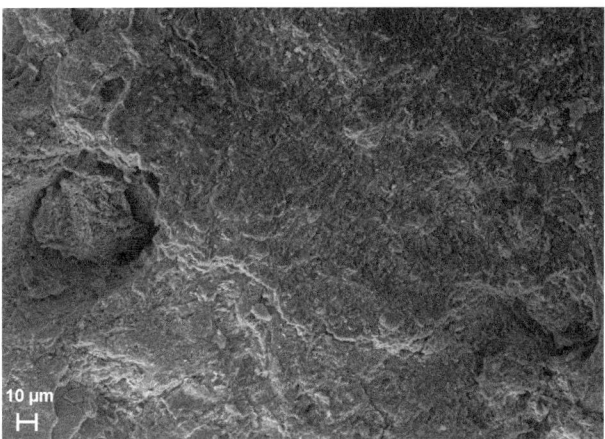

Fig. 10.11. Detail of mineralised leather before treatment (Image: © A. Fischer).

Fig. 10.12. After treatment: surface covered with fine needle-like crystals (Image: © A. Fischer).

operation of different states of preservation, systematic series of tests are required. Furthermore, it should be useful to compare different sequestrants. Nevertheless, it makes sense to complement further investigations with other research methods.

Examining Metal Conservation Treatments

We have little knowledge of how conservation treatments, developed for corroded metal artefacts, affect organic remains. For the long term stabilisation of artefacts made from iron, the alkaline sulphite method, established by Rinuy and Schweizer in 1982, is widely used. During the treatment, the objects remain in a strong alkaline solution tempered to 60 °C for several weeks. From the subsequent macroscopic examination, it is however evident that the fragile organic traces are damaged by the treatment. Experience has shown that, with large quantities of finds, it is hardly possible to maintain and store all metals with attached organic remains separately. A few attempts were made to consolidate or to cover the organic traces. Certain organic remains from the Lauchheim cemetery appeared stable enough and were treated in an alkaline sulphite solution. Preliminary investigations using VP-SEM microscopy visualised changes of the surface texture after the treatment. The surface of the iron and mineralised leather was covered by fine needle-like crystals. Figures 10.11 and 10.12 show the condition before and after treatment. Further investigations with complementary investigation techniques are planned.

The most common approach to the stabilisation of bronze artefacts is the inhibition of the active corrosion with benzotriazole (BTA). The use of this commercially well known chemical was suggested by Brinch Madsen in 1967. BTA molecules form polymer complexes with copper and copper compounds, acting as a physical barrier for vapours and gases. Edwards advises against the use of BTA if organic traces are adhering to the metal, because BTA crystallises and damages the mineralised organic material (Edwards 1989, 6). Preliminary tests using VP-SEM microscopy were carried out, examining organic remains treated with a 3% solution of BTA in ethanol. Edwards' doubts concerning the method are yet to be confirmed. In this case, further research will be carried out.

In connection with any metal conservation treatment, all conservators have to deal with the problem of consolidating the fragile organic remains. It is not possible to examine mineralised organic material after consolidating it with polymer resins, because the structure of the material and therefore the characteristic features at the microscopic level are coated. It should be investigated, whether volatile binding media (*e.g.* cyclododecane) are a viable alternative to other materials commonly used for consolidation. The development of using volatile binding media is based on the principle idea of consolidating fragile objects temporarily (Hangleiter *et. al.* 1995). Depending on the conservation purpose and the condition of the material, they can be applied in different ways and sublimated within a reasonably short time without leaving any residue. Series of tests using VP-SEM will allow a preliminary evaluation, if the diagnostic microstructures of organic remains are changed by volatile binding media.

The variety of observations made during and after treating metal artefacts is a strong incentive for several new research projects to be initiated. The visualisation at a microscopic level is a useful aid in assessing and understanding detrimental alternations and risks. Based on the VP-REM examinations, research, including further investigation methods, can be specified.

Conclusions

The examination of several archaeological finds recovered from soil blocks indicates that the application of the VP-SEM is very advantageous. Such non-destructive observation provides a thorough insight into organic remains, imaging the state of preservation of strongly disintegrated substances and preserved morphological characteristics, and thereby enables identification of the material. Much experience has been gained by examining waterlogged material *in situ*, even though drying is a limiting factor. In addition, previously unanswerable questions in conservation science may be resolved. Finally, it is possible to record and demonstrate the effects of various conservation techniques, such as the application of complexing agents, desalination methods or corrosion inhibitors on mineralised organic remains on artefacts.

Acknowledgements

I wish to express my thanks to: Christoph Krekel, head of the Laboratory for Archaeometry and Conservation Science; Gerhard Eggert, head of the Objects Conservation Programme; and the students of the Objects Conservation Programme at the State Academy of Fine Arts and Design in Stuttgart.

Bibliography

Chen, H. L. (1998) Preservation of archaeological textiles through fibre mineralisation. *Journal of Archaeological Science* 25, 1015–1021.

Banck-Burgess, J. (1999) Hochdorf IV. *Die Textilfunde aus dem späthallstattzeitlichen Fürstengrab von Eberdingen – Hochdorf (Kreis Ludwigsburg) und weitere Grabtextilien aus Hallstatt- und latènezeitlichen Kulturgruppen. Forschungen und Berichte zur Vor- und Frühgeschichte in Baden – Württemberg* 70. Stuttgart.

Bartel, A., and Knöchlein, R. (1993) Zu einem Frauengrab des sechsten Jahrhunderts aus Waging am See, Lkr. Traunstein, Oberbayern. *Germania* 71, 419–439.

Brinch Madsen, H. (1967) A preliminary note on the use of benzotriazole for stabilizing bronze objects. *Studies in Conservation* 12, 163–167.

Cameron, E. (1991) Identification of skin and leather preserved by iron corrosion products. *Journal of Archaeological Science* 18, 25–33.

Ebhardt-Beinhorn, C. (2003) Zur Trageweise des frühmittelalterlichen Amulettgehänges aus Gredingen-Großhöbingen, Grab 160. *Beiträge zur Erhaltung von Kunst und Kulturgut* 1, 55–68.

Edwards, G. (1989) Guidelines for dealing with material from sites where organic remains have been preserved by metal corrosion products. In *Evidence preserved in Corrosion products: A new field in artefact studies. United Kingdom Institute for Conservation of Historic and Artistic Works Occasional Papers* 8, 3–7.

Farke, H. (1992) Der Einsatz von Komplexon in der Textilkonservierung. *Arbeitsblätter für Restauratoren 2, Gruppe 10*, 176–178.

Fischer, A. (1997) Reste von organischen Materialien an Bodenfunden aus Metall – Identifizierung und Erhaltung für die archäologische Forschung. In K. W. Bachmann (ed.), *Schriftenreihe des Instituts für Museumskunde an der Staatlichen Akademie der Bildenden Künste Stuttgart*, 13.

Gillard, R. D., Hartman, S. M., Thomas, R. G., and Watkinson D. E. (1994) The mineralisation of fibres in burial environments. *Studies in Conservation* 39, 132–140.

Hangleiter, H., Jägers, E., and Jägers, E. (1995) Flüchtige Bindemittel. *Zeitschrift für Kunsttechnologie und Konservierung* 9, 385–392.

Haller, U. (2008) Gemälde im VP-REM: Beobachtungen zum Feuchteverhalten von Werkstoffen der Malerei. In *Beiträge zur 20. Tagung des Österreichischen Restauratorenverbandes*, 20–27.

Höhne, H. (1964) Präparation und Bestimmung vorgeschichtlicher Textilreste. *Neue Museumskunde* 7, 211–228, 306–319.

Janaway, R. C. (1983) Textile fibre characteristics preserved by metal corrosion: the potential of S.E.M. studies. *The Conservator* 7, 29–34.

Keepax, C. (1975) Scanning electron microscopy of wood replaced by iron corrosion products. *Journal of Archaeological Science* 2, 145–150.

Rast Eicher, A. (2002) Detailuntersuchungen an ausgesuchten Beigaben: Textilfunde. In A. Burzler, M. Höneisen, J. Leicht, and B. Ruckstuhl (eds), *Das frühmittelalterliche Schleitheim – Siedlung, Gräberfeld und Kirche*, 211–228, Schaffhausen.

Rast Eicher, A., and Burzler, A. (2002) Beobachtungen zur Tracht und Kleidung. In A. Burzler, M. Höneisen, J. Leicht, and B. Ruckstuhl (eds), *Das frühmittelalterliche Schleitheim – Siedlung, Gräberfeld und Kirche*, 372–399. Schaffhausen.

Rinuy, A., and Schweizer, F. (1982) Entsalzung von Bodenfunden in alkalischer Sulfitlösung. *Arbeitsblätter für Restauratoren 1, Gruppe 2*, 160–174.

Watson, J. (1988) The identification of organic materials preserved by metal corrosion products. In S. Olson (ed.), *Scanning Electron Microscopy in Archaeology*, 65–76. BAR International Series 452.

Wiesner, I., and Krekel, C. (In press) Low vacuum scanning electron microscopy of waterlogged archaeological leather. In *Proceedings of the 10th ICOM Group on Wet Organic Archaeological Materials Conference: Amsterdam, 2007*.

11 Textiles, Wool, Sheep, Soil and Strontium – Studying their Paths: a Pilot Project

by Karin Margarita Frei

Introduction

This paper summarizes the outcome of the recent study on the provenance of ancient textiles (Frei *et al.* 2009). Archaeological migration and trading studies in the last 25 years have brought archaeologists new and exciting answers based on strontium isotope analysis. It was Ericson (1985), who suggested the use of strontium isotopes in teeth and bones of archaeological human skeletal remains, in order to study dietary input from different geological environments. In his study, Ericson used strontium as an archaeometric parameter to determine the mobility or non-mobility of an individual, based on the geochemical signatures of strontium. The strontium isotopic system was, and still is, used in the fields of geochronology and geochemistry. Different rocks have different concentrations of strontium and isotope ratios that are characteristic. These ratios together with other trace elements provide a good indication of the geological characteristics of a certain area. Thus, the strontium isotope characteristics from the parent rock are passed on through the food chain without fractionation processes. In the pilot project presented in Frei *et al.* (2009), the strontium isotope system was taken a step further, in order to demonstrate that it is also applicable for the purposes of determining the provenance of modern wool and ancient textiles. Furthermore, we proposed a methodology that is an analytical protocol to determine contamination-free strontium isotopic signatures of sheep hair.

It is my aim to use this newly developed methodology to address questions regarding the possible trading routes of ancient textiles, and/or their raw materials.

Strontium in Hair

Wool is basically hair. It is widely accepted today "that hair can be considered a minor excretory organ, supporting its use as a biopsy material representing the body" (Attar *et al.* 1990, 477). As hair contains an elevated concentration of trace elements, it is a useful target for epidemiological studies and potential diagnostic considerations (see *e.g.* Attar *et al.* 1990). Consequently, the analysis of trace elements of human and animal hair is an important tool today to recognise environmental changes. This method has also been applied in the past on human hair from archaeological remains (see *e.g.* Wolfsperger *et al.* 1993).

Despite the promising prerequisites, there are not many studies based on the applications of the stable strontium isotope system to hair. This fact is probably due to the low concentration of strontium in hair: of a few parts per million (ppm) or less (Attar *et al.* 1990; Christian *et al.* 1997; Miekeley *et al.* 1998; Morita *et al.* 1986). Thus, the recovery of stable strontium from hair to obtain a precise mass spectrometric analysis is a difficult matter. Likewise, contamination by diagenetic processes is a delicate issue, especially when working with trace elemental analysis in archaeological samples. Therefore, the removal of post-depositional contamination is of crucial importance in order to obtain the primary isotopic signatures of the wool.

A pilot study recently conducted by von Carnap-Bornheim *et al.* (2007) aimed at determining strontium isotopic ratios in wool and leather samples from the archaeological remains from the Thorsberg peat bog (Germany), in order to ascertain the place of origin. The results and conclusions of von Carnap-Bornheim *et al.* (2007) are based on the assumption that the strontium isotopic composition of sheep hair reflects the bio-available strontium of respective grazing lands. Although this seems to be a fact for teeth and bones of modern and ancient humans/animals, it has not been proven for sheep hair yet.

By the examination of strontium signature relationships between modern sheep hair and the geological characteristics of respective grazing lands, it has been demonstrated that these assumptions are correct (Frei *et al.* 2009). Wool samples of modern sheep (preferably kept ecologically) and topsoils (A-horizon) from the respective pastures were collected from different places with different geological backgrounds. The samples originated from New Zealand, the Faroe Islands, the Shetland Islands, Norway, Sweden and Denmark. In addition to the modern wool samples, three important ancient textiles were chosen for this pilot study. The archaeological textiles were from Corselitze and Haraldskær in Denmark, and Gerum in Sweden. All the textiles have been recovered from peat bog environments, which are conducive to the preservation of organic materials. All samples were carefully mechanically and chemically decontaminated and analysed.

Results and Conclusions

The results of the analysis on modern sheep wool showed that there is an obvious and traceable strontium isotopic link between the grazing soil and the wool the sheep develops. Thereby, we can trace the wool's provenance and say if it is from a local or non-local source. It is important to mention however, that these studies are only possible when there are geological differences in the area of interest.

Strontium isotope signatures from the three ancient textiles from Corselitze, Haraldskær and Gerum, show that the raw material they were made of was from local sources, as they match the strontium isotopic signatures of the respective geological surroundings. Furthermore, the results of both ancient and modern wool demonstrated the importance of textile pre-treatment for decontamination purposes. As we are working with trace amounts, small post-depositional or environmental contamination can have a significant effect in the final strontium isotopic analysis, and thereby mask the true provenance. Previous work demonstrated that the lipid matter in hair acts as an open system in relation to environmental exposure (Attar *et al.* 1990). Consequently, the removal of the lipid matter in hair/wool, as well as the removal of dust particles (which often contain large amounts of strontium), is extremely important. This type of contamination can be removed chemically. Our method allows for an easy, rapid and controlled decontamination of wool, specially designed for minimal amounts of sample material (~30 mg).

These results form the necessary basis for future applications of the stable strontium isotopic system to ancient wool textiles for the purpose of provenance identification, and demonstrate the great potential of such analyses within the field of textile research.

Bibliography

Attar, K. M., Abdelaal, M. A., and Debayle, P. (1990) Distribution of trace-elements in the lipid and nonlipid matter of hair. *Clinical Chemistry* 36, 477–480.

Carnap-Bornheim, C. von, Nosch, M.-L., Grupe, G., Mekota, A.-M., and Schweissing, M. M. (2007) Stable strontium isotopic ratios from archaeological organic remains from the Thorsberg peat bog. *Rapid Communications in Mass Spectrometry* 21, 1541–1545.

Christian, J. M., Pielack, G., Freer, D. E., and Lee, M. J. (1997) Reference ranges for 34 trace elements in hair by inductively coupled plasma – Mass spectrometry (ICP-MS). *Clinical Chemistry* 43, 793.

Ericson, J. E. (1985) Strontium Isotope Characterization in the Study of prehistoric Human Ecology. *Journal of Human Evolution* 14, 503–514.

Frei, K. M., Frei, R., Mannering, U., Gleba, M., Nosch, M.-L., and Lyngstrøm, H. (2009) Provenance of Ancient Textiles – A Pilot Study Evaluating the Strontium Isotope System in Wool. *Archaeometry,* 51:2, 252–276.

Miekeley, N., Carneiro, M., and da Silveira, C. L. P. (1998) How reliable are human hair reference intervals for trace elements? *Science of the Total Environment* 218:1, 9–17.

Morita, H., Shimomura, S., Kimura, A., and Morita, M. (1986) Interrelationships between the concentration of magnesium, calcium, and strontium in the hair of Japanese schoolchildren *Science of the Total Environment* 54, 95–105.

Wolfsperger, M., Wilfing, H., Matiasek, K., and Teschler-Nicola, M. (1993) Trace elements in ancient Peruvian mummy hair: a preliminary report. *International Journal of Anthropology* 8, 27–33.

12 Not so much Cinderella as the Sleeping Beauty: Neglected Evidence of Forgotten Skill

by Ruth Gilbert

The aim of this paper is to draw attention to some textile remains, including important examples of shaped knitting, representing a type of garment overlooked by costume history and otherwise known in Britain only from a few mentions in documents. These fragments are undated finds from an 1888–1889 archaeological excavation of the Lindisfarne Priory buildings on Holy Island in Northumberland, on the coast of England just south of Berwick-upon-Tweed, which are kept at the English Heritage store at Helmsley in Yorkshire, England.

Woven, knitted and felted wool scraps were found in the ruins of the monastic buildings, although the circumstances are unclear. There is no question of the textiles being associated with a burial. The excavation report refers to "a solidified mass of soil, rags and rubbish" from a deposit dated between 1540 and 1603, which may be the source of these pieces (Crossman 1892, 234). There is a village on the island, the population in the early modern period being livestock farmers and fishermen. The priory was dissolved in 1536, and shortly after this the priory church and ancillary buildings were used as storage and workshops for naval victuallers, supplying the fleet maintained as defence against the Scots until the union of English and Scottish crowns in 1603 (Crossman 1892, 228). The castle garrison was maintained until 1819, with wives and children living with the soldiers, and there were subsequently other uses as coastguard station and militia headquarters (O'Sullivan and Young 1995, 14).

It is unlikely that the textiles date from the monastic use of the site. They may be offcuts from the remaking of second-hand clothing, either in the village or associated with the military presence. Crossman's description perhaps implies the rags having been found in a cesspit, which would be unsurprising (Crossman 1892, 234). There is also the possibility of textile refuse being used as wadding for gunnery: "Waddings is Okum, old clouts or straw, put after the powder and the Bullet" (Smith 1653, 66). "Old clouts" is an exact description of this archaeological assemblage of otherwise useless rags, conceivably bought from urban tailors in order to supply either ships or the guns at the castle.

In 1951 the textiles were examined by Elisabeth Crowfoot for what was then the Ministry of Works, now English Heritage. She carried out some conservation at that time and wrote a brief unpublished report, which came to my attention because she herself referred to it elsewhere (Crowfoot 1951). Having obtained a copy of the report, the importance of these pieces became clear and they were examined in the spring of 2007 nearly a 120 years after they were found. The information presented here is from my own observation of them, including a proposed reconstruction of a garment.

All the 16 knitted textiles show post-excavation damage. They are in poor condition and very dry and brittle, although now more safely mounted. They appear to have been fragmentary when deposited and some have darns. There is one patch, knitted from picked up stitches, on an unidentifiable fragment. All show some shaping or seam ribs and in one case a pattern of reverse loops (purl stitches). Three are identifiably the remains of stockings and another two may also be. The discussion here is of the seven pieces grouped by Crowfoot on the basis of similarity of the yarn and the fabric, which she described as "a knitted jerkin … ?17th century" (Crowfoot 1993, 50).

Of these seven pieces, 1a–1g, the largest piece, 1b, has a slightly different quality to the rest and may be a stocking, but has been left in the grouping. All seven pieces, are knitted of wool yarn consisting of two s-spun component yarns twisted Z. The count of the yarn is roughly 2/5 NM, 0.8–1mm in diameter.[1] The fabric is mainly in stockinet at an average gauge of 30 stitches and 40 rounds to 10 cm. None of this group were dyed but may include naturally pigmented fibres (Crowfoot 1951). They are now the reddish-brown typical of buried wool textiles.

The most easily identifiable garment pieces are 1c and 1g. Fragment 1c is a piece of the hem edge, approximately 14 cm high and 15 cm wide with welts to prevent the edge curling and neat paired decreases on either side of a seam rib (Fig. 12.1). Seam ribs in knitting are functional: they can be felt when working, alert the knitter when locating shaping and help in counting rounds. Their placing is determined by the shaping of the garment and will therefore resemble the placing of constructional seams. This rib may have started

Fig. 12.1. Piece 1c, Lindisfarne knitted fragments. Cast on edge with welt border, seam rib and decreases shown upright as worked (Photo: Ruth Gilbert, with thanks to English Heritage).

just above the border; the fabric is distorted and too brittle to attempt to re-align it.

Piece 1g is part of a sleeve, knitted round with increases and decreases for the elbow on either side of a seam rib which runs down the outside of the arm. It is 20.5 cm long and 13.3 cm across (26.6 cm round) at the widest. It was knitted down from the armhole. Piece 1a includes the armhole shaping, having fabric worked in two directions, that is, stitches were picked up from a selvedge and knitted at right angles to the original. This is common in the shaping of heels, but the size of this piece, about 30 cm wide, makes it too large for a stocking gusset. Two pieces of a less regular fabric show the same kind of shaping and may be from another similar garment (Crowfoot 1951). The pieces 1d, 1e and 1f are small and not readily identifiable fragments. 1b is a flat piece of fabric which may not be part of this garment, with a pattern described in the report as "a triangular flower in moss stitch and two leaves in purl stitch". This is shown on the garment diagram as knitted: if it were a stocking clock it would be worn the other way up.

The proposal is that these are pieces of a garment which Crowfoot called a 'jerkin' and would in the late 16th or 17th century have been called a 'knit waistcoat', a warm sleeved garment to wear under a more formal doublet, or indoors instead of a doublet or stiff bodice. Figure 12.2 shows conjectural reconstruction of the garment from the fragments, envisaged as similar to the wool night jackets in Scandinavian collections (Hazelius-Berg 1935; Kruse *et al.* 1988, 175). It is presumed to have been knitted in the round up to the armholes, the yokes knitted flat, probably with an opening on the chest, and the sleeves picked up and knitted down to the cuff. The method of making the armholes with retained stitches from the body is not known from Scandinavian examples but was used for the earliest surviving British garment, a little vest in the Museum of London, dated around 1540–1560.[2]

The other parallel in Britain is a cotton jacket in the Victoria and Albert Museum in London, presumed to be 18th century, which is hip length and shaped at the sides (Hart and North 1998, 185). This was knitted round and then cut down the front and hemmed. The shaping at the side on the V&A garment resembles piece 1c, although the decreasing is done not at the side seam, but where the gores meet the body.

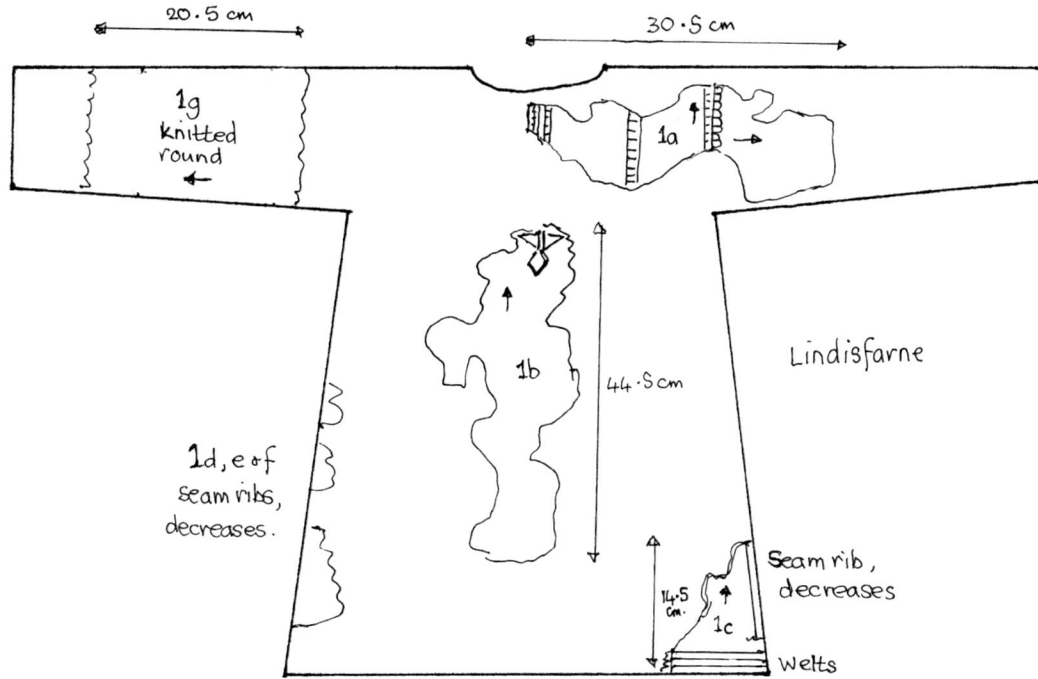

Fig. 12.2. Diagram showing conjectural reconstruction garment from Lindisfarne knitted fragments 1a–g (Drawing by Ruth Gilbert, with thanks to English Heritage).

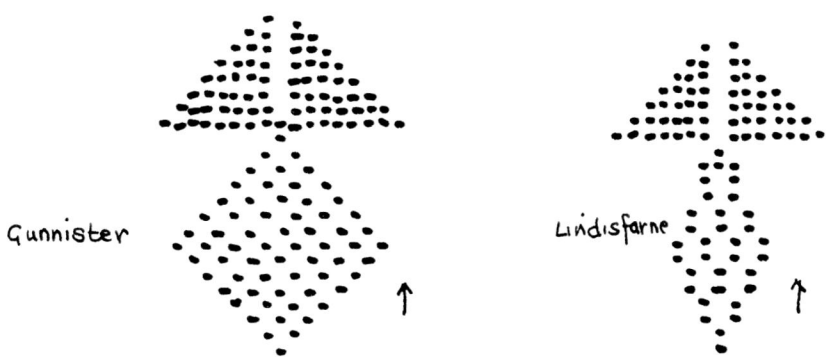

Fig. 12.3. Charts of stitch patterns. Dashes represent reverse loops (purl stitches), arrow indicates direction of work. Left: clock on Gunnister stocking, National Museums of Scotland NA 1043. Right: Lindisfarne piece 1b (Diagrams by Ruth Gilbert, Gunnister after Henshall and Maxwell 1952, 37; Lindisfarne with thanks to English Heritage).

From the mid-16th century in Britain, there was a trade in what are described as "knit waistcoats" (Willan 1962, 46). Wool knitted body garments called 'wylecoittis' were being made in Edinburgh in the second half of the 16th century and "knitt wastcoates" in Doncaster at the end of the 17th (Hey 1975, 25; Bennett 1981, 98). There are indications that seamen wore garments called waistcoats in the late 16th and early 17th centuries and that some of these were knitted (*Lord Brook* 1816, 15). The fragments described would appear to be the remains of such a garment.

Whether or not piece 1b is part of the same garment, it was deposited at the same time and is of a similar quality to the other scraps and can reasonably be assumed to be of around the same date. The pattern resembles stocking clocks, particularly one on a stocking from a burial at Gunnister in Shetland, *c.* 1690, the fabric of which is also very similar (Henshall and Maxwell 1952, 37.) Figure 12.3 shows charts of these patterns, which are very simply made with purl stitches. Clocks similar to this are also seen on two red silk stockings (not from the same pair) of the mid-17th century which belonged to King Charles X of Sweden, and on a yellow worsted stocking, also 17th century, excavated in Copenhagen (Ekstrand 1982, 176, 178; Kruse *et al.* 1988, 202).

The Scandinavian connection is not surprising. The trade in knitted goods and the similarity of surviving silk garments in Britain, Norway and Sweden has been noted (Hazelius-Berg 1935; Hoffman 1988). Knitted garments are easy to copy, and when technical detail of surviving garments is studied, it is apparent that simplicity of working was a criterion even when making high status garments. Once learned, patterns could be repeated and copied, and dating on the basis of the patterns cannot be exact. However, taking account of the yarn and fabric, the possible appearance of the garment and the circumstantial evidence, Crowfoot's dating to the 17th century seems reasonable. Dating might be assisted by examination of the woven textiles for comparison with securely dated assemblages such as those from Newcastle (Walton 1981; Walton 1983).

The point to be borne in mind is that whatever may or may not be deduced about the garments, these pieces prove the existence of the skill and technical knowledge necessary to make such things. There is very little record of knitting save obliquely of the enormous and lucrative trade in stockings. Knitters were not apprenticed, and do not feature in wage regulations. They appear in the main to have been women, and their skills must have been passed from one to another without written record. They were not taken account of during their working lives, fitting their knitting around their household work, and they have been overlooked by historians. The garments they made were worn out and discarded, and it is only by the examination of the few of their products that survive that their achievements can be recognised.

Acknowledgement

I would like to record my thanks to Claire Jones and Susan Harrison of English Heritage for their help and interest, and to Frances Pritchard for drawing my attention to Crowfoot's report.

Notes

1. The Nm count used in the text is an indirect system with constant weight. The Nm number is length in kilometres of a kilogramme weight of yarn; the higher the number, the finer the yarn. '5 Nm' describes a single yarn 5 kilometres of which weigh 1 kg. For plied yarn number and count of component yarns, 2/5 Nm, two yarns each of 5Nm, may also be expressed as 'resultant count', r2.5 Nm, arrived at by dividing the component yarn count by the number of components (Youngmark 1980).
2. This garment, 39.188/1, is on display in the medieval gallery in the Museum of London and can be seen at http://www.museumoflondon.org.uk/English/Events/Exhibitions/Permanent/medieval

Bibliography

Bennett, H. (1981) *Origins and Development of the Scottish Hand-Knitting Industry.* Unpublished PhD thesis, Edinburgh University.

Crossman, W. (1892) Recent Excavations at Holy Island Priory. *The History of the Berwickshire Naturalists' Club* 13, 225–240.

Crowfoot, E. (1951) Textiles from Lindisfarne, unpublished Ancient Monuments Laboratory Report 1509.

Crowfoot, E. (1993) Textiles. In S. Margeson (ed.), *Norwich Households: Medieval and Post-Medieval Finds from Norwich Survey Excavations 1971–78.* Norwich, Norwich Survey.

Ekstrand, G. (1982) Some Early Silk Stockings in Sweden. *Textile History* 13(2), 165–182.

Hart, A., and North, S. (1998) *Historical Fashion in Detail.* London, V & A.

Hazelius-Berg, G. (1935) Stickade Tröjor från 1600-och 1700-tolen. *Fataburen,* 87–100.

Henshall, A., and Maxwell, S. (1952) Clothing and Other Articles from a Late 17th Century Grave at Gunnister, Shetland. *Proceedings of the Society of Antiquaries of Scotland* 86, 30–42.

Hey, D., ed. (1975) *Doncaster People of Ten Generations Ago.* (Produced by the members of the Doncaster Local History Class held at Doncaster Museum and Art Gallery, Session 1974/75). Sheffield, University of Sheffield.

Hoffman, M. (1988) Of knitted "nightshirts" and detachable sleeves in Norway in the 17th Century. In I. Estham and M. Nockert (eds), *Opera Textilia Variorum Temporum: To Honour Agnes Geijer on her Ninetieth Birthday, 26 October 1988.* Stockholm, Statens Historiska Museum.

Kruse, A., Bogild Johannsen, B., Paludan, C., Warburg, L., and Østergaard, E. (1988) *Fru Kirstens Børn: To Kongebørns Begravelser I Roskilde Domkirke.* Herning, Poul Kristensen/Copenhagen National Museum.

Lord Brook's Life of Sir Philip Sidney, with a Preface etc. by Sir Egerton Brydges, Bart., (1816) Kent, printed at the private press of Lee Priory (1625).

O'Sullivan, D., and Young, R. (1995) *Lindisfarne: Holy Island.* London, Batsford/English Heritage.

Smith, J. (1653) *The Sea-Man's Grammar.* London (1627).

Walton, P. (1981) The Textiles. In B. Harbottle and E. Ellison (eds), An Investigation in the Castle Ditch, Newcastle-Upon-Tyne, 1974–6, *Archaeologia Æliana,* 5th series, 9, 190–228.

Walton, P. (1983) The Textiles. In E. Ellison and B. Harbottle (eds), Excavation of a 17th Century Bastion in Newcastle-Upon-Tyne. *Archaeologia Æliana,* 5th series, 11, 217–239.

Willan, T. S., ed. (1962) *A Tudor Book of Rates.* Manchester, Manchester University Press.

Youngmark, L. (1980) *Yarn Counts: Count Conversions and Calculations.* London, Handweavers Studio & Gallery.

13 Die Rekonstruktion des Vaaler Bändchens – ein archäologisches Kammgewebe aus Dithmarschen: Gemeinschaftsarbeit der Wollgruppe des Museumsdorfes Düppel, Deutschland

nach Annelies Goldmann und Eva-Maria Pfarr

Der Fundort des Bandes ist das Vaaler Moor in Dithmarschen im südwestlichen Teil Schleswig-Holsteins, wo die Teile des Bändchens beim Torfstechen im Jahre 1888 entdeckt wurden. Sie wurden von einem Lehrer, der den Wert des Fundes für die Geschichtsforschung erkannte (Schlabow 1976, 24), gerettet und kamen schließlich in das Textilmuseum in Neumünster. Der ehemalige Leiter dieses Museums, Klaus Tidow, bat uns, die ‚Arbeitsgruppe Wolle' des Museumsdorfes Düppel in Berlin, eine möglichst genaue Rekonstruktion des Bandes herzustellen.[1]

Gefunden wurden drei Fragmente: die Kette aus Schafwolle, der Schuss aus Ziegenhaar. Es war ein einfaches Kammgewebe – ein schmales Band, dessen Kette ursprünglich rot gefärbt war. Der Farbstoff konnte allerdings bisher nicht identifiziert werden. Diese Tatsache und die Verwendung von Ziegenhaar anstelle der üblichen Schafwolle unterscheiden diesen Fund von allen anderen aus der Umgebung.

39 Kettfäden ergaben eine Breite von *c.* 35 mm, 4 bis 5 Schussfäden eine Länge von 10 mm. Schuss und Kette waren in Z-Drehung gesponnen (Tidow and Walton 2001, 118 und 124). Die ehemalige Länge des Bandes konnte nicht ermittelt werden. Die Qualität des Gewebes war recht grob. Es ist bisher nicht gelungen mit Sicherheit zu bestimmen, aus welcher Zeit das Band stammt.

Im Folgenden werden die einzelnen Arbeitsschritte beschrieben:

Spinnen

Kette

Die Kettfäden wurden mit einer Handspindel gesponnen. Die Untersuchung der Kettfäden für das Band vom Vaaler Moor durch Penelope Walton ergab, dass diese aus weißer Schafswolle (*Medium fleece type*) von 1 bis 2 mm Durchmesser bestehen (Tidow und Walton 2001, 118).

Wir entschlossen uns, die Wolle der im Museumsdorf Düppel lebenden Skudden zu verspinnen. Die Skudden sind eine kleine und anspruchslose ursprüngliche Schafrasse, die in Ostpreußen und im Baltikum verbreitet war. Diese vor dem Aussterben gerettete, aber immer noch bedrohte Rasse wird in Düppel weitergezüchtet (Goldmann 1998, 223–242).

Die Wolle der Skudden ist weiß, wie auch die ursprüngliche Farbe der Kette des Bandes, die später überfärbt wurde.

Schuss

Die Schussfäden wurden mit einer Handspindel gesponnen. Der Schuss war laut Analyse Ziegenhaar mit einer natürlichen Pigmentierung von Dunkelbraun bis Schwarz. Das Vlies, das zum Verspinnen zur Verfügung stand, hatte lange glatte Deckhaare (≥ 15 cm) und stark verfilzte Unterwolle.

Zunächst wurde versucht, die Deckhaare zu verspinnen. Sie waren jedoch so glatt, dass es nicht gelang, sie zu einem Faden zu verbinden. Einen ersten Erfolg gab es, als den Ziegenhaaren Schafwolle beigegeben wurde; diese Mischung ließ sich zu einem Faden verspinnen, entsprach jedoch nicht der Analyse, denn es war kein Faden aus reinem Ziegenhaar. Um zum Spinnen geeignetes Ziegenhaar zu erlangen, wurden letztendlich die Deckhaare des Ziegenvlieses mit der Unterwolle gemischt.

Dazu wurden zwei Kämme mit langen Zinken benutzt: auf einen wurden abwechselnd je eine Lage Deckhaar und eine Lage Unterwolle gegeben und die Fasern dann durch mehrmaliges Kämmen vermischt. Bei diesem Arbeitsgang ergab sich eine Mischung, die sich mit der Handspindel zu einem Faden von etwa 2 mm Durchmesser verspinnen ließ. Mit diesem Schussfaden wurde das erste Band gearbeitet. Bei längerer Übung gelang es, dünnere Fäden mit einem Durchmesser von etwa 1 mm zu spinnen. Beide Stärken wurden bei den Webproben benutzt.

Färbepflanze	Ergebnis unterschiedlicher Farbaufzüge
Blutwurz (*Potentilla formentilla*)	Braun – Rot dunkel
Faulbaumrinde (*Ramus frangula*)	fahl, rötlich Braun / Gelb – Braun hell
Steinschlüsselflechte (*Parmelia omphalodes*)	Moorbraun – rötlich
Rotholz (*Caesalpina sappan*)	Kirschrot / Rotblau / Violett

Tabelle 13.1. Färbepflanzen und Ergebnisse unterschiedlicher Farbaufzüge.

Färben

Grundlage für die Farbversuche waren die Untersuchungen und die Berichte darüber von Penelope Walton Rogers. Die Kettfäden waren ursprünglich rot gefärbt. Es ist bisher nicht gelungen, den Farbstoff zu identifizieren, es könnte aber ein sattes Rot gewesen sein.

"On extraction, the dye was rich cherry red, whether in acid or basic conditions, although it was a more muted brownish red on the yarn" (Tidow und Walton 2001, 125). Auszuschließen ist jedoch eine Färbung mit Krapp, da der Farbstoff nach Walton jede zu diagnostizierende Charakteristik vermissen lässt.

Die Versuche, eine dem ursprünglichen Band möglichst ähnliche Färbung zu erzielen, führte im Museumsdorf Margarete Siwek durch. Zu allen Färbeproben benutzte sie Strangwolle.

In jahrelanger Arbeit hatte sie zunächst eine Reihe von Versuchen gemacht, um aus dem Ergebnis dieser verschiedenen Farbaufzüge die wahrscheinlichste Färbepflanze herauszufinden.

Nach diesen Vorversuchen schien Rotholz die am besten geeignete Färbepflanze zu sein. Das Rotholz oder auch Brasilholz lässt sich als Handelsartikel seit dem 9. Jahrhundert nachweisen (Ploss 1967, 31 und 55). Für das Färben benutzt wird das Kernholz dieses 3 bis 5 m hohen Baumes, der zu der Familie der Caesalpinen gehört; die Blätter sind zweifach gefiedert und haben eine oval längliche Form. Das rote Kernholz wird zu Spänen verarbeitet und kommt heute so in den Handel. Die färbenden Inhaltsstoffe sind Braselin (=3,4',5',7-Tetrahydroxy-2',3-methylen-neoflavan) und Gerbstoffe (Schweppe 1993, 412–415).

Erste Versuche, die vorgewaschene, aber ungebeizte Strangwolle zu färben, ergaben eine rotblaue Farbe. Ein besseres Ergebnis brachte dann die zweite Versuchsreihe mit Beize, die schließlich zur Rekonstruktion des Vaaler Bändchens benutzt wurde. Hier erfolgte die Färbung in zwei Arbeitsvorgängen:
1. Vorbeize der Schafwolle,
2. Färbung der Schafwolle mit Rotholzspänen.

Zu 1.
Die Beize bewirkt, dass die Wolle die Farbstoffe intensiver aufnimmt. Ein für Wolle häufig benutztes Mittel ist Alaun, das auch in diesem Fall verwendet wurde. Alaun ist ein weißes, in kristalliner Form gehandeltes Salz, ein Doppelsalz [K AL $(SO_4)_2$ 12 H_2O]. Die besten Ergebnisse wurden mit einer schwach konzentrierten Beize erzielt.

Abb. 13.1. Teil des Originalfundes „Vaaler Bändchen" (Foto: F. Peise).

Zu 2.
Nach den Vorversuchen wurde die Strangwolle nach der unter 1. angegebenen Vorbehandlung in dem Rotholzbad gefärbt. Der zweite Teil der Strangwolle wurde erst nach einer dreiwöchigen Gärung des Rotholzfarbbades gefärbt. Dieser Farbaufzug hatte das beste Ergebnis, ein sattes Kirschrot.

Über die Lichtechtheit der Färbung kann noch nichts Endgültiges gesagt werden. Einige Rotholzproben wurden über einen längeren Zeitraum dahingehend geprüft; Veränderungen des Farbtons konnten bisher nicht festgestellt werden. Interessant war, dass Nachfärbungen mit demselben Farbbad intensivere Farbnuancen brachten.

Weben

Nachdem diese Vorarbeiten zur Zufriedenheit abgeschlossen worden waren, konnte mit dem Weben begonnen werden. Zur Erinnerung: Es sollte ein Kammgewebe entstehen. Die Vorgabe für das Band war, mit 39 Kettfäden eine Breite von 3,5 cm zu erreichen, 4 bis 5 Schüsse sollten das Band um jeweils 1 cm verlängern.

Als Materialien für die verschiedenen Webproben wurden benutzt
für die Kette:
A: gebeizte Schafwolle (Skudde), 1. Färbung
B: gebeizte Schafwolle (Skudde), 2. Färbung (Nachfärbung)

für den Schuss:
C: Ziege, Durchmesser ca. 2 mm
D: Ziege, Durchmesser ca. 1 mm.

13 Die Rekonstruktion des Vaaler Bändchens – ein archäologisches Kammgewebe aus Dithmarschen

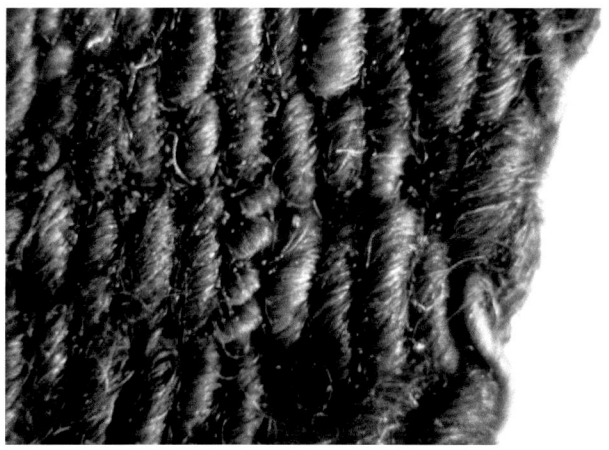

Abb. 13.2. Struktur des Bandes (Foto: F. Peise).

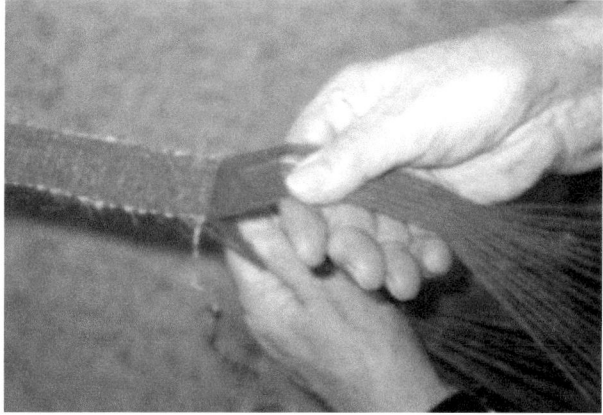

Abb. 13.4. Beim Weben des Bandes (Foto: R. Neumann).

Abb. 13.3. Rotholz – Baum mit Blatt, Blüte und Frucht und Spänen (Foto: E.-M. Pfarr unter Verwendung einer Zeichnung von G. Meier).

Abb. 13.5. Ergebnis des Rekonstruktion (Foto: E.-M. Pfarr).

Webprobe 1 bzw. Band 1

Für das 1. Band wurden Kette A und Schuss C verwendet und ein etwa 3 m langes Band gewebt. Es hatte bei der vorgegebenen Anzahl der Kettfäden allerdings nur eine Breite von 3 cm, obwohl die Kettfäden in der Breite nicht fest zusammengezogen wurden, so dass der Schuss noch zu sehen war. Das befriedigte nicht; einerseits war die Webprobe nicht besonders ansehnlich, andererseits – und das war der eigentliche Grund – war es beim Weben wesentlich mühseliger, eine gleichmäßige Breite herzustellen, wenn der Schuss nicht gleich fest angezogen wurde. Bei diesem Probeband mussten noch bei jedem Schuss die Kettfäden wieder auf die gewünschte Breite gezogen werden. Deshalb wurden mit den vorhandenen Materialien systematisch Webproben hergestellt.

Webprobe 2

Kette A – Schuss D

Der Schuss wurde fest angezogen. Insgesamt wurde das Band zu schmal, aber die Anzahl der Schussfäden pro cm stimmte.

Webprobe 3

Kette A – Schuss C

Das Band war noch immer zu schmal und nun stimmte auch die Anzahl der Schussfäden pro cm nicht. Die Folgerung aus diesem Versuch war, dass die Kettfäden offensichtlich zu dünn waren. Da jedoch keine dickeren gefärbten Kettfäden zur Verfügung standen, wurden für die weiteren Versuche zwei Fäden – ungezwirnt – statt des einen verwendet.

Webprobe 4

Kette B (doppelter Faden) – Schuss C

Die Ursache der veränderten Farbe der Kette ist, dass diese Kette aus der Strangwolle der 2. Färbung stammt. Das Material ist wie das der Kette A. Wie bereits erwähnt, wurde ein doppelter Kettfaden benutzt. Das Gewebe hatte jetzt die richtige Breite, jedoch zu wenige Schussfäden auf einen Zentimeter.

Webprobe 5 bzw. Band 2

Kette B (doppelter Faden) – Schuss D

Das Gewebe ist wie gewünscht fest, sowohl fest angeschlagen als auch angezogen. Es ist etwa 3,5 cm breit und die Anzahl der Schüsse auf 1 cm Länge ist zufriedenstellend.

Zusammenfassung

So wie dieses nach den verschiedenen Versuchen letztendlich hergestellte Band könnte auch das ursprüngliche Band aus dem Vaaler Moor einmal ausgesehen haben.

Anmerkungen

1   Weben: Ruth Neumann; Spinnen: Brigitte Freudenberg und Ruth Neumann; Färben: Margarete Siwek.

Literatur

Goldmann, A. (1998) Die Skudde, eine alte mittel- und osteuropäische Landschafrasse. *Textiles in European Archaeology. Report from the 6th NESAT-Symposium*, 233–242. GOTARC Series A/1. Göteborg.

Meier, G. (1994) *Pflanzenfarben – Forschung, Herstellung, Anwendung,* Dornach, Verlag am Goetheanum.

Ploss, E. (1967) *Ein Buch von Alten Farben*. München.

Schlabow, K. (1976) *Textilfunde der Eisenzeit in Norddeutschland.* Neumünster.

Schweppe, H. (1993) *Handbuch der Naturfarbstoffe*. Hamburg, Nikol Verlagsgesellschaft mbH und KG Hamburg.

Tidow, K., und Walton P. (2001) Recent Analysis of the Textiles from Bökener Moor and Vaaler Moor, Germany. In P. Walton Rogers, L. Bender Jørgensen und A. Rast-Eicher (Hg.), *The Roman textile industry and its influence. A Birthday Tribute to John Peter Wild*, 117–128. Oxford, Oxbow Books.

14 The Magdalensberg Textile Tools: a Preliminary Assessment

by Kordula Gostenčnik

Magdalensberg and Early Roman Noricum

When *Noricum* (located mainly in Austria, with a few small districts in the adjacent regions of southern Germany, northern Italy and Slovenia) became a Roman province during the reign of Emperor Claudius (AD 41–54), the south of the territory had been under Roman control for about 100 years already, as a result of Rome's interest in the richness of the region's metal deposits, especially iron and gold. In the north of Noricum, Roman influence increased considerably only when the frontier or *limes* along the river Danube was established by the Flavian emperors during the 2nd half of the 1st century AD.[1]

The oldest Roman town in Noricum was built *c.* mid-1st century BC on a mountain today called Magdalensberg, in the south of the region, 1058 m above sea level or 600 m above the surrounding plains (Fig. 14.1). Merchants from Aquileia, a Roman town and important trading centre on the shores of the northern Adriatic sea, set up an emporium in the Celtic *regnum Noricum* about three decades before the Roman military occupation of the Alps in 15 BC, to facilitate exchange and trade with the Mediterranean on a large scale, but also to start producing merchandise in their iron and bronze workshops, as metal processing was much easier and cheaper in the Alps than in north-eastern Italy. Since 1948, the town's centre with a forum, administrative centre, basilica, the baths and several adjacent buildings and quarters, a Roman fortification and temple on the summit and partly also the necropolis, have been excavated. At present, archaeological fieldwork concentrates on a rampart and terraces on the north-western slopes of the mountain.[2]

The advantages of the Magdalensberg excavations are clear: the town was inhabited for 100 years only, roughly between 50 BC and AD 50, and after it was abandoned, it was never reoccupied. The place remained untouched except for the building of a church on the summit in the 12th century and farmsteads nearby. When the town was abandoned, all the household supplies and usable building materials like timbers, roof tiles or marble decorations were carried away. As the town was built on natural and artificial terraces down the southern slopes of the mountain, much damage was caused afterwards by erosion and washing away of the mountain slope during the two millennia that followed.

The planning is typical for any Roman town of the Republic or the Empire (Fig. 14.1). The economic basis was wholesale trade, metal processing and, what has been understood only recently, the production of textiles (Fig. 14.2). What is unusual is the presence of metal processing workshops directly at, or near, the forum, and, furthermore, a workshop where gold ingots were cast. Two moulds for these ingots bear the owner's name, Emperor Caius Caligula (AD 37–41) himself (Piccottini 2001). Another interesting structure is a shopkeeper's warehouse, which was destroyed by fire in the AD 30s (Piccottini 1998, 61–75), during the final years of Emperor Tiberius (AD 14–37). The assemblage of the merchandise consists among many other goods of several hundred specimens of Roman tableware in *terra sigillata* and glass vessels, which melted due to the enormous heat, but also a bunch of 45 iron needles and several ring-distaffs in glass (Figs 14.5d, 14.13b).

The town's name was Virunum; this is attested by literary evidence which even fits the founding myth of the town that is connected to the killing of a wild boar by a brave man (*'vir unus'* therefore Virunum), as well as the fragment of an inscription (Dobesch 1997; Piccottini and Vetters 2003, 66). By mid-1st century AD, this earlier Virunum was abandoned and both name and inhabitants were transferred to a new town on the plains to the south-west of Magdalensberg, today called Zollfeld, some 10 km away, becoming the capital when Noricum received the legal status of a Roman province.[3]

The Magdalensberg Textile Making Equipment

Textile producing tools were made of a variety of raw materials such as wood, clay or loam, copper alloys, iron, bone, glass, or more precious materials like gold, silver, ivory or amber.[4] Among the Magdalensberg finds,[5] the latter, which are prestigious or highly symbolic artefacts (Trinkl 2004), are extremely rare. The most significant material is pottery recycled for the production of spindle whorls, which is freely available almost everywhere and by no means prestigious (Figs 14.6c–d).

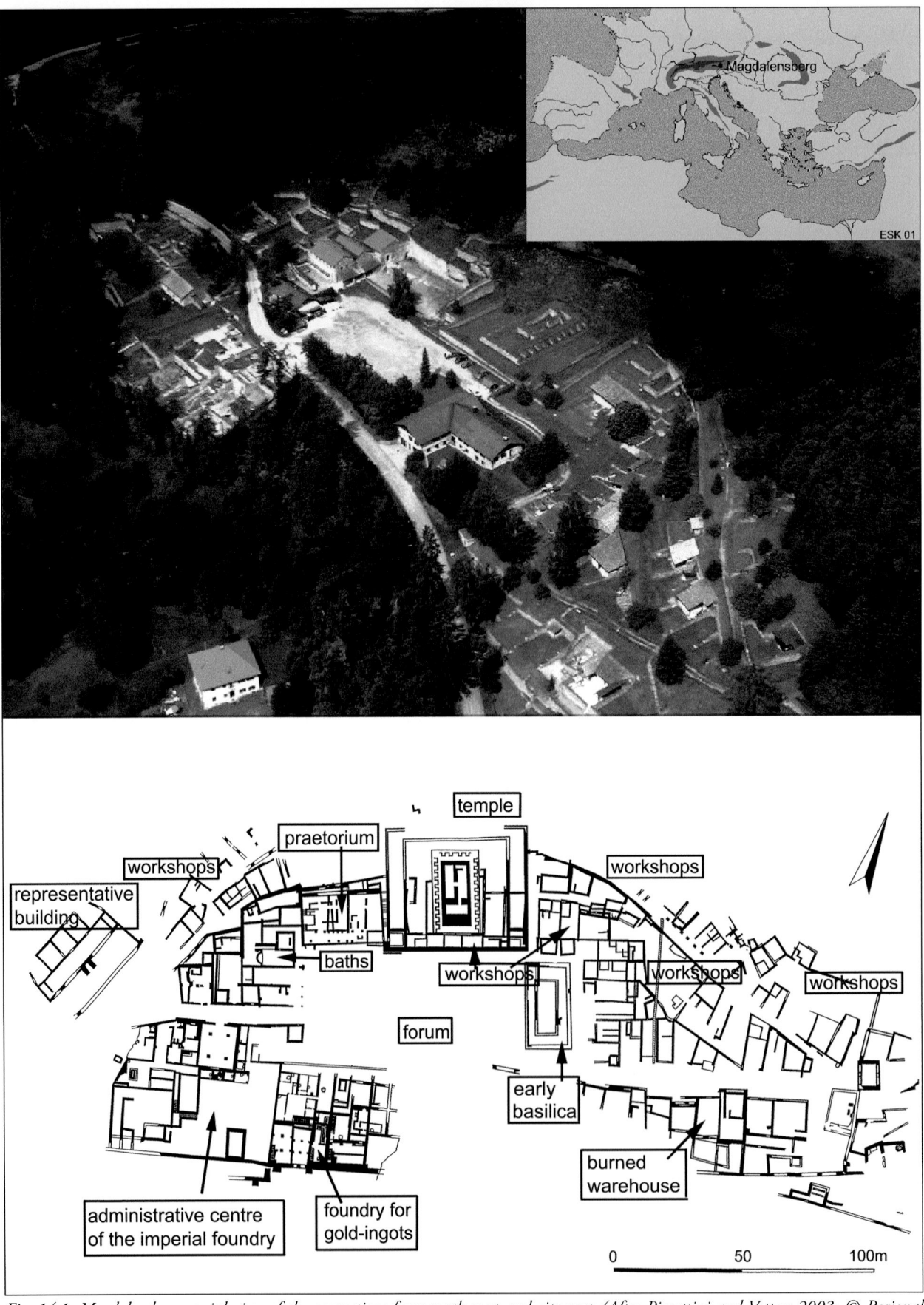

Fig. 14.1. Magdalensberg, aerial view of the excavations from south-west and city-map (After Piccottini and Vetters 2003; © Regional Museum Carinthia; addenda K. Gostenčnik).

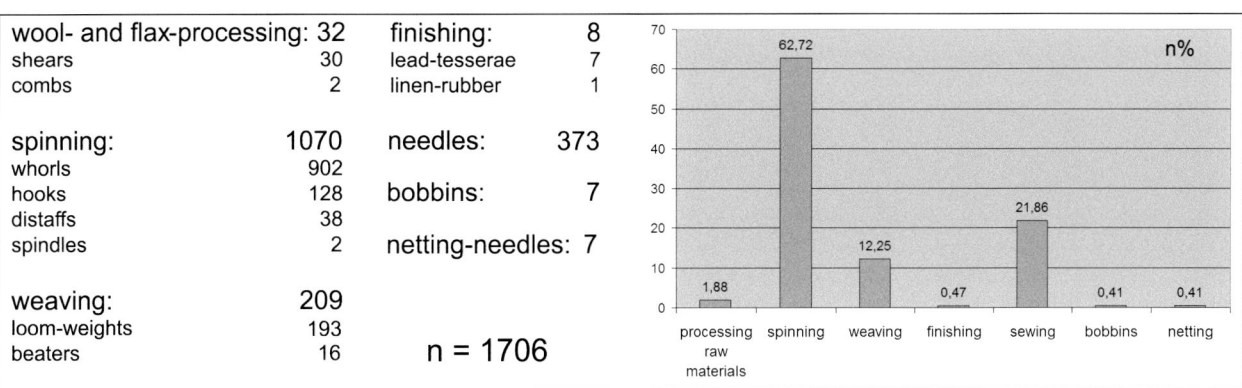

Fig. 14.2. Magdalensberg, textile equipment in statistics (© K. Gostenčnik).

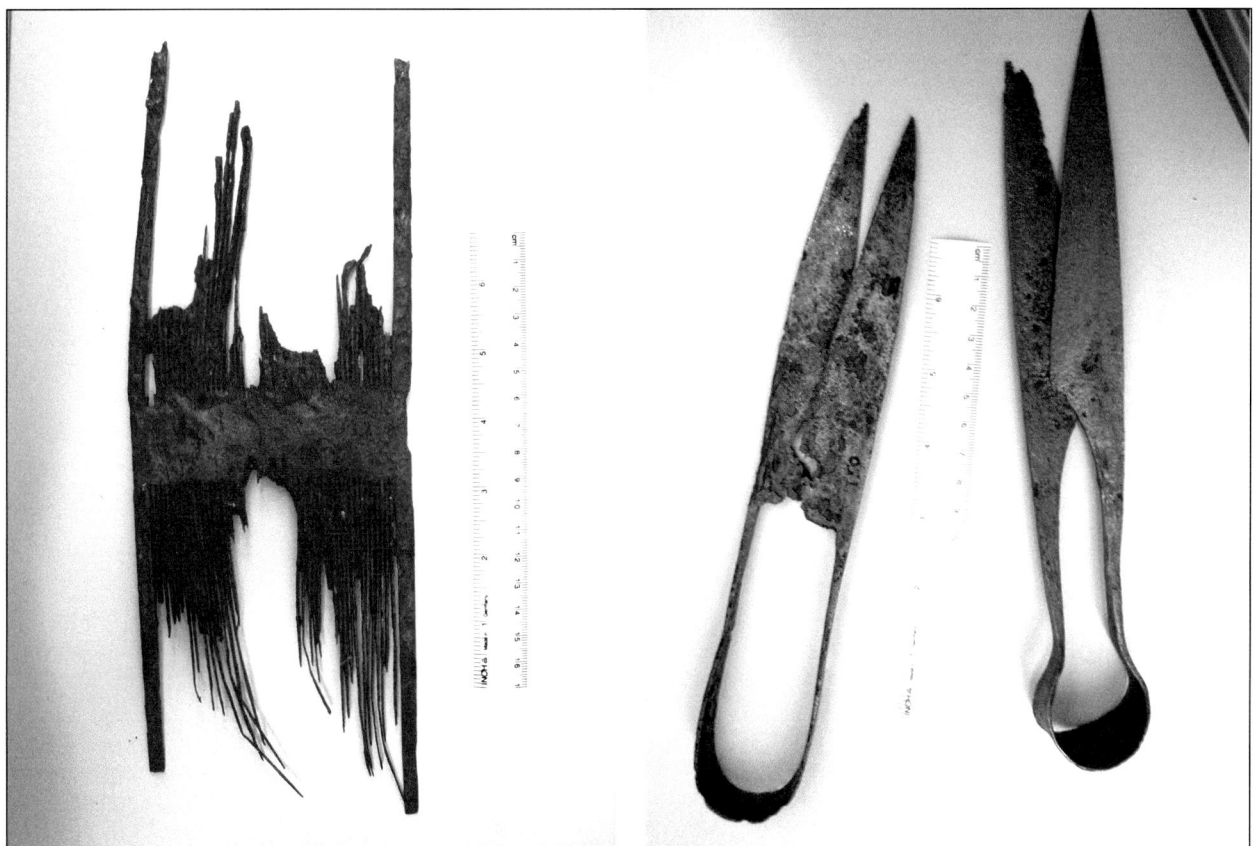

Fig. 14.3. Magdalensberg, wool- or flax comb; iron (© K. Gostenčnik).

Fig. 14.4. Magdalensberg, sheep-shears; iron (© K. Gostenčnik).

A few fragments of linen textiles have also survived.[6] Hints as to the availability of raw materials, mainly wool and flax, may be deduced both from archaeozoological and archaeobotanical sources from the site (Hornberger 1970; Werneck 1969), though they must have been produced outside the town.

The diagram in Figure 14.2 is a statistical summary of the Magdalensberg finds, 1706 items in total, without taking into consideration sets of loom-weights or hoards of needles or spindle-whorls. Apparently, the implements for spinning (62.72 %) and sewing (21.86 %) are the most important. Loom-weights are normally recorded as single finds only, except for one set of 18 and two sets of nine. Other artefacts which can be attributed to the chain of production are only marginally present, partly because items like shears or combs would not be preserved on a site like this, partly because raw materials like the lead tags (*tesserae fullonicae*) were recycled, and, primarily because items made of organic materials such as wooden tools (spindles, distaffs and looms) do not survive, except for a single wooden whorl.

Shears and Combs

Tools relating to the preparation of raw materials are not so numerous, which is by no means surprising considering the

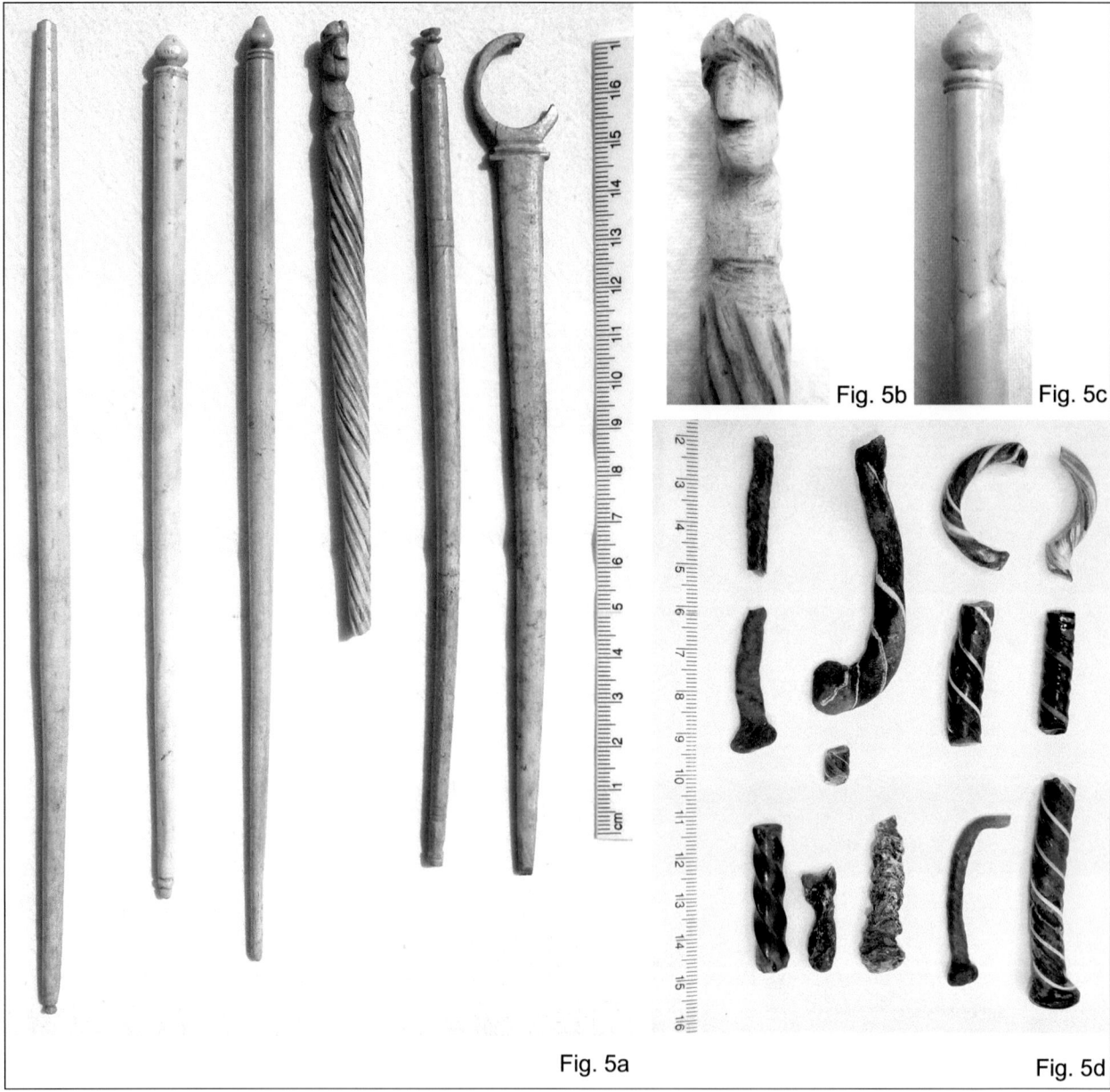

Fig. 14.5. Magdalensberg, spindle and distaffs, bone: 5a. bone spindle (left), distaffs and ring-distaff; 5b. bone distaff with a female bust, detail; 5c. pattern preserved on a distaff; 5d. glass ring-distaffs (Photos: © K. Gostenčnik).

fact that sheep breeding or flax cultivating were definitely done in the rural areas outside the towns. However, two iron combs (Dolenz 1998, Pl. 40, L66; Bitenc 2002) and 30 pairs of shears are recorded (Figs 14.3–14.4). Shears are not necessarily evidence of wool processing as they were multifunctional. Cropping shears (Wild 1970, Pl. XIIa) are not reported, however fulleries must have existed as is revealed from the evidence of lead *tesserae fullonicae* (Egger 1967). The absence of cropping shears might be due to the recycling of iron already in antiquity.

Spinning

Among the Magdalensberg finds, 25 distaffs, one spindle and five spindle-whorls of bone are preserved. The distaffs are especially interesting (Figs 14.5a–d), as the type with a small onion-shaped head is commonly found in female graves throughout the Roman Empire during the 1st and 2nd centuries AD (Gostenčnik 2005, 227–233) and similar prestige items are found in ivory as well. The bone tools are frequently dyed; some of the distaffs from Magdalensberg still have traces of red colour, and sometimes a pattern of spirals and dots is discernible (Fig. 14.5c). Thirteen fragments come from glass distaffs; most of these were excavated from among the destroyed merchandise in the burned warehouse SH/5 (Figs 14.1, 14.5d)

Typical for the Magdalensberg period are spindle-hooks (Fig. 14.7). Only two hooks with a tubular socket have been found, a type known from the Greek Classical Period onwards, while 126 spindle-hooks usually made from copper alloys have a twisted or sometimes plain shank; one is even preserved in iron. Previously, the only parallels considered

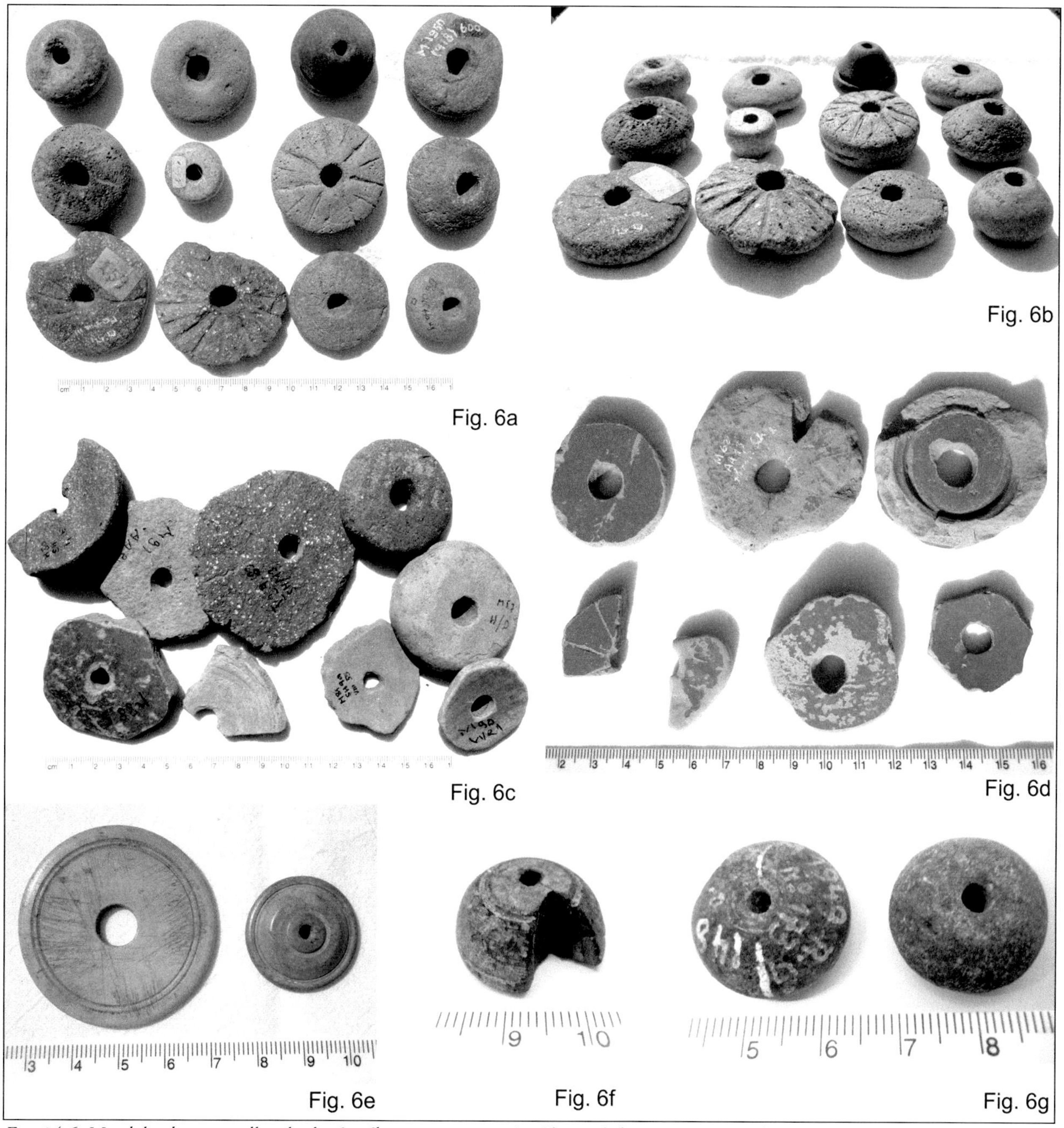

Fig. 14.6. Magdalensberg, spindle-whorls: 6a–6b. coarse pottery; 6c–6d. recycled coarse pottery and terra sigillata; 6e. bone; 6f. wood; 6g. amber (© K. Gostenčnik).

were those from Egypt (Gostenčnik 2001), but they are also present in Poland at least from the late pre-Roman Iron Age onwards and frequently recorded in Germanic graves in the Oder-Vistula region (Gostenčnik 2003). Later, during the Migration Period and Early Middle Ages, they were still in use; sometimes they were even made of silver. As attention is now drawn to these tools, they are published more frequently in southern Noricum, and therefore Magdalensberg is by no means the only place where they were in use.

Worked bone items also comprise 310 *stili* (Fig. 14.8) or writing-equipment of the Hellenistic and Republican periods and the Early Roman Empire (Gostenčnik 2005, 37–78).

Fig. 14.7. Magdalensberg, spindle-hooks: copper alloys (© K. Gostenčnik).

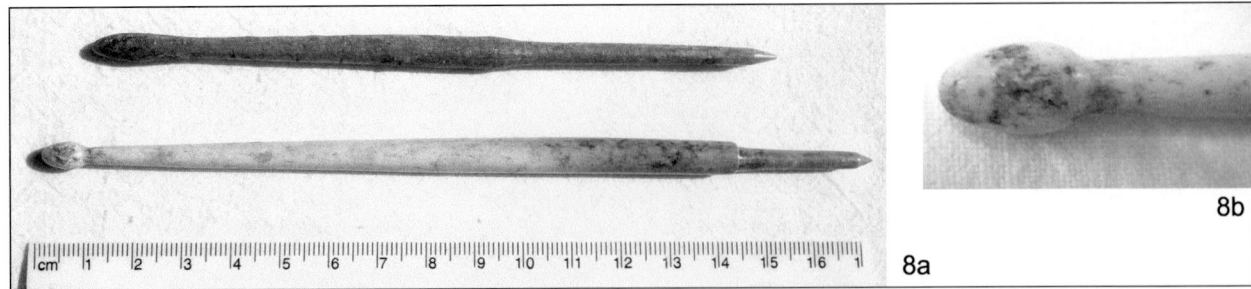

Fig. 14.8a–b. Magdalensberg, bone-stili; 8a main types; 8b detail with imprints of human teeth (© K. Gostenčnik).

They are frequently classified as spindles (Béal 1983, 151–162, although with reservation; Chazelles 2000); the types also exist in iron. However, there are no wear marks indicative of their use as spinning tools. Their heads very often bear the bite marks of human teeth (Fig. 14.8b), similar to chewing-marks on pencils.

Most of the 902 spindle-whorls from Magdalensberg (Fig. 14.6) were made from discarded coarse marble or quartz-tempered pottery, with the exception of a few in *terra sigillata*. A very limited number only was made from materials like bone, amber, or even wood and glass. Therefore, the most common shape among the Magdalensberg spindle-whorls is discoid, also among those in worked bone (Figs 14.6c–e). With the spindle-whorls made from clay, the outlines range from disc-shaped to conical and bi-conical (*cf.* Grömer 2004, prehistoric whorls). Only 36 out of 902 were not made from discarded pottery. As proven by hoards of half-finished whorls made from broken pottery, they were produced in bronze workshops, which is not unusual considering the fact that these were furnished with the necessary tools, that is hammer and drill. Therefore, we might even consider that these spindle-whorls were offered for sale. Bone-whorls, too, were manufactured locally, as is indicated by waste (Gostenčnik 2005, Pl. 54.4). The weights of the Magdalensberg spindle-whorls range from 3 to 81 g, the average weight being 8–30 g.

Weaving

Although different types of looms must have existed at Magdalensberg, only loom-weights survived (Figs 14.9a–f). Among the 193 loom-weights, the locally produced were made from two main raw materials, coarsely tempered loam and mortar. Of these, 87 were made of the latter material (Fig. 14.9b). Most weights (Fig. 14.10) are shaped like a truncated pyramid on a square base (Fig. 14.10.1–2) or like an egg (Fig. 14.10.5). Pyramids with a rectangular base are not so numerous (Figs 14.10.3–4), and as their fabrics are identical to those of tiles imported from Northern Italy, they, too, were definitely produced in the same workshops and imported to the site (*cf. e.g.* Mauné *et al.* 2005). Red and yellowish tiles and loom-weights are usually tempered with fire clay only; yellowish tiles are for instance typical for the tile *fabrica* owned by P. Catius Mato, whose workshops were located in north-eastern Italy; he sold a great many roof-tiles stamped with his name to Magdalensberg. The weight of the loom-weights ranges between 0.095 and 1.4 kg.

Weaving tools also include 16 bone beaters (Figs 14.11a–f). The oblong sword-beaters are of particular interest as the marks of wear (Fig. 14.11a) show the same traces as a wooden sword-beater used in experimental archaeology.[7] These beaters are also equipped with one or two dented edges (Béal 1983, Pl. 41.1323–1324) and should in that case be referred to as 'comb-beaters' (Fig. 14.11d). The fragment in Figure 14.11d shows two different kinds of teeth: short and narrow ones and long teeth with broader interstices. As the teeth do not show much wear, it is obvious that this tool had hardly been used.

Thanks to the graffito on a sword-beater from Trier (Bersu 1930, Pl. 16.5), we know that the Latin term for slender implements like the one in Figure 14.11e is *spatha* or sword-beater. There is always a groove at the narrow end, where threads may cut marks into the bone. Traces left by warp threads are also visible on the polished surfaces of those implements. The broader end is decorated with the relief depicting a fingernail, a hint to its usage instead of fingers to handle threads on the loom.

The graffito on one of the Magdalensberg sword-beaters (Fig. 14.11c) reads *Philargi* in genitive, a typical name of Greek origin for a slave. Two more implements of the same type from Augusta Raurica (present day Augst, Switzerland) and Lyon in France are similarly inscribed (Deschler-Erb 1998, Pl. 14.374; Béal 1983, Pl. 41.1323; in Lyon with initials only for *P(ublius) R(…)* in nominative or genitive, whereas the one from *Augusta Raurica* is indecipherable. A third example from Les Bolards in France bears an X (Sautot 1978, Pl. 19.9). The graffito on the above mentioned beater from Trier reads *L(ucii) Restituti spatha*, 'sword (or beater) owned by Lucius Restitutus'. It is interesting to note that in these three cases men are the owners and presumably also the users of the *spathae*, that is referring to men working at the loom (Moeller 1969).

Finishing

As previously mentioned, evidence for the existence of fullonicae at Magdalensberg is provided by the *tesserae fullonicae* (Egger 1967). A Roman fullonica does not necessarily full wool fabric only, but as Roman garments were brightly coloured and colours were prone to fading, garments were also dyed in those workshops. Moreover, the fullonica was the Roman dry cleaners and the establishment for the finishing of home-made cloth (Wild 1970, 79–88;

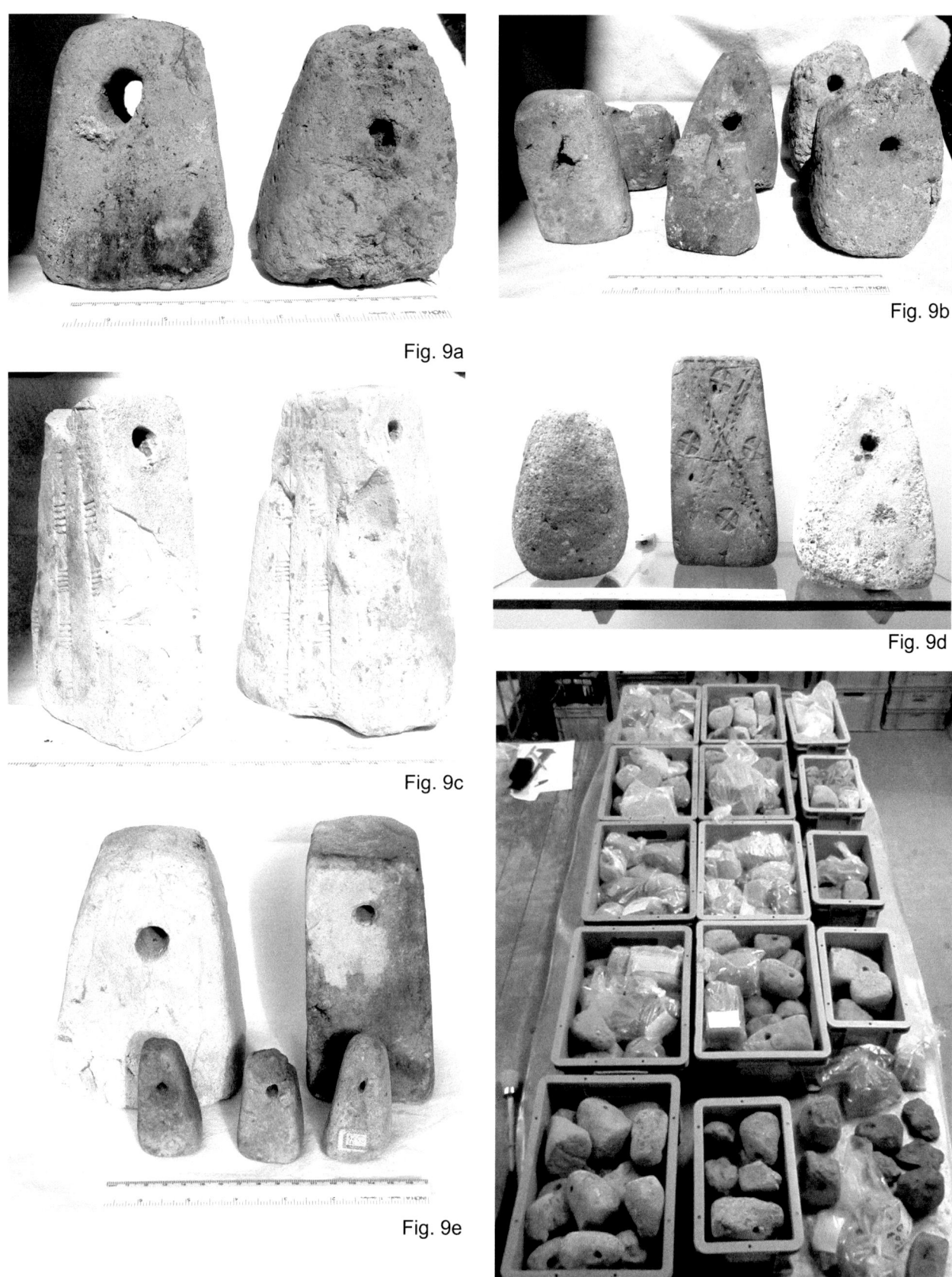

Fig. 14.9. Magdalensberg, loom-weights: 9a. loam; 9b. mortar; 9c. yellow tile-clay; 9d. three different types in clay and mortar; 9e. dimensions; 9f. overview (© K. Gostenčnik).

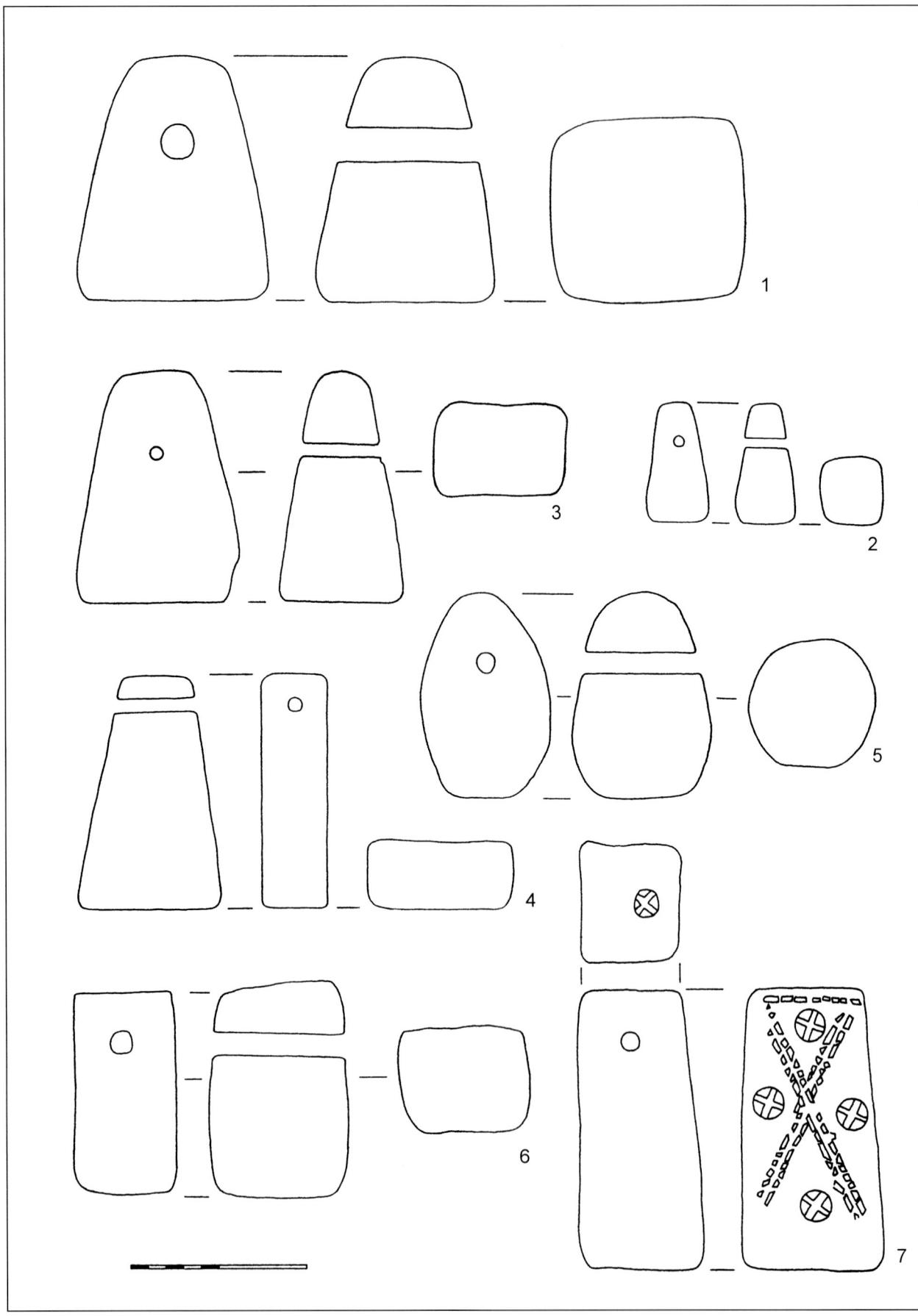

Fig. 14.10. Magdalensberg, main types of loom-weights: 1–2. pyramids on a square basis; 3–4. pyramids on a rectangular basis; 5. egg shaped; 6–7. cuboids (© K. Gostenčnik).

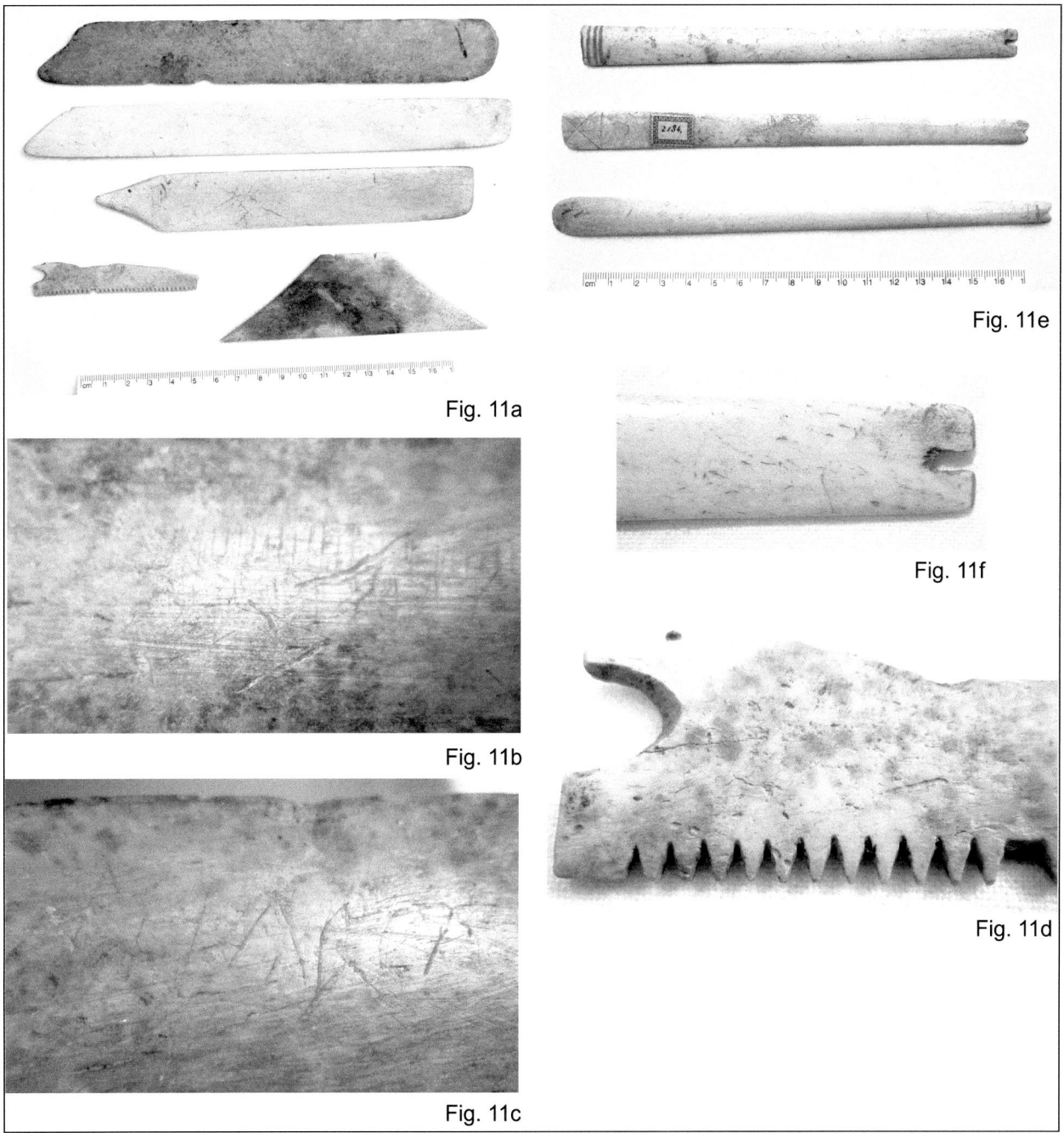

Fig. 14.11. Magdalensberg, beaters, bone; 11a. oblong sword-beaters, comb-beater, trapezoid beater; 11b. imprints of the warp; 11c. graffito with owner's name; 11d. comb-beater; 11e. slender sword-beaters; 11f. detail with notch and the grooves caused by threads (© K. Gostenčnik).

Uscatescu 1994; Borgard and Puybaret 2003; Wilson 2003; Flohr 2003). The contents of inscriptions on artefacts usually dubbed *tesserae fullonicae* differ greatly (Fig. 14.12a). We find prices for various tasks in the texts of the lead *tesserae*, for instance the prices for fulling or those for tailoring, the chemicals used for fulling, especially sulphur, the names of a variety of garments or the names of special patterns of garments, or hints to imports of textiles (Bonetto 2001). A hoard of nearly 200 of these *tesserae* from the 2nd century AD was found in Kalsdorf in the south-east of Noricum near the Roman town Flavia Solva attesting to the various tasks listed above (Römer-Martijnse 1990; 1993, 361–371, 402).

A rare piece is a linen smoothing stone made from dark glass (Fig. 14.12b), with traces of glossy polishing on the underside (Wild 1970, 84–85, fig. 76; Walton Rogers 1997, 1775).

Sewing

Regrettably, Roman needles in iron and copper alloys are rather neglected in archaeological literature. At Magdalensberg, 200 examples in iron and 172 in copper alloys are hitherto recorded. Roman needles in iron and copper alloys are very similar to those of today (Fig. 14.13a), with a groove

Fig. 14.12. Magdalensberg, finishing implements: 12a. lead tesserae fullonicae; 12b. linen rubber (Fig. 12a after K. Dolenz 2006, Figs. 2; 4; © Regional Museum Carinthia; Fig. 12b © K. Gostenčnik).

for the thread to be lowered into above and below the eye (Figs 14.13c–d), though specimens without these grooves are reported too. A bunch of 45 iron needles was among the debris of the burned warehouse SH/5 at Magdalensberg (Figs 14.1; 14.13b). These, too, are thin, with shanks 1–2 mm thick; their lengths are hard to establish, as they are all damaged, and the pieces vary from 2.5 to 11.8 cm. Roman needle cases or needle books are rare. One tubular needle case was excavated in Vindolanda with 28 thin iron needles approximately 7 cm long. A long piece of cloth was used as a needle cushion in Dura Europos with 17 iron needles about 5–6 cm long (Pfister and Bellinger 1945, Pl. 31, 293; Charpy 1993, fig. 47.02.13); their shanks are no more than 1.5 mm thick. These needles come closest to what we would expect to have been used for sewing Roman cloth.

On the other hand, the so-called bone needles are most common finds on Roman sites, although they are rare before mid-2nd century AD (Deschler-Erb 1998; Gostenčnik 2005). However, were they really used as needles? By comparing the eyes of numerous 'bone needles' with the signs of wear visible around the drilled holes of weaving tablets, *i.e.* grooves around the eyes caused by threads, raises an important question: why are there no such cuts or grooves caused from threads within the eyes of the 'bone needles' as well? Did those needles break or were they thrown away before any signs of wear occurred? A compelling answer, and this is pertinent to every single Roman 'bone needle' the world over with either one two or three eyes, is: they were never used as needles at all. The shanks range between three and six mm and when found in inhumation graves, they either appear around or below the skull or on the chest of the corpse (Gostenčnik 2005, 101–107, notes 409–410 and 412). Regrettably, too many of these 'bone needles' have been excavated in cremation-graves, *i.e.* contexts which leave no hints as to their intended purpose. On top of the eyes, where the thread is supposed to be pulled through permanently during sewing, no hint ever appears of either polishing or grooves caused from threads (Figs 14.13e–f). Almost every eye of the 'bone needles' has sharp edges, which would cause threads to break immediately. Furthermore, these artefacts are reported along with thousands of pins of various shapes in the sewer systems of Roman baths, indicating an alternative function.[8] Nevertheless, as an exception to the rule, Magdalensberg actually did produce one bone needle with carefully rounded and polished edges in the eye and a groove to sink the thread into (see Figs 14.13e–f).

Bobbins and Toggles

A small number of worked bones with an oblong shape provided with grooves on their smaller ends seem to have been bobbins for keeping threads (Fig. 14.14). One specimen, made of the metapodial of a pig, has a groove around the

Gostenčnik, K. (2006) Beinfunde aus Virunum – ein Überblick. In *Carinthia I* 196, 41–66. Klagenfurt, Geschichtsverein für Kärnten.

Gostenčnik, K. (2007) Bernsteinfunde vom Magdalensberg. In *Carinthia I* 197, 51–69. Klagenfurt, Geschichtsverein für Kärnten.

Gostenčnik, K. (Forthcoming a) Textilerzeugung in Virunum: Die Webgewichte aus dem Umfeld der Fullonica. In H. Dolenz and J. Polleres (eds), *Die Fullonica am nördlichen Stadtrand von Virunum. Textilproduktion und Textilverarbeitung im südlichen Noricum*. Kärntner Museumsschriften. Klagenfurt, Landesmuseum Kärnten.

Gostenčnik, K. (Forthcoming b) Ribs as a raw material among Roman bone artefacts from Virunum (southern Austria). In I. Sidéra (ed.), *Proceedings of the 6th conference of the ICAZ Worked Bone Research Group, Université de Nanterre/Paris X, September 2007*. Nanterre, Maison René Ginouves.

Grömer, K. (2004) Aussagemöglichkeiten zur Tätigkeit des Spinnens aufgrund archäologischer Funde und Experimente. *Archaeologia Austriaca* 88, 169–182.

Grömer, K. (2007) *Bronzezeitliche Gewebefunde aus Hallstatt – Ihr Kontext in der Textilkunde Mitteleuropas und die Entwicklung der Textiltechnologie zur Eisenzeit*. Wien, PhD Dissertation Universität Wien.

Gugl, C. (2004) Der Beitrag der Amphitheater-Grabungen 1998–2001 zur Stadtgeschichte Virunums im ausgehenden 3. und 4. Jahrhundert. In R. Jernej and C. Gugl (eds), *Virunum. Das römische Amphitheater. Die Grabungen 1998–2001*, 501–513. Archäologie Alpen Adria Vol. 4. Klagenfurt, Wieser.

Günther, R. (2000) *Matrona, vilica und ornatrix*. Frauenarbeit in Rom zwischen Topos und Alltagswirklichkeit. In T. Späth and B. Wagner-Hasel (eds), *Frauenwelten in der Antike. Geschlechterordnung und weibliche Lebenspraxis*, 350–376. Stuttgart, Metzler.

Hornberger, M. (1970) *Gesamtbeurteilung der Tierknochenfunde aus der Stadt auf dem Magdalensberg in Kärntnen (1948–1966)*. Naturkundliche Forschungen zu den Grabungen auf dem Magdalensberg Vol. 10. Kärntner Museumsschriften Vol. 49. Klagenfurt, Landesmuseum für Kärnten.

Jenkins, D. (ed.) (2003) *The Cambridge History of Western Textiles*, Vol. 1. Cambridge, Cambridge University Press.

König, G. G. (1987) Die Fingerkunkel aus Grab 146. In K. Roth-Rubi and H. R. Sennhauser (eds), *Verenamünster Zurzach 1. Ausgrabungen und Bauuntersuchungen*, 129–144. Veröffentlichungen des Instituts für Denkmalpflege an der ETH Zürich Vol. 6. Zürich, vdf Hochschulverlag AG.

Kurzynski, K. v. (1996) „… und ihre Hosen nennen sie bracas" *Textilfunde und Textiltechnologie der Hallstatt- und Latènezeit und ihr Kontext*. Internationale Archäologie Vol. 22. Espelkamp, Marie Leidorf.

MacGregor, A.(1985) *Bone, Antler, Ivory & Horn. The Technology of Skeletal Materials since the Roman Period*. London/Sydney, Croom Helm.

Martijnse, E. (1993) *Beschriftete Bleietiketten der Römerzeit in Österreich*. Wien, PhD Dissertation Universität Wien.

Mauné, S., Bourgaut, T., Lescure, J., Carrato, C., and Santran, C. (2006) Nouvelles données sur les productions céramiques de l'atelier de Dourbie à Aspiran (Hérault) (première moitié du 1er siècle apr. J.-C.). In *SFECAG Actes du congrès de Pézenas, 25–28 Mai 2006*, 157–188. Marseilles, SFECAG.

Moeller, W. O. (1969) The male weavers at Pompeii. *Technology and Culture* 10, 561–566.

Moeller, W. O. (1976) *The wool trade of ancient Pompeii*. Leiden, Brill.

Pfister, R., and Bellinger, L. (1945) *The Excavations at Dura Europos. Final Report Vol. 4, Part 2, The Textiles*. New Haven, Yale University Press.

Pfuhl, E., and Möbius, H. (1977) *Die ostgriechischen Grabreliefs* Vol. 1. Stuttgart, Theiß.

Pfuhl, E., and Möbius, H. (1979) *Die ostgriechischen Grabreliefs* Vol. 2. Stuttgart, Theiß.

Piccottini, G. (1977) Die Stadt auf dem Magdalensberg – ein spätkeltisches und frührömisches Zentrum im südlichen Noricum. In H. Temporini and W. Haase (eds), *Aufstieg und Niedergang der römischen Welt* Vol. II, Part 6, 263–301. Berlin/New York, Walter de Gruyter.

Piccottini, G. (1989) *Bauen und Wohnen in der Stadt auf dem Magdalensberg*. Studien zur Ur- und Frühgeschichte des Donau- und Ostalpenraumes Vol. 4. Denkschriften der Österreichischen Akademie der Wissenschaften, philosophisch-historische Klasse Vol. 208. Wien, Österreichische Akademie der Wissenschaften.

Piccottini, G., ed. (1998) *Die Ausgrabungen auf dem Magdalensberg 1980–1986*. Magdalensberg-Grabungsbericht Vol. 16. Klagenfurt, Geschichtsverein für Kärnten.

Piccottini, G. (2001) Norisches Gold für Rom. *Anzeiger der Österreichischen Akademie der Wissenschaften, philosophisch-historische Klasse* 136, 41–76.

Piccottini, G. (2002) Virunum. In M. Šašel-Kos and P. Scherrer (eds), *Die autonomen Städte Noricums und Pannoniens: Noricum*. Situla Vol. 40, 103–126. Ljubljana, Slovenska akademija znanosti in umetnosti.

Piccottini, G. and Vetters, H. (2003) *Führer durch die Ausgrabungen auf dem Magdalensberg*. 6th edition, Klagenfurt, Landesmuseum Kärnten.

Praschniker, C., and Kenner, H. (1947) *Der Bäderbezirk von Virunum*. Wien, Rudolf M. Rohrer.

Rast-Eicher, A. (2001) Roman textiles in Switzerland. In P. Walton Rogers, L. Bender Jørgensen and A. Rast-Eicher (eds), *The Roman Textile Industry and its Influence. A Birthday Tribute to John Peter Wild*, 84–90. Oxford, Oxbow Books.

Riederer, J. (1974) Römische Nähnadeln. In *Technikgeschichte* 41, 153–172. Berlin, Kiepert.

Roche-Bernard, G. (1993) *Costumes et textiles en Gaule romaine*. Paris, Errance.

Römer-Martijnse, E. (1990) *Römerzeitliche Bleietiketten aus Kalsdorf, Steiermark*. Denkschriften der Österreichischen Akademie der Wissenschaften, philosophisch-historische Klasse Vol. 205. Wien, Österreichische Akademie der Wissenschaften.

Šašel Kos, M., and Scherrer, P., eds (2002) *Die autonomen Städte Noricums und Pannoniens: Noricum*. Situla Vol. 40. Ljubljana, Slovenska akademija znanosti in umetnosti.

Sautot, M. C., ed. (1978) *Le cycle de la matière, l'os*. Dijon, Musée Archéologique.

Schwinden, L. (1989) Gallo-römisches Textilgewerbe nach Denkmälern aus Trier und dem Trevererland. *Trierer Zeitschrift* 52, 279–318.

Seitz, G. (1999) *Rainau-Buch 1. Steinbauten im römischen Kastellvicus von Rainau-Buch (Ostalbkreis)*. Forschungen und Berichte aus Baden-Württemberg Vol. 73. Stuttgart, Theiß.

Sievers, S. (1992) Knochen- und Geweihgerät. In F. Maier (ed.), *Ergebnisse der Ausgrabungen 1984–1987 in Manching*, 192–195. Die Ausgrabungen in Manching Vol. 15. Stuttgart, Theiß.

Trinkl, E. (2004) Zum Wirkungskreis einer kleinasiatischen *matrona* anhand ausgewählter Funde aus dem Hanghaus 2 in Ephesos. *Jahreshefte des Österreichischen Archäologischen Institutes* 73, 281–303. Wien, Österreichisches Archäologisches Institut.

Urban, O. H. (2000) *Der lange Weg zur Geschichte. Die Urgeschichte Österreichs. Österreichische Geschichte bis 15 v. Chr.* Wien, Ueberreuter.

Uscatescu, A. (1994) *Fullonicae y tinctoriae en el mundo romano.* Barcelona, Promociones y Publicaciones Universitarias.

Vetters, H. (1977) Virunum. In H. Temporini and W. Haase (eds), *Aufstieg und Niedergang der römischen Welt*, Vol. II, Part 6, 302–354. Berlin/New York, Walter de Gruyter.

Wächter, A. (1986) Untersuchungen von Textilfragmenten vom Magdalensberg. In H. Vetters and G. Piccottini (eds), *Die Ausgrabungen auf dem Magdalensberg 1975–1979*, 448–449. Magdalensberg-Grabungsbericht Vol. 15. Klagenfurt, Geschichtsverein für Kärnten.

Walton Rogers, P. (1997) *Textile Production at 16–22 Coppergate.* The Archaeology of York Vol. 17, Fasc. 11. York, York Archaeological Trust and The Council for British Archaeology.

Wedenig, R. (2008) Römische Webgewichte aus der Steiermark als Schriftträger. In G. Grabherr and B. Kainrath (eds), *Akten des 11. Österreichischen Archäologentages in Innsbruck 23.–25. März 2006*, 323–342. Ikarus Vol. 3. Innsbruck, Universität Innsbruck.

Werneck, H. L. (1969) *Pflanzenreste aus der Stadt auf dem Magdalensberg bei Klagenfurt in Kärnten.* Naturkundliche Forschungen zu den Grabungen auf dem Magdalensberg Vol. 10. Kärntner Museumsschriften Vol. 49. Klagenfurt, Landesmuseum für Kärnten.

Wild, J. P. (1970) *Textile Manufacture in the Northern Roman Provinces.* Cambridge, Cambridge University Press.

Wild, J. P. (1976) The *gynaecaea*. In R. Goodburn and J. C. Mann (eds), *Aspects of the Notitia Dignitatum. Papers presented to the conference in Oxford, December 13 to 15, 1974*, 51–58. BAR Int Ser 15. Oxford.

Wild, J. P. (1999) Textile manufacture: a rural craft? In M. Polfer (ed.), *Artisanat et productions artisanales en milieu rural dans les provinces nord-ouest de l'Empire romaine. Actes du colloque organisé à Erpeldange, mars 1999*, 29–38. Monographies Instrumentum Vol. 9. Montagnac, Éditions Monique Mergoil.

Wild, J. P. (2000) Textile Production and Trade in Roman Literature and Written Sources. In D. Cardon and M. Feugère (eds), *Archéologie des textiles, des origines au Ve siècle. Actes du colloque de Lattes, octobre 1999*, 209–213. Monographies Instrumentum Vol. 14. Montagnac, Éditions Monique Mergoil.

Wild, J. P. (2002) The Textile Industries of Roman Britain. *Britannia*, 33, 1–42.

Wilson, A. (2003) The archaeology of the Roman fullonica. *The Journal of Roman Archaeology* 16, 442–446.

Zimmer, G. (1982) *Römische Berufsdarstellungen.* Archäologische Forschungen Vol. 12. Berlin, Deutsches Archäologisches Institut.

15 Silk Ribbons from Post-Medieval Graves in Poland

by Dawid Grupa

Silk ribbons belonged in the past to the category of very precious and desirable luxurious objects, testifying the prestige and wealth of their owner. This long-lasting and beautiful craft had great impact not merely on their use as decorations on clothes, but also in cases where silk ribbons and tapes appeared as belts and headbands. They were used for various purposes depending on their flexibility, durability, length and width. This paper focuses on silk ribbons excavated in crypts and inhumation graves in churches in Toruń and Lublin, Poland. The ribbons presented here had been used as:

• scapular hangers
• gown ornaments
• coverings inside and outside coffins

Scapular hangers are plain bands in tabby weave of different density. Their width varies from 0.8 cm to 1.5 cm. They were sewn onto scapular bags with a thread of the same colour as the ribbon, and usually both ribbon and the scapular textile were of the same colour. The occurrence of silk scapulars in graves indicates an infringement of the rules on their wear and use, because in graves they should have been made of wool, and their colour defined membership in a particular fellowship (Szkopek 2005, 11–35). In the case of scapulars found in Toruń and Lublin, it is difficult to speak about scapular fellowship rules, because the objects were silk, and very often colourful, with ornaments made of gold wrap.

In contrast to scapular hangers, ribbons for gowns and coffins were small works of art. An example from Toruń is a brown band of tabby weave, warp density 60 threads per cm, and weft count of 28 threads per cm. Edges decorated with yellow thread made permanent striped ornaments in the shape of tiny rectangles arranged in straight lines. The length of the band is 44 cm, width 4.2 cm (Fig. 15.1).

Another ribbon in a natural silk colour has a geometric pattern – tiny rectangles repeated every 2 cm (Fig. 15.2). Its edges are finished in a ribbed weave with 16 wefts and 120 warp threads per cm. The band ends tied in a knot seem to suggest that it had probably been used as a belt. The total length is 88.0 cm, and width 3.2 cm. The dark brown ribbon with a total length of 88 cm and width of 3 cm was finished with two triangular cuts in it, protecting the band from frilling. Corrosion traces left by brass pins suggest that, it had been tied in a knot. The band has a tabby weave with a delicately marked edge 1 mm wide, where the weave changes into a little interlacing warp with two threads of weft. Moreover, the edge was ornamented by omitting 9 and 5 threads alternately, creating a very delicate final fringe (Fig. 15.3). The density of the warp is 56, the weft 36 threads per cm.

Fig. 15.1. A brown band of tabby weave from Toruń with edges decorated in yellow thread making permanent striped ornaments in the shape of tiny rectangles arranged in straight lines (Photo: D. Grupa).

Fig. 15.2. Ribbon in a natural silk colour with a geometric design (Photo: D. Grupa).

Fig. 15.3. The edge of a dark brown ribbon ornamented by omitting 9 and 5 threads alternately, creating a very delicate final fringe (Photo: D. Grupa).

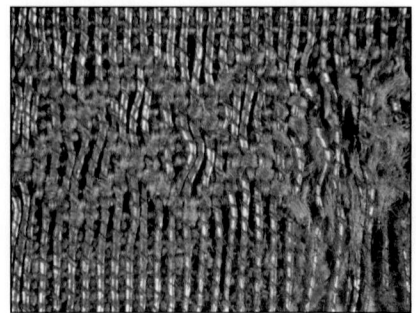

Fig. 15.6. A ribbon from the Toruń collection, with an ornament consisting of sequences of lozenges and triangles (Photo: D. Grupa).

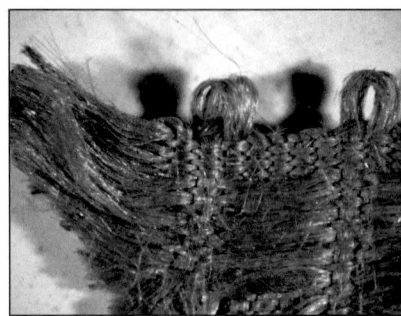

Fig. 15.4. A tape in a faded olive colour. Stripes made of sequences of yellow rectangles arranged in lines run along both edges. Its ends are toothed (Photo: D. Grupa).

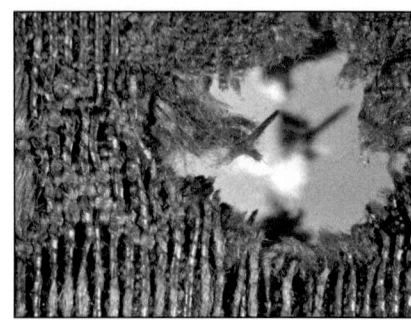

Fig. 15.7. A detail of Fig. 15.6. There are numerous holes left by iron threads fastening the tape to the coffin edges (Photo: D. Grupa).

Fig. 15.5. Detail of Fig. 15.4. Five brown threads of warp create a stripe running through the centre of the tape (Photo: D. Grupa).

Fig. 15.8. A very richly ornamented tape from Lublin decorated with tulips on a winding band, with the empty spaces filled in with a geometric pattern (Photo: D. Grupa).

The third ribbon analysed is a tape in a faded olive colour. Stripes made of sequences of yellow rectangles arranged in lines run along both edges. The tape is 2.5 cm wide and 77.5 cm long, in a tabby weave with a warp density of 79 threads and weft 25 threads per cm. Its ends are ragged. Five brown threads of warp create a stripe running through the tape's centre (Figs 15.4–15.5).

The fourth ribbon, 2.7 cm wide, of tabby weave, with a warp density of 26 threads and weft density of 21 threads per cm can be considered as part of a coffin's interior decoration. The pattern was created by introducing gold wrap as additional weft. The ornament consists of sequences of lozenges and triangles (Fig. 15.6). There are numerous holes left by iron threads fastening the tape to the coffin edges. The Toruń collection contains about 5 m of this type of band (Fig. 15.7).

An example of a very richly ornamented ribbon is a tape from Lublin, decorated with tulips on a winding band, with the empty spaces filled in with a geometric pattern. The composition was shaped by introducing an additional weft made with gold thread (Figs 15.8–15.9). Warp density is 48 threads and weft density 24 threads per cm. The edges are decorated with 3 mm long loops of gold threads. The width is 2 cm and the preserved length 57 cm.

Another ribbon from Lublin is 1.3 cm wide and its

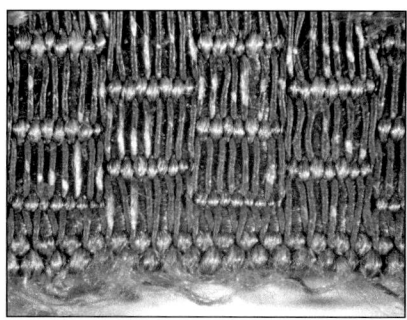

Fig. 15.9. Detail of Fig. 15.8. The empty spaces in between the tulips are filled in with a geometric pattern (Photo: D. Grupa).

Fig. 15.10. A ribbon from Lublin in a ground weave in olive and warp threads in a natural silk colour. The ornament presents stylized calyx. (Photo: D. Grupa).

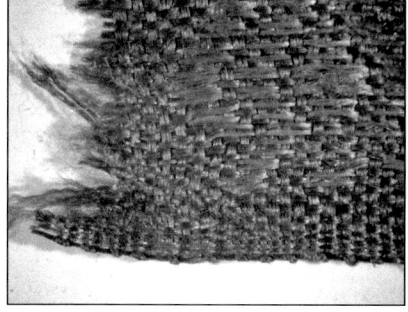

Fig. 15.11. A detail of Fig. 15.10. The ribbon bears traces of iron rivet and its edges consist of two warp threads in a natural silk colour. (Photo: D. Grupa).

preserved length is 6.8 cm; it bears traces of iron rivet. The ground weave has an olive colour and the pattern was made with warp threads in a natural silk colour. The ornament presents stylised calyx. Warp density is 39 threads, weft density 23 threads per cm. The edges consist of two warp threads in a natural silk colour (Figs 15.10–15.11).

Ribbons used as decoration on clothes are identified by means of traces of folds, pin holes, kinks or knots. Some of them also bear traces of thread remains and stitches that enable us to assume that they had been used to strengthen parts of garments, or to decorate them, as *e.g.* on cuffs or borders of collars.

Identifying ribbons and tapes serving as coffin coverings turned out to be difficult, as we have no evidence of how they were fastened to a coffin. However, the appearance of nail holes on some of them proved helpful in the identification process. Another argument for their use as coffin coverings was the length of the bands. These preserved silk pieces are rather long. An important characteristic is the fact that the tapes concealed imperfections in the decoration inside the coffin edges, when the decorations had not always been carefully finished.

The above descriptions show that, ribbons were of various quality and size. They were plain in most cases, with shiny satin weave giving them a delicate elegance. Ribbed silk ribbons were extremely varied. Few examples are richly ornamented with plant or geometric patterns. These patterns were rarely repeated, making each item unique. The decoration most often was made using gold thread. The perfection of the work indicates the high skill of the haberdashers and the composition of motives shows their sense of beauty while designing these small works of art. The ribbons presented here are not the only ones in the Toruń and Lublin collections, and the rich finds, their number and elaboration indicate their popularity with townsmen and aristocrats burying their dead within the church walls.

Ribbons are known from various archaeological excavations on Polish territory. They were used both in Catholic and Protestant burial ceremonies. In the burial crypt of Our Lady's Church in Kostrzyń upon Oder, a burial gown from the end of 17th or turn of the 18th century, sewn specially for that ceremony, was excavated. The dress front was decorated with a band with pinned green knots starting from the smallest to the biggest (Drążkowska 2005, 57). A grave shirt (dated to the 17th century), in which a one-year old boy had been buried in the parish church in Tworków was also decorated with ribbons sewn onto it (Drążkowska 2006, 214). The civic inventory list, too, contains a large amount of information about ribbons for the dead (*Inwentarze mieszczańskie* 1965, 380, 399).

Broad and very long ribbons could have served other purposes as well. Swaddling bands were found in a grave crypt in Our Lady's Church in Kostrzyń upon Oder. They were made of ribbed silk band, similar to those from Toruń. New born infants had been probably buried in them, with the bands wound tightly around the dead infant. It protected and decorated the plain clothes of the baby. The ribbon's ends were pinned in the front and a flat knot was placed in the middle (Drążkowska 2005, 56).

Tapes and ribbons were part of a haberdasher's production. Long tapes excavated in 1983 in the Assumption of Our Lady's Church in Toruń had been used for decorating the coffins. It is proven by traces and the occasional presence of haberdashery buttons (iron and brass), width of tapes, their considerable length, and remains stuck on the bottom of coffin covers. Tapes serving as decoration on the inside of coffin edges were also found during work in Płonkowo and Lublin (the material is in the process of being analysed at present). They were sometimes ornamented with gold thread, but were often equipped with fringes made of loose thread.

Civic inventories, apart from information on property bequeathed to families, contain notes concerning funeral ceremonies and their costs, and tapes for coffins: 'for 50 ells of coffin tapes-16zł, 14 gr, for coffin nails-17zł [...] for 50 ells of coffin tapes-10zł..." (*Inwentarze mieszczańskie* 1965, 43 and 90).

Small ribbons were commonly used in funerary elements for decorative or finishing purposes to emphasise the prestige of the deceased or his/her family. Both for aesthetic and practical reasons, tapes and ribbons, silk ones in particular, were used to create beautiful, rich decorations of burial clothes, coffins and separate ornamentations in the shape of wreaths and bunches, frequently combined with artificial or dried flowers. They decorated catafalques, walls in houses and churches. Funerals of those times were associated with theatrical performances and their splendour defined the social position of the deceased and his relatives. The higher the position of the deceased, the higher the costs of the funeral decorations and special clothes the family in mourning had to bear, frequently having to borrow money for that purpose. Ribbons were used in burial ceremonies of men, women and children. They were the sign of the fashion of that time.

Bibliography

Bartkiewicz, M. (1974) Odzież i wnętrza domów mieszczańskich w Polsce w drugiej połowie XVI i w XVII wieku. In Z. Kamieńska (ed.), *Studia i materiały z historii kultury materialnej, t. XLIX*. Wrocław.

Drążkowska, A. (2005) Konserwacja XVII- i XVIII-wiecznej dziecięcej odzieży grobowej. *Ochrona zabytków* 1, 55–66. Warszawa.

Drążkowska, A. (2006) Dziecięca odzież grobowa z XVII i XVIII wieku. *Kwartalnik historii kultury materialnej* 2, 211–220, Warszawa.

Drążkowska A., and Grupa, M. (2002) *Katalog tkanin jedwabnych pochodzących z krypt grobowych Archikatedry w Lublinie*. Lublin. Wojewódzki Oddział Służb Ochrony Zabytków.

Grupa M. (2005) *Ubiór mieszczan i szlachty z XVI–XVIII wieku z kościoła p.w. Wniebowzięcia Najświętszej Marii Panny w Toruniu*. Toruń.

Inwentarze mieszczańskie (1965) *Inwentarze mieszczańskie w wieku XVIII z ksiąg miejskich i grodzkich Poznania*, t.2 (1759–1793), wyd. J. Burszta, Cz. Łuczak. Poznań.

Kajdańska A., and Kajdański E. (2007) *Jedwab, szlakami dżonek i karawan*. Warszawa.

Szkopek T. (2005) *Szaty zbawienia – Pierwsza książka o wszystkich szkaplerzach*. Gdańsk.

16 Silks from Kwidzyn Cathedral, Poland

by Malgorzata Grupa

In 2007 in Northern Poland, an archaeologist, Dr Antoni Pawlowski, carried out an excavation (Pawłowski 2007) under the Kwidzyn Cathedral floor in search of the burial place of the medieval mystic, the Blessed Dorota from Mątowy. Two crypts were found in the presbytery, under the mensa. The first one was empty, whereas three male skeletons were found in the other one. One of them was of huge stature, the two others were much smaller and of delicate build. The lack of serious pathological changes in the bones, which can be connected with food deficiency, avitaminosis, anaemia, chronic infection or anatomical movement structure overloading caused by hard manual work, testify that these three persons may have belonged to the social elite of the 14th century (Landau 2008, 84–87).[1] Even their spines did not bear any traces of degeneration changes, common for older people who had undertaken hard manual labour (Landau 2008, 85).[2]

The second indicator, or rather the third one – as the circumstance of being buried in the presbytery is also a proof of belonging to the highest social groups of those times – is the fact that the men had been buried in silk clothes. Until now, 21 types of silk have been identified. Six of them are textiles of tabby weave of various densities. The others are textiles with geometrical and plant motifs (Fig. 16.1). Some of them were in two colours, *e.g.* the ground weave was red and the pattern in a natural coloured silk (Fig. 16.2). It is difficult to identify the types of gowns, as the textiles were excavated in small pieces. Two of these garments may have been tunics. Fragments, which can belong to the low-cut neck of the tunic – with edges finished with bands of gold and silver threads, were found.

A piece of textile excavated from under one of the skeletons was particularly interesting. It could have been a coat or rather a cloak, made of textile pieces cut in lozenges. One textile was in tabby weave, the other had stripes (Fig. 16.3), and they were sewn together alternately. It is the first example in Poland of a gown constructed in this manner. Can it be connected in some way with the family coat of arms of the buried men? This direction of studies seems promising, taking

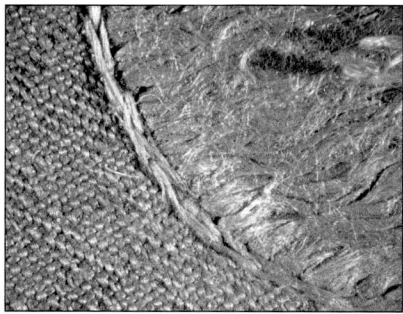

Fig. 16.1. Lozenge weave base of textile with embroidery in natural silk colour (Photo: D. Grupa).

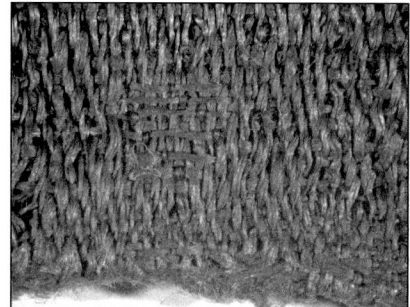

Fig. 16.2. Red textiles – a pattern shaped with natural coloured silk (Photo: D. Grupa).

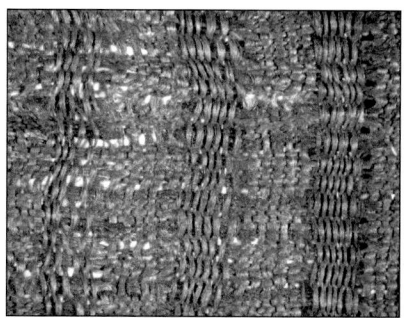

Fig. 16.3. Textile with stripes of three colours (Photo: D. Grupa). Reproduced as a colour plate on page 303.

into consideration the colours of textiles and rules of the medieval dyeing trade (Pastoureau 2006, 141–148).

Another interesting element is a band sewn on a piece of tabby textile, in the shape of a cross. It has two recurring ornaments made of gold thread. First is a bird with its wings outstretched, possibly an eagle. The other is a spray of flowers. Each ornament section is separated from the other with a plain fragment of textile resembling gauze.

In analysing the entire textile collection, the question of who could have been buried in such expensive clothes, in so rich an environment, arises. There are two hypotheses: It could be the burial places of three Grand Masters of the Teutonic Order, Werner von Orseln, Rudolf Konig von Wattzau and Henryk von Plauen. Excavations to locate their burials have been carried out for years (Pawłowski 2007). Although the rules of their Order forbade such luxury, these rules had been created for the lower ranks and not for the superiors; therefore, they still could be the burials of the Grand Masters themselves. Another possibility is that it could be the burial place of the Pomeranian bishops, as Kwidzyn was their property and their burials naturally had priority there. However, it is difficult to answer this question at this stage of the enquiry, although it continues to puzzle scholars.

Notes

1   Dendrochronological studies of the coffins were made by Prof. Tomasz Ważny from the University of Nicolaus Copernicus in Toruń. The first coffin was made of wood cut in 1325, the others came from the end of 14th century.
2   Anthropological analysis was made by Dr. Alicja Drozd and Dr. Tomasz Kozlowski from the Anthropology Department of the University of Nicolaus Copernicus in Toruń.

Bibliography

Landau M. (2008) Kod Wielkiego Mistrza, *Wprost* 16, 84–87.
Pastoureau M. (2006) *Średniowieczna gra symboli*. Warszawa.
Pawłowski A. (2007) *Sprawozdanie z badań w prezbiterium Katedry w Kwidzynie*. Wojewódzki Oddział Służb Ochrony Zabytków. Gdańsk.

17 Norwegian Peat Bog Textiles: Tegle and Helgeland Revisited

by Sunniva Wilberg Halvorsen

Migration Period graves in Norway have left us with a large collection of textiles especially famed for beautiful and well preserved tablet weaves. Less well known, and far fewer, are textile finds from peat bogs. This article presents a new investigation of the two peat bog textile finds from the Migration Period: Tegle and Helgeland.[1]

Tegle and Helgeland are both located in Rogaland, in south-western Norway. Rogaland was heavily populated in the Migration Period, AD 400–575 (Myhre 1983, 160). Both finds are located in the vicinity of Early Iron Age farms. The Tegle farm is situated in the Jæren area, a very flat and agriculturally rich region. Tegle lies by a lake, which stretches through a landscape of several wealthy Early Iron Age finds. Close by is a place called Frøyland, or Freya's land. This name appears in the vicinity of Helgeland, too. Additionally, the name Helgeland might by interpreted as 'holy land'. There is quite a distance between the two finds. The Helgeland bog is situated in the bottom of a fjord, at a highland plateau above Helgeland, the innermost farm in the valley.

Both finds consist only of textiles and fibres, and seem to have been intentionally deposited. The nature of the textiles in the find is very diverse. Both were found during peat cutting in the interwar period. The finds have been discussed by earlier scholars, primarily from a technical point of view (Dedekam 1924; von Walterstorff 1928; Hoffmann 1964). The new investigation was aimed at taking a more holistic view of the finds, considering the material from both a technical and a contextual angle. Some of the results of these new investigations are presented here.

Tegle

The Tegle find was discovered in 1921, and published by Hans Dedekam in 1924. The find consists of a cloth bag, with several different artefacts inside: a warp, a sprang tube, a fringed band, a fine twill textile fragment, unspun wool, yarn, threads of twisted hair, and a bone needle which is now lost (Figs 17.1–17.3). It is ^{14}C dated to AD 445–545 (Halvorsen forthcoming).

Best known is the warp, which is made with a tablet-woven starting border (Fig. 17.2). The weft of the tablet weave was drawn out in loops, creating the warp threads. The same technique was applied when producing the 206 cm long band with a fringe on one side (Fig. 17.3 top). Both the twill fragment and the sprang tube have different variations of functional tablet-woven borders. The warp and the tablet-woven boarder have provided important insight into understanding of warp-weighted loom. In 1928, Emilie von Walterstorff demonstrated how the find could illuminate Iron Age weaving technology (von Walterstorff 1928). Marta Hoffman developed this research further and, in 1959, she presented a new technical analysis of the Tegle find (Hoffman and Trætteberg 1959). She later applied this research to a collection of modern warp-weighted looms from the Nordic countries. This resulted in her seminal PhD thesis: *The warp-weighted loom: Studies in the history and technology of an ancient implement* (Hoffmann 1964). Since then, the Tegle find has been, and still is, frequently referred to in publications on prehistoric weaving.

As the find is important for understanding weaving technology, it has been much discussed by scholars. The aim of the present study was to focus on other aspects of the find, such as the tube made in sprang technique (Fig. 17.3 right). It has often, and without much discussion, been referred to as a stocking. Margrethe Hald also suggested it might be a loose sleeve (Hoffmann and Trætteberg 1959, 45; Hald 1962, 19). I wished to gain a closer understanding of its function. Sprang is a very flexible technique, but tablet weaving is not. As the sprang is 40 cm long, and the tablet-woven edges are quite narrow, 26 cm in circumference, this would indicate whom it was made for and how the sprang tube would fit. The question was turned into an experiment, with the help of the Historical-Archaeological Experimental Centre, Lejre in Denmark. A number of people of varying sizes, ages and both sexes tried on a reconstituted version of the sprang piece.

It turned out that very few adults could fit it over the ankle, but several women could use it as a loose sleeve. It was the opposite for the children, who could easily use it as a legging, but not as a sleeve. The best fit was for pre-adolescent children around 10 years old. The result might not be directly

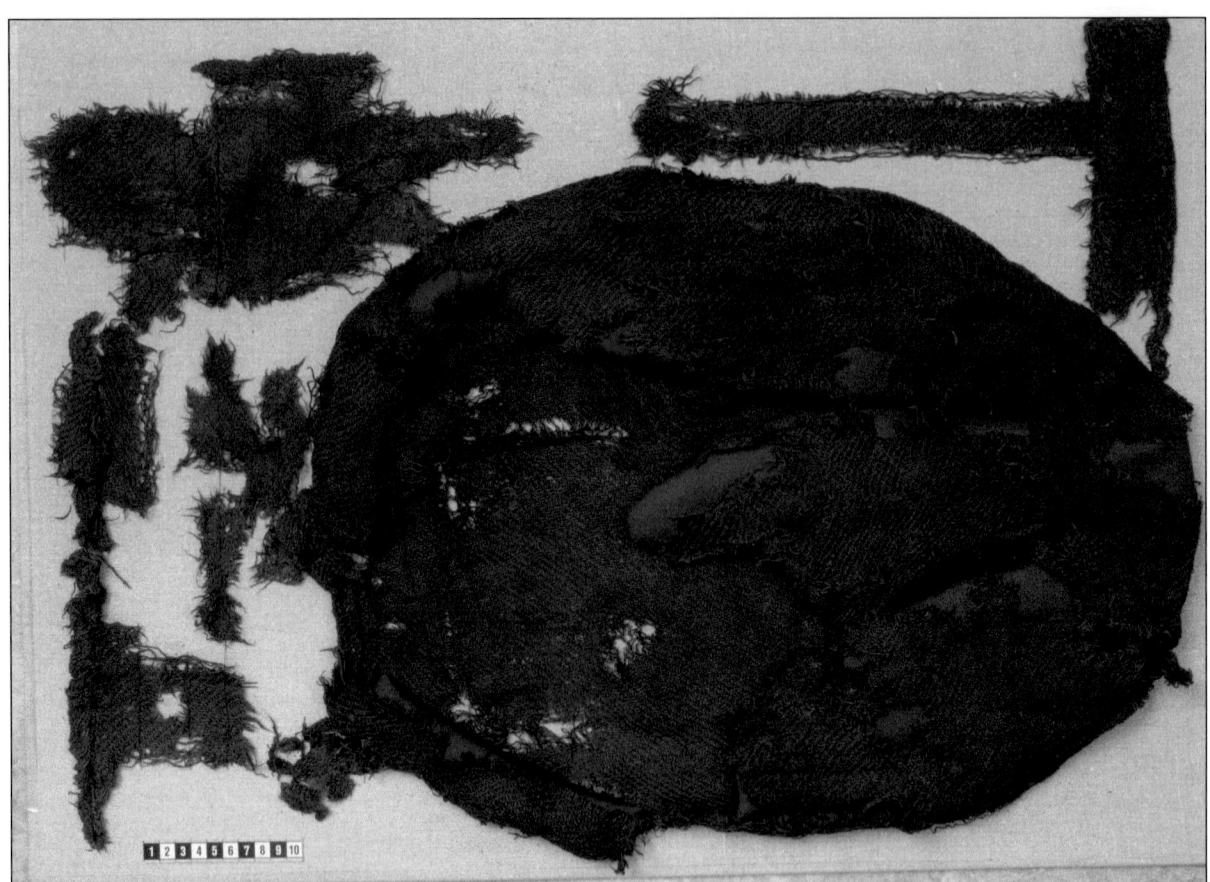

Fig. 17.1. The bag from the Tegle find and some of its contents: the bag (© Arkeologisk Museum i Stavanger).

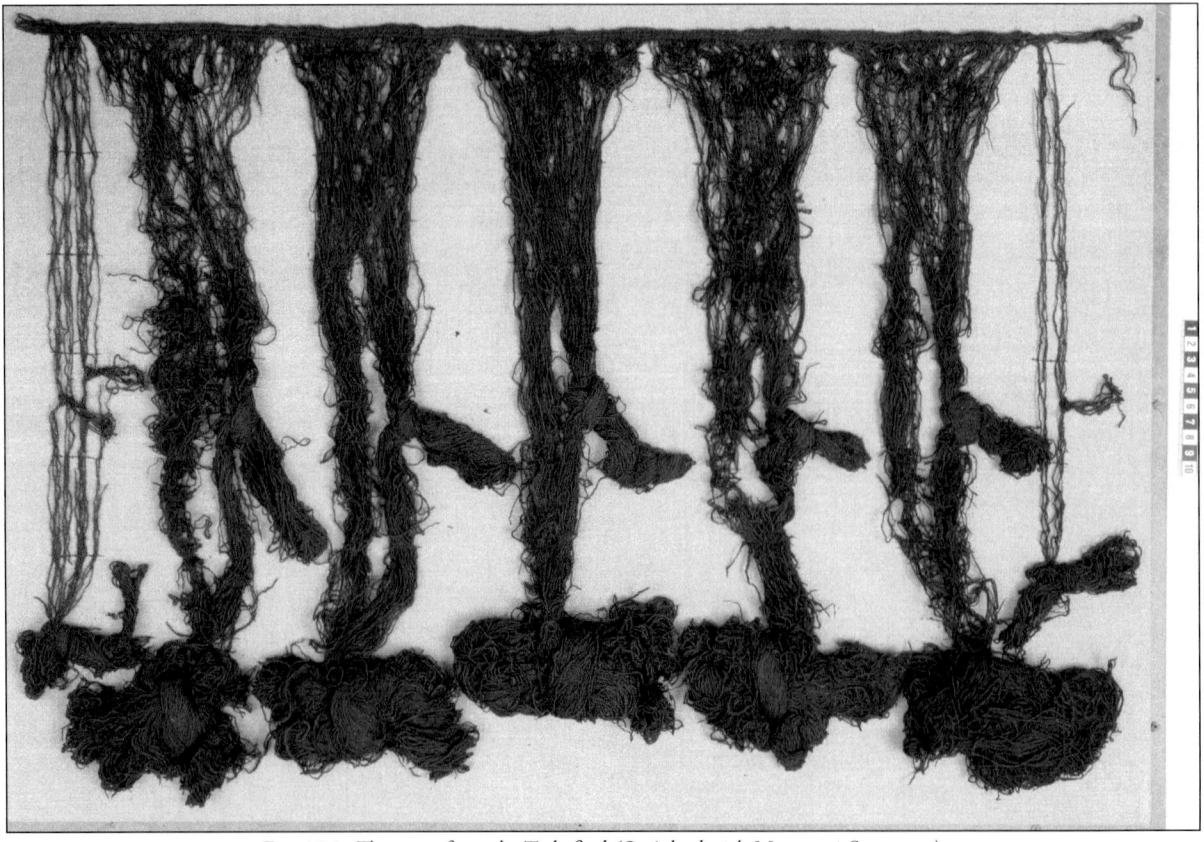

Fig. 17.2. The warp from the Tegle find (© Arkeologisk Museum i Stavanger).

Fig. 17.3. The Tegle find contents: top left – fringe; bottom left – twill fragment; right – sprang tube (© Arkeologisk Museum i Stavanger).

applicable to people of the Iron Age, as different lifestyles would have affected bone and muscle size. However, other finds indicate that the conclusion is valid. A sprang item has probably been found at the feet of a girl in a Roman period grave in Fallward, in northern Germany (Kadereit 2005, 15). From Roman York comes another sprang tube, 51 cm long and 38 cm around the opening, which is probably more suitable for a grown up (Henshall 1951). My conclusion is that the Tegle sprang item was probably a legging, used by a relatively young person. It is also worth noting that Migration period costumes seem to have had long sleeves,[2] although there can of course be variations according to age, status, workload and weather.

Along with the textiles, the Tegle find contained different qualities of yarn, as well as unspun wool. A few twisted threads identified earlier as woollen, seem to be made of hair, possibly human. It is interesting to note how much the quality and thickness of the yarn varies. There is a light, plied yarn which is smooth and even, but the rest of the yarns are quite coarse and unevenly spun. Compared to the fine quality of the threads in the fringe and the twill fragment, this might indicate that the different textiles and yarns have been made by more than one person. I suggest that the yarn might have been spun by a somewhat unskilled person, possibly the owner of the sprang tube. One possible explanation for the varying yarn quality is that the sprang tube could have belonged to a child, who made some of the yarn. A further plausible explanation for the qualitative variations is low wool quality.

The bag itself has been somewhat neglected, since in previous studies the objects it contained were of more interest. Although it is coarsely made, it is still a very interesting piece, being the only known Iron Age bag in Norway. It is sewn together with large stitches and gives a somewhat sloppy impression. The varying size of the pieces adds to this impression. Random stitching suggests that it was also repaired. A patched up part at the back indicates that the bag was made of a secondary-use textile, but the few loose pieces that exist today were torn off when it was discovered. The drawing illustrates how they probably fit together, according to twill-lines, seam types and edges (Fig. 17.4).

The fabric of the bag is a 2/2 twill with 8 z-spun threads per cm in both directions. The bottom is elliptical, 12 × 36 cm. The bag measures 52 cm lengthwise. One side is made of an oval piece of fabric; the other side is made of three pieces of different dimensions. The bottom and side seams and the long seam in front are sewn in a thick thread and long running stitches. The loose piece in the left corner of the figure 17.1 was sewn with smaller stitches, but with the reverse side of the seam outwards, showing the raw edges of the fabric. The bottom and sides have no edging, but the longest front seam does. The opening is only partly edged and lined. One of the loose fragments from the bag has a woven tubular selvedge. The occurrence of different kinds of edging in the woven fabrics of a closed find like this might indicate that choice of edging technique was based on the intended use of the fabric, rather than on weaving traditions.

The find from Tegle was probably intentionally deposited in the bog. The find consists of a collection of new and old objects. The sprang item has been worn and repaired. The fringed band has been attached to something, and it still has pieces of sewing thread. The twill item was probably worn out before it was placed in the bag. These three objects are also of fine quality. The fringe and the twill fragment even have traces of dyes (Vanden Berghe *et al.* 2009). The new and unfinished objects, the warp and the yarns, are of a far lower and more varied quality. It seems to be a selection of personal items, as both the clothing and the unfinished items must have been closely connected to the person who owned and made them. The unfinished textiles were probably intimately connected to the craftsperson, as that person's skills would be reflected in the products.

The Tegle find is famed for its tablet-woven warp. The warp certainly testifies to a specific handicraft process, but also to the craftsperson. The yarn bears the fingerprint of the spinner. As all the objects inside the bag would need to be transformed to gain economic value, I suggest that it is the social and personal value from the items' past lives that imparted the material real value. I would suggest that the different artefacts inside the bag have ascribed value and meaning for the people, places and social situations connected to them.

Helgeland

The textiles from the bog at Helgeland are less famous, but still of great interest. Found during peat cutting, and partly excavated by an archaeologist, textiles were unearthed between 1929 and 1932. Bjørn Hougen published the find in 1933, and discussed it in more detail in 1935 (Hougen 1933; 1935, 83). Textiles were found in a 20 × 1 m wide area of the Helgeland bog. Several earlier reports on textile finds in the bog indicate that the bog contained quite a few textiles, spread out over a vast area. Except for two glass beads, only textiles and fibres have been found in the bog. The find contained fragments of a broad tablet-woven band, several fragments of woven textiles, a little piece of unspun wool, some thick wool threads and a bundle of hair. There were no reports of human remains other than a lump of human hair wrapped with twisted braids. One of the now lost finds was described as a wrapped up bundle of fabric, fastened with a metal pin. The fact that the textiles were wrapped might indicate that they were deposited without any connection to human remains. Braids cut from human hair have been found in bogs in Norway, Denmark and Holland (Brøndsted 1958, 277; Ingstad 1961, 33; van der Sanden 1999, 219). The vast area of textile finds, the description of the lost find and the lack of human remains make it difficult to interpret Helgeland as being connected to burials. It rather indicates that textiles were deposited in the bog on their own. The find is ^{14}C dated to AD 425–535 and the tablet-woven band and the woven twill fragment D differ only by five years (Halvorsen forthcoming). Although ^{14}C provides the date when the wool was taken from the sheep, it may indicate that the textiles were deposited within a rather short span of time.

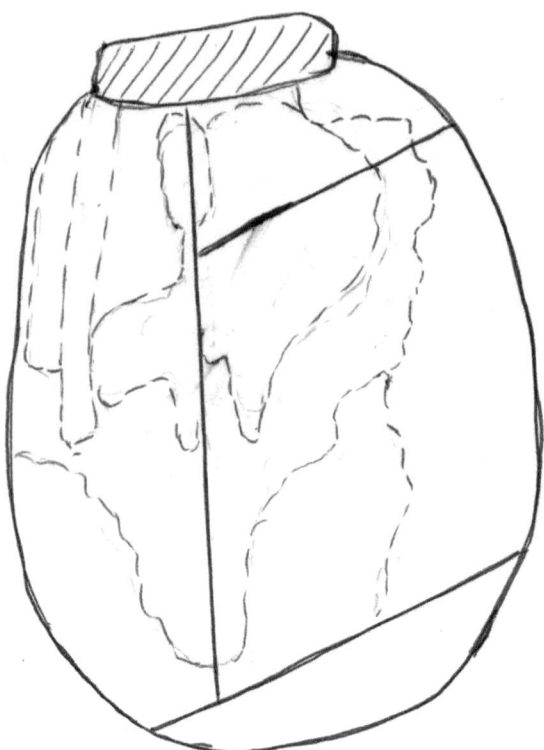

Fig. 17.4. A sketch of how the Tegle bag was stitched together. The fully drawn lines indicate seams, while punctured lines indicate the preserved fragments (Drawing: S. Halvorsen).

The woven fragments seem to have come from different items of clothing. As the textiles were primarily unearthed by peat diggers, the relationship between the different fragments is unknown. The fragments were grouped A–S according to the way they were found in the same lumps of peat, but these groups seem to contain fragments of different qualities. The woven textiles all fall within the overall quality of Migration Period textiles. They are 2/2 twills, with 8 to 16 threads per cm, most often 12 to 14 threads per cm. Fragment C is proved to have traces of the blue dyestuff indigotin probably from the plant woad (Vanden Berghe *et al.* 2009).

Most of the fragments are rather small, and it is not known whether textile variations are due to disturbances in the bog, or to real differences in quality. This makes it difficult to determine how many different textiles there were originally. Based on a comparison of the fabric qualities and variations in sewing techniques, I find that there must be at least seven different textiles. At least three of these must be from clothing, as they have a neck opening and two different sets of gores (Fig. 17.5).[3] Additionally, there are a few fragments with sewn edges, and two with tubular selvedges. The new investigation of the woven fragments also made it clear that one small fragment (P), interpreted by Hougen (1933) as a weft-faced fabric, is instead a 3 cm broad warp-faced band.

The Helgeland bog also contained what is probably the best preserved tablet weave from the Early Iron Age in Norway (Fig. 17.6). The band stands alone in the Norwegian Migration Period, both in terms of its dimensions and technique. Migration Period bands are primarily patterned either by

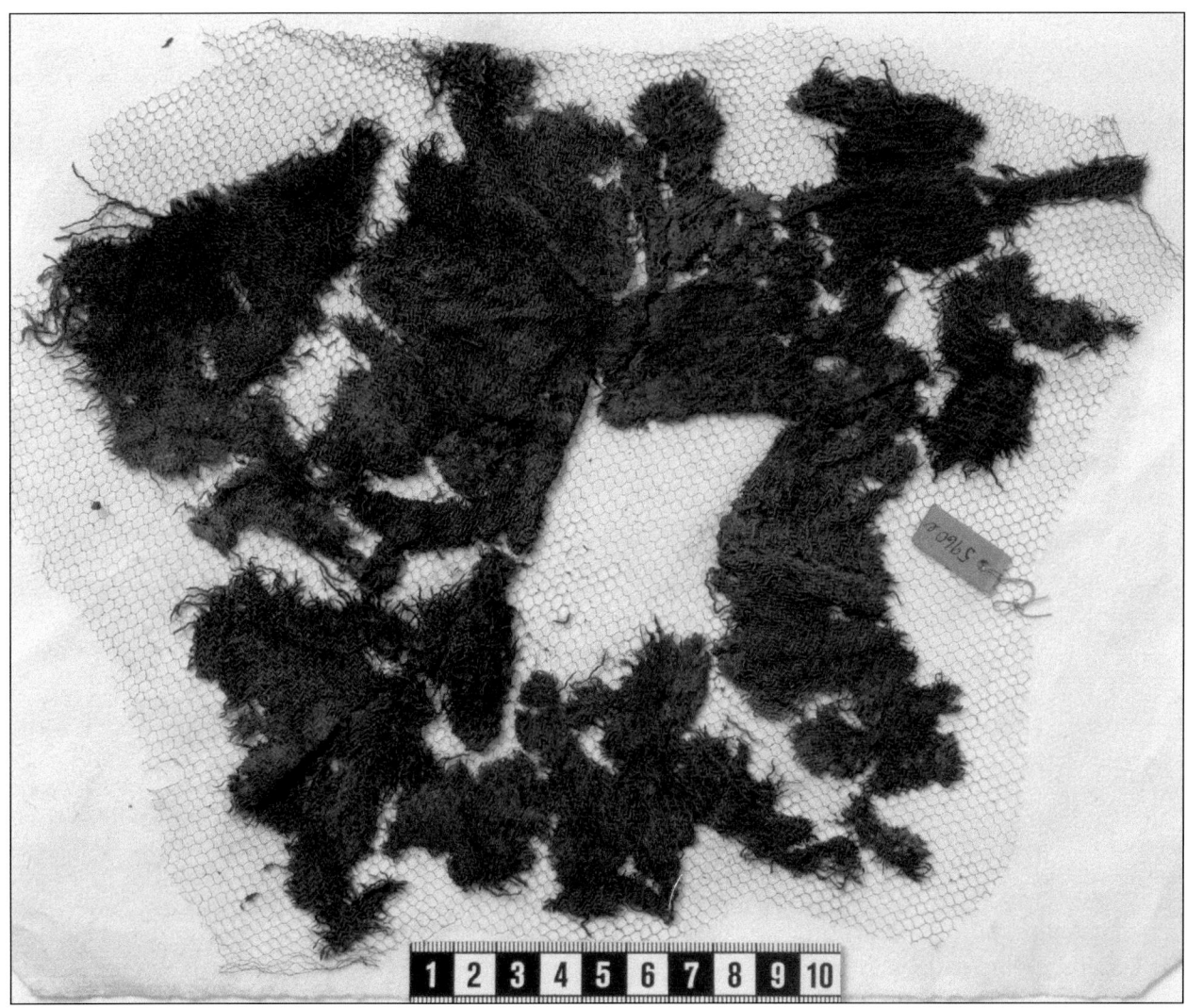

Fig. 17.5. One of the woven fragments (E) with inserted gores, from the Helgeland find. (© Arkeologisk Museum i Stavanger).

individually turning of the tablets, or by adding a decorative pattern weft of horsehair in a tapestry-like technique (Nockert 1991). The Helgeland band is made completely of wool; it is 10–11.5 cm wide, with a maximum of 81 tablets. Stitches indicate that one edge has been attached to another object. The band is patterned with a yellow pattern weft, in 9–14 cm wide rectangles, with plain patterned rectangles of 2–3 cm in between. The ground weave is simple warp-twisting technique, with paired tablets in a fishbone pattern. The ground weave looks purple-reddish, though only yellow dye components have been detected in these threads (Vanden Berghe *et al.* 2009). An impressive number of rather large fragments are preserved. The preserved fragments indicate a band at least 250 cm long. This measurement is based on fitting all the fragments into the smallest possible length, based on the sizes of the continuous patterned and blank rectangles, together with the twisting direction of tablets and occurrence of weaving mistakes (Fig. 17.6).

When the band was first analyzed, Bjørn Hougen divided the fragments into groups A and B, describing them as possibly two bands (Hougen 1933, 60–64). However, the same weaving mistake, that of tablets twisting the wrong way in certain places, is repeated in both band groups on the same tablets. I therefore find that these fragments are most likely from one band, or at least one warp. The most notable difference between A and B groups is in the preservation. The fragments grouped as B are more disintegrated.

The first publication of the Helgeland find left some doubt as to how the band had been constructed (Hougen 1933). A few scholars have commented on it (Geijer 1972, 273; Nockert 1991, 83; Ræder Knudsen 1996, 37; Collingwood 1996, 250), but it had not been further investigated or discussed. Part of my project was to understand which techniques had been applied to make the band. The approach was to combine a new analysis of the fragments with test weaving of the possibilities that had been proposed earlier. Bjørn Hougen (1933; 1935) proposed that the bands were either brocaded or embroidered (1933, 61–62). When Hougen's notes were published posthumously in the volume on the Oseberg textiles (Christensen and Nockert 2006, 129), it became clear that he had changed his view on the technique in the Helgeland band, considering it to be brocaded soumak.

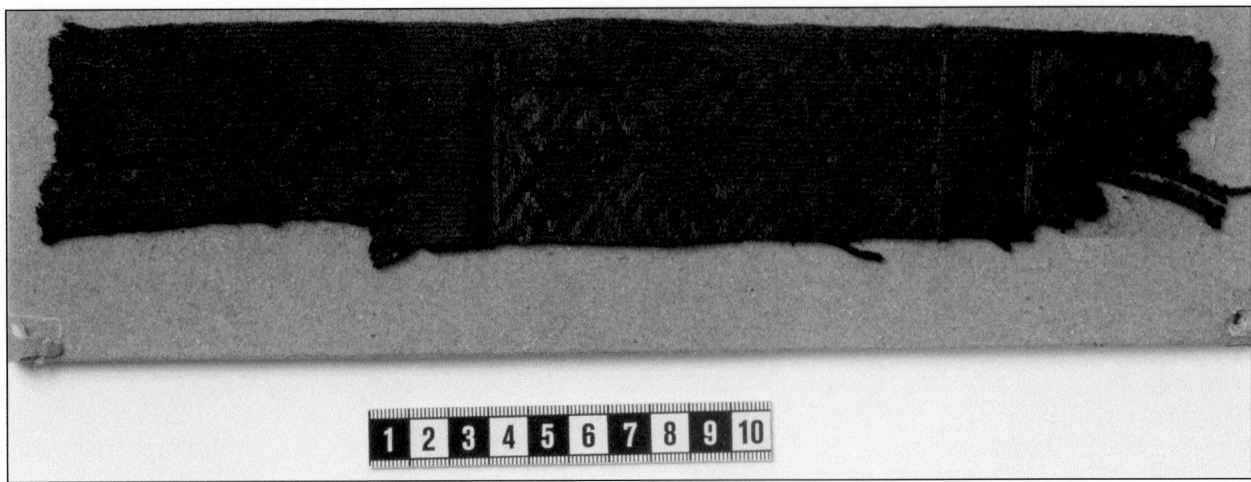

Fig. 17.6. One fragment (Af) from the broad tablet weave from the Helgeland find (© Arkeologisk Museum i Stavanger).

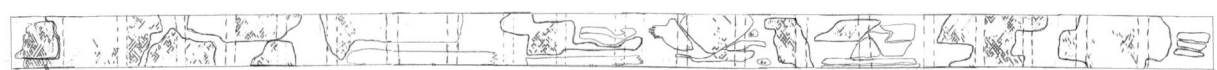

Fig. 17.7. An illustration of how all the fragments in the Helgeland band might fit together, to create the shortest possible band. Each patterned area is approximately 10 cm long and each blank area is about 2 cm long (Drawing: S. Halvorsen).

The new analyses included drawings of each pattern weft thread in each fragment, with variations such as weaving mistakes, stitches and number of tablets. Registering how the pattern thread behaves on the back of the band turned out to be an important key to understanding the technique. Unfortunately, the pattern thread is hardly ever preserved either in the front or in the back of the band. This is clearly due to decomposition, not the technique. There are a few places where both sides are completely preserved. The pattern is somewhat the same on both sides, but is far more irregular on the reverse, with the pattern thread in different angles and even crossing over itself.

Test weaves of the different proposed techniques were compared to the new analyses of the band. Bjørn Hougen's first idea of ordinary brocading (Hougen 1933, 61), where the pattern weft follows the ground weave when not visible on the front side, had to be discarded because the pattern weft does not show on the reverse with this technique. His second idea of brocading (Hougen 1933, 62), where the pattern thread floats on the back when not patterning the front, and thus forms a negative of the front side pattern, was also discarded. The back of the original band does not show a negative of the front pattern. Hougen (1933, 62) also thought there were indications of the pattern having been embroidered onto the band after it was woven. Although this might be a possible way of doing it, it would be far more time-consuming than weaving a pattern. The pattern threads run parallel with the ground weft, and this is hard to achieve by inserting it after the band is woven. Furthermore, as I have observed that the pattern thread runs in the shed in some places, the theory of it being embroidered does not hold true.

Hougen's notes on brocading with soumak (Christensen and Nockert 2006, 129) indicate a technique where the pattern thread passes over the front threads, then backwards beneath them, then up and forwards again. It has a lot in common with embroidery, but is done while weaving. The pattern thread in the Helgeland band does clearly turn backwards, but there are flaws with this technique as well. It can only be used to cover whole areas, if the thread is not put in the shed between the blank spaces in the pattern. The band has a pattern thread that lies in the shed in a few places, but it does not seem to be the rule everywhere. I made two more tests, where I used several discontinuous pattern threads to build the different parts of the pattern. Merely weaving with discontinuous pattern wefts was not successful either, as the pattern thread did not lie in the shed, and the reverse side seemed too perfect. The last test was done by letting several discontinuous threads weave soumak, sometimes floating in the back, and sometimes lying in the shed. This seems to be quite close to the original. The band thus seems to be brocaded with soumak, with several discontinuous pattern threads following different parts of the pattern. Where the pattern thread needs to move between areas in the pattern, it is put either in the shed or floats at the back, depending on what is most convenient and how wide the leap is. It is thus a combination of different techniques. The band was not made with the thought of having a nice looking reverse side. As I see it, the same pattern could have been accomplished in a far simpler and faster way. The reasons for using this more time-consuming technique may be due to some colour variations that are not visible anymore. It can also be rooted in the social framework of the handicraft, or how the craftsperson used to think of pattern weaving.

Conclusions

The textile finds from the Helgeland bog were scattered over a wide area. We do not know to what extent textiles have been discarded by earlier peat diggers, or how the preserved textiles relate to one another. It is not known whether there were one or several instances when textiles were deposited. As in the case of all bog finds, there are too many unknown variables. These factors make it difficult to draw conclusions on the prehistoric setting that led to the textiles being deposited. The Tegle find, being a closed find, provides us with slightly more information. We know the items were deposited together, at the same time. Although bog finds are difficult to interpret, these two finds have given us some important knowledge. Apart from being of great importance to our understanding of textile technology of the Migration Period, these two finds are significant for what they can tell us of textiles as social objects. The fact that they were deposited, probably due to a mixture of their economic, biographic and symbolic value, tells us that textiles and textile production were meant to be more than merely keeping warm and looking good. These are textiles that have meant something and had their own stories. The textiles have probably been deposited in the bogs due to the importance ascribed to them.

Notes

1. This paper is based on an MA thesis in archaeology written at the University of Bergen in 2007: *Myrfunn av tekstiler – en ny undersøkelse av funnene fra Tegle og Helgeland.*
2. There are several finds of wrist clasps in female graves from the Migration Period. One example comes from a female grave at Døsen, Hordaland: B6090, grave 2.
3. The fragments of S5960D, E and K seem to belong to a gored piece of clothing, as S5960E has a gore. S5960D was ^{14}C dated to AD 530-535 (Halvorsen forthcoming).

Bibliography

Brøndsted, J. (1958) *Danmarks Oldtid. Bind 2: Bronzealderen.* København, Gyldendal.

Christensen, A. E., and Nockert, M. eds, (2006) *Tekstilene.* Osebergfunnet, bind IV. Oslo, Universitetets Oldsaksamling.

Collingwood, P. (1996) *The Techniques of tablet weaving.* Arkansas, Robin & Russ.

Dedekam, H. (1924) Et tekstilfund i myr fra romersk jernalder. *Stavanger Museums Aarshefte* 1921–24, 3–29, 1–57.

Geijer, A. (1972) *Ur textilkonstens historia.* Stockholm, Gidlunds.

Hald, M. (1962) *Jernalderens Dragt.* København, Nationalmuseet.

Halvorsen, S. (Forthcoming) Dates and Dyes – New test-results for the finds from Tegle and Helgeland, Norway. *ATN* 49.

Henshall, A. (1951) Note on an early stocking in "sprang" technique found near Micklegate bar, York. *Annual Report & Transactions of the Yorkshire philosophical society* 1950, 22–24.

Hoffmann, M., and Trætteberg, R. (1959) Teglefunnet. *Stavanger museums årbok* 1959, 41–60.

Hoffmann, M. (1964) *The warp-weighted loom: Studies in the history and technology of an ancient implement.* Oslo, Universitetsforlaget.

Hougen, B. (1933) Helgelandsfunnet. Et myrfunn av tekstiler fra eldre jernalder. *Stavanger Museums Årshefte* 1930–1932, 55–75.

Hougen, B. (1935) *Snartemofunnene: Studier i folkevandringstidens ornamentikk og tekstilhisorie.* Norske oldfunn, bind 7. Oslo, Kulturhistorisk museum, Universitetet i Oslo.

Ingstad, A. S. (1961) Votivfunnene fra nordisk bronsealder. *Viking* 25, 23–50.

Kadereit, F. (2005) Das Madchengrab der Fallward: vorläufiger bericht. In F. Pritchard and J. P. Wild (eds), *NESATVII, Northern Archaeological Textiles, Textile Symposium in Edinburgh, 5th–7th, May 1999,* 12–16. Oxford, Oxbow Books.

Ræder Knudsen, L. (1996) *Analyse og rekonstruktion af brikvævning.* Unpublished thesis. Konservatorskolen.

Myhre, B. (1983) Beregning av folketall på Jæren i yngre romertid og folkevandringstid. In *Hus, gård och bebyggelse. Föredrag från det XVI nordiska arkeologmötet, Island 1982,* 147–164.

Nockert, M. (1991) *The Högom find and other Migration Period textiles and Costumes in Scandinavia.* Archaeology and environment 9, Högom part II. Umeå, Umeå universitet.

Sanden, W. van der (1999) Wetland archaeology in the province of Drenthe, the Netherlands. In B. Coles (ed.), *Bog bodies, sacred sites and wetland archaeology,* 199–222. Exeter, WARP.

Vanden Berghe, I., Gleba, M., and Mannering, U. (2009) Towards the identification of dyestuffs in Early Iron Age Scandinavia peat bog textiles. *Journal of Archaeological Science.*

Walterstorff, E. von (1928) En vevstol och en varpa. *Fataburen* 1928, 143–159.

18 Smooth and Cool, or Warm and Soft: Investigating the Properties of Cloth in Prehistory

by Susanna Harris

'Studies of materiality cannot simply focus upon the characteristics of objects but must engage in the dialectic of people and things' (Meskell 2005, 4).

A number of researchers have looked at the significance of the properties of cloth to understand their suitability to environment and function (*e.g.* Rast 1990, 125; Barber 1991, 15; Rast-Eicher 1997, 303). This research is a good basis and has potential to be developed further. In this paper I investigate the physical, chemical and aesthetic properties of linen, wool and lime bast fibres, and the structure of knotless netting, woven textiles and twining that were used to make cloth from the Neolithic to Bronze Age in the Alpine region of Europe. Through these results I look at examples of how these cloth types may have been used and valued in these societies.

Properties and Materials

The original idea for this research came from a conversation with a social anthropologist. She pointed out that while archaeologists are excellent at dealing with the technology and production of cloth, they are not as good at dealing with cloth as a material, and the social importance of materials in terms of *materiality* (*e.g.* Küchler 2003; Küchler and Were 2005). This investigation of materials poses particular problems for archaeologists examining cloth in prehistoric societies. Usually, preserved fragments of cloth are fragmentary, fragile and decayed and do not retain their original properties. To overcome this problem and understand the properties of these materials, archaeologists need to analyse the preserved fragments, and compare the results with modern examples. The analysis of preserved fragments is currently carried out to a high standard at many sites following standard cataloguing systems (*e.g.* Walton and Eastwood 1988; Bazzanella *et al.* 2003). The results from these analyses are highly suited for identifying the properties of the archaeological materials.

However, the identification of properties is only one part of a materials analysis; the relationship between people and materials or *materiality* is equally significant (Meskell 2005, 4). Through everyday encounters, people associate ideas with the surfaces and structures of materials such as cloth in complex and subtle ways (Küchler and Were 2005, 198). This occurs through a combination of factors including the performance of cloth based on its properties, and the way people interact with it, and associate meaning with this relationship. To take an everyday example, doctors around the world wear a white coat. The colour of this garment is a selected property, as white is believed to show up dirt and is associated with hygiene and cleanliness. However, the actual significance of this material is more than this. Through a combination of the colour and cloth type, the shape of the garment and the context in which it is worn, the doctor's white jacket is imbued with beliefs about the wearer's ability to heal the sick. In this example, a material is deemed appropriate due to some of its properties, but takes on meaning that is more than a sum of these. While a *materiality* approach to materials is arguably more difficult to research in prehistoric archaeology than social anthropology, it is necessary to ensure that an investigation of materials does not limit itself to investigating properties. Therefore, a materials approach should see these materials as surfaces that people engaged with as socially understood materials.

Archaeological Evidence of Cloth from the Neolithic to Bronze Age in the Alpine Region

The majority of preserved cloth fragments in the Alpine region are made of plant fibres and come from the waterlogged contexts of lake dwellings, and belong to sites dating from the early 4th to mid-2nd millennium BC. Excavation reports of these sites identify a rich variety of cloth constructions including twined cloth, woven textiles, knotted netting, knotless netting and woven basketry; the raw materials used were often tree bast and flax plus unmodified fibres from grasses and rushes (Winiger 1981, 57–64, 148–171; Rast-Eicher 1997, 302–310; Körber-Grohne and Feldtkeller 1998). Other important sources of preserved cloth include the frozen Iceman dating to the late Neolithic/Copper Age (Egg 1992, 35–100) and the mainly wool woven textiles from the Middle to Late Bronze Age galleries of the Hallstatt salt mines, Austria (Grömer 2005).

18 Smooth and Cool, or Warm and Soft: Investigating the Properties of Cloth in Prehistory

Properties	What is this?
Abrasion resistance	Resistance to flexing, compression, twisting, rubbing; variables including type of abrasion, pressure, speed, tension
Air permeability	The readiness with which air can pass through the cloth
Dimensional stability	The extent a fabric retains its original dimensions subsequent to manufacture
Drape	The way a cloth hangs under its own weight
Elastic recovery	Force applied to extend below the breaking point and then allowed to recover
Elongation	Force applied so it extends and eventually breaks
Fibre fineness	Mass per unit length of fibre
Flammability	Behaviour when in contact with a flame
Handle	Subjective properties assessed by touch and feel such as smooth, rough, limp, stiff, drape
Insulation	Heat loss by conduction and convection
Lustre	Reflection of light
Prickle	Caused by coarse and stiff fibres protruding from the surface
Regain	Weight of water in a material expressed as a percentage
Resistance to biological attack	Resistance to microorganisms
Tensile strength	Maximum tensile force when extended to breaking point
Tickle	Caused by fabric hairiness
Twist	The number of turns per unit length, direction measured as S or Z
Water absorption	Two determining factors; the speed of water uptake and the quantity
Water repellency	The prevention or delay or water penetration or absorption
Windproofing	Resistance to wind penetration by coating or using a tight weave
Yarn fineness	Weight per unit length of yarn

Table 18.1. List and description of a selection of industrial tests (After Saville 1999; Airoldi 2000, 21–33; Wulfhorst 2001, 9–10).

Flax and tree bast from indigenous lime, oak, willow and elm were important raw materials in the Neolithic (Rast 1995, 149). As a raw material, tree bast is less represented in the Bronze Age (Rast-Eicher 2005, 127; Médard 2005). Although wool is rarely preserved in the lake dwelling, other lines of evidence suggest it was probably an important raw material in the Bronze Age, whereas flax seems to become less important in this period (Rast-Eicher and Reinhard 1998, 285; Schibler 2005, 153; Rast-Eicher 2005, 127–128). In terms of cloth construction techniques, twining was common in the Neolithic with many variations known from the Neolithic lake dwellings of the Alpine region (Vogt 1937, 12–32; Rast-Eicher 1997, 307–308; Cardon 1998, 17–18). Knotted netting and variations of knotless netting are known throughout the Neolithic, Copper Age and Bronze Age but are not known continuously in all areas (Rast-Eicher 1997, 305; Cardon 1998, 17–18). Plain weave was the preferred weave structure in the Neolithic; the earliest appearance of twill weave dating to the Beaker period or early Bronze Age (Rast-Eicher 2005, 124–128).

Sources of Comparative Evidence

The investigation of a materials approach depends on the identification of the archaeological cloth remains to understand the raw materials, thread diameter, thread count, cloth structure and other attributes. Once this information is established, it is then possible to compare the ancient cloth types with modern or historically known cloth types. Textile industry tests that measure the properties of different cloth types are a useful source for archaeologists to compare with ancient cloth types. Such industrial tests measure and investigate an extensive range of physical, chemical and aesthetic properties. Some of the properties tested for are outlined in Table 18.1, with a short description of their meaning.

However, these comparisons should be used with the following reservations in mind. First, hand processing as practised in prehistory may create different effects to modern mechanical processing; for example, industrial tests on sheep wool do not take into account the presence of lanolin on the fibres. Second, some raw materials have changed since prehistory. For example, Neolithic flax stems were only c. 30 cm in length (Körber-Grohne and Feldtkeller 1998, 137) and therefore shorter than modern plants. Similarly, Bronze Age sheep fleece was coarser and hairier than in later periods (Ryder 1969, 500–501). Experimental archaeology and the modern ethnographic or historical accounts of craftspeople are useful for understanding how non-industrial processes affect the properties of cloth and to understand fibres and fabrics that are rarely encountered in the present day, such as knotless netting and lime tree bast (Table 18.1).

Results – the Fibres

Flax Fibres

The properties of flax fibres are outlined in Table 18.2. Cool, crisp and smooth to the touch with its excellent ability to absorb moisture, such as body sweat (Needles 1981, 62; Airoldi 2000, 30–34), the properties of flax fibres show how suitable they are for summer clothing. This summer clothing aspect of linen has come up in interpretations of woven linen (Barber 1991, 14–15). In addition, with a handle that is comfortable close to the skin, woven linen can be used for undergarments as part of a layered costume, suitable for any time of year. However, the fineness or coarseness of linen depends on the quality of the fibres. The short (*tow*) fibres produce coarser

Flax fibres	
Physical properties	Strong
	Good tensile strength
	20% stronger when wet
	Standard regain 12%
	Good heat conductivity
	Good water absorption
	Rigid fibre, creases on bending
	Break under repeated flexing
	Low elongation at break, but fairly elastic at low elongations
	Stable shape and size
	Resists abrasion
	Highly inflammable
Chemical properties	Good resistance to insects and micro organisms
	Only susceptible to mildew in extremely moist conditions
	Slow degradation by sunlight
	Resists acids, bases, chemical bleaches
Aesthetic properties	Dull fibre but becomes more lustrous if beaten (*beetling*)
	Natural colour: white, golden yellow, silver grey
	Accepts dyes, but the application of a mordant improves fastness
Handle	Soft
	Cool
	Crisp
	Smooth

Table 18.2. Properties of flax fibres (After Kornreich 1952, 11–17; Needles 1981, 60–62 and 73; Puliti 1987, 21–22; Airoldi 2000, 12–35).

cloth, which was historically used for sacks, work cloth and towels; the long (*line*) fibres produce a more lustrous, stronger, smoother cloth which was used for fine clothing and bedding (Chandler 1995; Mott and Tomasoni 2000, 15, 206). Besides the significance of linen cloth for clothing, the diverse properties make flax useful in other ways.

Flax fibres were one of the strongest fibres available to people in prehistoric Europe. In addition, its resistance to abrasion and chemical attack was probably useful in cloth for working tools and equipment. Not only strong, but increasing in strength when wet, it was a very suitable fibre for the knotted fishing nets that are excavated from the Neolithic lake dwellings (Körber-Grohne and Feldtkeller 1998, 135–137). Other properties are that it resists decay from mildew and does not loose its shape when wet (Needles 1981, 62; Airoldi 2000, 34–35). The appearance of cloth made from flax is also interesting as although naturally dull, flax fibres become lustrous when they are beaten (*beetling*) or smoothed (Needles 1981, 62; Airoldi 2000, 34). As I understand from experienced weavers, this can occur also through extensive wear. This aesthetic property brings to mind the attention researchers have given to the colour and shiny, luminous surface of metals in the Copper Age (Keates 2002, 111). One negative property of flax fibres is their flammability (Needles 1981, 62). So much so that historically in Britain, the waste from preparing cellulose plant fibres (scutching and breaking debris) was sold as fuel (Evans 1985, 23). Although the burnt layers in the prehistoric lake dwelling settlements cannot be attributed to the presence of flax fibres and linen cloth, their presence shows that it is likely that they contributed to these blazes.

Wool Fibres

The properties of sheep's wool fibres are outlined in Table 18.3. Wool cloth is often associated with winter clothing. This is supported by its excellent insulating properties, warm feel, and ability to absorb nearly 40% of its weight in water (Needles 1981, 88) and still feel dry and warm (Chandler 1995, 205). In contrast to plant fibres, wool has a low to moderate strength, with decreased strength when wet (Needles 1981, 88). An elastic fibre, it is even more elastic when wet, but will return to its normal shape and size except in very humid conditions (Needles 1981, 88). In many contexts, wool's stretch, resistance to flexing and ability to absorb shocks compensates for its lack of strength (Kornreich 1952, 12–14).

These qualities were possibly exploited in the Hallstatt Bronze Age salt mines where coarse rags of fulled wool textiles may have been used to carry the mined salt (Grömer 2005, 20; Reschreiter 2005, 13). As the salt would have been a dry filling, wool's weakness and over-elasticity when wet was probably not important. The salt mines are an interesting context to evaluate, as here many of the textile fragments appear to be reused from clothing, showing how cloth of the same type was valued for different properties depending on the context of use. For example, it probably did not matter that wool is a good insulator or good at taking dyes when reused to make containers. Another compelling reason to have wool in the salt mines rather than linen or other plant fibres could have been its resistance to fire. This may well have been useful in the confined environment lit by burning wooden spills (Barth and Lobisser 2002, 15). As mentioned above, linen is highly flammable, which may be why it is rare in the mines.

The stiffness of wool depends on the fineness of the

	Wool fibres
Physical properties	Low to moderate strength
	Weaker when wet
	Good heat insulator due to low heat conductivity and bulkiness
	Wool degrades and chars on heating
	Burns very slowly even in contact with a flame
	Elastic
	Good stretch and recovery except in very moist conditions
	Standard regain 13-18%
	Highly absorbent: can hold nearly 40% of its weight in water
	Resists repeated flexing
	Absorbs shocks
	Fairly abrasion resistant
	Will felt if agitated in warm water
	Slow drying
	Stiffness will vary according to breed and diameter of individual fibre
Chemical properties	Susceptible to attack by moths
	Quite resistant to mildew
	Resistant to acids
	Vulnerable to bases, even in low dilutions
	Slow degradation and yellowing in contact with sunlight
Aesthetic properties	Readily dyed and good colourfastness
	High to moderate lustre
	Natural colour: white, yellowish, reddish-brown, black
Handle	Warm
	Soft, moderate or rough
	Drapes well

Table 18.3. Properties of sheep wool fibres (After Kornreich 1952, 10–17; Needles 1981, 88–90; Puliti 1987, 11; Airoldi 2000, 12–35; Wulfhorst 2001, 11).

	Lime bast fibres
Physical properties	Stronger than elm or oak bast, particularly if prepared without retting
	47% stronger when wet
	Low water absorption
	Limited swelling when wet
	Lightweight
	Low extensibility
	Low resistance to wear
	Floats on water
	Quick drying*
Chemical properties	Resistant to attack by moths *
	Resistant to decay
Aesthetic properties	Natural colour: light to medium golden brown*
Handle	Retted lime bast is soft

*Table 18.4. Properties of lime bast fibres (After Myking et al. 2005), *observations from own experiments working with lime bast fibres.*

individual fibres, which therefore affect the handle. When spun into thread, coarser fibres can be uncomfortable to the skin, producing what industry calls 'tickle' (hairiness) and 'prickle' (coarseness) (Saville 1999, 232). Before and during the Bronze Age, wool contained a mixture of fine underwool and hairy kemp fibres (Ryder 1969, 500–504; Rast-Eicher 2005, 27) and was therefore hairier, stiffer and coarser than modern specialized fleece.

Tree Bast Fibres

Tree bast is extracted from the inner bark of lime, willow, oak and elm.[1] The species of tree bast fibres have different properties and provide a range of natural colours from nearly white to dark brown (Körber-Grohne and Feldtkeller 1998, 156). However, the information on the properties of these fibres concerns mainly lime as this has been subject to industrial testing.

The properties of lime bast are outlined in Table 18.4. A strong fibre, lime bast is particularly interesting in its reaction to water. It is substantially stronger when wet than dry and is resistant to decay. Lime has a low extensibility, floats and due to its low water absorption does not swell in contact with water (Myking *et al.* 2005, 69–70). From my own experiments, I found that lime bast dries quickly, presumably because of the low water absorption. Undoubtedly, these are good properties for fishing equipment, but would also be a good choice of material for shoes, floor or wall coverings, clothing and containers by people living and working in wet environments, such as the Alpine lake dwellings. Historically,

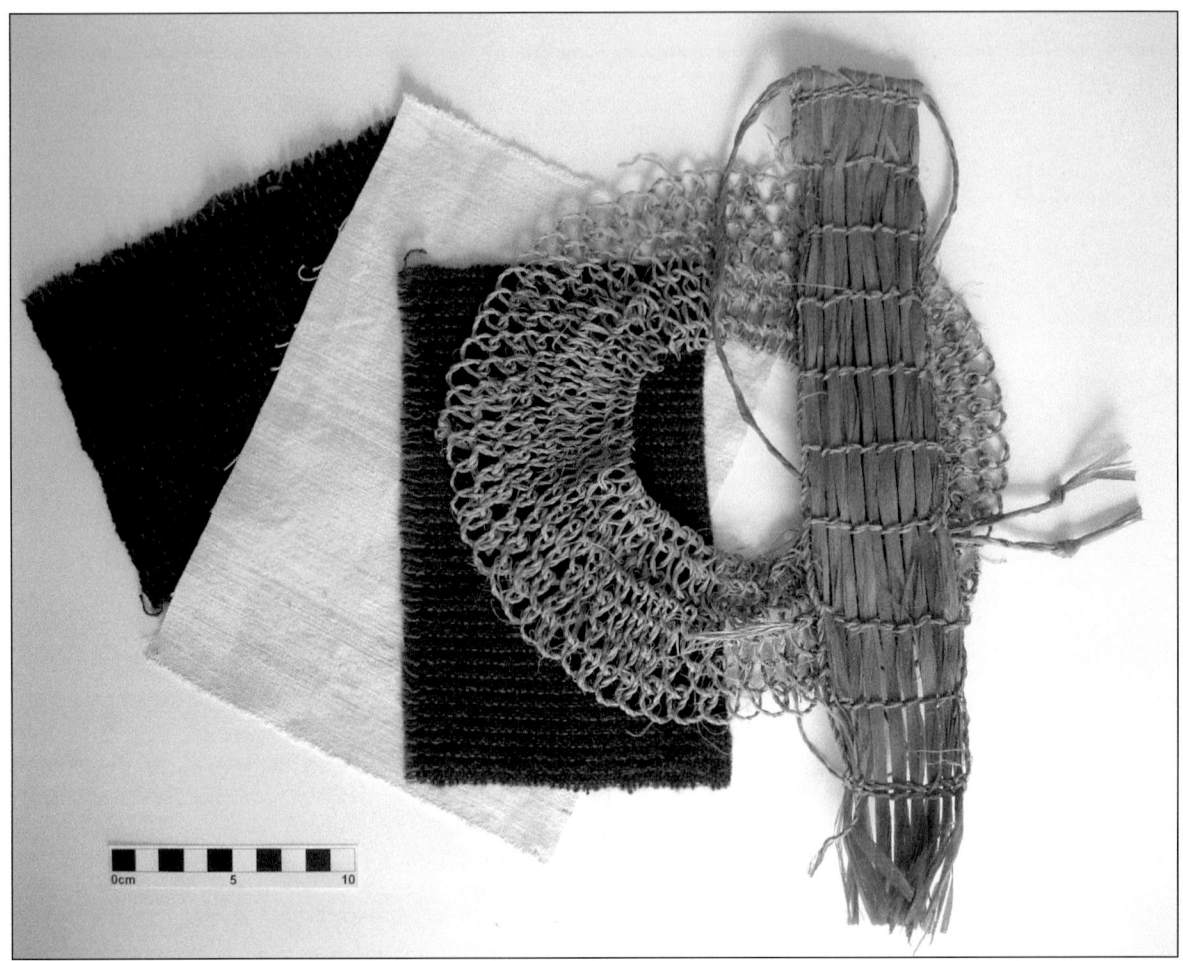

Fig. 18.1. Modern samples of cloth types from left to right: twill weave sheep's wool, plain weave linen, plain weave sheep's wool, open looping with single twist from lime tree bast, twining from lime tree bast (Photo: © S. Harris).

lime bast rope was considered soft to handle in industries where manual work was carried out without gloves (Myking *et al.* 2005, 70). However, this softness depends on the fineness of fibres; the finest fibres lie close to the wood, whereas those extracted from near the bark are noticeably coarser. On the negative side, lime bast is prone to wear, making it less durable than other fibres.

Results – Cloth Structures

The properties of the finished cloth depend on the properties of the fibres, the thickness and spin of the threads (tight or loose, single or plied) and the way they are interworked. With this in mind, each fragment of cloth can be considered for its individual merits based on the technical analysis of the original preserved fragment. In the following section, I look at some general ideas of how different structures (Fig. 18.1) affect the properties of the finished cloth.

Knotless Netting

Looped cloth types, such as knotless netting, are the most flexible and elastic cloth types; the extendibility and firmness depends on the looping method and mesh width (MacKenzie 1991, 128–129; Seiler-Baldinger 1994, 11). Combined with strong fibres and thread, knotless netting provides a structure with no distinct direction of maximum strength; this is unlike textiles where the maximum strength is in the direction of the warp or weft (MacKenzie 1991, 132–133; Saville 1999, 154). Examples from prehistoric Europe are often open looping (*e.g.* Winiger 1981, 190–191, taf. 76.2 and 3). Such open looping is strong, lightweight, flexible, expandable, permeable and see-through. These properties make it suitable for bags carrying heavy loads and stretching round awkward shapes. Some knotless netting archaeological artefacts are interpreted as possible bags (Winiger 1981, 190).

In Papua New Guinea, knotless netting bags (*bilums*) are associated with women and women's labour. As well as expandable, 'strong and capable of hard work', the open looping means that people can see the contents of the bag, which in turn reveals the owner's capacity to contribute to society (MacKenzie 1991, 129–136). Such permeable and see-through properties of open knotless netting are in contrast to dense cloth structures. These properties may be important in the way they can conceal or reveal their contents.

Twined Cloth

In the Neolithic, twining was used to produce a rich variety of cloth types of different thread thickness, warp and weft spacing. Although mainly spun, plaited threads are also employed as the passive element, as are fronds of tree bast or grasses. In some cases, the threads are tightly packed creating a dense structure; in others, they are widely spaced creating gaps in the structure; some are recognised as sieves (Körber-Grohne and Feldtkeller 1998, 144). Some examples are covered with tufts, known as pile (Rast-Eicher 1997, 308). This wealth of variation indicates the skill involved in manipulating the cloth structure to control the properties of the finished product. Here I consider some examples of how twined cloth has been used across the world as a way to understand how structure relates to properties.

With widely spaced warps and wefts, twining can produce an open construction that is lightweight, permeable and see-through. As mentioned above, such structures seem to have been used as sieve bottoms in the Neolithic (Körber-Grohne and Feldtkeller 1998, 144). Open twining was (and is) used for fish traps and large containers in Australia and North America (Aboriginal people of Jumbun 1992, 20–24; Fienup-Riordan 2005, 55–57, fig. 2). By contrast, closely twined warps and wefts produce a dense and solid cloth. In New Zealand, closely twined capes made from plant fibres covered with narrow strips of dog skin were reputed to be strong enough to withstand a 'spear thrust'. On the basis of this property they were worn by warriors and were highly valued (Roth 1923/1979, 50–51, pl. XIX). Twined cloth made from thick threads has the ability to insulate, cushion and absorb shocks. This combination of warm, lightweight and insulating properties is recognised in the interpretation of the large twined grass item found with the Copper Age Iceman, identified as a cape or mat (Spindler 1995, 144–145; Reichert 2006, 9).

Examples of hats, shoes and large pieces that may be used for capes or mats show the use of twining for clothing in the Neolithic (Feldtkeller and Schlichtherle 1987, 78–80). However, the types of garment that could have been produced are more extensive than this. In North America, twined cloth from grasses, tree bast and other plant fibres were (and are) used for garments such as capes, coats, socks, boots, mittens to protect from the cold, as mats to sit and sleep on and as covering to protect fragile pottery (Turner 1998, 32, 68, 109, 145; Fienup-Riordan 2005, 54 58). Twining with a pile surface provides a water resistant surface as the tufts encourage the water to run away (Rast-Eicher 1997, 308). Such tufted surfaces could also provide warmth; the Maori of New Zealand made rain cloaks out of twining with pile, using coarse plant fibres that were described as impervious to rain and also warm (Roth 1923/1979, 46–48).

Aesthetically, twined cloth has a distinctive texture and drape; a stiff structure with poor drape, it falls in flat sheets rather than fine gathers (see Turner 1998, 123; Anawalt 2007, 348, figs 562–564). Archaeologists note the thick, furry appearance of twining with pile and its similarity in appearance to fur (Feldtkeller and Schlichtherle 1987, 78–79; Rast 1995, 150). In terms of visual properties, twined cloth from fine thread is quite distinct from twining with thick threads; it is worth noting that, historically, on the Northwest coast of America, fine close-twined cloth was highly valued and exchanged and worn in the potlatch (Gillow and Sentence 1999, 64; Anawalt 2007, 352).

Woven Textiles

The properties of woven textiles are affected by the fineness of the threads, the number of threads per centimetre (the thread count), the way the threads were spaced on the loom (the set), the weave structure (*e.g.* plain weave or twill), and the post loom processing (the finish). A number of these attributes are recorded in the regular cataloguing of archaeological textiles (Walton and Eastwood 1988). During some periods of the Neolithic, the structure of linen textiles is noticeably uniform (Rast 1995, 149). At other times, there is more variation in thread count, thickness and set (*e.g.* Bazzanella *et al.* 2003, 161–172; Grömer 2005, 28–32). Woven textiles made of fine threads such as the examples of plain weave linen from the lake dwellings are flat and thin and would have draped well. Balanced plain weave drapes well and is good for non-tailored clothing, although by comparison twill will drape better and is more pliable than plain weave (Chandler 1995, 132). A weft or warp faced cloth (*reps*) will be more pliable in one direction than another (Chandler 1995, 120–121, 132). Twill has a slightly more textured surface and is particularly noted for its flexibility, however this is relative; looped cloth types such as knotless netting are more flexible (Chandler 1995, 132).

Although weaving patterns, dyes and finishes such as fringes are the most obvious sources of decoration in prehistoric textiles (Barber 1994, ch. 3), these would not have been the only way that value and meaning was associated with the visual appearance of cloth. The appearance of cloth without decoration known from everyday situations is also a significant visual statement. In this, the smooth, flat, thin properties of woven textiles are distinctive and would have contrasted with the twined or netted cloth structures, although in some cases fine twining appears very similar to weaving (Rast-Eicher 2005, 123).

In many cases, it is assumed that woven textiles were used for clothing. Yet, taking note of historical examples, we should remember that textiles were used as sacks and sheets in agricultural work, for bedding and towels, as cloths for rubbing dishes and floors as well as shirts, skirts and underwear (Mott and Tomasoni 2000, 15). Therefore, when we find fragments of woven textiles, they may have had any number of uses.

Discussion

Through the combination of raw materials, processing methods, thread type, cloth structures and finish, each fragment of archaeological cloth would have had multiple properties. This makes the task of understanding materials complex in several ways. Properties of a material that were important in one context of use, such as colour or absorbency, may have been irrelevant in another. This also makes it difficult to understand which properties were valued and which were of secondary significance. How far were fineness and the ability to conceal important from the Neolithic to Bronze Age in contrast to cloth types that were thick and cushioned or see-through? Neither should we expect that properties were used in optimal ways. Flammable fibres may have been used in pyrotechnical activities and coarse cloth may have been worn close to the skin. In addition, the exploitation of properties can be contradictory, showing how difficult it is to separate cultural beliefs from properties. For example, historically in Britain there are contradictory accounts as to whether light or dark fishing nets were more effective on the basis of their invisibility to fish (Geraint Jenkins 1974, 79). However, through understanding the materials better, we are better able to approach these debates.

By looking at the range of properties of fibres, threads and cloth, it is possible to expand the range of possible uses of the fragments of cloth found in excavation. This expands the potential role of cloth beyond 'textile' research. For example, the potential use of dense twining as armour to protect against piercing and cutting suggests a relationship between cloth and weapons; the resistance of wool to a naked flame suggests its use in pyrotechnical industries, or the strength of linen textiles for sacks, harvesting and food collection.

To the more regularly cited properties such as insulation, strength and thickness, I have added aesthetic properties such as texture, drape, lustre, colour and the ability to conceal or reveal. This is significant in appreciating that even when not specially decorated or dyed, cloth would have been an aspect of visual culture in past societies; something that can be considered the aesthetics of the everyday. In this way, the range of cloth types at any one time would have represented a visual norm in past societies; the characteristic drape of clothes, the texture of cloth covers, the area of the body a cloth was expected to conceal or reveal. The aesthetic of cloth surfaces and structures may also have drawn comparison with other material surfaces. There are some hints towards these relationships, textured pottery surfaces that appear like textiles or other cloth structures, the tufted surface of twining with pile that resembles fur or the lustre of beaten linen textiles and metals. This approach is not new to archaeologists; as mentioned above, the colour and luminosity of metals in the Copper Age is seen as part of their value in addition to the properties of cutting and durability. Such an approach to cloth is also necessary.

In this paper I have approached some of the most common fibres and cloth structures in the Alpine region from Neolithic to Bronze Age; there are more types to examine. Another approach could be to investigate individual fragments in a site context and chart the range of properties held by different cloth types at a particular time and place. It would also be interesting to consider change and continuity in the materiality of cloth from the Neolithic to Bronze Age, alongside change and continuity in the technology of cloth production.

Conclusions

With exceptions, archaeologists have focused on understanding techniques and technology above materials. Yet, the material surfaces and structures of cloth are as much an indication of social values and meaning as any other item of material culture such as housing, pottery and stone tools. A materials approach is therefore worth developing to understand the role of cloth in past societies.

The investigation of a materials approach depends on the accurate analysis of the preserved cloth. Fortunately, the standard cloth cataloguing system offers a ready resource, including the identification of raw materials, thread diameter, thread count, cloth structure and other attributes. These factors can then be compared with modern samples, reports of craftspeople, and experimental archaeology to understand the original properties of cloth, before the decay and degradation resulting from the preservation processes. From this knowledge, it is then necessary to evaluate these materials in the context of the societies they belonged to. This helps understand how cloth types may have been used, and why they were used in particular ways. In addition, as an aspect of visual culture, the aesthetic properties of fibres and cloth structures bring to attention the everyday aesthetic of cloth for clothing, housing and equipment in prehistoric societies.

Acknowledgements

This research was completed as a British Academy Post-doctoral Fellow at the Institute of Archaeology, University College London. Thanks to Professor Susanne Küchler who suggested I turn my research in the direction of the materiality of cloth, Jasper Chalcraft and Salvatore Notaro who read earlier versions of this paper and made useful suggestions. I am grateful to the organisers of NESAT X for allowing me the opportunity to present this paper and to the many participants who discussed these ideas with me at the conference.

Note

1   Although Médard questions whether oak fibres were actually used for textiles, or, if in the fibre analysis they have been confused with elm bast (Médard 2005, 101).

Bibliography

Aboriginal people of Jumbun and Helen Pedley (1992) *Aboriginal life in the rainforest* (compiled and photographed by Helen Pedley). Peninsula Region, Cairns (Queensland), Queensland Department of Education.

Airoldi, G. (2000) *Il tessuto ordito e trama ad intreccio ortogonale.* Lipomo (Como), Centrocot S.p.A.Editore.

Anawalt, P. R. (2007) *The Worldwide History of Dress.* London/New York, Thames and Hudson.

Barber, E. J. W. (1991) *Prehistoric Textiles. The Development of Cloth in the Neolithic and Bronze Age with Special Reference to the Aegean.* Princeton, Princeton University Press.

Barber, E. J. W. (1994) *Women's Work: The First 20 000 years. Women, Cloth and Societies in Early Times.* London & New York, W. W. Norton & Co.

Barth, F. E., and Lobisser, W. (2002) *Das EU-Projekt Archaeolive und* das archäologische Erbe von Hallstatt. Wien, Naturhistorischen Museum in Wien.

Bazzanella, M., Mayr, A., Moser, L., and Rast-Eicher, A. (2003) Schede [Catalogue]. In M. Bazzanella *et al.*, (eds), *Textiles: intrecci e tessuti dalla preistoria europea,* 133–289. Provincia Autonoma di Trento, Servizio Beni Culturali Trento.

Cardon, D. (1998) Neolithic Textiles, Matting and Cordage from Charavines, Lake of Paladru, France. In L. Bender Jørgensen and C. Rinaldo (eds), *Textiles in European Archaeology: Report from the 6th NESAT Symposium, 7–11th May 1996,* GOTARC Series A, Vol.1, 3–21. Göteborg University, Department of Archaeology, Göteborg.

Chandler, D. (1995) *Learning to Weave.* Loveland (Colorado), Interweave Press.

Egg, M. (1992) Die Ausrüstung des Toten. *Jahrbuch des Römisch-Germanischen Zentralmuseums Mainz* 39, 35–100.

Evans, N. (1985) *The East Anglian Linen Industry: Rural Industry and Local Economy 1500–1850.* London, Gower, The Pasold Research Fund.

Feldtkeller, A., and Schlichtherle, H. (1987) Jungsteinzeitliche Kleidungsstücke aus Ufersiedlungen des Bodensees. *Archäologische Nachrichten aus Boden* 38–39, 74–84.

Fienup-Riordan, A. (2005) Tupigat (Twined Things): Yup'ik Grass Clothing, Past and Present. In J. C. H. King, B. Pauksztat, and R. Storrie (eds), *Arctic Clothing of North America – Alaska, Canada, Greenland,* 53–61. London, The British Museum Press.

Geraint Jenkins, J. (1974) *Nets and Coracles.* Newton Abbot and London, David and Charles (Holdings) Ltd.

Gillow, J., and Sentance, B. (1999) *World Textiles. A visual guide to traditional techniques.* London, Thames & Hudson.

Grömer, K. (2005) The Textiles from the prehistoric Salt-mines at Hallstatt (Die Textilien aus dem prähistorischen Salzbergwerk von Hallstatt). In P. Bichler *et al.* (eds) *Hallstatt Textiles: Technical Analysis, Scientific Investigation and Experiments on Iron Age Textiles,* 17–40. Oxford, Archaeopress.

Keates, S. (2002) The Flashing Blade: Copper. Colour and Luminosity in Northern Italian Copper Age Society, In A. Jones and G. MacGregor (eds), *Colouring the Past: The Significance of Colour in Archaeological Research,* 109–125. Oxford, Berg.

Körber-Grohne, U., and Feldtkeller, A. (1998) Pflanzliche Rohmaterialien und Herstellungstechniken der Gewebe, Netze, Geflechte sowie anderer Produkte aus den neolithischen Siedlungen Hornstaad, Wangen, Allensbach und Sipplingen am Bodensee. In *Siedlungsarchäologie im Alpenvorland V,* 131–189. Stuttgart, Konrad Theiss Verlag.

Kornreich, E. (1952) *Introduction to Fibres and Fabrics, their Manufacture and Properties.* London, The National Trade Press Ltd.

Küchler, S. (2003) The Poncho and the Quilt: Material Christianity in the Cook Islands. In C. Colchester (ed.), *Clothing the Pacific, Photography by Glenn Jowitt,* 97–118. Oxford and New York, Berg.

Küchler, S., and Were, G. (2005) *Pacific Pattern.* London, Thames & Hudson.

MacKenzie, M. A. (1991) *Androgynous objects: string bags and gender in central New Guinea.* Chur, Harwood Academic Publishers.

Médard, F. (2005) Les textiles préhistoriques. Anatomie des écorces et analyse des traitements mis en œuvre pour en extraire la matière textile. In P. Della Casa and M. Trachsel (eds), *WES'04. Wetland Economies and Societies, Proceedings of the international conference in Zürich, 10–13 March 2004.* vol. 3, 99–104. Zürich, Publication of the Schweizerisches Landesmuseum Zürich.

Meskell, L. (2005) Introduction: Object Orientations. In L. Meskell (ed.), *Archaeologies of Materiality,* 1–17. Oxford, Blackwell Publishing.

Mott, A., and Tomasoni, R. (2000) *Filo da torcere: le fibre tessili e la loro lavorazione nella tradizione trentina.* Trento, Museo degli Usi e Costumi della Gente Trentina si San Michele all'Adige.

Myking, T., Hertzberg, A., and Skrøppa, T. (2005) History, manufacture and properties of lime bast cordage in Northern Europe. *Forestry*, 78:1, 65–71.

Needles, H. L. (1981) *Handbook of Textile Fibres, Dyes, and Finishes.* New York/London, Garland STPM Press.

Puliti, M. (1987) *Le Fibre Tessili Naturali-Artificiali-Sintetiche: Nozioni Elementari ad Uso degli Istituti Statali d'Arte.* Firenze, Istituto Statale d'arte di Firenze.

Rast, A. (1990) Jungsteinzeitliche Kleidung. In *Die ersten Bauern* 123–126. Zürich, Schweizerisches Landesmuseum Zürich.

Rast, A. (1995) Le vêtement néolithique. In A. Gallay (ed.), *Dans les Alpes, à l'aube du metal. Archéologie et bande dessinée. Ouvrage publié à l'occasion del'exposition: Le Soleil des Morts,* 149–153. Sion, Musée cantonal d'archéologie et bibliothèque municipale.

Rast-Eicher, A. (1997) Die Textilien. In J. Schibler *et al.* (eds), *Ökonomie und Ökologie neolithischer und bronzezeitlicher Ufersiedlungen am Zürichsee: Ergebnisse der Ausgrabungen Mozartstrasse, Kanalisationssanierung Seefeld, AKAD/Pressehaus und Mythenschloss in Zürich,* Band A, Text. 300–328. Zürich, Direktion der Öffentlichen Bauten des Kantons Zürich, Hochbauamt, Abt. Kantonsarchäologie.

Rast-Eicher, A. (2005) Bast before Wool: the first textiles. In P. Bichler *et al.* (eds), *Hallstatt Textiles: Technical Analysis, Scientific Investigation and Experiments on Iron Age Textiles* 117–131. Oxford, Archaeopress.

Rast-Eicher, A., and Reinhard, J. (1998) *Textile et vannerie,* 285–291. Basel, Verlag Schweizerische Gesellschaft für Ur- und Frühgeschichte.

Reichert, A. (2006) Umhang oder Matte? Versuche zur Rekonstruktion des Grasgeflechts des 'Mannes aus dem Eis'. *Waffen- und Kostümkunde. Zeitschrift für Waffen- und Kleidungsgeschichte,* 48:1, 1–16.

Reschreiter, H. (2005) Die prähistorischen Salzbergbaue in Hallstatt und ihre Textilreste (The prehistoric Salt-mines at Hallstatt and its Textile remains). In P. Bichler *et al.* (eds), *Hallstatt Textiles: Technical Analysis, Scientific Investigation and Experiments on Iron Age Textiles,* 11–16. Oxford, Archaeopress.

Roth, H. L. (1923/1979) *The Maori mantle. Reprinted from the original limited edition published by the Bankfield Museum, Halifax, England, together with 'The Maori mantle': a review (1924) by Sir Peter Buck (Te Rangihiroa).* First published 1923. Carlton (Bedford), Ruth Bean.

Ryder, M. L. (1969) Changes in the fleece of sheep following domestication (with a note on the coat of cattle). In P. Ucko and G. W. Dimbleby (eds), *The domestication and exploitation of plants and animals,* 495–521. London, Gerald Duckworth & Co.

Ryder, M. L. (1993) Skin and wool remains from Hallstatt. *Circea* 10:2, 69–78.

Saville, B. P. (1999) *Physical testing of textiles.* Cambridge, Woodhead Publishing in association with The Textile Institute.

Schibler, J. (2005) Bones as the key for reconstructing the environment, nutrition and economy of the lake-dwelling societies. In F. Menotti (ed.), *Living on the Lake in Prehistoric Europe. 150 years of lake-dwelling research,* 144–161. Abingdon, Routledge Taylor and Francis Group.

Seiler-Baldinger, A. (1994) *Textiles: A Classification of Techniques.* Washington (D.C.), Smithsonian Institute Press.

Spindler, K. (1995) *The Man in the Ice; The preserved body of a Neolithic man reveals the secrets of the Stone Age*. London, Phoenix.

Turner, N. J. (1998) *Plant Technology of First Peoples in British Columbia*. Vancouver, University of British Columbia Press in collaboration with the Royal British Columbia Museum.

Vogt, E. (1937) *Geflechte und Gewebe der Steinzeit*. Basel, E. Birkhäuser & Cie.

Walton, P., and Eastwood, G. (1988) *A Brief Guide to the Cataloguing of Archaeological Textiles*. 4th ed. London, Institute of Archaeology.

Winiger, J. (1981) *Feldmeilen Vorderfeld: der Übergang von der Pfyner zur Horgener Kultur*. Frauenfeld and Basel, Verlag Huber & Verlag Schweizerische Gesellschaft für Ur- und Frühgeschichte.

Wulfhorst, B. (2001) *Processi di lavorazione dei prodotti tessili*. Italian Edition. Milan, Tecniche Nuové.

19 Oriental Influences in the Danish Viking Age: Kaftan and Belt with Pouch

by Anne Hedeager Krag

Introduction

This paper focuses on the Danish finds of archaeological objects, which can be related to oriental dress, such as the kaftan, in the Viking Age. These finds are linked to the well-preserved kaftans from this period, which have been found in Moshchevaya Balka on the Silk Road in the northern Caucasian region (Ierusalimskaja 1996, 47).

The kaftan was originally a garment characteristic to Central Asia. Archaeological sources and depictions reveal that the kaftan influenced the costume of the European nobility in the Late Iron Age and the Viking Period, from about AD 550 to 1050, often through Byzantine fashion (Hedeager Krag 2004, 81). The belt and belt-mount were a part of male dress, and were worn over a kaftan. In the Byzantine Empire, the belt was part of the official dress, and functioned as an important diplomatic gift. There was a detailed code that determined the material and quality of manufacture that were suitable for each recipient.

Approximately 10 km west of Kolding in Jutland, Denmark, in a locality named Dollerup, a small fitting was found in a posthole of a farm building (Jensen 1991a, 12; 1991b, 5). It was made of bronze and formed by a stamping technique with a pattern of palmettes in a heart-shaped frame, placed around a rectangular hole. The piece, which is quite small, only 2.7 cm by 2.7 cm, was riveted onto leather (Fig. 19.1).

Remains of leather could still be seen on the back of the fitting. Similar fittings, sometimes with a comparable ornamentation, have been found in Eastern Europe, as well as in Hungary (Fig. 19.2), on the coast of the Baltic Sea and at Birka in Sweden (Hedenstierna-Jonson and Holmquist Olausson 2006, 56). Dollerup is one of the most westerly sites where such fittings have been found.

It is most likely that the fitting was fastened to the flap of a leather bag, and functioned as a fastening in connection with a strap. This strap extension would have bound the bag to the belt. There could also have been other fittings mounted onto the strap, besides the belt and the bag, so the fitting should be seen as a detail of a greater whole.

A pendant belt pouch decorated with metal mounts was an important adjunct to the oriental sword-belt, from which it hung. The number of mounts on the pouch, the type of decoration, their size and metal type, illustrated the rank and the social status of the bearer. This type of pouch most likely had its origin in the East, in the Russian Viking towns; its shape finds parallels among the Hungarian sabretaches

Pouches (sabretaches) from Rösta in Ås parish

Fig. 19.1. Dollerup fitting (Photo: Museet på Koldinghus, Kolding).

Fig. 19.2. Belt pouch, a so-called italics, reconstructed according to finds in a Hungarian cavalry grave; notice the middle fitting's similarity to the piece from Dollerup (Drawing © Anne Sofie Gräslund).

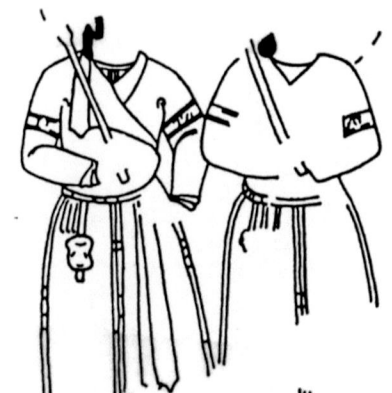

Fig. 19.3. Soldiers in kaftans with italics on an Afghani wall painting from c. AD 1000 (Drawing © Ingmar Jansson).

[IME] •ΧΑΙΡΕ ΕΝΔΟΞΕ•ΚΥΡ
ΙΒΑΝΗ ΠΡΟΤΟΣΠΑΘΑΡΙΕ•ΑΝΘΗΣ ΣΥ[Ν] ΝΕΟ-
ΤΗΣ ΦΑΙΔΡΥΝΗ[C]•

Fig. 19.4. The Greek inscription from the tablet-woven band from Moshchevaya Balka. It translates: "All hail glory protospatharios Sir Ivanes! May you be well in your young days. May you have courage…"

(Hedenstierna-Johnson and Holmquist Olausson 2006, 56). This type of belt-mounts emerged between AD 500 and 600, worn by kaftan-clad cavalrymen who came from Central Asia (Fig. 19.3).

In Central Europe, this type of fitting continued to be worn through the centuries until the Viking Age, still in connection with the cavalrymen, in whose graves they have been found. The fittings which most clearly and convincingly resemble the find from Dollerup belong to the first half of the 10th century AD. For instance, in the Ladby grave in Denmark, small buttons made in passementerie in gold and silver thread, as well as gold-thread embroidery, which belonged to a kaftan were recovered (Hedeager Krag 2004, 83, fig. 2, no. 501). In grave EK, Kaagarden, Langeland in Denmark, oriental upper belt accessories were found. It was a belt mount with oriental floral pattern and negative relief on the reverse (Grøn *et. al.* 1994, 129). They were found in great numbers in Gnezdovo, which was one of the main centres of the old Russian State situated in the Middle Dniepr Region (Eniosova 1999, 1099). Both the Ladby grave and the grave from Kaagarden are dated to the 10th century AD.

Belts and Mounts as an Expression of Warrior Rank and Status

The composite belt of oriental type functioned presumably as a symbol of rank. Within the cultural groups and in the areas where these types of belts were originally worn, there are clear indications that the belt expressed a hierarchical ranking system (Hedenstierna-Jonson and Holmquist Olausson 2006, 65). When the belt was removed from its original context and taken to Scandinavia, this symbolic value was almost certainly altered. Still, in the West, there was a clear connection between the warrior's rank and status and the wearing of a belt; this must have been well known to the Scandinavian warrior elite during the Viking period. However, the belt was only part of a larger context along with the warrior's weapon, all seen as part of his dress, the totality of which showed his military rank (Hedenstierna-Johnson and Holmquist Olausson 2006, 66).

Moreover, the Frankish royal courts, which had such great influence on the Scandinavian courts, had long presented diplomatic gifts of belts decorated with mounts – an item of costume they borrowed from Byzantium. An example of the belts' symbolic importance is obvious from a passage in the *Annals of St. Bertin* for the year AD 838. Here, it is mentioned that when the Emperor Charlemagne granted his son, Charles the Bald, an area in Neustria (in present-day France), a sword belt was also bestowed on the boy (Hedenstierna-Jonson and Holmquist Olausson 2006, 54). As mentioned earlier, the kaftan with belt and belt-mount constituted part of male attire among noblemen in Europe from about AD 550 to 1050. (See also Zanchi in this volume).

The Finds from Moshchevaya Balka

The cut of the kaftan is typical of male garments found in the burial site known as Moshchevaya Balka in the Northern Caucasus in present-day Abkhasia, a province in Russia. The burials are dated to AD 600–800, and optimal conditions for preservation have yielded 16 complete kaftans (Ierusalimskaja 1996, 47–54). The kaftans are cut on the straight grain, with a closely fitting upper part and a wide lower part, without a fastening, but with side slits made for greater convenience on horseback. Very long fasteners would have been attached to the braided trimmings. One of the kaftans is trimmed with fur, and the whole kaftan was lined with squirrel fur. Covered with splendid Persian silk, this kaftan was probably intended for a chief. A mythological animal, known as a *senmurv* is portrayed in a medallion set against a green background (Ierusalimskaja 1996, 248). Other kaftans from the same burial place are made of pieces of Sogdian, Byzantine and Chinese silks.

Another unique find from Moshchevaya Balka is a piece of tablet-woven belt. No other similar finds of woven silk from the Viking Age are known (Ierusalimskaja 1996, 251–252.) The belt is made of silk and woven using 37 tablets with four threads in each. Each tablet has two red and two yellow s-twisted silk-threads. The weft is s-twisted beige silk, with 15 threads per cm. The belt could have other functions, for example a headband or a ribbon that wrapped the sword, but it is quite unique, and we do not know other examples of this type.

The tablet-woven belt has a Greek inscription (Fig. 19.4), which includes the term *Protospatharios*, and is made in Constantinople, during the 8th–9th centuries AD. Two pieces of the belt are preserved in Russian museums today. One piece, 42 cm long and 1.1 cm wide, was excavated by N. Kaminskij in 1979, and belongs to the regional museum in Stavropol. A smaller piece, only 4.5 cm long, was discovered by N. Kaminskij in 1982, and is today in the Hermitage, in St. Petersburg (Ierusalimskaja 1996, 17).

This tablet-woven belt is of great significance for the history of the northern Caucasus. The title of *Protospatharios* was a high military rank (Ierusalimskaja 1996, 251). The belt belonged to a person called Ivanes, and he was presumably sent by the Byzantine Emperor to Moshchevaya Balka. As the two pieces of the tablet-woven belt were found in different places (Ierusalimskaja 1996, 251), perhaps they were distributed to different persons. Perhaps the belt had been cut into pieces. The title *Protospatharios* commanded respect in the imperial hierarchy in Byzantium. It was introduced in AD 718 and, as a military rank, it was comparable to brigadier general in the AD 900s. During the AD 1000's, the designation was altered to the rank of a colonel, especially after the reign of Byzantine Emperor Basil II (AD 976–1025). The Norse King, Harald Hardarada (AD 1047–1066) was accorded the title *Protospatharios* by the Byzantine Emperor during his stay in Constantinople. Harald Hardarada ("The Ruthless") was the last of the great Viking warriors, and he served as a mercenary leader fighting for Russian princes. He later served in the Byzantine emperor's elite Varangian Guard, posted to Bulgaria, Sicily and Syria, before he returned to Norway in AD 1045 and was accepted as King (Hall 2007, 98).

At first, Harald held the rank of *manglabites,* which was usual for soldiers of the imperial guard. The term *manglabites* derives from the Latin '*manuclavium*'. As a literal meaning of Harald Hardarada's title, Constantine VII Pophyrogennetos (AD 908–959) identified this rank with the title of *protospatharios*. But later Harald received the rank of *spatharocandidatos*, which was not as high a rank as *protospatharios*. According to Philotheos (AD 899), it was an ordinary rank given to foreign friends of the Empire (Bibikov 1996, 208). The rather moderate titles given to Harald Hardarada in the Byzantine hierarchy suggests that he did not hold an extremely strong position at the Byzantine Court.

It has been suggested that the garments from Moshchevaya Balka, including kaftan and belt equipment belong to an East Russian style (Hägg 1984; 1991). This dress fashion appeared in Kiev in the multicultural environment of the princely court, where Persian, Khazarian, Scandinavian and Byzantine impulses created an eclectic, aristocratic culture. The dress itself and its various elements were aimed at displaying its wearer's high social rank. The main types of garments found in Birka, Sweden were a version of the *scaramangion*, a kaftan-like ceremonial garment used at the imperial court at Constantinople (Hägg 1984, 218). The special purpose of these garments is stressed, in some cases, by elements with Christian content. For example, the small pockets made of silver threads, and other pockets made of silk, decorated with the figure of a stag (Geijer 1938, Taf. 35, 36) are explained as reliquary-capsulae (Hägg 1984, 213–214). A small St. Andrew's cross made of gold or silver threads adds to the Christian symbolism of these high-status garments (Duczko 1996, 194).

International research on the background of the Byzantine kaftans refers directly to Constantine VII Pophyrogennetos (AD 908-959) and his *Book of Ceremonies*, where he names the production of the kaftan, the *scaramangion*, in the imperial workshops in Byzantium (Kondakov 1924, 13). According to this contemporary source, the kaftan was fashionable at the Byzantine court in the AD 900s, the same century as the Dollerup find in Denmark. We cannot be certain whether the item with oriental influence from Dollerup belonged to a Christian, but we can see the person who may have owned this item as a member of the local Danish nobility, who adopted a special dress fashion in order to show his social status.

Development of dress in the Scandinavian Viking Age was generally influenced by new trends during and after the transition to Christianity, and the result was that the upper class was inspired by the Byzantine style after AD 700. Examples of kaftans and other unusual richly decorated clothes have been found in Viking Age Scandinavia, where they can all be linked to the highest social class (Hedeager Krag 2007, 241).

Centre of Power and Trade

An important commercial and political centre in the late 9th and 10th century AD is Gnezdovo, which is situated near Dniepr, *c.* 12 km. west of present city of Smolensk, Russia. Slavs and Scandinavians, warriors and craftsmen, merchants and farmers, all lived in peace in Gnezdovo. The "golden age" of Gnezdovo was in the 10th century, when the town became one of the major centers controlling the route from the Varangians (including the Vikings) to the Greeks (*i.e.* the Byzantines)(Eniosova 2007, 199). It was the place, where the North was connected with the South of Europe, and in many graves from Gnezdovo, Scandinavian objects are combined with Russian and Byzantine objects.

The Byzantine Official Dress

The styles of dress worn at the court and by officials in Byzantium have their origin partly in the dress worn by the Roman statesmen and partly in eastern regions of Central Asia including Persian areas (Boucher 1996, 147). The Byzantine upper class, like its Islamic counterpart, wore boots, trousers, tunic and kaftans of silk with buttons and a cape with sleeves. These dress pieces derive from more ancient costumes in Central Asia.

Scaramangion, the Persian cavalry kaftan worn with trousers, was made of varying colours, such as purple or gold (Pilz 1984, 24). Of all the materials available, silk was the most expensive, and it signified high status. Silk was imported in large quantities at great expense from China via the Silk Road to Byzantium, but from the AD 500s, the Byzantines started to produce silk themselves, although its import continued. In AD 568, an imperial workshop was established where costumes of silk with decorative bands of precious metals were produced for both the court and private persons. Silk functioned as a form of currency in diplomatic circles and as presents for distinguished visitors (Geijer 1994, 156).

The Byzantine Empire was a trendsetter in many areas, from style of dress to diplomacy, in both eastern and western Europe for several centuries. Graeco-Roman culture was practised and developed, and its forms and means of expression moved to lands directly connected to the empire, as well as more remote areas such as Scandinavia. For example, the court of the Franks as well as the court in Kiev both adopted Byzantine dress, and from here the style spread to the north. The well-travelled Vikings came naturally in direct contact with this dress-style during visits to the empire.

Byzantine Silks in the Viking Age

Byzantine silks have been found in several places in Scandinavia, often in the same localities as the buttons of silver and gold thread. A survey of the Scandinavian graves of the Viking Age shows that 26 contain silk. Seven of these graves are in Denmark, six in Norway and 13 in Sweden (Hedeager Krag 2005, 30, fig. 6.2). This survey does not include the silk items from Birka in Sweden, where about 60 graves containing silk have been recorded. With a few exceptions, including Chinese silk at Birka, all of the Scandinavian Viking Age silks were weft-faced compound twill, also known as satin. Such silk fabrics were produced predominantly in the silk-weaving workshops of Byzantium and the Islamic eastern Mediterranean up to the 12th century AD (Geijer 1994, 257).

Conclusions

The oriental belt is known from several finds in Denmark, and it functioned as a symbol of rank. At the same time, there is a clear connection between warrior rank and status and the wearing of a belt; this was known to the Scandinavian warrior elite during the Viking period (Hedensterna-Johnson and Holmquist Olausson 2006, 66). The Frankish royal court had for a long time presented diplomatic gifts of belts decorated with mounts – a custom itself borrowed from Byzantium. Silk from Byzantium was presumably used in the garments with oriental belts. Furthermore, the Danish examples with buttons of silver and gold threads and belt accessories show a dress style, which was apparently modeled on an oriental dress fashion, demonstrated by the kaftans of Moshchevaya Balka from Caucasus.

There has been a close link between Scandinavia and Russia since the 9th century AD. The trade route from the Varangians to the Greeks, which connects northern Russia with Byzantium, was established at the end of the 9th century AD when Gnezdovo on the Dnieper, in the Smolensk region of Russia, was founded. From the AD 900s close links are seen, and these indicate that also earlier there were connections. In the literary sources, there are examples of Russian counts going into exile in Scandinavia only to return later (Bibikov 1996, 207). Harald Hardarada is an example of a Scandinavian, who was in Kiev for a period, and he also served the Byzantine emperor, acquiring the title *protospatharios,* a title of honour, also known from an inscription on a tablet-woven silk belt from Moshchevaya Balka, dated to the 8th–9th century AD.

Traditional scholarship of the Viking Age has often focused on the interests of the Danes and Norsemen in the West, and those of the Swedes in the East. However, continuing finds support the theory that both the Danes and the Norsemen also had contacts to the East. The belt-mount from Dollerup is merely one example that attests to this.

Bibliography

Bibikov, M. (1996) Byzantinoscandica. In K. Fledelius (ed.), *Byzantium, Identity, Image, Influence. XIX International Congress of Byzantine Studies University of Copenhagen, 18–24 August 1996,* 201–211. Copenhagen, Eventus Publisher.

Boucher, F. (1996) *A History of Costume in the West.* London/New York, Thames & Hudson.

Duczko, W. (1996) Viking Sweden and Byzantium – an Archaeologists Version. In K. Fledelius (ed.), *Byzantium, Identity, Image, Influence. XIX International Congress of Byzantine Studies University of Copenhagen, 18–24 August 1996,* 193–200. Copenhagen, Eventus Publisher.

Eniosova, N. (1999) Manufacturing Techniques of Belt and Harness Fittings of the 10th Century AD. *Journal of Archaeological Science (1999)* 26, 1093–1100.

Eniosova, N. (2007) Viking Age Gold from Old Rus. Cultural interaction between east and west. In U. Fransson, M. Svedin, S. Bergerbrant, and F. Androshchuk (eds), *Archaeology, artefacts and human contacts in northern Europe,* 175–179, Stockholm Studies in Archaeology 44, Stockholm University. Stockholm.

Geijer, A. (1938) *Birka III. Die Textilefunde aus den grabern.* Uppsala.

Geijer, A. (1994) *Ur textilkonstens historia.* Stockholm, Tidens Forlag.

Grøn, O., Hedeager Krag. A., and Bennike, P. (1994) *Vikingetidsgravpladser på Langeland.* Langelands Museum.

Hall, R. (2007) *Exploring the World of the Vikings.* London/New York, Thames & Hudson.

Hedeager Krag, A. (2003) Herskersymboler i dragten fra Danmarks yngre jernalder og vikingetid. In A. Hedeager Krag (ed.), *Dragt og magt,* 62–77. Copenhagen, Museum Tusculanums Forlag.

Hedeager Krag, A. (2004) New light on a Viking garment from Ladby, Denmark. Priceless invention of humanity textiles. In J. Maik (ed.), NESAT VIII *Acta Archaeologica Lodziensia* 50/1, 81–86.

Hedeager Krag, A. (2005) Denmark-Europe: dress and fashion in Denmark's Viking Age. In F. Pritchard and J. P. Wild (eds), *NESAT VII,* 29–35. Oxford, Oxbow Books.

Hedeager Krag, A. (2007) Christian Influences and Symbols of Power in Textiles from Viking Age Denmark. Christian Influence from the Continent. In C. Gillis and M-L. B. Nosch (eds), *Ancient Textiles. Production, Craft and Society,* 237–243.Oxford, Oxbow Books.

Hedensterna-Jonson, C., and Holmquist Olausson, L. (2006) *The Oriental Mounts from Birka's Garrison.* Antikvariskt arkiv 81. Stockholm, Kungl. Vitterhets Historie och Antikvitets Akademien.

Hägg, I. (1984) Birkas orientaliska praktplagg. *Fornvännen* 78, 204–223.

Hägg, I. (1991) *Die Textilefunde aus der Seidlung und aus den Gräbern von Haithabu.* Berichte über die Ausgrabungen in Haithabu 29. Naumünster, Karl Wachholtz.

Hägg, I. (2003) Härskarsymbolik i Birkadräkten. In A. Hedeager Krag (ed.), *Dragt og magt,* 14–27. Copenhagen, Museum Tusculanums Forlag.

Ierusalimskaja, A. A. (1996) *Die Gräber der Moscevaja Balka. Frühmittelalterliche Funde an der Nordkaukasischen Seidenstrasse.* München, Editio Maris.

Jensen, A.-E. (1991a) Lokale vikinger. *SKALK* 6, 12–15.

Jensen, A.-E. (1991b) Dollerup i vikingetiden. Den første totalundersøgte enkeltgård fra dansk vikingetid. *Museet på Koldinghus, Årbog 1990–1991,* 5–17.

Kondakov, N. P. (1924) Les Costumes orientaux á la Cour Byzantine. *Byzantion. Revue Internationale des Études Byzantines I,* 7–49. Paris.

Pilz, E. (1984) Vad vet vi om ämbetsdräkten i Bysans? *Bulletin/Svenska kommitté för bysantinska studier* 2, 23–27.

20 A Study of Two Medieval Silk Girdles: Eric of Pomerania's Belt and the Dune Belt

by Viktoria Holmqvist

Introduction

Very little has been published about the extraordinarily well-preserved medieval silk girdle popularly known as 'Eric of Pomerania's Belt' kept at the National Museum in Copenhagen, Denmark. Poul Nørlund describes the weave as technically complex, and refers to a study of the girdle carried out by Elisabeth Budde-Lund (Nørlund 1937, 318). Apparently, this study was never published and no record of it exists in the National Museum's archives.[1] Nørlund's article does not include a detailed analysis of the belt's weave structure, but he partially dismisses the claims that it could be tablet-woven by indicating that it lacks the usual distinguishing features of that technique. Instead, he tentatively describes the weave as a double-sided, patterned twill with a supplementary pattern weft of spun silver gilt thread (Nørlund 1937, 318). Ilse Fingerlin also deals with Eric of Pomerania's Belt in *Gürtel des hohen und späten Mittelalters* (Fingerlin 1971), but does not elaborate on the weave structure either. Like Nørlund, she discusses the belt fittings from the famous Dune hoard from Gotland, Sweden, in conjunction with Eric of Pomerania's Belt, but neither of them makes much of the woven belt fragments that were also found there (Nørlund 1937, 322; Fingerlin 1971, 53).

However, after having studied[2] both Eric of Pomerania's Belt and the Dune Belt fragments, I have reached the conclusion that although these two girdles display both superficial similarities *and* differences, their ground weave is actually of the same type and also very unusual. Nørlund's description of it as a double-sided, patterned twill does not recount the entire story.

Unlike Eric of Pomerania's Belt, the Dune fragments do have the warp-twined edges that are usually associated with tablet-weaving, while the rest of the weave does not show any obvious signs of warp-twining. My examination of the edges suggests that they were made with octagonal tablets each carrying eight threads, rather than the prevalent square tablets with four threads. Thus far, I have only been able to find one other case of eight-thread warp-twining documented in a medieval textile: a 15th century silk ribbon (Vial 1971).

Eric of Pomerania's Belt

Eric of Pomerania's Belt is extraordinarily well preserved and although many of the threads are worn away, it is still possible to make out the different patterns and colours (Fig. 20.1). A table of the complete analysis can be found in Appendix A. The main portion of the warp threads appears a uniform beige, but Nørlund has established that their original colour was red (Nørlund 1937, 318). In fact, a couple of these threads have actually kept their colour and can be seen in various places throughout the belt. At first glance, these very bright red threads might appear to be later additions, but they are fully integrated in the weave and must therefore be original. The rest of the warp threads are blue and green and the colours are especially vivid on the reverse side of the belt. The thin silver gilt lamella of the supplementary pattern weft is almost completely missing. In most places, only the silk core remains, although it too, is badly worn. The silver has oxidised, leaving behind a blackish crust on the belt even where the thread itself has disappeared. The warp threads that the metallic weft used to cover are also mostly missing in these places, but their ends can still be seen with the help of a microscope.

To establish the principles of the weave and to document all the different patterns of Eric of Pomerania's Belt, I worked from close-up photographs I had taken during my examination of the belt. It is made up of 26 pattern sections, not counting the partial sections at either end, that are half-covered by the chape of the buckle and the strap-end respectively. Each section is approximately 6.3 cm in length and separated from the next section by a bar-shaped belt-mount. There are 27 belt-mounts altogether and most of them are set with semi-precious stones that match the colours of the girdle. The pattern sections all follow the same overall design, starting and ending with zigzag diagonals with a central, diamond-shaped motif in-between (see Appendix B for pattern drawings of the entire belt). The sections change colour in a fixed sequence, which is repeated throughout the belt in the following manner (counting from the strap-end to the buckle):

- Section one: Green with red pattern elements
- Section two: Red with blue pattern elements
- Section three: Blue with red pattern elements
- Section four: Red with green pattern elements
- Section five: Green with red pattern elements
- The pattern repeats.

The colours of the patterns are mirrored on the reverse side of the girdle; green pattern elements are blue on the back and blue ones are green. Red pattern elements are red on both sides.

Although the overall design of each pattern section is the same, most of the details of the patterns are different, with the exception of four designs which are repeated once, more or less identically (Appendix B, nos. 2 and 24, 3 and 19, 4 and 20 and 6 and 18). The patterns are mainly variations of the swastika shape, but there are also a couple of zoomorphic patterns with a deer-like creature and a bird.

The belt is double-woven and the ground weft and the pattern weft both go round in the edges in round-weave fashion. The pattern weft is visible on the reverse side of the belt too, making it truly double-sided. As may be seen in Figure 20.2, the weft threads are broken in several places, allowing a glimpse of the two-layered structure of the weave. A closer examination of the weave reveals that each coloured diagonal seen in the patterns is made up of five threads, the middle one being of a different colour from the others. The four main threads make warp-floats *over two, under one* in twill-fashion (2/1), with the differently coloured one making a short float *over one*, as in the schematic drawing in Figure 20.3. It is difficult to reconstruct the course of the badly worn pattern weft in any detail, but it complements or accentuates the woven pattern. The weaver appears to have used the differently coloured thread in the middle of the diagonals as a guide when picking the pattern shed – the metallic weft regularly covers the two threads *to the right* of the short float (see Fig. 20.4). This would make picking the pattern shed a relatively quick procedure that can be repeated on both sides of the belt identically, without having to count threads or referring to a pattern.

It is understandable why Nørlund describes Eric of Pomerania's Belt as a double-sided, patterned twill; that is exactly what it looks like. Attempting to weave it as such, with the green and blue threads working in pairs and the red threads doing the same to give the colour combinations seen in the belt, will produce a piece that looks very much like the original, except when you turn it over. In Eric of Pomerania's Belt, the reverse side (*i.e.* the bottom layer of the double-weave) is slightly skewed in relationship to the front (the top layer). Some of the warp threads appear to shift positions sideways, which is not usual in an ordinary interlaced woven structure. It is difficult to see this, because the threads are so tightly packed, and it only became apparent to me after I had woven a full-scale sample that did *not* show this subtle shift.

The Dune Belt

Unlike Eric of Pomerania's Belt, the Dune Belt is not in one piece (Fig. 20.5). It consists of 30 silver gilt belt-mounts, a

Fig. 20.1. The so-called 'Eric of Pomerania's Belt' (Photo: V. Holmqvist, courtesy of The National Museum, Copenhagen).

Fig. 20.2. The edge of Eric of Pomerania's Belt – the broken weft reveals the two-layered structure (Photo: V. Holmqvist, courtesy of The National Museum, Copenhagen).

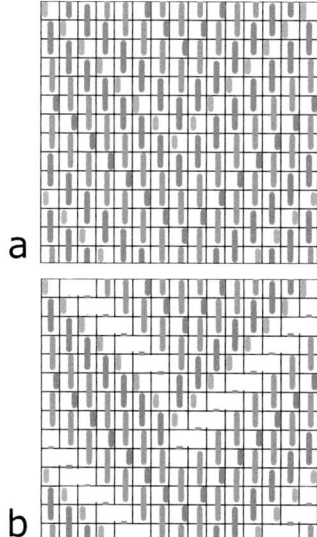

Fig. 20.3. a. Schematic drawing of the basic pattern of Eric of Pomerania's Belt; b. with the supplementary pattern weft drawn in (Drawing: V. Holmqvist).

buckle and a strap-end, which are displayed in the 'Golden Room' in the National Museum of Antiquities in Stockholm, Sweden. Some of the mounts are still attached to fragmentary pieces of the woven silk girdle and there are also five loose fragments which are not on display (a table of the complete analysis may be found in Appendix A). The belt, or what is left of it, was found as part of a large treasure hoard on the island of Gotland, Sweden. It is believed to have been buried around the year AD 1361 when the Danish king Valdemar invaded the island. According to a numismatic study, a coin found in the hoard is minted by Winrich von Kniprode, Grand Master of the Teutonic Knights in 1351–1382, thus it cannot have been buried before the year 1351 (Lannby and Rispling 2006, 61).

The poor state of preservation of the girdle makes it difficult to reconstruct the pattern with any certainty, but it appears to have consisted of multi-coloured diamonds, almost like Eric of Pomerania's Belt, but probably less elaborate.

In places, the weft has a slightly pinkish tint and some of the warp threads might have been blue or green. On some of the fragments, warp-twined edges are preserved, which suggests with a high degree of certainty that the Dune Belt is tablet-woven. Interestingly, however, the main portion of the weave shows the same type of double-woven structure as Eric of Pomerania's Belt, with the characteristic twill-like warp-floats *over two, under one* and the short floats *over one*. In the Dune Belt, the warp threads are pulled together tightly by the weft, which causes the short floats to disappear almost completely under the long ones. This, and the fact that the Dune Belt does not have a supplementary pattern weft, account for much of the differences in the appearance of the two girdles.

On both the front and the back, the warp-twined cords of the Dune fragments have the *appearance* of ordinary four-thread warp-twining produced by square tablets. However, it is not possible to achieve this effect with the single shed created by square tablets *in a double-woven structure*, as is the case here. Standing the tablets on their points to create the two sheds required would not help either, as the result would look very different from four-thread warp-twining. Hexagonal tablets threaded with six threads would produce the correct look in both layers, but the broken weft on one of the fragments reveals that the twining warps move from the top to the bottom layer in a manner, which is consistent with the use of octagonal tablets and eight threads (Fig. 20.5). To my knowledge, the only other documented occurrence of eight-thread warp-twining is in an Italian silk ribbon from the 15th century, analysed by Gabriel Vial. The edge-cords of this band do not appear to have the same straightforward twining as the Dune Belt, but, according to Vial, they may nevertheless be produced with octagonal tablets (Vial 1971, 74).

Weaving with Octagonal Tablets – Reconstructing a Forgotten Technique?

The warp-twined edges of the Dune Belt make it reasonable to expect that the rest of the girdle is then also tablet-woven, probably with octagonal tablets. The same interpretation would also be valid for Eric of Pomerania's Belt, since the

Fig. 20.4. The largest woven fragment of the Dune Belt (Photo: V. Holmqvist; Courtesy of The National Museum of Antiquities, Stockholm).

two girdles share the same type of basic weave structure. The question is, how does one achieve this weave with tablets?

The warp-twined edges of the Dune Belt are easily produced by continually turning octagonal tablets with eight threads in one direction. The four warp-twined cords that are preserved at either edge of some of the Dune fragments indicate that the edge tablets were threaded alternately in S and Z. When it comes to reconstructing the rest of the

Fig. 20.5. a. Side-view of one of the Dune fragments; b. the same fragment with the path of the twining warp threads and the wefts drawn in (Photo: V. Holmqvist; Courtesy of The National Museum of Antiquities, Stockholm).

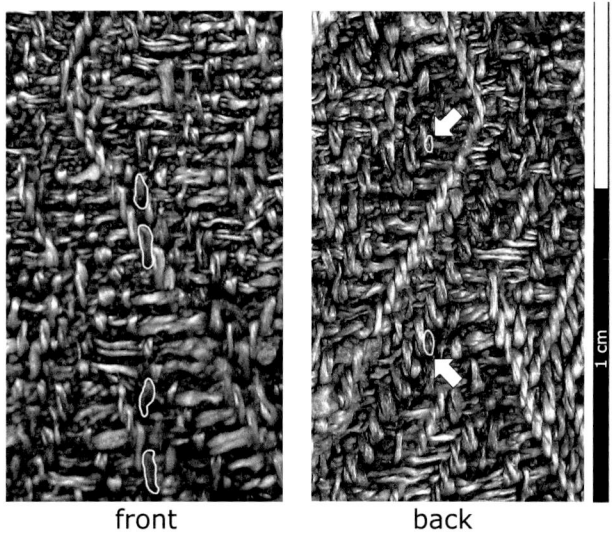

Fig. 20.6. The red thread visible in Eric of Pomerania's Belt, circled in white (Photo: V. Holmqvist; Courtesy of The National Museum Copenhagen).

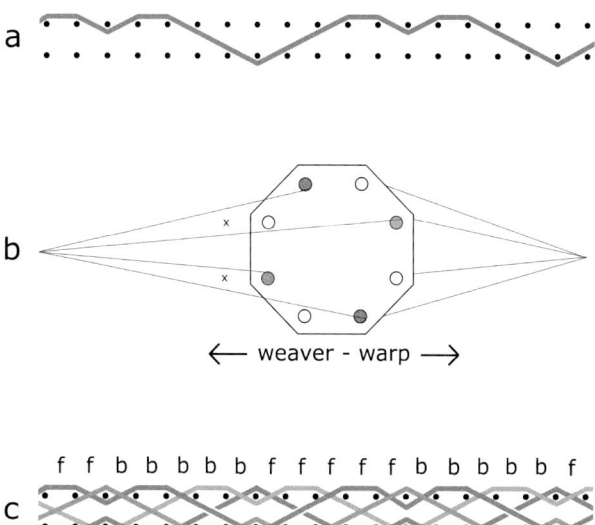

Fig. 20.7. a. Cross-section showing the course of the red thread from Fig. 20.6; b. S-threaded octagonal tablet with the sheds marked with 'x'; c. the turning sequence needed to weave the basic pattern of Eric of Pomerania's Belt (f – forwards; b – backwards). The cross section also shows the slight warp-twining that shifts the threads sideways between the two layers (Drawings: V. Holmqvist.)

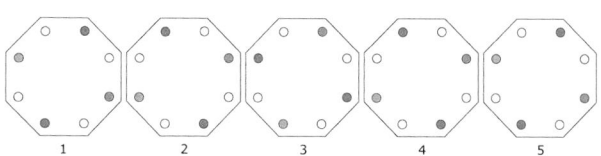

Fig. 20.8. The 'staggered' starting position of the threads in the octagonal tablets (Drawing: V. Holmqvist).

weave, it is easier to look for clues in the better preserved Eric of Pomerania's Belt. In pattern section 14 (counting from the strap-end to the buckle), one of the bright red threads mentioned earlier is present, and, because it stands out among its faded counterparts, it is possible to follow its course through the weave. As can be seen in Figure 20.6, its main function is to be part of the diagonal pattern in the top layer, but it also disappears down to the bottom layer to make short floats on the reverse side. The cross section in Figure 20.7a shows how the red thread moves between the wefts of the double-woven structure.

Fig. 20.9. Detailed diagram of pattern section 20, Eric of Pomerania's Belt. The reconstructed position of the supplementary pattern weft is indicated by transparent yellow (Drawing: V. Holmqvist). Reproduced as a colour plate on page 303.

An S-threaded octagonal tablet with the colours red, blue, red and green in every other hole, as in Figure 20.7b, can be used to reproduce the movement of the red thread by making one-eighth turns, following the turning sequence outlined in Figure 20.7c. This will automatically control the other threads in a way that matches the structure of the original belt. The weft is passed twice before the tablet is turned; first through the top shed and then through the bottom shed. To recreate the simplest pattern element of Eric of Pomerania's Belt, *i.e.* two differently coloured diagonals, every tablet is turned five times *forwards* (away from the woven band) and five times *backwards* (towards the woven band). However, to create the overlapping, twill-like floats, the tablets' starting position, and therefore also their turning sequence, need to be slightly staggered (see Fig. 20.8), so the tablets cannot be turned together. They have to be turned in groups, depending on where in the five forwards/five backwards-sequence they appear. The principle is roughly the same as that of other complex tablet-weaving techniques, such as double-faced 3/1 broken twill (*cf.* Collingwood 1996, 208–218).

Fig. 20.10. Turning direction of the tablets for pattern section 20. One square represents one octagonal tablet and one 1/8 turn. White squares indicate forward turns; grey squares indicate backward turns (Drawing: V. Holmqvist).

Figure 20.7c shows the slight warp-twining that occurs when a thread passes from one layer to the other – this explains the sideways shift between the front and back in Eric of Pomerania's Belt mentioned earlier. The threads actually do cross, but the warp-twining is more or less cancelled out by the constant shift between forward and backward turns. This is why the patterned weave of the girdles does not immediately appear to be tablet-woven. In the case of Eric of Pomerania's Belt, it is only in the transition between two pattern sections that the warp-twining will remain. When a section is finished, every tablet is given a *quarter* turn backwards to bring the colours into the right position for weaving the next section. It takes four whole pattern sections for the octagonal tablet to make one complete turn, so the common tablet-weaving problem with warp-twining building up behind the tablets is kept to a minimum.

It should be noted that it is also possible to achieve this weave structure with square tablets carrying four threads. The tablets have to be used standing on their points to make two sheds, and some tablets must be left idle every pick to create the long, twill-like floats. This increases the complexity of the turning sequence significantly and the eight-thread warp-twining of the Dune Belt still needs octagonal tablets to work. A combination of square and octagonal tablets could of course be used, but the difficulty of working with tablets of different shapes on the same warp makes it a less appealing alternative.

A comparison between the detailed analysis of pattern section 20 of Eric of Pomerania's Belt (Fig. 20.9) and the schematic drawing of the turning direction of the 85 tablets needed to weave it (Fig. 20.10), shows that the intricate patterns will follow a very regular and straightforward system if woven with octagonal tablets. In other words, the tablets move in units of three to weave the zigzag diagonals that begin and end each pattern section, and in units of five to produce the two-coloured patterns. When a diagonal changes direction in the pattern, the tablets' turning direction is reversed. This, coupled with the considerable skill of the medieval weaver, could possibly account for the lack of apparent weaving mistakes in the belt. Furthermore, I believe that a practised weaver, well versed in the technique outlined here, would be able to weave a girdle like Eric of Pomerania's Belt or the Dune Belt without referring to a detailed pattern.

The top and bottom sheds produced by octagonal tablets are very small and run the risk of becoming tangled and unclear, especially when weaving with fine threads and many tablets. However, there is a way to make the work a little easier, which can be found in several medieval depictions of the Virgin Mary weaving (see, for instance, Collingwood 1996, 31). In these images, the loom generally consists of two upright beams with a horizontal cross bar. The warp is wrapped around the poles and the whole weave is tilted on its side. Collingwood argues that this is not necessarily a product of aesthetic convention or the inability to draw it otherwise, but a depiction of real weavers at work (Collingwood 1996, 30), and I concur – weaving with the band facing sideways makes a great deal of sense if the shed is small. It means that one can look at the shed from above as one weaves and make sure it is clear when passing the weft. Furthermore, tablets with empty holes, as in the technique just described, are unbalanced and have a tendency to move out of position by themselves, but when the warp faces sideways, they are stacked neatly on top of one another and kept in place between turns.

Having the weave set up on its side takes a little getting used to, but it offers another definite advantage when weaving a double-sided belt – the weaver can see both sides of it without having to flip it over. This is especially useful when adding the supplementary pattern weft in Eric of Pomerania's Belt. Since every pattern shed needs to be picked both on the front *and* the back, it saves a lot of trouble not having to twist the belt 180 degrees ever so often to do it.

Weaving as an Analytical Tool

The advantages of an experimental approach to solving a problem as that posed by the intriguing weave of the two girdles are obvious. Both Nørlund's and my own initial assumption that Eric of Pomerania's Belt was not tablet-woven, simply because it did not seem to contain any warp-twining, was proven wrong by weaving a sample and comparing it to the original. It is relatively easy for mistakes and misconceptions to slip into a two-dimensional analysis of a complex weave structure; after all, it is called complex for a reason! What appears to work on paper does not always translate well into reality, where the threads suddenly move in three dimensions. By using practical weaving experiments to test the validity of an analysis, it is possible to identify and avoid mistakes, at least to a certain degree. Of course, using what is basically a craft skill as an analytical tool demands a great deal from the weaver in terms of evaluating one's own handicraft, as well as being well acquainted with historical techniques and research. When used together with a sound, critical approach, weaving as an analytical tool can play a valuable role in the examination of archaeological textiles. It offers what seems to be a more thorough analysis, and even, in this case, the added bonus of the discovery of an unusual, and, as far as can be ascertained, previously undocumented tablet-weaving technique.

By comparing Eric of Pomerania's Belt and the Dune fragments and weaving several samples, it has been possible to reconstruct a potential technique for weaving the complex structure of these two girdles. Apart from explaining the look of both girdles, using octagonal tablets makes it possible to weave the varied and colourful patterns of Eric of Pomerania's Belt with a minimum of effort once the technique has been fully mastered.

Acknowledgements

I would like to thank The National Museum in Copenhagen and The National Museum of Antiquities in Stockholm for being very helpful and giving me access to Eric of Pomerania's Belt and the Dune Belt.

Notes

1 Vivian Etting, personal communication 2007.03.29.
2 I use a simplified version of Lise Ræder Knudsen's system for drawing tablet-woven structures, which involves drawing the warp threads as they are seen on top of the weave (Ræder Knudsen 1996, 14). I use graph paper to do this, so each row of squares represents a weft and the drawn 'stitches' the warp. For the technical aspects of tablet weaving, I use the terminology outlined by Peter Collingwood. For in-depth discussions on this and tablet weaving in general, see Collingwood 1996.

Bibliography

Collingwood, P. (1996) *The Techniques of Tablet Weaving*. McMinnville, Robin & Russ Handweavers, Inc.
Fingerlin, I. (1971) *Gürtel des hohen und späten Mittelalters*. München Berlin, Deutschen Kunstverlag.
Lannby, M. G., and Rispling, G. (2006) Duneskattens mynt, myntstycken och besvärjelser. *Svensk numismatisk tidskrift* 3, 60–64.
Nørlund, P. (1937) Det saakaldte "Erik af Pommerns Bælte". In *Från*

stenålder till Rokoko. Studier tillägnade Otto Rydbeck, 317–323. Lund, Gleerups förlag.

Ræder Knudsen, L. (1996) *Analyse og rekonstruktion af brikvævning.* 2. dels opgave, Konservatorskolen, maj 1996.

Ugglas, C. R. af (1936) *Gotländska silverskatter från Valdemartågets tid.* Ur Statens historiska museums samlingar 3, Stockholm.

Unionsdrottningen. Margareta I och Kalmarunionen (1996) Exhibition Catalogue. Copenhagen, National Museum.

Vial, M. G. (1971) Un ruban de velours tissé "aux cartons". *Bulletin de Liaison du CIETA*, 34, 54–74.

Appendix A. Analysis sheets of Eric of Pomerania's Belt and the Dune Belt fragments. (Photos: V. Holmqvist; Courtesy of The National Museum of Antiquities, Stockholm).

Analysis Sheet

Item:	'Eric of Pomerania's Belt'
Function:	Girdle
Museum:	The National Museum (Nationalmuseum), Copenhagen, Denmark
Inventory Number:	27
Find location/Provenance:	Found in a storage room in Fredensborg Castle, Denmark, in the late 18th century. Attributed to Eric of Pomerania by mistake.
Date of item:	*Girdle and belt mounts*: early 13th – early 14th century *Buckle and strap end*: mid-15th century (Fingerlin 1971; Nørlund 1937)
Description/Technical data:	One-weft double-weave with twill-like warp-patterning by floats *over two, under one*, together with short floats *over one*. Supplementary weft-patterning with silver gilt thread. The weave is most likely produced with octagonal tablets, each carrying four threads (green, red, blue, and red). The girdle is divided into 26 sections by both the belt mounts and the pattern. The designs of four of the pattern sections are repeated once, all the other sections have unique patterns. The designs are mainly variations on the swastika-shape, with the exception of a couple of zoomorphic patterns; a deer and a bird.
Width:	3 cm
Length:	178 cm
Warp:	approx. 113 threads/cm; silk with a slight S-twist, red, blue and green. The red is very faded (beige).
Weft:	approx. 30 threads/cm (15 per layer); ground weft: red silk; pattern weft: three strands of spun silver gilt thread, but very little remains of the thin metal lamella.
Literature:	Fingerlin, I. (1971) *Gürtel des hohen und späten Mittelalters*. München Berlin, Deutschen Kunstverlag. Nørlund, P. (1937) 'Det saakaldte "Erik af Pommerns Bælte"', in: *Från stenålder till Rokoko. Studier tillägnade Otto Rydbeck*. Lund, Gleerups förlag, 317-323. *Unionsdrottningen. Margareta I och Kalmarunionen*. (1996) Exhibition catalogue, Nationalmuseum, København, 386.

Analysis Sheet

Item:	Dune Belt fragments
Function:	Girdle
Museum:	The Museum of National Antiquities (Statens Historiska Museum), Stockholm, Sweden
Inventory Number:	6849:68
Find location/Provenance:	Dune, Dalhems sn, Gotland, Sweden
Date of item:	Mid-14th century
Description/Technical data:	Five fragments. Only four were examined (one was missing).

One-weft double-weave with twill-like warp-patterning by floats *over two, under one*, together with short floats *over one*. Warp-twined edges, made by 4? octagonal tablets carrying eight threads. The pattern is most likely woven with octagonal tablets carrying four threads.

The fragments are all various shades of brown, but have originally been multi-coloured. Further fragments of the belt are still attached to the belt mounts and displayed in the museum's 'Golden Room'. Width & length: see photos below.

Warp: silk, with a slight S-twist; some threads still have a faint blue-green tint

Weft: approx. 26 wefts/cm (13 per layer); silk, slightly thicker than the warp, double?; some threads are light pink and have probably once been red

Literature: Fingerlin, I. (1971) *Gürtel des hohen und späten Mittelalters*. München Berlin, Deutschen Kunstverlag.

Nørlund, P. (1937) 'Det saakaldte "Erik af Pommerns Bælte"', in: *Från stenålder till Rokoko. Studier tillägnade Otto Rydbeck*. Lund, Gleerups förlag, 317-323.

Ugglas, C. R. af (1936) *Gotländska silverskatter från Valdemartågets tid*. Ur Statens historiska museums samlingar 3, Stockholm.

128 Viktoria Holmqvist

Appendix B. Drawings of all the complete pattern sections from Eric of Pomerania's Belt, numbered from the strap-end to the buckle. (Drawings: V. Holmqvist). Reproduced as a colour plate on page 304.

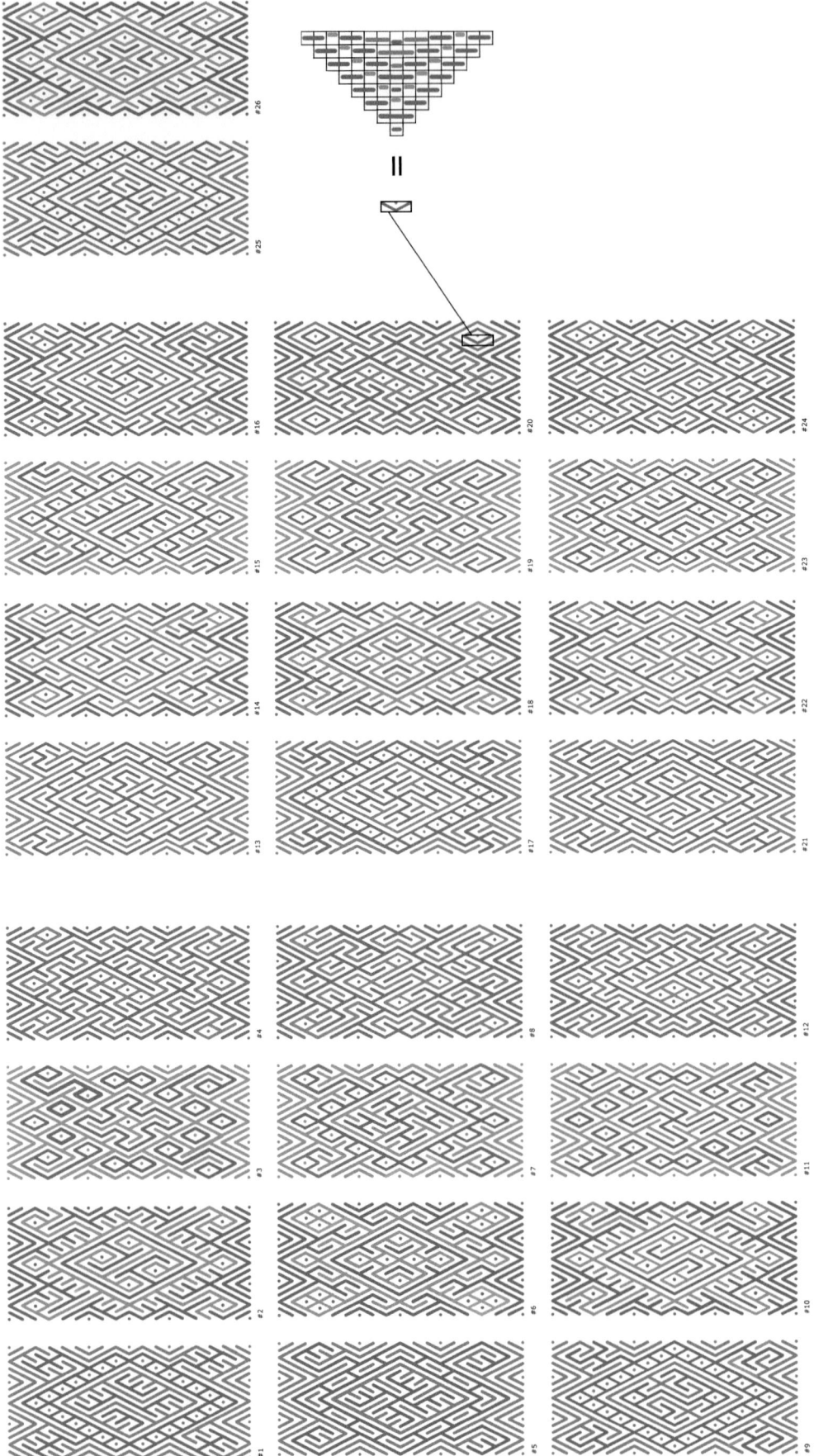

21 Nettle and Bast Fibre Textiles from Stone Tool Wear Traces? The Implications of Wear Traces on Archaeological Late Mesolithic and Neolithic Micro-Denticulate Tools

by Linda Hurcombe

Introduction

Unlike the other papers in this volume, my starting point is not the textiles themselves, but a particular kind of stone tools. Wear analysts have long been concerned that many of the plant wear traces that they identify need not be connected with subsistence activities but may instead be part of the processing of plants as the raw material for crafts (Juel Jensen 1994; Hurcombe 1998, 2000a, 208b; Aranguren and Revedin 2001; Beugnier and Crombé 2007). In particular, this paper suggests that a tool type known variously as a micro-denticulate or serrated edge flake, has its best wear comparisons with experiments used in the processing of plant bast for producing fine fibres. The flint tool itself is distinctive due to small indentations on the working edge (Fig. 21.1). It is also characterised by an intense silica-rich polish often visible macroscopically (Fig. 21.2). This polished area extends for perhaps 1.5 cm only along the tool edge and it is usually located on a straight or inwardly curving edge. Under the microscope the wear traces show that it was used not in a sawing motion, but in a light scraping motion. The asymmetry of the wear traces with the greatest wear on the ventral surface along with the distribution of the striations and polish clearly show a light scraping action. The study of the microwear on these serrated stone tools constitutes a 20-year old functional puzzle, with most authors considering some form of plant-processing task a possibility (Bocquet 1980; Vaughan and Bocquet 1987; Juel Jensen 1994; Silva and Keeley 1994; Beugnier 2007). In recent years, more experiments with plant bast fibres have allowed a closer match to the archaeological wear traces to be presented to a microwear audience (Hurcombe 2007b; 2008b), but as the implications are that these tools are for fibre production processes with no specific ethnographic counterparts, an overview is presented here to a textile readership.

Fig. 21.1. A serrated edge tool showing the fine serrations formed by fine retouch located on the dorsal edge (© L. Hurcombe).

might be some micro-scars on the tips of the teeth, especially where these are pointed, the polish is not extensive and does not go into the bases of the denticulations. Experimentation has shown that activities like stripping leaves or shredding fibres have to be ruled out as in all cases these experiments showed a strong polish at the base of the denticulations. The polish is also asymmetrical on the dorsal and ventral aspects and the action suggested by the polish distribution is very much a light scraping motion at a shallow angle. The distribution of the polish along the edge also suggests that the material contacted is limited in its lateral extent; certainly, no more than 1.5 cm as a main contact point. The extent and development of the polish is also different between the dorsal and the ventral surface. The ventral is by far the stronger set of wear traces indicating it was the main contact zone.

The function of this tool, therefore, needs to answer a series of questions. First amongst them, in such a distinctive edge, is why serrate? What advantage does serration give over a non-serrated edge? In the contemporary world, most of our examples of serrated edges are for sawing. However, there are examples of serrated edges being used as scraping tools for defleshing hides. In this case, the serrations stop the tool digging into the hide surface but instead allow it to scrape off the smallest amount of material from the surface without adversely affecting the underlying fibres of the hide. It is perhaps this aspect of serration which micro-denticulate tools also employ. Experimentally, it can be shown that a serrated edge skims the surface, but does not dig in when drawing it down a plant in a light scraping action. The same edge, non-serrated, tends to dig into the underlying wood or pith and give more of a light whittling action. Therefore, experimentation shows that the serrated edge keeps the scraping action very light with a minimal amount of material shaved from the item being scraped.

Another question is why is the extent of the gloss so limited? Here the argument must be that the contact material surface and the nature of the task simply restrict the amount of the edge used. Thus, such tools could not be used effectively on large items; they would have to be used on some small portion of it and the prevalence of straight or inwardly curving edges shows itself to be very suitable for scraping curved items. In a variety of experiments, the serrated tool edge, especially where it is straight or slightly concave in plan view, was found to be ideal for scraping small rounded surfaces such as plant stems and roundwood.

Furthermore, in assessing this 20-year-old functional puzzle, the task has to be one that would require tools such as these to be so numerous. On many sites, they are numerically on a par with other scrapers. The task suggested must also be one which explains the geographically widespread nature of this tool's distribution – Britain and Denmark and the circumalpine region. Ideally, the suggested use should also explain why it is a period-specific tool. This tool is, in Britain, present in the early Neolithic and declines markedly in the later periods of the Neolithic. In Denmark, micro-denticulate tools are known from Ertebølle Mesolithic contexts as well as Neolithic ones (though some tools have the serration scars on the ventral face rather than the dorsal) (Juel Jensen 1994,

53–57). In Britain, the tool is found on a wide range of site types; for example, the Sweet Track flints include some and there are examples known from long barrows and causewayed enclosures. Also, the wear traces strongly suggest that these are long-lived tools and that the wear on their surfaces perhaps represents days or weeks of wear. Again, this suggests that the activity associated with these tools is a very significant part of the activities of the societies where this tool is found in such prevalent numbers. Thus, if the task is associated with a suite of plants, they have to be a group whose distribution coincides with that of the tools. And finally, the tools themselves can have a variety of different weights and shapes. Some of them are naturally backed with cortex, while others have deliberate backing. Some examples occur on blades that are two-edged, so both sides of the straight edges have serrations. Others are on quite broad thin flakes. In many cases, there seems to be no deliberate backing or shaping such as may be the case if these tools were hafted. Rather the reverse, so the suggested function must also take account of the use of these tools in the hand.

Given all of these points, the key issues for the experimental programme is the operational chains for processing plants for cordage and fibre production. Essentially, the operational chain can be reduced to a list of main tasks, some of which can be performed in slightly different sequences.

Figure 21.3 does not present the full range of tasks and their variations that we have investigated. As a part of this *chaîne opératoire,* there could also be water used in some of these processes, and possibly also chemicals such as the wood ash solution. The key point about this sequence working backwards through the *chaîne opératoire* is that not all raw materials used in fabric production need be twisted and those that are twisted can be rolled on the thigh or twisted using a spinning hook. The alignment of fibres needed for these methods of twisting need not involve the same techniques used to prepare fibres for spinning with a spindle. Furthermore, the fibres spun with such techniques might open up other processing options earlier in the *chaîne opératoire*. In a larger scale of production, water retting of plant material in bulk for later breaking, scutching and hackling might make more sense, even though these processes can damage the fibres and produce wastage to some extent. However, a different scale of production might favour a system where more investment in earlier phases of the *chaîne opératoire* resulted in fibres being sufficiently aligned for minimal processing before being thigh or hook spun, or indeed spindle-whorl spun. These ideas have originated from a practical engagement with these early processing phases.

What the archaeological stone tool wear traces suggested was a very light scraping action on plant materials. Why such an action might be useful or how it would fit into some task is not known. A wide range of tasks was attempted with this tool; not all of these are reported here. Instead, the focus here is on the set of activities which seemed to make most sense of the wear traces in relation to a task. The key issue is to identify the most promising general task and then examine the 'best fit' examples within the genre.

Figure 21.4 shows the ample use of such a serrated tool for

Sequence	Task
1.	harvest the plant materials in their prime; seasonal implications
2 (or 5)	dry them prior to use; seasonal implications
3 (or 4)	remove the fibres from the inner pith or wood *(process can involve retting but need not)*
4 (or 3)	remove the epidermis or outer bark layers *(process can involve retting but need not)*
5 (and/or 2)	dry them prior to use; some seasonal implications but fibres processed manually in this stage will now dry quickly and take up less storage space
6	if fibres are to be twisted, align and prepare them in ways suitable for the different ways of twisting
7	twisting (on the thigh, with a spinning hook, with a spindle)
8	ply or cable if necessary
9	make into nets, baskets, fabrics *etc.*

Fig. 21.3. Generalised and simplified early stages in the plant fibre chaîne opératoire *for producing material culture with more variations possible (© L. Hurcombe).*

Fig. 21.4. Using the tool to scrape the outer bark from willow bast fibres while still on-the-wood (© L. Hurcombe).

manual stripping of the outer bark of willow, which can then be peeled away cleanly from the inner wood in the spring and early summer when the sap is still running. This is a relatively rapid manual technique for stripping the unwanted material away from the desired bast fibres. In 2006, a variety of tree bast fibres were processed in this way. Such experiments also suggested that nettles could perhaps be scraped to remove all of the fibres from their woody stems and epidermis. This was tried and was successful to a limited extent. The fibres came away very easily; however, they came away in a tangled mass using this technique and thus they added to the job of aligning the fibres and sorting them for spinning, and also they still had the epidermis attached, which, if a tight spin is wanted, is not ideal. Here, the work of an expert weaver, Anne Batzer at the Land of Legend Lejre, Centre for Historical-Archaeological Research and Communication in Denmark

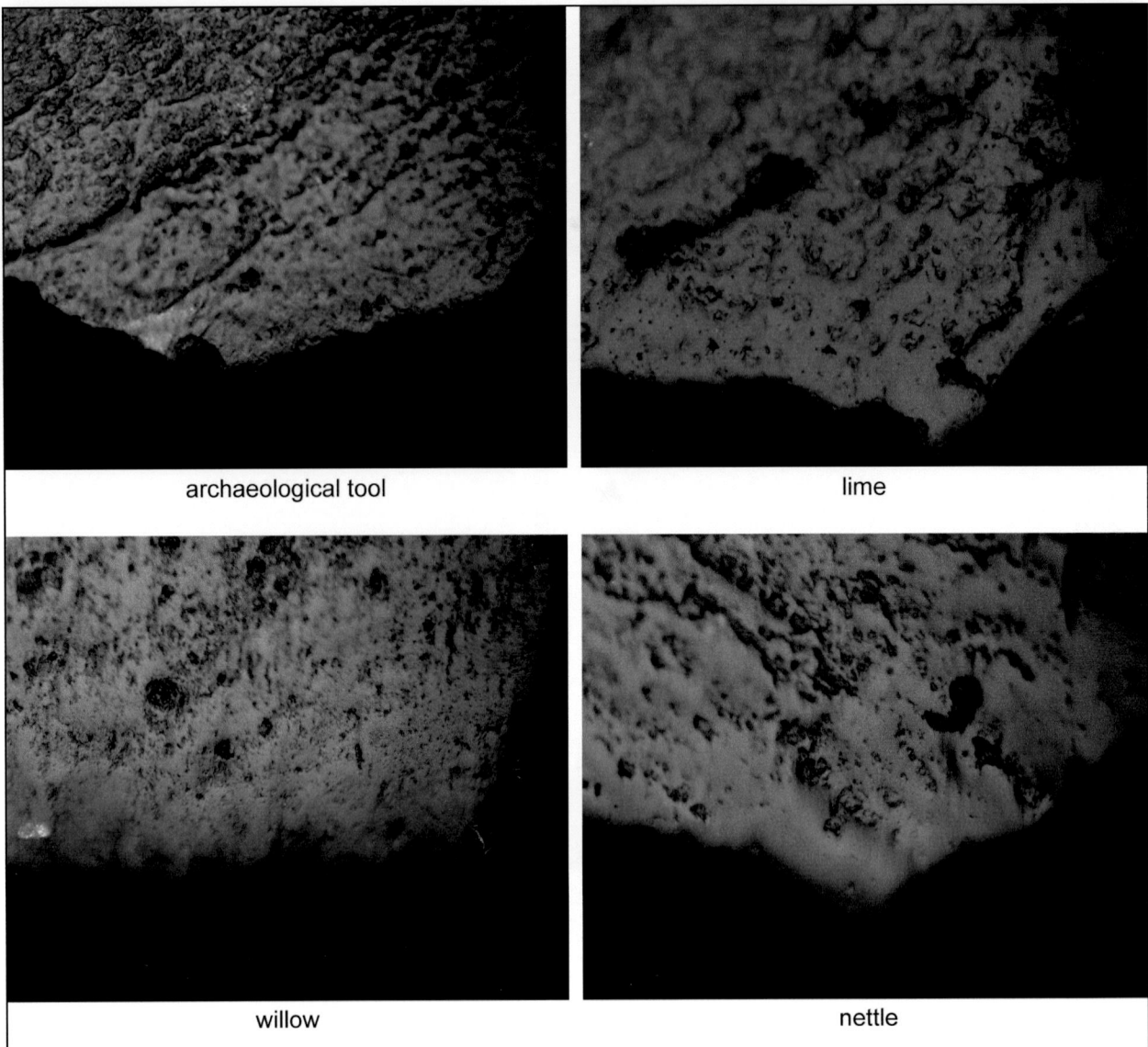

Fig. 21.5. Wear traces (original magnification 200×) on the ventral surface of an archaeological tool and three experimental tools used for 11–12 hours to scrape the outer bark off lime, willow and nettle (© L. Hurcombe).

was invaluable in showing that it would be impractical to aid one part of the *chaîne opératoire* by a particular technique if it caused a great deal of more work in another part of the *chaîne opératoire*.

A comparison of the tools used to scrape the outer bark off lime and willow shoots and saplings for three hours did show some positive points in common but did not match the archaeological tool wear traces in all aspects. These conclusions were drawn by using microscopic analyses at 50, 100, 200 and 500×. The outer bark stripping tasks showed the same polish distribution overall at magnifications of 50× and 100× and the same distribution over the micro-topography, as well as similar microscopic edge damage on some pointed tips. However, there were slight differences at 200× and 500× magnification in the fine polish textural details. The closest of these matches was actually provided by scraping the fibres and outer bark off the nettle together to the woody pith, but it was still not a good match at the higher levels of magnification.

Furthermore, although the distribution of wear traces was right, it was obvious that the wear traces on the experimental tools used for three hours were not as well developed as those on the archaeological examples. The following year (2007), another set of experiments was conducted with a standard duration of 11–12 hours. The wear on these experimental tools is matched against the archaeological wear in Figure 21.5. The distribution of wear traces was correct and the tool proved very effective at such tasks, so the ergonomics of the task in all three cases was appropriate, but the specific contact material seemed not to be correct. However, two aspects of the experiences were important. Firstly, Anne Batzer showing the need to keep the fibres aligned and, secondly, the experience of removing the outer bark from the tree bast fibres by scraping on-the-wood and only then removing the bast fibres from the wood. Based on these practical insights, the latter *chaîne opératoire* was extended to nettle and found to work in much the same way as for the tree bast fibres. It

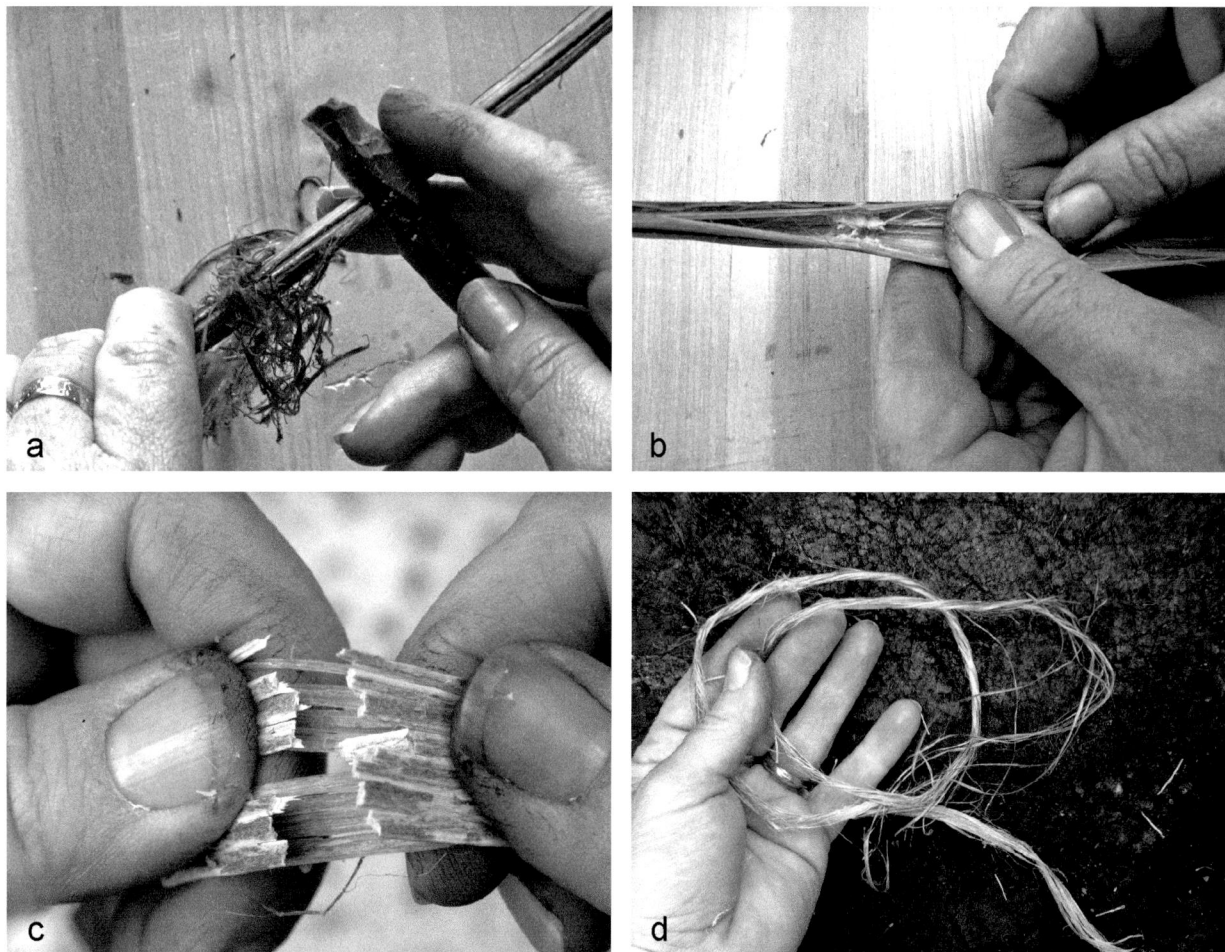

Fig. 21.6. A new chaîne opératoire *for extracting fine fibres from nettles manually using the stone tool. The nettle stem is stripped of leaves, then the outer bark is scraped off using the tool. After careful splitting of the stem and thorough drying, the fibres are carefully peeled away from the woody pith in good alignment ready for use in thigh rolling or hook spinning (© L. Hurcombe).*

was found that a very light scraping action down the nettle stem once the leaves had been stripped off could skim off most of the epidermis from the fibres and the stem could then be split and allowed to dry (Fig. 21.6). It was found that, several days of drying was best, and at this point the split nettle stem could be broken in half and the fibres, with care, extracted in one long section. The fibres produced were judged by the expert spinner and weaver to be long and good material for spinning up. Thus, it is possible to produce good fibres for fine work without using water as part of the *chaîne opératoire*, but instead stripping both epidermis or outer bark and inner pith or wood from the fibres manually. This has the advantage that the fibres come off in a long hank, and, where the nettles grow to about shoulder height, it is possible to get off *c.* 1 m of spinnable fibres. From 18 nettle stems of similar height and growing in the same conditions, the manual method described produced average weight of 0.744 g of spinnable fibres per nettle stem. The range was 0.45–1.30 g. Very little time need then be spent aligning the fibres. Although these lengths could be processed and cleaned further, they can also be used as they come off the stalk. In this manner, approximately 7 minutes of scraping per nettle stem, splitting the stem in perhaps 2 minutes, drying it for several days, and then spending a further 4 to 6 minutes stripping the fibres from the pith gives a total active processing time per stem of approximately 15 minutes plus drying time.

The wear traces produced by using the tool on nettle stems to skim off the outermost layers only down to the bast fibres produced the closest comparisons to date to the archaeological traces. Across all the lower scales the match was excellent, at 200× and 500× magnification there were good matches of spread and extent over the microtopography and in terms of general brightness, but, there are very slight textural differences which suggest there is still scope for further refining the experiments. Minor variations are planned for summer 2008.

It is also possible to split fresh nettle stems and strip off the fibres and outer bark together by hand. This can be called the Mears method since Ray Mears gives a good account of this technique (Mears 2005). Such material can be spun up for string but, for finer material, it might be desirable to remove the outer bark. Our experiments showed that it was possible to do this by lightly beating or hand manipulating (jiggling) the hank once dry, but these techniques were invariably time-consuming (10–20 minutes) and the fibres were lightly damaged or misaligned during these processes. As part of the

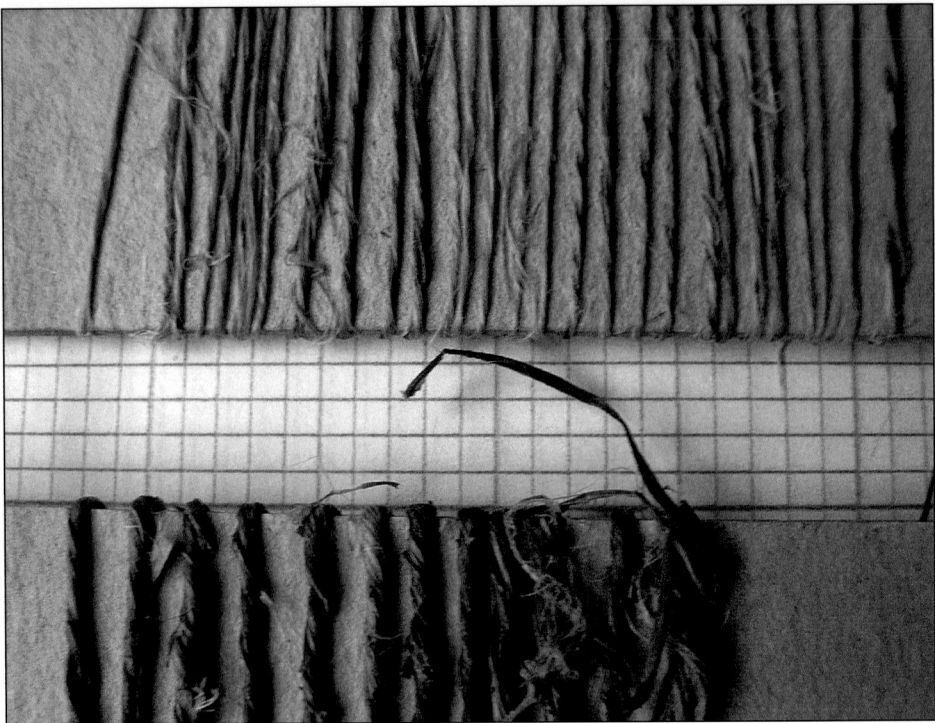

Fig. 21.7. The difference made by the using the tool for 6–8 minutes extra processing time per stem: the fibres below have been hook spun after manually stripping outer and inner bast together from nettle stems, the fibres above have been hook spun after using the tool to remove the outer layer from the fibres whilst on-the-stem and only then stripping the remaining inner bast fibres from the stem (© L. Hurcombe).

chaîne opératoire approach adopted by the project, the fibres from all the variations attempted were twisted up by Anne Batzer using a spinning hook. The Mears method and the Hurcombe method of nettle fibre processing were directly compared. The Hurcombe *chaîne opératoire* is illustrated in Figure 21.6. It can best be described as follows: cut the nettle stem or pull the nettle; strip off the leaves; lightly scrape with a serrated tool to remove most of the epidermis ensuring that the fibres are not cut by the tool in use, but are skinned; the stems are then split carefully in half and flattened and left to dry; and lastly the pith is carefully allowed to fall away as the fibres are stripped from the stems – the fibres come off well aligned and with a slight twist and easily form a small coil. In the images shown of this process, it is only the serrated tool which survives in the archaeological record. Figure 21.7 shows a direct comparison between the fibres hook spun from the Mears method and the Hurcombe method using the tool. Thus for approximately 7 minutes extra work per nettle stem with the serrated stone tool, a much finer product is obtained.

Conclusions

To go back to the questions posed at the start for this tool type, although a variety of rushes, reeds and tree bast fibre extraction processes have been the subject of experimentation, the best match for the archaeological wear traces at low power is scraping the outer bark off woody stems or saplings prior to stripping the fibres from the wood/pith. At higher magnifications, using the tool in such a way on nettle stems is the best match to date, but further experimentation may offer refinement of this idea. The experimental work has convincingly demonstrated that serrated edges are well-suited to such tasks. In addition, experiments have demonstrated that nettle fibres can be manually processed without using water to ret the stems.

To return to some of the larger questions, the nettle stem is small, the nettle grows widely across temperate Europe, and the processing of fine fibres in such a manner would require large commitments of time on a seasonal basis each year and explain the ubiquitous occurrence and numerous quantities of this tool in the archaeological record. The changing fashions of fabrics or changes in the *chaîne opératoire*, perhaps to either concentrate on flax as the raw material for fibres or to water-retting as a process for extracting plant fibres, could individually or together explain the dwindling nature of this once-prevalent tool type. The woody nature of the stem of the nettle in particular gives a reason why this particular *chaîne opératoire* works in a way that it would not on thinner or softer plant stems such as flax and rush materials. The larger question may yet be that if this tool is a marker for fine fibre production then it strongly suggests that, at this moment in prehistory, there was an explosion in the production of fine fabrics. The media of fibre-based material culture gives many possibilities for varying the pattern, colour/contrast, weave and style, as well as the design of bags and clothing, making this sphere of material culture a rich means of expressing personal and community identities, and it is perhaps this aspect which might drive the 'moment in time' aspect of the serrated tools. If the process in which they are used is based on fibre production for nets and cordage, there is no reason for it to have such a 'moment in time' character, but if they are seen as tools used in the manufacture of fine products expressing personal and cultural identity, this ties in with other cultural phenomena of the Neolithic.

Perspectives

After these experiments, it is possible to look again at some of the archaeological sites where such tools are prominent and where there is better preservation than usual, to see tantalising glimpses of what could be glaring evidence for fabric production in the earlier Neolithic in temperate Europe. As previously argued (Hurcombe 2000a), Etton, a causewayed enclosure in East Anglia, UK, is one such site (Pryor 1998, and especially reports by Nye and Scaife 1998; Taylor 1998). It seems to have many examples of cordage, basketry and fibre style tool production signatures. There are wood heeled awls, the antler comb (combs prevalent on these types of site in Britain could be for hackling or part of the production process for stripping leaves), the flat bone pieces could be scutching implements or be part of the fine processing removing the final part of the epidermis, and the y-shaped sticks, previously suggested as suitable for lifting hanks of retting or dying fibres from pools or vats (Hurcombe 2000a), could instead be seen as belt distaffs to aid in the production of spinning up such long fibres. In addition, Etton has a hank of plant material identified currently as flax, but with no secure reasoning given. It cannot help but be suspected that this might yet prove to be nettle. There are also internal pits with nettle macrofossil remains prevalent, despite the problems of nettle survival. Taken altogether, what started out as a very simple question – what is the function of this tool? – has entered a wider discussion on perishable material culture. The ideas formed over the years of experimentation with a variety of plant materials now strongly suggest that these tools are associated with fine fabric production, on present evidence most likely to be associated with nettles.

Acknowledgements

I would like to thank the Leverhulme Trust for funding a pilot phase of the 'organics from inorganics' project, and Land of Legend Lejre for funding further experimental work on bast fibre processing and to those who acted as assistants in the experiments at Lejre: Peter Groom, Mandy Pike, Emily Pike, Tine Schenck, and Hannah Simons. Marianne Rasmussen and the weavers Anne Batzer and Ida Demant were especially helpful and the 'textiles workshop' run by Lejre has enormously aided this work by creating opportunities for discussions with other experts such as Linda Mårtensson Olofsson and Anne Riechert. Discussions with other wear specialists have also been very fruitful and I thank Patty Anderson, Helle Juel Jensen and Annelou van Gijn (and others from Leiden) for recent and longstanding discussions. Exeter University has supported this research and the production of this article has been assisted at Exeter by Adam Wainwright, Sean Goddard and Mike Rouillard.

Bibliography

Alfaro, C. (1992) Two Copper Age tunics from Lorca, Murcia (Spain). In L. Bender-Jørgensen and E. Munksgaard (eds), *Archaeological Textiles in Northern Europe: Report from the 4th NESAT Symposium, 1–5 May 1990 in Copenhagen,* 20–30 Tidens Tand Nr. 5. Copenhagen, Konservatorskolen det Kongelige Danske Kunstakademi.

Alfaro Giner, C. (1984) *Tejido y Cestería en la Península Ibérica: Historia de su Técnica e Industrias desde la Prehistoria hasta la Romanización.* Bibliotheca Præhistorica Hispana Vol. XXI, Madrid, Instituto Español de Prehistoria.

Andersen, S. (1987) Tybrind Vig: A submerged Ertebølle settlement in Denmark. In J. Coles and A. Lawson (eds), *European Wetlands in Prehistory,* 253–280. Oxford, Clarendon.

Aranguren, B., and Revedin, A. (2001) Interprétation fonctionnelle d'un site gravettien à burins de Noailles. *L'Anthropologie* 105, 533–545.

Bender Jørgensen, L. (1990) Stone-Age Textiles in North Europe. In P. Walton and J. P. Wild (eds), *Textiles in Northern Archaeology. NESAT III: Textile symposium in York 6–9 May 1987.* London, Archetype.

Bender-Jørgensen, L. (1992) *North European Textiles until AD 1000.* Aarhus, Aarhus University Press.

Bender-Jørgensen, L. (1994) Ancient costumes reconstructed: a new field of research. In G. Jaacks and K. Tidow, (eds), *Archäologische Textilfunde – Archaeological Textiles, Textilsymposium Neumünster, 4–7 May 1993 (NESAT V),* 109–113. Neumünster, Textilmuseum Neumünster.

Bernick, K. (1998) Stylistic characteristics of basketry from Coast Salish area wet sites. In K. Bernick (ed.), *Hidden Dimensions: the Cultural Significance of Wetland Archaeology,* 139–156. Vancouver, University of British Columbia.

Beugnier, V. (2007) Préhistoire du travail des plantes dans le nord de la Belgique. Le cas du Mésolithique ancien et du Néolithique final en Flandre. In V. Beugnier and Crombé (eds), *Plant Processing from a Prehistoric and Ethnographic Perspective,* 23–40. Oxford, BAR Int. Ser. 1718.

Beugnier, V., and Crombé, P., eds. (2007) *Plant Processing from a Prehistoric and Ethnographic Perspective.* Oxford, BAR Int. Ser. 1718.

Bocquet, A. (1980) Le microdenticulé, un outil mal connu: essai de typologie. *Bulletin de la Société Préhistorique Française* 77:3, 76–85.

Bocquet, A. (1994) Charavines il y a 5000 ans. *Dossiers d'Archéologie* 199, 1–104.

Cardon, D. (1998) Neolithic textiles, matting and cordage from Charavines, Lake of Paladru, France. In L. Bender Jørgensen and C. Rinaldo (eds), *Textiles in European Archaeology: Report from the 6th NESAT Symposium, 7–11th May 1996 in Borås,* 3–22. Göteborg. Department of Archaeology, Gothenburg University.

Catling, D. and Grayson, J. (1998) *Identification of Vegetable Fibres.* London, Archetype.

Davy, D. (2007) Aroumans (*Ischnosiphon spp., Marantaceae*). Vannerie et symbolisme en Guyane française. In V. Beugnier and Crombé, P. (eds), *Plant Processing from a Prehistoric and Ethnographic Perspective,* 101–119. Oxford, BAR Int. Ser. 1718.

Deur, D., and Turner, N. J., eds. (2005) *Keeping it Living: Traditions of Plant Use and Cultivation on the Northwest Coast of North America.* Washington, University of Washington Press and Vancouver, University of British Columbia Press.

Dreyer, J., Drayling, G., and Feldmann, F. (1996) Cultivation of stinging nettle Urtica dioica (L) with high fibre content as a raw material for the production of fibre and cellulose: qualitative and quantitative differentiation of ancient clones. *Journal of Applied Botany* 70, 28–39.

Dufraisse, A. (2008) Firewood management and woodland exploitation during the Late Neolithic at Lac de Chalain (Jura, France). *Vegetation History and Archaeobotany* 17, 199–210.

Dunsmore, S. (1985) *The Nettle in Nepal: a Cottage Industry.* Surbiton, Land Resource Development Centre.

Favre, P., and Jacomet, S. (1998) Branch wood from the lake shore settlements of Horgen Scheller, Switzerland: evidence for economic specialization in the Late Neolithic period. *Vegetation History and Archaeobotany* 7, 167–178.

Grieve, M. (1931) *A Modern Herba.* London, Jonathan Cape.

Grosse-Klee, E. (1997) Schichtkorrelation der Grabungen Mozartstrasser, Kanalisationssanierung Seefeld, AKAD/Pressehaus, Kleiner Hafner und Utoqua. In C. Brombacher, E. Gross-Klee, H. Hüster-Plogmann, S. Jacomet, A. Rast-Eicher and J. Schibler (eds), *Ökonomie und Ökologie Neolithischer und Bronzezeitliche Ufer-Siedlungen am Zürichsee Band A: Text,* 20–29. Zürich, Kommunikation-Verl.

Gustafson, P. (1980) *Salish Weaving.* Vancouver, Douglas and McIntyre.

Hald, M. (1980) *Ancient Danish Textiles from Bogs and Burial.* Copenhagen, National Museum of Denmark.

Harris, S. (2007) Investigating social aspects of technical processes: cloth production from plant fibres in a Neolithic lake dwelling on Lake Constance, Germany. In V. Beugnier and P. Crombé (eds), *Plant processing from a prehistoric and ethnographic perspective,* 83–100. Oxford, BAR Int. Ser. 1718.

Hurcombe, L. M. (1998) Plant-working and craft activities as a potential source of microwear variation. *Helenium* 34, 201–209.

Hurcombe, L. M. (2000a) Plants as the raw materials for crafts. In A. Fairburn (ed.), *Plants in Neolithic Britain and Beyond,* 155–173. Oxford, Oxbow/Neolithic Studies Group.

Hurcombe, L. M. (2000b) Time, skill and craft specialisation as gender relations. In M. Donald and L. M. Hurcombe (eds), *Gender and Material Culture in Archaeological Perspective,* 88–109. London, Macmillan.

Hurcombe, L. M. (2007a) *Archaeological Artefacts as Material Culture.* London, Routledge.

Hurcombe, L. M. (2007b) Plant processing for cordage and textiles using serrated flint edges: new chaînes opératoires suggested by combining ethnographic, archaeological and experimental evidence for bast fibre processing. In V. Beugnier and P. Crombé (eds), *Plant processing from a prehistoric and ethnographic perspective,* 41–66. Oxford, BAR Int. Ser. 1718.

Hurcombe, L. M. (2007c) A sense of materials and sensory perception in concepts of materiality. *World Archaeology* 39, 532–545.

Hurcombe, L. M. (2008a) Organics from inorganics: using experimental archaeology as a research tool for studying perishable material culture. *World Archaeology* 40, 83–115.

Hurcombe, L. M. (2008b) Looking for prehistoric basketry and cordage using inorganic remains: the evidence from stone tools. In L. Longo and N. Skakun, (eds), *Prehistoric Technology 40 Years Later: Functional Studies and the Russian Legacy.* Oxford, Archaeopress, BAR Int. Ser.

Jacomet, S. (2004) Archaeobotany: a vital tool in the investigation of lake dwellings. In F. Menotti (ed.), *Living on the Lake in Prehistoric Europe: 150 Years of Lake-Dwelling Research,* 162–177. London, Routledge.

Jacomet, S. (2006) Plant economy of the Northern Alpine lake dwellings – 3500–2400 cal. BC. *Environmental Archaeology: the Journal of Human Palaeoecology* 11:1, 65–85.

Jacomet, S., Brombacher, C., and Dick, M. (1989) *Archäobotanik am Zürichsee: Ackerbaum Sammelwirtschaft und Umwelt von Neolithischen und Bronzezeitlichen Seefersiedlungen im Raum Zürich. Ergebnisse von Untersuchungen Pflanzlicher Makroreste der Jahre 1979–1988.* Zürich, Füssli.

Juel Jensen, H. (1994) *Flint Tools and Plant Working: Hidden Traces of Stone Age Technology.* Aarhus, Aarhus University Press.

Kirby, R. H. (1963) *Vegetable Fibres: Botany, Cultivation and Utilization.* London, Leonard Hill.

Kooistra, L. (2006) Fabrics of fibres and strips of bark. In L. P. Louwe Kooijmans and P. F. B. Jongste (eds), *Schipluiden: a Neolithic Settlement on the Dutch North Sea Coast c. 3500 CAL BC,* 253–259. Analecta Praehistorica Leidensia 37/38. Leiden, Faculty of Archaeology, Leiden University.

Küchler, S., and Were, G., eds (2005) *The art of clothing: a Pacific experience.* London, UCL Press.

Kuoni, B. (1981) *Cestería Tradicional Ibérica.* Barcelona, Ediciones del Serbal.

Maier, U. (1999) Agricultural activities and land use in a Neolithic village around 3900 BC, Hornstaad-Hörnle IA, Lake Constance, Germany. *Vegetation History and Archaeobotany* 8, 87–94.

Maier, U., and Vogt, R. (2001) *Siedlungsarchäologie im Alpenvorland VI: Botanische und Pedologische Untersuchungen zur Ufersiedlung Hornstaad-Hörnle IA.* Stuttgart, Theiss.

Marles, R. J., Clavelle, C., Monteleone, L., Tays, N., and Burns, D. (2000) *Aboriginal plant use in Canada's northwest boreal forest.* Vancouver, UBC Press.

Martial, E., and Médard, F. (2007) Acquisition et traitement des matières textiles d'origine végétale en Préhistoire: l'exemple du lin. In V. Beugnier and Crombé (eds), *Plant processing from a prehistoric and ethnographic perspective,* 67–82. Oxford, BAR Int. Ser. 1718.

Mears, R. (1990) *The Survival Handbook.* Oxford, Oxford Illustrated Press.

Mears, R. (2002) *Bushcraft.* London, Hodder and Stoughton.

Mears, R. (2005) Making string from nettle, skills section. *Bushcraft* [DVD]. London, BBC.

Médard, F. (2000) *L'Artisanat Textile au Néolithique: l'Exemple de Delley-Portalban II (Suisse) 3272–2462 avant J.-C.* Préhistoires 4. Montagnac, Monique Mergoil.

Nye, S. and Scaife, R. (1998) Plant macrofossil remains. In F. Pryor (ed.) *Etton: Excavations of a Neolithic Causewayed Enclosure near Maxey, Cambridgeshire 1982–7,* pp. 289–300. English Heritage, London.

Owoc, M. A., and Gilligan, S. (2006) Prehistoric String Theory: Perishable Impression Design, Typology and Significance in the Southern British Bronze Age. Paper presented at the Society for American Archaeology Annual Meeting, April 2006, San Juan, Puerto Rico.

Pryor, F. (1998) *Etton: Excavations at a Neolithic Causewayed Enclosure near Maxey, Cambridgeshire 1982–7.* London, English Heritage.

Rast, A. (1990) Die Verarbeitung von Bast. In *Die ersten Bauern: Pfahlbaufunde Europas/Forschungsberichte zur Ausstellung im Schweizerischen Landesmuseum und zum Erlebnispark / Ausstellung Pfahlbauland in Zürich, 28 April bis 30 September 1990,* 119–122. Zürich, Schweizerischen Landesmuseum.

Rast-Eicher, A. (1992) Neolitische Textilien im Raum Zürich. In L. Bender-Jørgensen and E. Munksgaard (eds), *Archaeological Textiles in Northern Europe, Report from the 4th NESAT Symposium, 1–5 May 1990 in Copenhagen,* 9–19. Tidens Tand Nr. 5. Copenhagen, Konservatorskolen det Kongelige Danske Kunstakademi.

Rast-Eicher, A. (1997a) Die Textilen. In C. Brombacher, E. Gross-Klee, H. Hüster-Plogmann, S. Jacomet, A. Rast-Eicher and J. Schibler (eds), *Ökonomie und Ökologie Neolithischer und Bronzezeitliche Ufer-Siedlungen am Zürichsee Band A: Text*, 300–328. Zurich, Kommunikation-Verl.

Rast-Eicher, A. (1997b) Die Textilen. In C. Brombacher, E. Gross-Klee, H. Hüster-Plogmann, S. Jacomet, A. Rast-Eicher and J. Schibler (eds), *Ökonomie und Ökologie Neolithischer und Bronzezeitliche Ufer-Siedlungen am Zürichsee Band B: Datenkatalog*, Tabellen D 62–360. Zürich, Kommunikation-Verl.

Rast-Eicher, A. (2005) Bast before wool: the first textiles. In P Bichler, K Grömer, R. Hofmann-De Keijzer, A. Kern and H. Reschreiter (eds), *Hallstatt Textiles: Technical Analysis, Scientific Investigation and Experiment on Iron Age Textile*, 117–131. Oxford, Archaeopress, BAR Int. Ser. 1351.

Rast-Eicher, A., and Schweiz, E. (1994) Gewebe im Neolithikum. In G. Jaacks and K. Tidow (eds), *Archäologische Textilfunde – Archaeological Textiles, Textilsymposium Neumünster, 4–7 May 1993 (NESAT V)*, 18–26. Neumünster, Textilmuseum Neumünster.

Reinhard, J. (2000) Textiles et vannerie. In D. Ramseyer (ed.), *Muntelier / Fischergässli: un Habitat au Bord du Lac de Morat (3895 à 3820 avant J.-C.,* 200–205. Archéologie fribourgeoise 15. Fribourg, Editions Universitaires Fribourg Suisse.

Saville, A. (2002) Lithic artifacts from Neolithic causewayed enclosures: character and meaning. In G. Varndell and P. Topping (eds), *Enclosures in Neolithic Europe,* 91–105. Oxford, Oxbow Books.

Schlichtherle, H. (1990), *Die Sondagen 1973–78 in den Ufersiedlungen Hornstaad-Hörnle I. Siedlungsarchäologie im Alpenvorland I: Forschungen und Berichte zur Vor und Frühgeschichte in Baden-Württemberg*, 36. Stuttgart.

Silva, R. J. and Keeley, L. H. (1994) "Frits" and specialized hide preparation in the Belgian Early Neolithic. *Journal of Archaeological Science* 21, 99.

Spindler, K. (1995) *The Man in the Ice*. London, Phoenix.

Stewart, H. (1973) *Indian Artefacts of the Northwest Coast.* Seattle, University of Washington Press.

Stewart, H. (1977) *Indian Fishing on the Northwest Coast.* Vancouver, Douglas and McIntyre.

Stewart, H. (1984) *Cedar: Tree of Life to the Northwest Coast Indian.* Vancouver, Douglas and McIntyre.

Taylor, M. (1998) Wood and bark from the enclosure ditch. In F. Pryor (ed.), *Etton: Excavations at a Neolithic Causewayed Enclosure near Maxey, Cambridgeshire 1982–7,* 115–160. London, English Heritage.

Turner, N. J. (1998) *Plant Technology of First Peoples in British Columbia.* Vancouver, University of British Columbia Press.

Turner, N. J. (2004) *Plants of Haida Gwaii.* Winlaw (British Colombia), Sono Nis Press.

Vaughan, P., and Bocquet, A. (1987) Première étude fonctionnelle d'outils lithique Néolithiques de village de Charavines Isere. *L'Anthropologie* 91, 399–410.

22 Construction and Sewing Technique in Secular Medieval Garments

by Katrin Kania

While medieval clothing and fashion have attracted students and researchers for a long time, the techniques of their making have often been neglected. Between the studies of archaeological textiles, most often centred on fabrics and materials, and the studies of historical costume, usually based on text and contemporary art as main sources, there is a large gap. The well-published theories about the development of clothing and fashion spring mainly from costume history and rarely take archaeological finds into account.

An attempt to trace the development of medieval garments based on the archaeological evidence was the subject of my PhD thesis. It was set up in three parts: a catalogue of garments; an analysis of the extant pieces with the additional use of text and picture sources; and finally an experimental section to try out the reconstruction of the medieval tailor's procedures and methods. The following text presents a summary of the main results of this thesis.[1]

Part I – Catalogue of Extant Garments

Since it was to serve as the basis for the other two parts, the first step in the appraisal of medieval clothing and the techniques used was to compile a comprehensive catalogue of objects that can provide information on tailoring techniques and clothes development. The catalogue of extant textiles consists mainly of complete garments or large fragments from different sources: archaeological finds, relics and other extant garments from across Europe (excluding the very east of Europe as well as Asia; see Fig. 22.1). The garments date from about AD 500 to AD 1500, though there are fewer garments

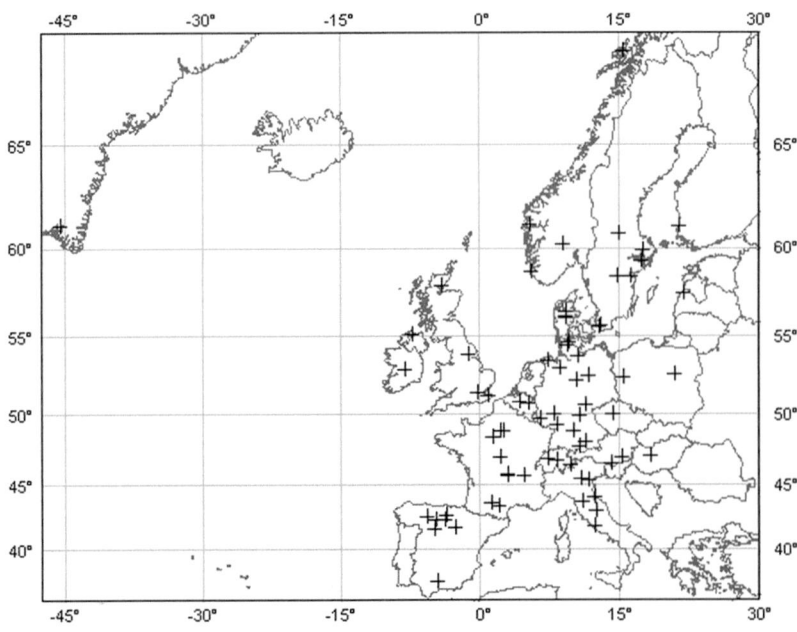

Fig. 22.1. Map with garment finds. (K. Kania, plotted with PanMap Software).

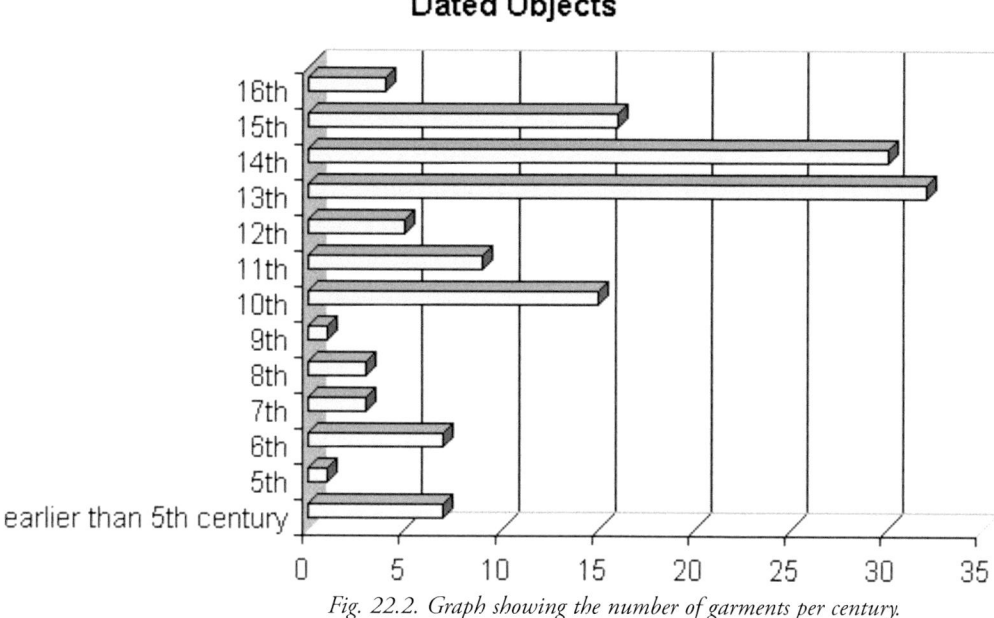

Fig. 22.2. Graph showing the number of garments per century.

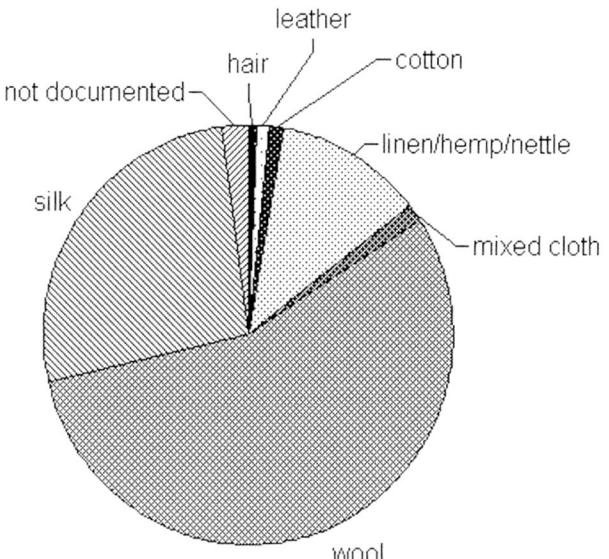

Fig. 22.3. Graph showing the proportions of the different materials constituting the garments.

in the first half of this period, from AD 500 to AD 1000 than in the latter 500 years. Headdresses and shoes were excluded from the catalogue, since shoemaking is a different trade and already well researched, and most headdresses are not sewn and thus not comparable to body garments. All other kinds of garments – undergarments, tunics and dresses, overgarments like mantles and leg- and footwear from cloth – are listed and described. Including some few earlier and later garments as comparison, the catalogue consists of 173 pieces altogether, with about one fifth of them not dated more precisely than 'medieval' or 'probably medieval'. There are only 15 objects from the 5th to 9th centuries, 24 garments date to the 10th and 11th centuries, 37 are from the 12th and 13th centuries, 46 pieces from 14th and 15th centuries, 11 are earlier or later than that, and the rest is undated (compare Fig. 22.2). The most common materials, as would be expected, are wool and silk, since both animal fibres stand up better to the usually acidic soil environment; vegetable fibres and other materials occur only rarely (see Fig. 22.3).

To make the comparison between the garments in the catalogue easier, the pieces were classified into four categories: undergarments, garments, overgarments and leg- or footwear, and sorted chronologically within the categories. The undergarments are defined as clothes that would not be an acceptable outer layer of clothing in public; wearing only undergarments would be followed by social repercussions. Garments are the expected outer layer when being in public, and, finally, overgarments can be added to serve as an additional protection against cold, dirt or bad weather, or to serve as an added representative layer, increasing the value of the outfit. While the assignment of garments to the various categories must be based on visual and textual sources, ambiguities often do not allow for exact certainty on, for example, whether a certain piece was an undergarment or a garment. The status of such a garment might also depend on the social environment in which its wearer moves (see Vavra 1988 on problems of categorizing garments). Despite these problems – and surely some of the choices in the sorting will merit a debate – the categories were introduced because of much easier comparison between pieces: with this set-up, tunics that are chronologically close to each other can be compared without the need to skip over the entries of leg-wear, mantles and shirts in between.

Altogether, there are 21 undergarments (18 of them dated), 79 tunics and dresses (55 of them dated), 35 overgarments (all dated) and 38 pieces/pairs of legwear (31 of them dated).

This number of garments, although assembled with great

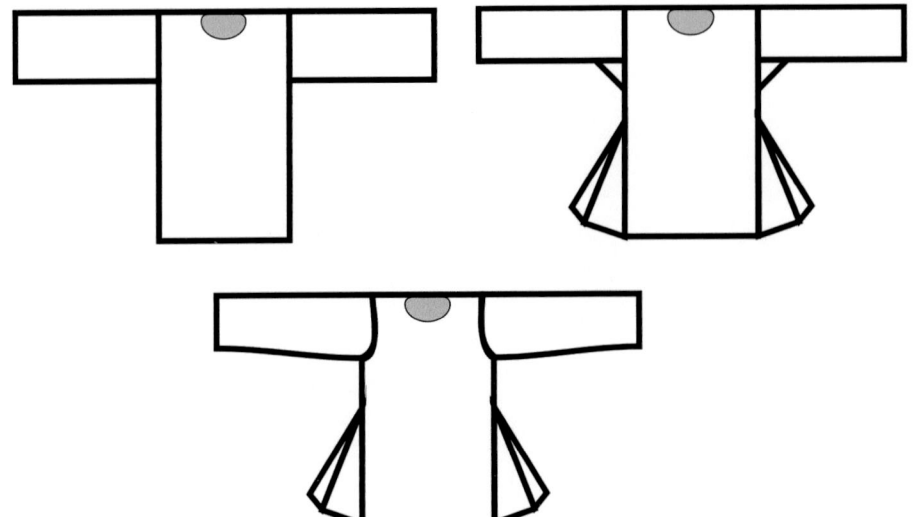

Fig. 22.4. Sleeve forms in use during the Middle Ages: Top left tunic with simple rectangular sleeves, top right sleeves with underarm gussets, bottom tunic with armscye and sleeves with curved sleeve head (Drawing: K. Kania).

care to find as many pieces as possible, is a paltry amount upon which to base a whole continent's development during a time span of 1000 years. Moreover, the very wide range – from elaborate, tight-fitting garments to the simplest cut conceivable – makes giving any certainties or any exact dates of first occurrence highly risky. However, while it is not possible to find the date or the place where a new development in the cut and construction of garments took place, we can try to comprehend why and how the different developments occurred.

Part II – The Analysis

Garments that are woven directly on a loom seem to be the first type of clothing in the extant pieces, long before the Middle Ages. Making garments woven to shape signifies a great deal of complicated work in setting up the loom for just one piece of clothing, and this complex manufacture can be made easier by weaving separate rectangles that are sewn together to form the body part and sleeves of a tunic, and easier again by weaving a large cloth and cutting the pieces from this, as can be seen on a tunic from Martres-de-Veyre, dated to the 2nd century (Roche-Bernard and Ferdière 1993, 11–12). Clothing woven to shape remains common until at least the 3rd century AD. However, elaborate cut already did exist at that time, as can be seen by the example of the trousers from Thorsberg in North Germany, dated to c. AD 175 (Schlabow 1976; Müller 2003, 70, footnote 26). These trousers can be reconstructed as very tight, form-fitting leg wear, displaying even the ripple of muscles through the cloth, while still granting its wearer full freedom of movement (Kania 2007). With this piece of tailoring in mind, it should be safe to assume that form-fitting, well-tailored clothing could have existed in the Early Middle Ages as well, even if textual and iconographic sources suggest the existence of tight garments only from the High Middle Ages onwards.

The discernible types of tunics in the Early and High Middle Ages are only few, and they are all quite simple. The most basic tunic consists of a long rectangle, folded in half and fitted with a hole in the middle to accommodate the head of the wearer, and two smaller rectangles that form relatively wide, straight sleeves. This corresponds to the form of garments woven to shape or garments made from separately woven rectangles (Fig. 22.4, top left). With a tunic cut like this, the sleeves have to be wide to prevent the ripping of the joint under the arm, where sleeve and body part of the tunic meet. In a wide sleeve, the arm can move without taking the fabric with it and thus the stress on the underarm seam is reduced. For tighter sleeves, some other means to prevent ripping is needed. The common practice for this is inserting an underarm gusset, a square piece of fabric that sits between the sleeve and the body of the garment (Fig. 22.4, top right). These underarm gussets are a common part of simple-sleeved tunics in the Middle Ages, and they can be found in undergarments as late as the 19th or even 20th century. Since these gussets are squares with a side length of usually around 10–15 cm, weaving them separately would not have been very practical. Thus it can be assumed that tighter sleeves with underarm gussets only came in use when the garment parts were cut from cloth.

With or without underarm gussets, the seam between sleeve and body part runs across the upper arm in both cases. For tight garments with tight sleeves, this placement of the connecting seam causes problems, since the arm is hindered in its movements by the seam joint. The solution for this problem is to set back the seam so that it is now placed right on top of the shoulder and does not get moved along with the arm. This can be done by removing some fabric at the shoulder part from the body rectangle – a procedure that would initiate the development of the armscye and also of its counterpart, the curved sleeve-head (Fig. 22.4, bottom). These two together would greatly improve the fit of the garment on the upper torso and make much tighter sleeves possible. This typical medieval way of setting the sleeve into the garment, with

Fig. 22.5. Comparison between tunic with riding slit with no additional middle-gores (left side) and with the riding slit in middle-gores (right side) (Photo: K. Kania)

the low part of the sleeve-head to the front of the garment, where the armscye is deepest, would provide tension to the front panel of the garment. This would emphasize the breast of the wearer, and naturally was especially eye-catching when used on women's garments.

Until the 13th century, only three types of garments can be found in the archaeological evidence. The first type consists of a rectangular front and back panel and simple rectangular sleeves, with or without gusset added (as in Fig. 22.4, top left). To allow sufficient ease of movement, the seams have to be left open from the hips downward, resulting in two side slits. For a garment closed all around, gores are needed to make movement possible. These gores are inserted into the sides of the tunic instead of leaving them with slits (as

Fig. 22.6. Mi-parti garment, schematic drawing (Drawing: K. Kania).

Fig. 22.7. Quarter-cut garment, schematic drawing (Drawing: K. Kania).

in Fig. 22.4, bottom). Both of these simple forms are not suitable for horse-riding, however: the side-slits of the first tunic will uncover the legs of the wearer, and the moderately wide tunic with gores will ride up the thighs when the legs are spread apart, also uncovering most of the legs. The solution is the riding-slit, an additional slit cut into the front and back of the tunic.

With just a simple slit cut into the front and back rectangle of the second type of tunic, the garment will function as intended for horse-riding, by covering the thighs of the wearer when on horseback. However, the slit in the tunic will open when walking, too, and the garment might even gape open at the slit when standing. This would not make the tunic a very good, warm and practical garment for everyday use, which might explain why there is no archaeological evidence at all for that version. The solution to the 'gaping problem' is, again, to insert gores that will add the necessary width. Instead of cutting the riding-slit into the front and back panel itself, a triangular gore is inserted there, and the riding-slit transferred into the middle-gores. In this manner, the riding-slit will remain hidden when walking or standing, and the legs are well covered when riding. Figure 22.5 shows the difference in drape and fall between a tunic with just the slit cut into front and back panel (left) and with the gores inserted (right). Tunics with gores and riding-slit can be found several times in the archaeological evidence, for example, in the tunic from Moselund (Østergård 2004, 135, 141), while the version without gores, but with a slit, is not documented at all. Together with the fact that there are no medieval illustrations showing the characteristic gap of a slit with no gores, this can lead us to assume that a tunic form with a slit, but no gores has probably never been in use.

With the altered fall and drape of the tunic that middle-gores will bring in comparison to a tunic without middle-gores, we can suppose that even people not needing a riding slit could have wanted to give the impression of higher status by inserting middle-gores into their tunics. Examples of this can also be found in the archaeological evidence; the tunic from the Bocksten find is one of them (Nockert 1997).

These three types of tunics – no gores but side slits, side gores, and side and middle-gores with or without riding slit – remain the basis for all garments until the very end of the 12th or the beginning of the 13th century. At this time, a change in the tailoring concept can be traced back to the appearance of the closed helmet in the military sector. These face-concealing helmets make recognition of individuals on a battlefield difficult or even impossible, leading to the development of heraldry as a means of distinguishing between persons. However, heraldry and marking the shield with the arms of a person are not sufficient for distinguishing or collating all the persons on a battlefield. A very easy way to show where a person belongs is to use the garments, since these are worn by everybody anyway. For the additional distinction, different colours were used: colours that matched the coat of arms of the corresponding knight or lord.

This is achieved by making *mi-parti* garments, *i.e.* clothes that are divided vertically into two colours (Fig. 22.6). These *mi-parti* clothes, tailored in the two main colours of a person's coat of arms, can be used to show a relationship between the wearer of the heraldic arms and the other wearers of his colours.

While at first glance, only a small additional seam in the front and back are needed to make *mi-parti* garments, this would lead to a completely new concept of garments. Since both front and back need to be cut from differently-coloured cloth, the tailor has to mentally divide the front and back of the garment, and he also needs to change colours on the sides of the garment to show the same sequence: Colour A – Colour B from left to right in the back of the garment. This inevitably leads to a conception of the clothing in quarters – 'front left, front right, back left, back right' – instead of 'front, back, gores at the sides, middle-gores front and back' as before. In turn, this new concept leads to the development of a new cutting scheme of garments, the quarter cut (Fig. 22.7). While the old style of making garments can still be found in the Late Middle Ages, quarter-cut garments, like the gown of Queen Margareta (Geijer, Franzén and Nockert 1994) (Fig. 22.8), are a common style in the 14th and 15th

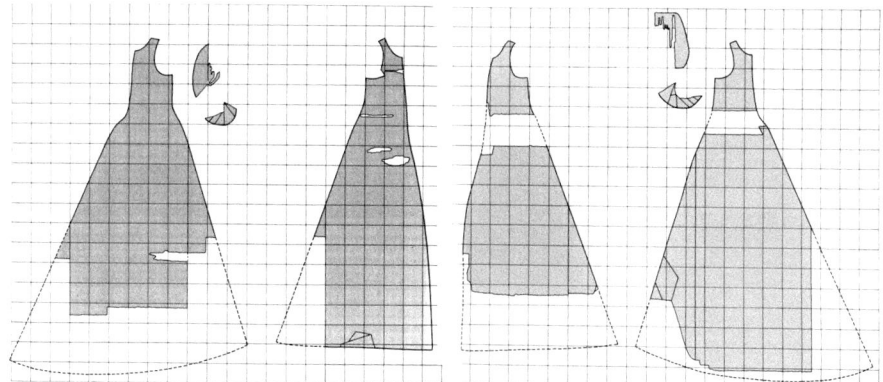

Fig. 22.8. Pattern of the gown of Queen Margareta, 15th century (After Geijer, Franzén and Nockert 1994, 22).

century and, basically, are still in use today, since a seam in the middle back and an opening in the middle front can be found on almost all modern blouses, jackets and blazers.

The additional seams in the middle front and middle back of the body now allow even tighter, more form-fitting garments than could be achieved before. Previously, the dress could be tightened by using side lacings between the front and back panel and gores from the hips downward. With the new quarter-cut concept, the lacing can be placed into the front of the dress or tunic, and shaping the garment to the body can be done on all four seam lines, not only on the sides. Using firm cloth as a lining to add stability to the dress, it is now even possible to some extent to shape the soft parts of the body, especially to lift the female breast to the desired high position on the torso, as can be seen on pictures like Jean Fouquet's depiction of the Virgin Mary, dating to the 1450s (Boucher 1987, 205).

The quarter cut also allows tighter garments for men, usually with a front closure. These male garments are the short, padded doublets that were probably developed for use under the armour. Padded garments have a long history of use as protective layer under the actual armour; not only do they protect the wearer from skin irritations caused by abrasions due to metal or hard-boiled leather on skin, but they also dampen any impact on the armour because of the padding. The short doublet emerges at a time when metal braces for the legs have become quite common in armouring, making a long padded garment for protection of the legs unnecessary and maybe even hindering movement. The tightness of the doublets also corresponds to the tighter, more form-fitting body armour that becomes increasingly more prominent during the 13th century. This military item – short, lightly padded and very form-fitting – then crosses over into civilian dress. Its sudden appearance in plain view has led to speculations about a hard break in clothing development in the past, but then unprecedented shortness and tightness can be explained by the evolution of the short doublet under the armour.

In depictions of short, tight doublets in civilian use, we can now see an elongation of the hose. While the longer tunics allowed the hose to end at mid-thigh, the doublet that is only slightly longer than hip-length necessitates hose that reach up to the hips. These longer hose now are attached to eyelets in the hem of the doublet with short strings (called *points*), while the old-style hose are tied to the belt holding up the *braies*, the wide, voluminous breeches worn under the long tunic. Since the breeches now no longer have any supportive function for the other legwear, they can be reduced to the bare minimum needed to cover the male genitals, resulting in underwear strongly resembling modern boxer shorts or even tanga slips. Those much reduced undergarments for men can be seen on late medieval and early modern illustrations (Mütherich and Dachs 1987, Table 72).

As this brief summary shows, there is a continuous line of development in clothing. The gradual development of garment construction can be traced in all areas of clothing before, throughout and beyond the Middle Ages. The changes that occur seem to spring from a practical requirement or from a development in another, *e.g.* military, context. Changes in one part of dress will influence other parts of dress and the other gender's dress as well. While there is often no possibility of dating the first appearance of a new type of garment, or of a new way of cutting and assembling a garment, due to the scarcity of archaeological sources, it is possible to grasp a *terminus post quem* when considering iconographic and textual evidence. The great range of variety in garment forms adds to the danger of misinterpretation: the simplest cuts for basic, non-representative garments can be found throughout all ages, and this has to be taken into account when judging the complexity or the possibilities of past tailoring.

Part III – Experimental Reconstruction

In recent years, experimental archaeology has grown and developed a great deal. Experiments as a means to evaluate reconstructed processes or the properties and uses of artefacts are widely accepted in archaeology. Compared to those in other fields, experiments with textiles and clothes have some additional limitations, since the properties and wearability of a garment depend on its proper fit, and evaluation therefore is less objective than in other approaches in experimental archaeology. Nevertheless, experimenting with medieval garments can provide valuable new ideas and insights.

To understand the construction of a garment, it is very helpful to know how its pattern came into being. Furthermore, if a sustainable idea of how medieval garments

Fig. 22.9. Dress of St. Elizabeth of Thuringia, Church St. Martin, Oberwalluf. The dress is housed in a neo-gothic shrine (Photo: K. Kania).

looked on their owners is needed, it is necessary to look at garments worn on the body they were tailored for, not at copies of extant garments that probably do not properly fit the model wearing them. For this, in turn, the reproduction of the methods used by the medieval tailor to make a length of cloth into the garment his customer desired is necessary.

Reconstructing these tailoring techniques can only be done by experimenting, based on the well-researched patterns of the extant garments. A few preliminary assumptions need to be taken into account as well. For instance, the modern method of complex constructing on large sheets of paper cannot have been used in the Middle Ages or before, as they lacked suitable, cheap material. Complicated mathematics for the pattern making can most probably be ruled out, since before the early Modern Age, there is no evidence for scaled measuring tapes that would be needed for this. Thus, a simple yet efficient way of cutting the pattern pieces is needed. Another common trait of the preserved medieval garments is the consequent use of simple geometric shapes as the base of every pattern piece, greatly reducing the amount of waste material when cutting. This economy also has to be taken into account when trying to reconstruct the tailoring procedures.

Starting from these cornerstones, it is possible to reconstruct a method of cutting and making garments that could very well have been used in the Middle Ages. In modern tailoring, the pattern of the garment is constructed on paper, transferred onto the fabric, and all the pieces are cut out at once. In contrast, the pattern pieces in the reconstructed method are cut out as geometric shapes in single groups and obtain their final shape by fitting right on the body of the garment's future wearer.

Using this reconstructed method results in garments that

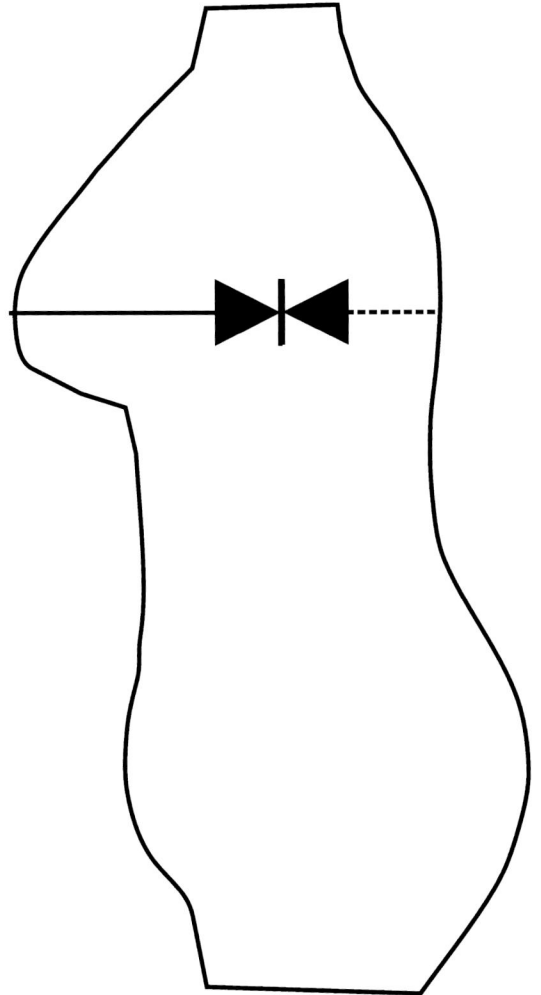

Fig. 22.10. Measuring width of front and back panel, from middle of torso to middle of torso (Drawing: K. Kania).

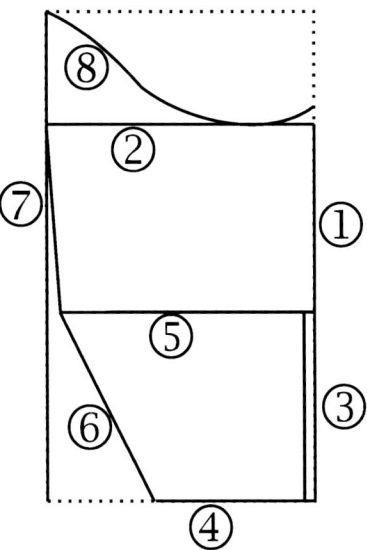

Fig. 22.11. Measurements for making the sleeve. Needed are the measurements for length of arm, straight and bent at 90° at the elbow, length of wrist to elbow, circumferences of hand/wrist, elbow bent at 90°, and upper arm (Drawing: K. Kania).

are cut efficiently and are very wearable and ergonomic. The reconstructed procedure is described below using an example of a simple female dress. A corresponding historical example is the gown of St. Elizabeth of Thuringia (Grönwoldt 1977), shown on Figure 22.9.

First of all, the width of the front and back panel are found by measuring from the middle of the torso to the middle of the torso on the other side (Fig. 22.10). This will automatically result in a wider front than back panel. The length of the two panels is determined by measuring from the person's shoulder to the floor and then adding the desired extra length for the typical floor sweeping dress with a little train plus some allowance for rounding off the hem and hemming. Typically, the back panel will thus be longer and narrower than the front panel.

Both panels are then cut from the length of fabric, following the grain. The top centre of each panel is marked with a small vertical cut, about 1 cm in length in the back panel and about 10 cm long in the front panel. The two pieces are centred on each other, and starting from the top corners of the back panel, they are then basted together on the top, leaving enough free space in the middle for the wearer's head. The piece can now be put on, enabling the tailor to take the shoulder slant right from the shoulders of the wearer by tracing the shoulder line with a piece of chalk. The shoulder seams will run along this marked line, starting from the point where the top of the shoulder meets the arm and ending at the edge of the neck-opening. The shoulder seam is then put in before size and position of the armscye are determined: It runs from the outer edge of the shoulder seam downward and then curves to the outside of the front panel, taking the shape of a halved letter U. The lower end of the armscye, the point where it meets the edge of the front panel, can be found by pulling the edge of the panel towards the back of the body and very slightly upwards. This puts tension on the fabric, pressing it to the body in the bust region. Gripping the panel edge at different heights when pulling, the fabric will conform better or worse to the bust form. The point where it will best emphasise the curves of the body marks the lower end of the armscye. With the sleeve set in, thus, the completed dress will accent the bust line by gently following the curve of the body.

The sleeve itself is the most complicated part of the garment, and more measurements are needed for making the sleeve pattern than for the rest of the dress combined. It is, however, possible to make up the pattern of the sleeve without resorting to pre-made templates or to mathematics; the schematic drawing in Figure 22.11 shows such a sleeve with reference lines drawn in. The wearer's arm is measured to get the length of the arm from armpit to wrist (1), length from the back of the shoulder to the wrist with the elbow at a 90° angle (7), the circumference of the elbow in that position (5), circumference of the upper arm at the shoulder (2), the circumference of the hand (4) and the distance between elbow and wrist (3). From these measurements, the sleeve can be constructed; the curved sleeve head has to be

23 Tiny Weaving Tablets, Rectangular Weaving Tablets

by LiseRæder Knudsen

Introduction

A project on tablet weaving in the Danish Iron Age took place under the aegis of the Danish National Research Foundation's Centre for Textile Research in Copenhagen. Its aim was to collect all evidence of tablet weaving in Denmark dated until *c.* AD 400 (the Early Roman Iron Age) (Ræder Knudsen forthcoming). At the same time, the very limited evidence of weaving tablets from this period was closely examined. Weaving tablets have been found in three different places in Denmark (Hald 1980, 225; Nielsen 2000a, 215; Sørensen forthcoming) and they are all quite unusual compared to general expectations of what a weaving tablet ought to look like. One is rectangular and the tablets from the two other find places are very small.

This led to the question: could the size and shape of these three tablets be due to practical considerations, and does their appearance represent evidence of a special working method or a special task? These issues are the focus of this paper.

Tablets of the Early Roman Iron Age in Denmark

From the Early Roman Iron Age in Denmark, there are finds from 3 different sites. The first comes from Dejbjerg (Fig. 23.1), where an intact tablet and fragments of others were found in a clay vessel close to two ceremonial wagons in a

Fig. 23.1. Dejbjerg tablets (NM 4746). Size of the tablet c. 4.8 cm × 5.5 cm (Photo: Lise Ræder Knudsen with permission from The National Museum, Copenhagen).

Fig. 23.2. Iron comb from Sejlflod with 6 tablets preserved in the corrosion. Left: the positive X-ray image; right: the same X-ray with one tablet and comb outlined. Size of tablet 3.4 cm × 3.4 cm (X-ray: Per Hadsund, Aalborg Museum).

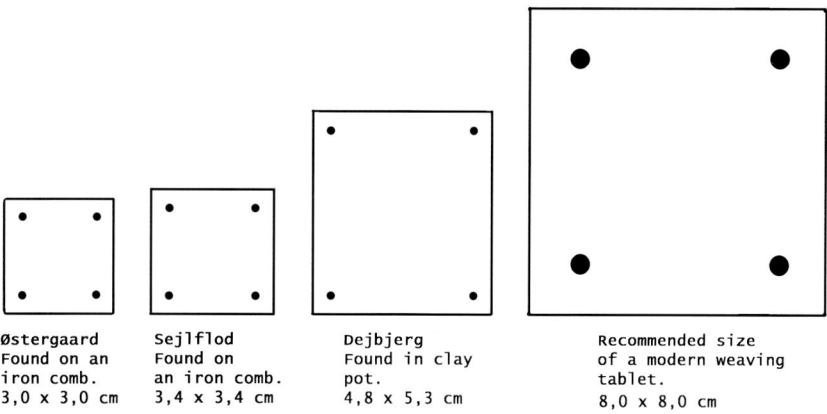

Fig. 23.3. Comparison between the size of the three preserved tablet from the Early Iron Age in Denmark and the recommended size of a modern weaving tablet (Drawing: L. Ræder Knudsen).

peat bog. This find is very famous and there are pictures of these tablets in many books on tablet weaving. Previously, the Dejbjerg tablets were dated to the Pre-Roman Iron Age on the basis of the wagons (Hald 1980, 225), but recently, ^{14}C dating indicates that they date to AD 90–270 and therefore it must be concluded that the clay vessel and tablets belong to another deposition in the bog than the wagons do (Egebjerg Hansen 1996, 211). The intact tablet measures 48 × 55 mm. It is 3–5 mm thick, the holes in each corner are quite small and they are placed 2.5–3.5 mm from the corner. The material is ash wood.

The second find is Sejlflod (Fig. 23.2), where a grave (ZZ) was found with grave goods including 15 glass and amber beads, probably a fibula, clay vessels, an iron rivet indicating the presence of a bone comb and an iron comb (Nielsen 2000, I, 215). On the iron comb, embedded in the corrosion, an intact tablet and fragments of other five tablets are preserved. The thickness of the tablets is about 2 mm and they measure 34 × 34 mm. The material used is sycamore wood. The holes are quite tiny, measuring a little less than 3 mm. The tablets can be clearly seen on the x-ray image of the comb.

The last example is an as yet unpublished find from Østergaard in Southern Jutland, where an even smaller weaving-tablet of 30 × 30 mm was also found embedded in the corrosion of an iron comb (Sørensen forthcoming).

The above mentioned weaving tablets found at the three different sites all seem unusual compared to tablet-weaving equipment from later periods, when a square tablet with an edge of 50–60 mm would be the standard: one tablet is rectangular and the tablets from the two other contexts are very small and both are preserved in the corrosion of iron combs (Fig. 23.3). Three finds seem to be a rather limited amount from which to draw conclusions. However, taking published material from other countries into consideration may yield some more information.

Weaving Tablets from Other Countries

In Norway, a very interesting weaving tablet dated to the Early Roman Iron Age was found at the site of Mæle, Osterøy, Hordaland (Bergen Museum inventory number B6981 I n). It is of exactly the same size as the tablet from Sejlflod and it was also found in the corrosion of an iron comb.

From El Cigarralejo in Spain, another very interesting and early find dated between 400 and 375 BC has been recorded. Several carbonised tablets measuring about 30 × 30 mm made of boxwood were found in grave 200. The tablets have a thickness of about 1 mm and rather large holes placed at some distance from the corners (Hundt 1968, 192–193). Even if the tablets have lost some of their volume during

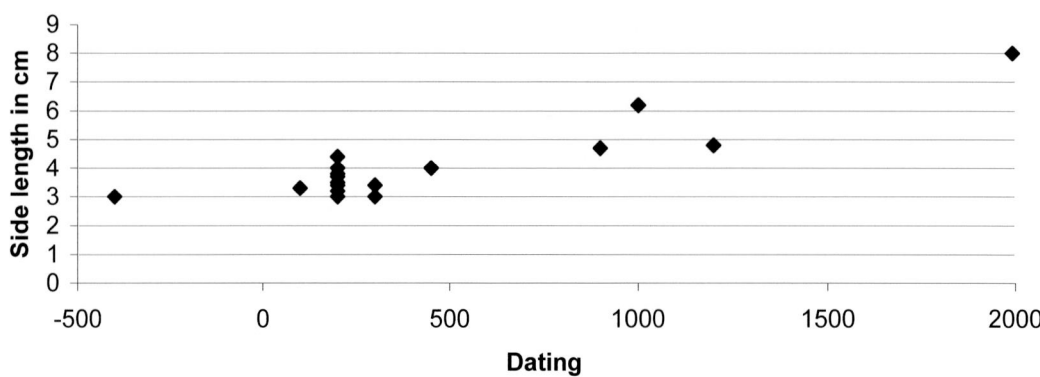

Table 23.1. Side length of square weaving tablets compared with the dating. Efforts have been made to collect all published literature describing weaving tablets dated up to AD 400. Only examples of weaving tablets dated to later periods are listed.

the carbonisation process, their original size must have been very small. The hard boxwood is most suitable for tablets as it is dense, flexible and can be polished giving a soft and smooth surface. Moreover, it is possible to make the tablets very thin and strong at the same time. In the same grave, a considerable amount of different textiles made of flax were found. Many of the fragments had tablet weaves integrated in the fabric and these are presumably starting- or side borders. One of the borders was woven using 33 tablets and the tablets were threaded three left, three right and so on (Hundt 1968, 191).

In the Musée du Cinquentenaire in Brussels (Musées Royaux d'Art et d'Histoire), there is a set of 25 weaving tablets found most likely at Antiöe in Egypt and dated to the 4th to 5th century AD (Crowfoot 1924, 100). Other tablets presumably found at the same location are kept at the Louvre, Paris. The tablets are about 40 × 40 mm and made of sycamore wood. Some of them still bear saw or rasp marks on their flat surfaces, and their sharp pointed corners and lack of smoothness suggest that they were never used. Others are very smooth and worn, and there are marks of heavy use around the holes (Collingwood 1982, 15 and 24). The marks of use are most obvious on the diagonal lines of the tablets suggesting they were mostly used in a position when the tablet loom has two sheds.

Several weaving tablets are known from Roman sites in Germany and England. Wild describes some 40 weaving tablets found at 28 different sites (Wild 1970). Of these, 23 are of triangular shape and 17 are square. All but one are made of bone and the remaining are bronze. The square tablets measure between 30 and 44 mm, the majority being less than 40 × 40 mm. Many of them have grooves around the holes indicating heavy or modest wear (Wild 1970, 140–141). Some of these tablets have engraved patterns of lines and circles on one side. The marking of one side of the tablet may serve as a practical help when creating a pattern weave, where the threading of the tablets changes during the weaving process (Behrens 1925, 46–47).

Furthermore, early evidence of tablets in Roman London was found at Upper Walbrook Valley, dated to around the 1st century AD. Bone waste of bovine shoulder blades was found with triangular saw marks. Evidence of about 70 triangular bone tablets could be found in the waste. Prichard concludes that, these are a hint of a relatively "professional level" of workshop production of weaving tablets (Prichard 1994, 157–160).

Tablets of a later date seem often to be larger. A square weaving tablet found in Trondheim, Norway dated with some certainty to the 10th century AD measures 62 mm (Christie 1985, 56). Fifty-two square tablets found in the Oseberg burial also in Norway (9th century AD) measure 47 × 47 mm (Christensen and Nockert 2006, 143). A tablet from a burial in Lund in Sweden dated to around AD 1200 measures 48 × 48 mm and has a runic inscription (Hald 1950, 227). However, very small square weaving tablets dating from the 9th to 12th centuries AD have been found in the Curonian parts of Latvia and Lithuania. Measuring from 11 mm to 27 mm, they are made of thin bronze plate and were found together with other miniature weaving equipment and most certainly used in necklaces. They are considered votive objects and are too small for real use (Götze 1907, 489–493; *Latvijas PSR arheoloģija*, 190–191).

Conclusions on the Evidence of Small Weaving Tablets

We started out in Denmark looking at the three existing samples of weaving tablets preserved from the Early Iron Age. A comparison with square weaving tablets found in Europe and Egypt shows us that we find the smaller weaving tablets in the earlier periods, whereas weaving tablets dated later are larger (Table 23.1).

What at first sight seemed to be unusual tools, because we compare them with our knowledge of tools throughout time, are in fact quite common for this period. This realization led to the following question: Are the tablet-woven textiles different in the Early Iron Age compared to later periods?

Fig. 23.4. Making a starting border for the warp-weighted loom using tiny tablets of 3.4 cm × 3.4 cm (Photo: © L. Ræder Knudsen).

Here my work for the Danish National Foundation's Centre for Textile Research in Copenhagen allowed me an ideal opportunity to answer this question, as all possible evidence of tablet-woven textiles in Denmark dated to the Early Iron Age or earlier had recently been collected. At first, the work on the early tablet-woven borders was estimated to be rather easy, as only a few and rather simple tablet-woven textiles from this period were known from the literature. During the work, however, more than 80 different tablet weaves from about 35 different locations were analyzed. Usually, tablet weaving is seen primarily as a band-weaving technique, which could be used as an integrated part of a fabric, for instance in a starting border for the warp of a warp-weighted loom. However, my research shows that we have to reconsider this approach. Among the tablet weaves from Danish Early Iron Age bogs and graves, we have approximately 50 well enough preserved fragments for analysis. Out of these, only approximately 14% were woven as independent bands sewn on a fabric or used as bands. The other 86% of the fragments were integrated parts of fabrics, such as starting, side or finishing borders. This indicates that, in Denmark, tablet-weaving began as a helping technique for weaving on a loom or finishing a fabric (Ræder Knudsen forthcoming).

As the weaving tablets seem to be smaller during the same period, this may indicate that small weaving tablets and integrated borders could be connected. Could it be possible that it is easier to keep small tablets hanging on the side of the loom while weaving, as they are lighter? Would larger tablets be heavier and have a tendency to slide down the warp threads? And would small weaving tablets be problematic when weaving a starting border?

Experiment using Tiny Tablets

An experiment was carried out to answer the above questions. A warp was prepared using a thin, wool single weaving thread of 6 m per g and small tablets of 34 × 34 mm made of sycamore with a thickness of 2 mm (Fig. 23.4). Three tablets were used to weave the starting border, and the warp of a narrow tabby fabric somewhat like the leg wrappers F.S.3692 from Thorsbjerg was set up (Schlabow 1976, 89 and abb. 230). The fabric was about 12 cm wide and the starting border was woven using three tablets and it continued down along the one side of the fabric going from being a starting border to a side border. In the opposite side, three sycamore tablets of 58 × 58 mm (thickness of 2 mm) were used for the other side border. Using the small weaving tablets in the preparation of the starting border caused no problems (Fig. 23.5).

Fig. 23.5. Preparing a warp with tiny tablets of 3.4 cm at one side border and tablets of 5.7 cm at the other side border (Photo: © L. Ræder Knudsen).

Starting the weaving and opening the shed was easy, but the difference of working between the two different sizes of tablets was immediately obvious. The small tablets kept their position whereas the larger tablets were sliding down the warp, when it was handled to give a new shed. The larger tablets served their purpose, but had to be moved up to the correct position one or more times for each shed (Fig. 23.6).

The weight on each thread was 12 g. To prevent the larger tablets from sliding down the warp, the weight on each border thread was doubled. This made the tablets a little more stable, but caused the threads of the outer tablet to be pulled apart.

The set up of the experiment had some weaknesses, as not only the weight of the tablets, but for instance, also the smoothness of the tablets, the quality of the fibre and yarn, the weight on each thread, the size of the holes, and the weaver's experience using the warp-weighted loom, would have had an impact on the result of the experiment. Yet, the result was very clear: weaving was easier using the small tablets on one side of the fabric compared to the larger tablets on the other side of the fabric.

The interpretation of the experiment

The surface and the material of the two different sizes of tablets used in the experiment were the same. The only real difference was the size and the weight of the tablets.

Fig. 23.6. When using tablets, as in this experiment, the tiny tablets are easier to work with. It is perfectly possible to use the larger tablets but they have a tendency to slide and tangle up. The smaller shed of the tiny tablets does not present a problem (Photo: © L. Ræder Knudsen).

The volume (height × length × width) and weight of the different tablets were compared. The weight of the different tablets was estimated using the density of the modern tablets. All weight measurements provided the use of the same kind of wood (sycamore). The approximate weight of a tiny tablet from Østergaard is 1.3 g; Sejlflod is 1.7 g; and the tablet from Dejbjerg is 5.6 g. The wooden tablets used in the experiment were copies of the Sejlflod tablet at the left side, and at the right side of the warp, tablets weighing 5.8 g with a side length of 5.7 cm. The recommended modern tablet with a side length of 80 mm weighs 9.4 g (Fig. 23.7).

My conclusion is that in the Early Iron Age in Denmark, small weaving tablets were a specialized tool for making integrated tablet-woven borders in connection with weaving fabrics on some kind of loom. Margrethe Hald (1950/1980) and Lise Bender Jørgensen (1986) both reached the conclusion that during the Early Iron Age in Denmark, the warp-weighted loom replaced the loom used earlier, the loom with a tubular warp. It is obvious to combine the interpretation of the small weaving tablets with the warp-weighted loom, as the emergence of the warp-weighted loom and tablet weaving seem to be simultaneous.

Rectangular Weaving Tablets

The completely preserved weaving tablet from Dejbjerg is rectangular and measures, as mentioned earlier 48 × 55 mm (Hald 1980, 226). One could imagine different reasons why this tablet is not square as would be expected. The difference in shape could be caused by the long period in the peat bog, but the difference between the two edges is more than 12% and this exceeds the dimensional change of wood due to moisture. Another possibility is that during conservation, the tablet could have been put together of two different fragments. It is more than 125 years since the excavation took place and the restoration principles and the documentation were different at that time. A visual examination of the whole tablet shows that there might be a joint in the middle, which does not fit 100%. This could indicate that the tablet was restored when

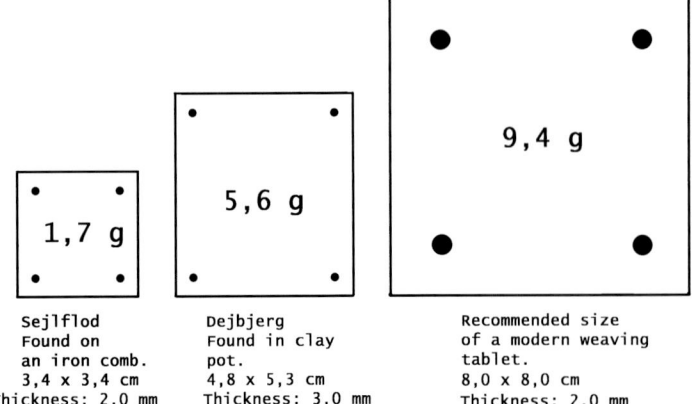

Fig. 23.7. Comparison of the size and weight of preserved weaving tablets from the Danish Early Iron Age and the recommended size of a modern tablet if they were made of the same kind of wood (Photo: © L. Ræder Knudsen).

Fig. 23.8. At the textile reconstruction workshop at Moesgård Museum, tablets painted blue at one edge and red on the opposite edge helped turn the tablets in the right direction, when working on complicated patterns involving the individual turning of each tablet (Photo: © L. Ræder Knudsen).

Fig. 23.9. Using rectangular tablets makes it very easy to see if one out of a larger bundle of tablets has not been turned correctly. It would be useful when making broad tablet-woven borders of Iron Age cloaks (Photo: © L. Ræder Knudsen).

it was found. The joint is filled with a wood-coloured material and it is not possible to see if the two fragments originally were part of the same tablet or not. This requires a more thorough investigation, for instance, comparing the growth rings of the wood. A third possibility is that the shape was due to poor craftsmanship. Knowing the highly developed skills of woodwork in the Iron Age, this explanation does not seem likely. The fourth possibility is that the shape of the tablet is intentional.

Although they are rare, the rectangular tablet from Dejbjerg is not the only example of a type from prehistory. As mentioned earlier, a tiny, rectangular tablet of 32 × 35 mm was found at Fenchurch Street in London (Pritchard 1994, 157). The difference of the edges of this tablet is only 3 mm. One could wonder how large the difference between the two sides should be to make the tablet rectangular and to make any difference in the weaving process compared to square tablets. When only one tablet is preserved, it is impossible to say, as it will depend on the irregularities of the other tablets. If a bunch of tablets are very regular, even a difference of 1 mm would be visible.

In the 'Black Earth' in Birka, Sweden, a rectangular bone tablet was found measuring 42 × 37 mm (Geijer 1938, Pl. 39). It is dated to about AD 900 (Collingwood 1982, 25). From the settlement of West Heslerton, North Yorkshire in England, another bone tablet measuring 36 × 30 mm was found (Walton Rogers 2007, 35).

Tablet-woven bands with patterns are not very common in the Early Iron Age. In the following period (AD 400–500), we have a large number of very fine and highly developed patterned tablet weaves, for instance, the animal frieze from Evebø-Eide in Norway (Collingwood 1982, 283). Until recently, we did not know how these patterns were made, and only intense analyses and work with reconstructions have revealed some of the working methods (Collingwood 1982; Hansen 1990). During this process, it became clear that some kind of marking of the tablets is of great help, when some tablets must be turned forwards and some backwards in a complicated pattern. During Egon Hansen's textile reconstruction work at Moesgård Museum during the early 1980s, we painted one edge of the tablet red and the opposite edge blue, leaving an unpainted area in the middle. When putting the dark threads in the blue side and the lighter threads in the red side, it was possible from outside the bundle of tablets to see the colours of the warp threads inside the bundle (Fig. 23.8). Weaving a copy of the animal frieze would have been very difficult without this help.[1]

What if the rectangular shape of a tablet served the same purpose, as our colouring of the tablets? In a bundle of tablets, it would be easy to see the difference between tablets standing on their short edge and tablets standing on their long edge.

Experiment using Rectangular Tablets

An experiment was carried out to answer the above question. Some 20 rectangular tablets were produced and a warp set up. It was not immediately obvious that, the rectangular shape of the tablets were useful when making pattern-woven tablet borders. However, the work is rather complicated and a great deal of practice is needed to become familiar with the tools. Therefore, further work is needed to produce more reliable information.

Yet a new idea occurred: what if the rectangular tablets were used for producing the broad warp-twined tablet borders as seen on cloaks from, for instance, Vrangstrup and Donbæk? These cloaks and the Dejbjerg tablets are dated to the same period, whereas the vast majority of pattern-woven tablet weaves are dated to after AD 400. The find from Vrangstrup has a tablet border woven to the eyelets of the cloak fabric and thus the tablet weaving is integrated invisibly in the cloak fabric (Ræder Knudsen 1998, 79–84). The find from Donbæk was made in the same way and here the border needed more than 150 tablets (Hald 1980, 88 and 92; Ræder Knudsen forthcoming).

Therefore, a new experiment was set up to try to answer this question. A warp was made using 20 rectangular tablets and very thin double wool thread. The reconstructed rectangular tablets had their holes very near the corners similar to the

Dejbjerg tablets, and this was slightly problematic as the thread tangled up more easily. However, the rectangular shape turned out to be useful when making certain that all the tablets are turned correctly. When more than 150 tablets are turned, it would be easy to miss one, but the preserved broad borders in the archaeological textiles are very regular and without any mistakes. The rectangular shape of the tablets would make it very easy to see any tablets which were not turned correctly (Fig. 23.9). Thus, it is my conclusion, based on this experiment that this might precisely be the purpose of the rectangular shape.

Conclusions

In Denmark, tablets are preserved from three sites dated to the Early Iron Age. Two of these sites contained very small tablets found on iron combs and the third is a medium size rectangular tablet. My aim was to investigate if the shape of these early weaving tablets from Denmark could have a practical purpose.

The conclusion to the practical experiments is that tiny, lightweight weaving tablets are very useful for making integrated tablet-woven side borders or starting borders for the upright loom. Heavier tablets can be used, but make the work a little more difficult.

It is not possible to come to any definitive conclusions about the rectangular tablets, which is probably due to my lack of experience in weaving with this shape of weaving tablets. However, one thing was certain: using rectangular tablets makes it very easy to see if all the tablets are turned when making a simple warp-twined band, where all the tablets are turned together. It is my conclusion that making broad warp-twined cloak borders may precisely be the purpose of the Dejbjerg tablets.

Acknowledgements

I very much wish to thank Anne Birgitte Sørensen of the Museum Sønderjylland for the opportunity to mention the weaving tablet from Østergård even if it is not yet published. I am also grateful to Aalborg Museum and The Danish National Museum for the opportunity to analyse their material and take photographs. Handmade tablets were provided by Michael Højlund Rasmussen. I would also like to thank Irita Žeiere of the Latvian National History Museum for the translation of a Latvian reference.

Note

1 Personal experience of making copies of ancient tablet weaves in Egon Hansen's textile reconstruction workshop at Moesgård 1982–1983.

Bibliography

Behrens, G. (1925) Brettchenweberei in römischer zeit. *Germania* IX, 45–47.
Bender Jørgensen, L. (1986) *Forhistoriske textiler i Skandinavien*. København.
Christensen, A. E., and Nockert, M. (2006) *Oseberg Funnet IV Tekstilene*. Oslo.
Christie, I. L. (1985) *Brikkevevede bånd i Norge, Levende tradition og glemte teknikker*. Oslo.
Collingwood, P. (1982) *The Techniques of Tablet Weaving*. London.
Crowfoot, G. (1924) A tablet woven band from Qua ei-Kebir. In *Ancient Egypt, Part IV*.
Egebjerg Hansen, T. (1996) Et jernalderhus med drikkeglas i Dejbjerg, Vestjylland. *Kuml* 1993–1994, 211–237. Århus.
Geijer, A. (1938) *Birka III. Die Textilfunde aus den gräbern*. Uppsala.
Götze, A. (1908) Brettchenweberei im Altertum. *Zeitschrift für Ethnologie*, Vol. 40.
Hald, M. (1980) *Ancient Danish Textiles from Bogs and Burials*. København.
Hald, M. (1950) *Olddanske Tekstiler*. København.
Hansen, E. (1990) *Brikvævning, historie, teknik, farver, mønstre*. Højbjerg.
Hoffmann, M. (1974) *The Warp-Weighted Loom*. Oslo.
Hundt, H.-J. (1968) Die verkohlten Reste von Geweben, Geflechten, Seilen, Schnüren und Holz-geräten aus Grab 200 von el Cigarralejo. *Madrider Mitteilungen* 9, 187–205.
Ilkjær, J. (1993) Die Gürtel. Bestandteile und Zubehör. *Illerup Ådal 3–4*. Århus.
Latvijas PSR arheoloģija (1974) Riga.
Nielsen, J. N. (2000a) Sejlflod, eisenzeitliches Dorf in Nordjütland. Katalog der Grabfunde. Band I: Text und Pläne. *Nordiske Fortidsminder. Serie B. 20:1*. København.
Nielsen, J. N. (2000b) Sejlflod, eisenzeitliches Dorf in Nordjütland. Katalog der Grabfunde. Band II: Abbildungen und Tafeln. *Nordiske Fortidsminder. Serie B. 20:2*. København.
Pritchard, F. (1994) Weaving Tablets from Roman London. In G. Jaacks and K. Tidow (eds), *Archaeological Textiles. Textilsymposium Neumünster 4.–7.5.1993*. NESAT V, 157–161. Neumünster.
Ræder Knudsen, L. (1998) An Iron Age Cloak with Tablet-woven Border: a New Interpretation of the Method of Production. In L. Bender Jørgensen and C. Rinaldo (eds), *Textiles in European Archaeology. Report from the 6th NESAT Symposium 7–11th May 1996 in Borås*, 79–84. GOTARC Series A, Vol. 1. Göteborg, Göteborg University.
Ræder Knudsen, L. (forthcoming) Tablet woven bands from Bogs and Burials of the Early Iron Age. In U. Mannering and M. Gleba *Designed for Life and Death*.
Schlabow, K. (1976) *Textilfunde der Eisenzeit in Norddeutschland*. Göttinger Schriften zur Vor- und Frühgeschichte. Neumünster.
Sørensen, A. B. (forthcoming) En yngre romertids grav fra Østergård med usædvanligt udstyr. *Symposium Jarplund 2008. Archäologie in Schleswig 12*.
Walton Rogers, P. (2007) *Cloth and Clothing in Early Anglo-Saxon England, AD 450–700*. CBA Research Report 145. York, Council for British Archaeology.
Werner, J. (1990) Eiserne Wollkämme der jüngeren Kaiserzeit aus dem freien Germanien. *Germania* 68, 608–611.
Wild, J. P. (1970) *Textile Manufacture in the Northern Roman Provinces*. Cambridge, Cambridge University Press.

24 Warriors' Clothing in the *Rigspula*

by Annika Larsson

Introduction

The paper presents results of my PhD thesis *Klädd krigare. Skifte i skandinaviskt dräktskick kring år 1000* (Warriors' Clothing: The shift in Scandinavian Costume *c.* 1000), defended at the Department of Archaeology and Ancient History at Uppsala University in 2007. The thesis is a comparative interdisciplinary investigation of textile sources covering the Viking Age and the Scandinavian Early Middle Age. By tracing changes in the male dress code, it was possible to identify a significant societal change in eastern Scandinavia around AD 1000.

Traditionally, the transition from the Viking Age to the Scandinavian Medieval Age is ascribed to AD 1050, however, my hypothesis is that the change from pre-Christian to Christian time is clearly reflected in changes in dress codes at least 50 years earlier. The pre-Christian dress of markedly oriental influence associated with the Viking Age trade centre Birka was replaced by a Christian medieval dress code of western European influence when the town of Sigtuna was established after Danish model around AD 970. The research results were obtained by tracing various kinds of textile codes, containing dress codes as well as handicraft traditions, trade routes, norms and values pertaining to previous societies (Larsson 2007).

As a textile archaeologist, one often resorts to interpretations based on very small textile fragments preserved in soil. Texts, pictures, textile implements, folklore dress and ancient terms thus become essential tools when attempting to reconstruct a complete dress. It is rarely – or rather never – possible to make a completely accurate reconstruction. It is not only the lack of important parts of the dress that hampers our understanding, but to an even larger extent the influences from our contemporary norms and ideals. Every researcher interprets previous societies based on contemporary society and his/her own pre-recognitions.

This is, thus, my quite radical re-interpretation of the dress code in the Old Norse poem *Rigspula*, seen from the interdisciplinary historical perspective of a modern textile archaeologist. The conclusion is that the Jarl's dress as described in the *Rigspula* belongs to a high-born Christian male warrior. In my opinion, previous interpretations indicating that this dress belonged to a pagan woman are based on national-romantic views from the 19th century. Such interpretations are, however, hardly relevant for the medieval manuscript *Codex Wormianus*, in which *Rigspula* was written down, since it was written on parchment around AD 1300.

What follows is an overview of the re-interpretation of the relevant part of the text concerning the dress code of the jarls in the *Rigspula*. For more detailed studies and argumentation, the reader is referred to the above mentioned thesis (Larsson 2007).

Archaeology and Old Norse Sources

Viking Age dresses are – among other sources – depicted on carved stones and wall hangings. Some textile terms can also be traced linguistically all the way back to the time of runes. There are several archaeological textile fragments preserved from the time, thanks to a rich burial tradition where metals, in for example, weapons and costly dress accessories, have conserved the textile materials. Despite these sources, it has been difficult to obtain a clear representation of Viking Age dress. The images on our retinas are probably influenced most by the national romantic paintings from the 19th century, where ideals from antiquity are mixed with Nordic peasant romanticism.

During the 19th century, industrialism presented a threat to the old and familiar societal order. Preservation of the agrarian society was thus high on the agenda for all art forms. Not only was painting influenced by peasant romanticism, but also literature. Interest in the history of nations flourished, and direct descriptions of the heroic Viking and his pagan world were seen in Old Norse poetry. The origin of what was genuinely 'Nordic' was to be found in Viking Age society.

It is not uncommon for archaeologists who work with the Viking Age to use the Old Norse texts in order to explain the archaeological finds and their context. The fact that the texts were written on parchment as late as final decades of the 13th century, several hundred years later than the Viking Age, is

Fig. 24.1. Reconstructed warrior from the 11th century, dressed in dark blue shining 'mail armour' (bláserkr). The archer's sleeves are reinforced by cloth (strauk of ripti sterti ermar) (Photo: A. Larsson).

Fig. 24.2. Duke Harold Godwinson wearing the folded cloak trailing at the back, hanging from a clasp on the chest: Kinga var á bringu, siðar slæður *(Drawing: A. Larsson, after the Bayeux tapestry).*

generally not considered as a significant problem. Also, the texts were translated and interpreted as late as during the 19th century, when the disintegrating agrarian society was a role model. Besides, archaeology as a science did not exist then (*e.g.* Frykman and Löfgren 1979, 61). Furthermore, it should be noted that poetry was a genre with established ideals already during medieval times. There are thus several filters to pierce when attempting to interpret Old Norse texts.

Medieval Ideals

The Old Norse *Rigspula* is a rich source of textual descriptions of textiles and dress: such as those of jarls, peasants and slaves. My own research has focused on the dress of the higher classes in society, primarily because those environments have left the most archaeological remains for us to examine. Comparisons can be made not only between archaeological finds and texts, but also between a rich material of, for example, pictures, coin portraits and religious and linguistic expressions.

The results of my research show that a significant societal shift took place around AD 1000, and that this shift is expressed directly in costume and fashion. My conclusion is that it is Christian norms that are expressed in the dress from around AD 1000. I make a claim that the dress of the jarl as described in the *Rigspula*, written in the *Codex Wormianus* around AD 1300, should be associated with a male Christian warrior, and not a several hundred years older Viking Age female as previously claimed.

According to my interpretation, the *Rigspula Poem* is a Christian expression, since newborns are baptized, and marital faithfulness through marriage rituals is prescribed in ways comparable to provincial laws from the same time. Marriage and baptism are two of the fundamental sacraments of Christianity. On these grounds, I would like to attribute the content of the text to Christian times, or more exactly between AD 1000 and the time it was written down around AD 1300.

In order for the reader to follow my arguments concerning the interpretation of the jarl's dress in the *Rigspula*, there follows below the actual text corresponding to verses 28–29 (Dronke 1997) with modern interpretations. First, comes the original Old Norse text from the Rigspula describing the dress of the jarls (*Codex Wormianus folio 61v*), next the established traditional interpretation of the text since the 19th century for which I have chosen the most recently published English version, verses 28–29 (Dronke 1997), and finally, my new interpretation of the text, expounded in my thesis (Larsson 2007).

The Rigspula	Codex Wormianus, verses 28-29,	fol 61v, line 4-6, Dronke 1997	New interpretation, Larsson 2007
	sat húsgumi	The master of the house sat	The master of the house sat
	ok sneri streng	and twined a bowstring,	and twined a bowstring,
	álm of bendi	arched a bow of elm,	with flexible elm wood
	ǫrvar skepti	set shafts to arrows,	he set shafts to arrows.
	en húskona	while the mistress of the house	The mistress of the house
	hugði at ǫrmum	studied her arms,	was occupied with *his* arms,
	strauk of ripti	stroked the fine linen,	with strips of ribbed cloth
	sterti ermar	tightened the sleeves.	she strengthened *his* sleeves.
	keisti fald	High curved her head-dress,	She folded an imperial cloak
	kinga var á bringu	a coin-brooch was on her bosom,	from the clasp on *his* chest,
	síðar slæður	a trailing robe she wore,	and shaped a long trail
	serk bláfán	a bodice blue-died	over *his* blue-shining armour

Fig. 24.3. King Olof Skötkonung in Sigtuna struck coins around AD 1000 after English patterns, which depicted dresses in the same regal fashion as King Ethelred (Drawing: A. Larsson).

Clad Warrior

According to Old Norse terminology, *bláserkr* represents 'mail armour' (Heggstad *et al.* 1997), which means a warrior's coat of blue- or black shining steel (Fig. 24.1). Old Norse alliterations to describe expensive metals were very common – gold shone red like fire while steel shone dark blue. Let us thus assume that the *Rigspula* describes a high-born Christian warrior clad in mail armour, *serk bláfán*, rather than a woman in blue linen dress.

A rich source of information on early medieval mail armour in Western Europe is found in the Bayeux tapestry, where a couple of hundred fully-clad warriors are depicted (Rud 1983, Wilson 2004). When clad in armour, these warriors have reinforced sleeves and leg coverings, which are also clearly illustrated in reconstructions in the Bayeux Museum (Fig. 24.2). The same dress details are clearly illustrated in other contemporary depictions of Christian warriors, among others Jesus Christ himself in a Frankish depiction from AD 800 (Fig. 24.4), where he, clad in shining blue mail armour, aims/throws a spear at sin and paganism (*Stuttgart Psalter fol. 23*). St. George, too, is depicted around AD 1000 in several instances, fighting, clad in mail armour and reinforced sleeves (see Hartmann 1971).

Fig. 24.4. The Frankish version of Jesus Christ dressed as a warrior in bláserkr, *tied or reinforced sleeves and leg coverings, and a red cloak worn with a clasp on his chest. AD 830 (Stuttgart Psalter fol. 23).*

Considering the sleeves, the text states that the mistress of the house was occupied with strengthening them with some cloth (Dronke 1997, 168 verse 28), but it does not say to whom the sleeves belonged. The translators of the text have simply assumed that she was occupied with her own sleeves, but there is no linguistic support for this assumption. The text could just as well refer to how the mistress of the house was occupied with her husband's, the warrior's, sleeve arrangements (*en húskona hugði at ǫrmum strauk of ripti sterti ermar*).

In direct connection to the segment about the sleeves in the *Rigspula*, it is emphasized that the high-born husband, the jarl, twined a bowstring, arched a bow of elm and set shafts to arrows (Dronke 1997, 168 verse 28). It is thus not an overly bold assumption to think that the man was an archer, and as such could be assumed to have reinforced sleeves as part of his armour. My interpretation is, thus, that after her husband had taken care of his weapons, the housewife helped him by reinforcing his sleeves with strips of cloth. As a good Christian wife, she supported her husband in his undertakings (see *The Bible. Proverbs of Solomon: The Virtuous Woman*).

Grammatically, the sequence *strauk of ripti* corresponds to the preceding *álm of bendi*, or "flexible elm wood". Reasonably, *strauk of ripti* should be interpreted according to the same linguistic pattern, and thus represents "strips of ribbed cloth". I argue in my thesis that the textile term *ripti* corresponds to wool cloth, which has been claimed previously (Fritzner 1867). In this case, where only the warrior's clothing is examined, the particular material used in the strips is of subordinate importance, for which reason I leave this detail for the present.

Cloak with Clasp on the Chest

The armoured jarl is not fully equipped without a cloak. Let us thus continue with the original text in the *Codex Wormianus*: *Keisti fald kinga var á bringu siðar slæður*.

Returning to the Bayeux tapestry, to the depiction of Jesus Christ as warrior, or to St. George with his lance, we find another unmistakable common attribute, the arrangement of the cloak. All these powerful men wear a cloak, open at the front, thrown back over the shoulders and fastened with a clasp on the chest (English 2004, 373) – in contrast to subordinate men, who either do not wear a cloak at all, or wear it with a clasp on the right shoulder. See also *Konungs skuggsiá*; *Why one should not wear a cloak in front of the King*.

Kinga var á bringu, states *Rigspula*, which means that the clasp was on the chest. *Siðar slæður* – "the train was long". Which train? The train of the cloak, of course. *Fald* means fold, folded, covered, and should in this case refer to the cloak. Falk (1919) in his *Altwestnordische Kleiderkunde* highlights a later meaning from AD 1600 Iceland, where *fald* means a kind of female headgear. This interpretation seems less relevant for Old Norse terminology in the *Rigspula*. There are no indications that the *Rigspula* is of Icelandic origin, but rather it is Danish (Johansson 1998). The term *keisti* means imperial, regal. *Keisti fald* in the *Rigspula* most probably refers to a cloak worn in a royal fashion, properly folded from the clasp on the chest (*kinga var á bringu*), and thrown over the back in a long train (*siðar slæður*).

The same dress codes were manifest in the entire Christian area, in imperial courts as well as among loyal upper ranks. Every man of rank thus carried his cloak over the back, folded from the clasp on the chest so that the cloak formed a train. This is clearly depicted on contemporary portraits on coins – Sven Forkbeard, Canute the Great, Olav the Holy, King Olof Skötkonung (Fig. 24.3) *et al.* – all from the period around AD 1000 and with English coin portraits as inspiration. This attribute of power was also used in the Byzantine court, and it is depicted on Russian coins.

Divine Mission

After a comparison of the Old Norse *Rigspula* description of the dress of the jarl with other contemporary sources and artistic expressions, like the Bayeux tapestry, biblical illustrations and coin depictions, it seems reasonable to assume that the dress described in the text refers to a male Christian warrior. The jarl's dress comprised of chainmail armour of shining black-blue steel. The sleeves were reinforced with wrapped strips of cloth. To the equipment also belonged a cloak worn according to the manner of Christian royalty – starting from a clasp on the chest, then thrown back to trail over the back with distinct parallel folds.

The present interpretation contradicts the traditional interpretation, which claims that it is a pagan Viking Age female dress that is described. From a textile- and dress historical perspective, the description of the dress in the *Rigspula* indicates a continental male medieval ideal, with depictions of Jesus Christ, the son of God as a model. The high-born men had a divine mission to fight for the good of Christianity against pagan evil. The symbolic codes of the dress were expressed both in real terms in art and literature, as well as in political propaganda. The *Rigspula* should probably be placed within such a genre, but expressed in *fornyldirslag* – typical for the Edda.

Despite the passing of two hundred years since the Old Norse texts were translated to modern languages, and despite sciences and source criticism long since having entered the academic world, the influence of 19th-century art on Old Norse literature has not been questioned in relation to material culture. In modern times, when the written word easily takes precedence over other types of sources, this might be the reason for such an un-critical approach. I have been able though, to demonstrate the importance of material culture for interpretation of the older texts.

A new interpretation of the dress in the *Rigspula* is of significant importance not only for research in dress- and textile archaeology, but also for related disciplines such as Old Norse and medieval history. In this case, a relevant interpretation of the dress codes can help in determining the age and cultural context of textiles. The investigation unambiguously shows that interdisciplinary research is necessary if we aim to arrive at results suitable for comparison between disciplines. My results also indicate that the knowledge of our history has to be updated continuously with new research in order not to stagnate. Every new epoch gives research alternative tools in order to see with fresh eyes.

Bibliography

The Bible. Proverbs of Solomon: The Virtuous Woman.
Codex Wormianus. Rigspula. 14th Century manuscript.
Dronke, U. (1997) *The Poetic Edda. Mythological Poems* II. Oxford.
English, B. (2004) The Coronation of Harald in the Bayeux Tapestry. In P. Bouet, B. Levy and F. Neveux (eds), *The Bayeux Tapestry: Embroidering the Facts of History. Proceedings of the Cerisy Colloquium (1990).* Caen, Presses universitaires de Caen.
Falk, H. 1919. *Altwestnordische Kleiderkunde*. Oslo.

Fritzner, J. (1867/1973) *Ordbog över Det gamle norske Sprog*. Band III. Oslo Bergen Tromsö.

Frykman, J., and Löfgren, O. (1979) *Den kultiverade människan*. Lund.

Hauttmann, M. (1971) *Die Kunst des frühen Mittelalters*. London.

Heggestad, L., Hödnebö, F., and Simensen, E. (1997) *Norrøn ordbok*. 4th ed. Oslo.

Johansson, K. G. (1998) *Rigspula och Codex Wormianus. Textens funktion ur ett kompilationsperspektiv*. Alvísmál 8.

Konungs skuggsjá. (1983) Norrøne tekster 1. Oslo.

Larsson, A. (2007) *Klädd Krigare. Skifte i skandinaviskt dräktskick kring år 1000*. Opia 39. Uppsala.

Rud, M. (1983) *Bayeux Tapetet*. København.

Stuttgart Psalter, fol. 23. Illustration von Psalm 91:13, Würtembergische Landesbibliothek.

Wilson, D. M. (2004) *The Bayeux Tapestry. The complete tapestry in colour*. London.

25 Potential and Limitations of the Application of FTIR Microscopy to the Characterization of Textiles excavated in Greece

by Christina Margariti, Dinah Eastop, Georgianna Moraitou and Paul Wyeth

FTIR microscopy has been applied to the analysis of textiles excavated in Greece for the purpose of material identification and characterization of the type of preservation. While FTIR studies alone may not always prove conclusive, the outcome, depending on the condition and the method of preservation of the textiles, is an invaluable part of a more comprehensive analytical investigation.

The Particularity of Textiles excavated in Greece

Buried textiles are directly affected by their environment. A burial environment, *i.e.* the product of the prevailing conditions and the properties of other materials present, could be favourable or unfavourable to the preservation of textiles. Within Greece, conditions are generally considered to be unfavourable; nevertheless, recent work has confirmed that numerous textile finds have survived (Spantidaki 2004, 67–68; Moraitou and Margariti 2005). This seems to be due to specific preservation mechanisms: bibliographical and archival research (through DCAMM's General and Sampling Archives) has indicated the presence of a metal, the absence of oxygen, incomplete burning and inhumation burials. The unfortunate corollary in many cases is that the organic textile fibres may be transformed to delicate pseudomorphs or obscured by severe encrustation. In spite of their fragile nature, excavated textile finds are an invaluable source of information for the archaeologist and textile historian and therefore are worthy of conservation. The development of an appropriate strategy first demands non-destructive identification of the constituent materials.

Fourier Transform Infrared Spectroscopy and its Application to Textiles

Fourier Transform Infrared (FTIR) Spectroscopy has been successfully applied to the identification (*e.g.* Derrick *et al.* 1994, 59; Garside and Wyeth 2003, 269) and further characterisation of the physical condition of textile fibre (*e.g.* Garside *et al.* 2005, 90). Its value to the study of excavated mineralised textile fibres has also been demonstrated (Jakes and Sibley 1989, 240; Gillard *et al.* 1994, 132; Chen *et al.* 1996, 219). The characteristic FTIR spectrum (typically recorded from 4000 to 700 cm^{-1}; Skoog *et al.* 1998, 404) represents the absorption of specific infrared frequencies by particular bonds or groups of atoms within a molecule. Absorbance peaks due to functional groups, such as amides and hydroxyls, which show characteristic frequencies, may allow classification (Derrick *et al.* 1994, 11); however, the 'fingerprint' region (around 1400–700 cm^{-1}), which encompasses the frequencies of complex molecular vibrations, is particularly useful for more specific material identification (Skoog *et al.* 1998, 411).

Analysis – Fourier Transform Infrared Spectroscopy and its Application to the Samples

Small fragments of excavated textile finds from Greece were analysed under an FTIR microscope in reflectance mode. The samples were too thick for transmission spectroscopy and the requisite non-destructive analysis precluded any specimen preparation, such as pressing. The drawbacks are the resultant reduced quality of the spectra due to scattering, the mix of specular and diffuse reflectance, which can cause confusion, and the limitation to surface rather than bulk analysis. The parallel study of model reference samples proved an essential aid to interpretation.

Case Studies and the Relevant Reference Samples

The three case studies presented below were selected to represent the different methods of preservation and the variety of textile fibres.

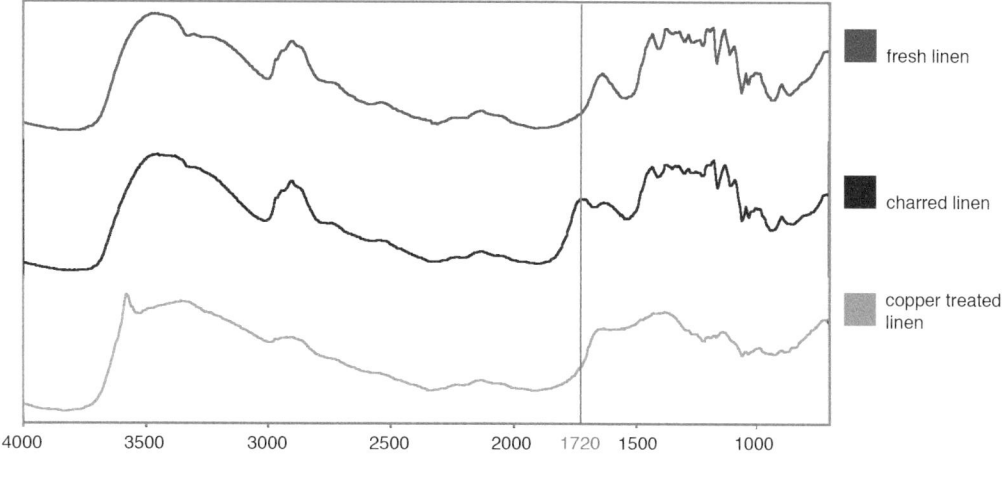

Fig. 25.1. Microscope FTIR spectra, reflectance mode, of the linen reference samples: fresh, charred and treated with copper (starting from top).

Case Study No. 1

A c. 18th century AD textile from an inhumation burial and a recent excavation. Stereomicroscopy suggests silk fibres.

Case Study No. 2

A 13th century BC charred textile from an old excavation. Documentation contemporary to the excavation identifies the fibres as wool, although the method of analysis is not mentioned.

Case Study No. 3

A 6th century BC textile preserved on a copper vessel from a recent excavation. Visual observation under the stereomicroscope indicates the presence of cellulosic, possibly bast fibres.

Preparation of the Reference Samples

Scoured new linen, wool and silk fibres were used for reference. An attempt was made to model the different types of preservation. Charring was achieved by placing fibres in a test tube over an open flame; a glass-fibre plug limited air ingress inhibiting complete combustion. Mineralised fibres were prepared by successive immersion in 1 mol dm^{-3} copper sulphate and 0.5 mol dm^{-3} sodium hydroxide solutions, followed by brief rinsing in purified water and air drying. Fresh fibres were taken as appropriate references for the inhumation burial.

Experiments

A Perkin-Elmer FTIR Spectrum One instrument attached to an AutoIMAGE microscope was used for the analyses. Spectra were acquired by accumulating 32 scans with a resolution of 4 cm^{-1} over the range 4000 to 700 cm^{-1}. Three different areas of each sample were analysed. Since these triplicate spectra were similar, they were synthesised to give an average one. Also the spectra of warp and weft fibres, were similar in each case. In the case of the mineralised fibres, spectra were collected from areas coated with copper oxidation products. Grams 8.0© software (Thermo Fisher Scientific) was used for spectral manipulation.

Results and Discussion

Reference fibres

LINEN (Fig. 25.1)

The reflectance spectrum of fresh linen shows a characteristic pattern for the cellulosic fibre, although some peaks are shifted by up to 20 cm^{-1} from the values quoted for other acquisition methods (e.g. Garside and Wyeth 2003, 270–273). Upon charring, the signature peaks are still evident and there is an additional absorbance at 1720 cm^{-1}, indicative of some oxidation to carbonyl moieties. Copper oxo-hydroxide mineralisation appears to mask the linen spectrum.

WOOL (Fig. 25.2)

In contrast, the spectra of fresh and copper mineralised wool are quite similar. There are three dominant peaks in the region 1700–1200 cm^{-1}, which may be assigned as amide I, II and III, typical of proteinaceous material (Skoog *et al.* 1998; Garside *et al.* 2005, 93, 410–413; Derrick *et al.* 1994, 181), although these appear at 40 to 60 cm^{-1} higher than for a standard ATR-FTIR spectrum, for example. The quality of the spectra precludes more detailed analysis. Charring improves the spectral quality to the extent that absorbance due to C-H stretches are now clearly evident (3070–2875 cm^{-1}); oxidation of the fibre is again evident (1780 cm^{-1}), and perhaps the generation of nitrile species (2060 cm^{-1}).

SILK (Fig. 25.3)

The spectra for the three silk references are of much better quality. In all cases, the amide I, II and III peaks are obvious

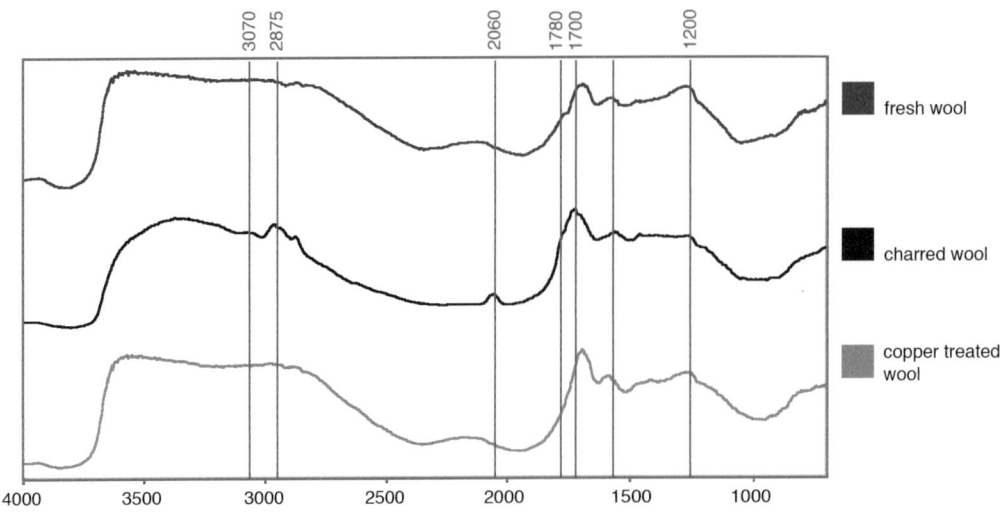

Fig. 25.2. Microscope FTIR spectra, reflectance mode, of the wool reference samples, fresh, charred and treated with copper (starting from top).

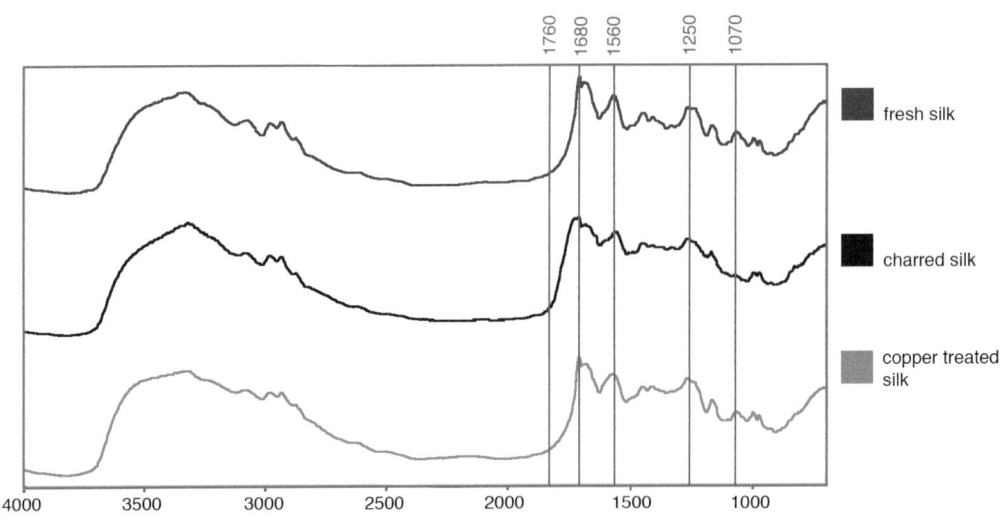

Fig. 25.3. Microscope FTIR spectra, reflectance mode, of the silk reference samples, fresh, charred and treated with copper (starting from top).

(1680, 1560, 1250 cm^{-1}), and there is additional detail. As with wool, treatment with copper solution has not masked the characteristic silk spectral pattern. Although the charred material is still clearly identifiable there are some discernible changes, *e.g.* a new broad shoulder around 1760 cm^{-1}, consequent upon oxidation, and loss of intensity around 1070 cm^{-1}, suggesting removal of hydroxyl side groups (*e.g.* from serine and threonine).

Case Studies (Fig. 25.4)

Case Study No. 1: Textile Fibres from an Inhumation Burial

The sample is clearly identifiable as silk from the spectral pattern, although the spectrum suggests that the silk is probably rather degraded, *e.g.* the amide peaks are somewhat broadened and the hydroxyl side-chain related peak at 1070 cm^{-1} is insignificant. Warp and weft fibres gave similar spectra suggesting that both are silk.

Case Study No. 2: Charred Textile Fibres

Spectra of the warp and weft of the charred specimen are again the same. Definitive spectral categorisation will require comparison with a more comprehensive graded set of carbonised fibres; nonetheless, the general spectral pattern is suggestive of wool, and it appears that cellulosic fibres can be ruled out.

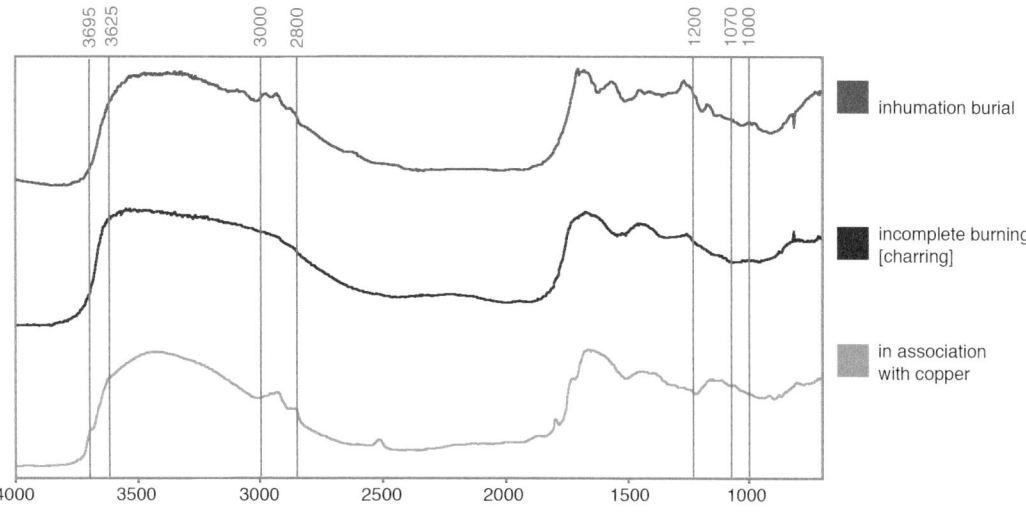

Fig. 25.4. Microscope FTIR spectra, reflectance mode, of the samples from the case studies. Case study No. 1: inhumation burial; Case study No. 2: incomplete burning; and Case study No. 3: in association with copper (starting from top).

Case Study 3: Textile Fibres Impregnated with Copper Oxides

Facile spectral identification may be hindered by mineral impregnation; there are, for example, peaks at high wavenumber (3695, 3625 cm^{-1}), which are suggestive of inorganic constituents. However, there are organic signatures, too, as evidenced by the νC-H peaks in the region 2800–3000 cm^{-1}. While conclusions are again tentative, the spectrum may not be inconsistent with a partially oxidised cellulosic material, which would account for the significant peaks in the range 1000–1200 cm^{-1}, where there are characteristic vibrations due to the cellulose chain. The copper-preserved specimen warp and weft give the same spectrum.

Conclusions

Fibre identification of textiles excavated in Greece using FTIR microscopy is not a straightforward issue. It was possible to unambiguously identify the original fibres in only one of the three case studies; tentative conclusions were drawn for the other two. There are perhaps two main reasons for this: the method of preservation and the age of the sample. The specimen successfully categorised was from a relatively recent inhumation burial.

Our research suggests that spectral acquisition by reflection is obligatory for FTIR analysis of excavated finds, if the specimen is to be recovered intact. Scanning for 32 times with a resolution of 4 cm^{-1} over the range 4000 to 700 cm^{-1} is adequate, although the spectra are generally of poor quality. The examination of at least three areas of a specimen is then essential to ensure representative analysis and the recording of the best quality spectrum possible.

Some caution is required in spectral interpretation, not least because bands are shifted by 20 cm^{-1} or more from their normal positions (as quoted for other acquisition modes). The fibre spectra are further affected by the method and degree of preservation; the fibre spectra may become masked and may alter due to deterioration. The need for a reliable spectral reference set is of utmost importance. Where mineralisation, adventitious soiling or consolidation is suspected, it may be possible to subtract such additional contributions from the fibre spectra, although if there is heavy contamination, other analytical methods of investigation, such as optical and electron microscopy may be required. First aid treatments, which could interfere with the analysis, such as consolidation *in situ*, although poorly documented were not uncommon for earlier textile finds.

In certain circumstances, FTIR reflectance spectroscopy alone can serve to identify excavated fibres using only a restricted reference data set; in other cases, a more comprehensive library of spectra for reference models may be necessary or other techniques may also need to be brought to bear. Further FTIR spectroscopic work is also required with a wider range of references to enable the assignment of the preservation type; this is not immediately apparent. However, currently, even if the FTIR spectral analysis is inconclusive, it may still yield useful information from a conservation viewpoint. Simply knowing that there is still organic material present should enable the conservator to undertake a more appropriate conservation strategy.

Acknowledgements

We would like to thank the following: Emma J. C. Richardson, PhD student, The Textile Conservation Centre, for her continuous help with the FTIR spectrometer and GRAMS 8.0© software; Andreas Koutouvalas, Art Director, 'Fox Design' Athens, Greece, for his invaluable help with image processing; Ioanna Papantoniou, Director, Peloponnesian Folk Art Museum, for offering the fabrics for the reference samples; Nell Hoare, Director, The Textile Conservation Centre; Nikolaos Minos, Director, Directorate of Conservation of Ancient and Modern Monuments.

Bibliography

Chen, H. L., Jakes, K. A., and Foreman, D. W. (1996) SEM, EDS and FTIR Examination of Archaeological Mineralised Plant Fibres. *Textile Research Journal* 66:4, 219–224.

Derrick, M. R., Stulik, D., and Landry, J. M. (1994) *Infrared Spectroscopy in Conservation Science*. Los Angeles, The Getty Conservation Institute.

Garside, P., Lahlil, S., and Wyeth, P. (2005) Characterisation of Historic Silk by Polarised Attenuated Total Reflectance Fourier Transform Infrared Spectroscopy for Informed Conservation. *Applied Spectroscopy*, 90–95. Society for Applied Spectroscopy.

Garside, P., and Wyeth, P. (2003) Identification of Cellulosic Fibres by FTIR Spectroscopy: Thread and Single Fibre Analysis by Attenuated Total Reflectance. In D. Scott *et al.* (eds) *Studies in Conservation* 48:4, 269–275.

Gillard, R. D., Hardman, S. M., Thomas, R. G., and Watkinson, P. E. (1994) The Mineralisation of Fibres in Burial Environments. *Studies in Conservation* 39, 132–140.

Jakes, K. A., and Sibley, L. R. (1989) Evaluation of a Partially Mineralised Fabric from Etowah. In Y. Maniatis (ed.), *Archaeometry, Proceedings of the 25th International Symposium*, 237–244. Amsterdam, Elsevier.

Moraitou, G., and Margariti, C. (2005) Excavated Archaeological Textiles in Greece. Past, Present and Future. In *Postprints from the 2nd International Symposium of Textiles and Dyes of the Ancient Mediterranean World, 24–26 November 2005*, 165–167.

Skoog, D. A., Holler, F. J., and Nieman, T. A. (1998) *Principles of Instrumental Analysis*. 5th edition. USA., Brooks/Cole Thomson Learning Publications.

Spantidaki, Y. (2004) Ancient Textiles. In S. Panelis (ed.), *Corpus* 66 (December), 66–73. Athens, Stavros Panelis Publications.

Taylor, N. (2004) How does FTIR work? In N. Taylor (ed.), *The Internet Journal of Vibrational Spectroscopy*, 5, Edition 5. Toronto, Canada, John Wiley & Sons, Ltd.

Tímár-Balázsy, Á., and Eastop, D. (1998) *Chemical Principles of Textile Conservation*. Oxford, Butterworth-Heinemann.

26 Evidence of War and Worship: Textiles in Roman Iron Age Weapon Deposits

by Susan Möller-Wiering

Introduction

From Roman times, more than two dozen sites are known in southern Scandinavia and neighbouring areas where weapons were deposited in lakes, which later became bogs. Sometimes, there are only a few pieces. In other cases, there are thousands of weapons and parts of bits of personal equipment. Four of the largest deposits are Vimose on the Danish island of Fyn, Illerup Ådal in Northern Jutland, Nydam in the south of the same peninsula and Thorsberg still a little more further south. Rather large areas in Thorsberg, Nydam and Vimose were already excavated in the 1850s–1860s by the Danish archaeologist, Conrad Engelhardt (1863; 1865; 1869). Further work was done in Nydam primarily in the 1990s (Jørgensen and Petersen 2003, 258–259), and already some years earlier, mainly in the 1970s–1980s, Illerup Ådal was investigated (Ilkjær 2003, 47). The hypothesis put forward already 140 years ago that the weapons and personal belongings found in these large deposits were once owned by foreign warriors who were defeated by the local or regional inhabitants, may still be the most probable one, although other possibilities are also discussed (Ilkjær 2003, 59–60; Lund Hansen 2003, 85–86). After the battle, the victorious survivors sacrificed these treasures to their gods.

Methodology

Vimose, Illerup Ådal and Nydam are bogs with a basic chemistry in which iron was well preserved, while the conditions were unfavourable for textiles. In Thorsberg, the situation is the opposite: wool textiles have survived in a remarkably good condition, while the iron has disintegrated because of the bog's acidic character. The material is therefore very diverse.

Already Conrad Engelhardt mentioned textiles in connection with weapons. For example, he found some spearheads as a closed find in Nydam and he described them as wrapped in wool cloth (1865, 4). An idea arose that, during the ritual after the battle, some sacking material could have been used for collecting and depositing the weapons and other equipment.[1] This hypothesis was the point of departure for the current enquiry, which – up to date – has consisted of the following three steps:

1. The examination and documentation of the textile material on a microscopic scale, *i.e.* with magnification, which is the basis for all conclusions.
2. The comparison and interpretation within each site. Besides a comparative analysis of the textiles themselves, it includes a closer examination of the weapons and other finds as well as an interpretation of their spatial distribution. Detailed maps of the sites are of great importance at this point; these are, however, only available for Illerup Ådal and Nydam.
3. The comparison between the sites.

What should follow, but has to be left to the future, is the comparison on a still smaller scale, *i.e.* including material from other sites and regions as well as from other types of finds, particularly from graves. This would offer the possibility to find out more about issues like provenance and status.

All results will be published in a monograph (Möller-Wiering forthcoming). The aim of this article is to present some rather general technical aspects and to focus on the question of what the textile material may contribute to the reconstruction of the rituals which took place after the battle.

Vimose

Vimose is the oldest of the four sites. A little more than 50 textile finds were registered here. This material covers the surface of iron weapons, mainly lanceheads and spearheads. Apart from four tablet weaves,[2] the spinning directions and weaving types of 40 items could be determined. Of these, 90% are plain 2/2 twills in z/z spinning. The other 10% are tabbies in z/z spinning. These are the only two variants determined in Vimose. The textiles are heavily mineralized, thus concealing most details of the fibres. Seemingly original, non-mineralized fibres were observed in a few instances. They look like unpigmented wool but no fibre analysis was carried out. The yarns seem to be well prepared and the weaves

Fig. 26.1. Vimose, two layers of a 2/2 twill in z/z used for wrapping a lancehead (24721), scale in mm (Photo: S. Möller-Wiering).

are homogeneous in two respects. Firstly, the weaving was accomplished carefully. Secondly, most of the twill fragments are so much alike that they might derive from only one cloth with about 15–12 threads per cm. If so, it is likely that it was a cloak that was cut for this purpose because of the required size for the amount of wrapping. Three out of the four tabbies may also originate from one and the same fabric, which had a thread count similar to the twill.

In any case, the fabrics do not appear low quality and do not support the idea of sacking material. However, it is obvious that the weapons were indeed wrapped, as Conrad Engelhardt observed already in 1864 in Nydam and a year later in Vimose (1869, 3–4). Sometimes, it is clear that two or more objects were wrapped together. Many other pieces, however, were treated individually: the textile adheres to both sides of a blade or socket and – since the shafts are missing – sometimes even covers the opening to some extent. A very interesting phenomenon in this context is the orientation of the textiles: they follow the axes of the weapons, *i.e.* the thread system with the lower thread count lies parallel to the long axis (Fig. 26.1). This is particularly true for the sockets. On the blades, the direction may become a little oblique but usually, it is quite straight even there. This feature may be found on almost every individually wrapped lance and spear. Furthermore, it is not only the thread systems that follow

a certain pattern but also, apart from few exceptions, the oblique twill line is oriented in the same manner. On a closer examination of these exceptions, it emerged that in several cases, they are part of a distinct stratigraphy: there are two layers of the same fabric, with the typical orientation directly on the iron and the mirrored one on top of the other (see Fig. 26.1). Seemingly, the textiles were laid double before wrapping. In a few other cases, it was impossible to decide whether the textile with an opposite twill line represents a second layer or not, although the orientation of the thread systems was again the same. After all, it seems that somebody took his – or her? – time to prepare each item very carefully for the rituals.

Only a few textiles were found on weapons others than lances and spears. One of them is a tabby on a shield boss (Fig. 26.2). The fabric is marked by straight lines or cuts, many parallels to which were found in Illerup Ådal.

Illerup Ådal

Over 100 items with textiles were examined from Illerup Ådal: primarily shield bosses and lanceheads, but also swords, knives, axes and other objects. The spectrum is wider in Illerup than in Vimose, both in terms of the types of weapons and the types of textiles. The largest group of more than one third

Fig. 26.2. Vimose, tabby in z/z with cuts used for wrapping a shield boss (24767), scale in mm (Photo: S. Möller-Wiering).

of the total consists of 2/2 diamond twills in z/s spinning. Together with the plain variants, the twills in z/s make up a little more than 50%. 2/2 twills in z/z also exist in two variations: as diamond twills and plain ones, which together account for almost 30%. The rest encompasses spin-patterned twills and tabbies in z/z spinning. As in Vimose, all the textiles are of good quality and seemingly made of wool.

The site maps reveal that, sometimes, different fabrics were found on neighbouring weapons, thus pointing to individual wrapping. In other cases, there is a stratigraphy of different textiles on one and the same shield boss (Fig. 26.3). This possibly indicates individual wrapping first, followed by creating bigger packages. Quite often, the textiles, particularly on the shield bosses, are characterized by straight lines, seemingly cuts, like on the aforementioned example from Vimose. These cuts are analogous to the damage seen on the weapons themselves. Research has shown that they mainly derive from the rituals after the battle, not from the battle itself (*e.g.* Petersen 1995, 24). The cuts in the textiles therefore indicate that they were treated in the same way as the weapons. This would signify that the textiles would have belonged to the same people as the weapons did.

Nydam

In the case of Nydam, the archaeological material connected with textiles is more diverse. As a parallel to Vimose and Illerup Ådal, one might expect that the majority of fabrics were wrapping material for weapons. Conrad Engelhardt's observations of this phenomenon were mentioned earlier. However, except for very few items, the weapons have only very faint traces of textiles preserved and in too small a quantity and poor condition to give any technical description. This is true for the material excavated in the 19th century, as well as for the recently found objects. Nevertheless, traces of textiles were observed on practically every piece of a random test including some dozens of weapons, mainly lanceheads and spearheads, but also a few shield bosses. One may therefore assume that generally, the situation in Nydam is comparable to Illerup and Vimose.

Another group of material consists of 25 wooden boat fragments.[3] The textiles derive from recently excavated parts of the well known 'Nydam boat' which is on display in the *Archäologisches Landesmuseum*, Schloss Gottorf, Schleswig, Germany. The ship is, in fact, one of four boats that were sacrificed on the site. The textiles lay between the planks to keep the ship watertight. Sometimes, real fibres are left,

Fig. 26.3. Illerup Ådal, stratigraphy of a tabby in z/z beneath a plain 2/2 twill in z/z used for wrapping a shield boss (RGZ), scale in mm (Photo: S. Möller-Wiering).

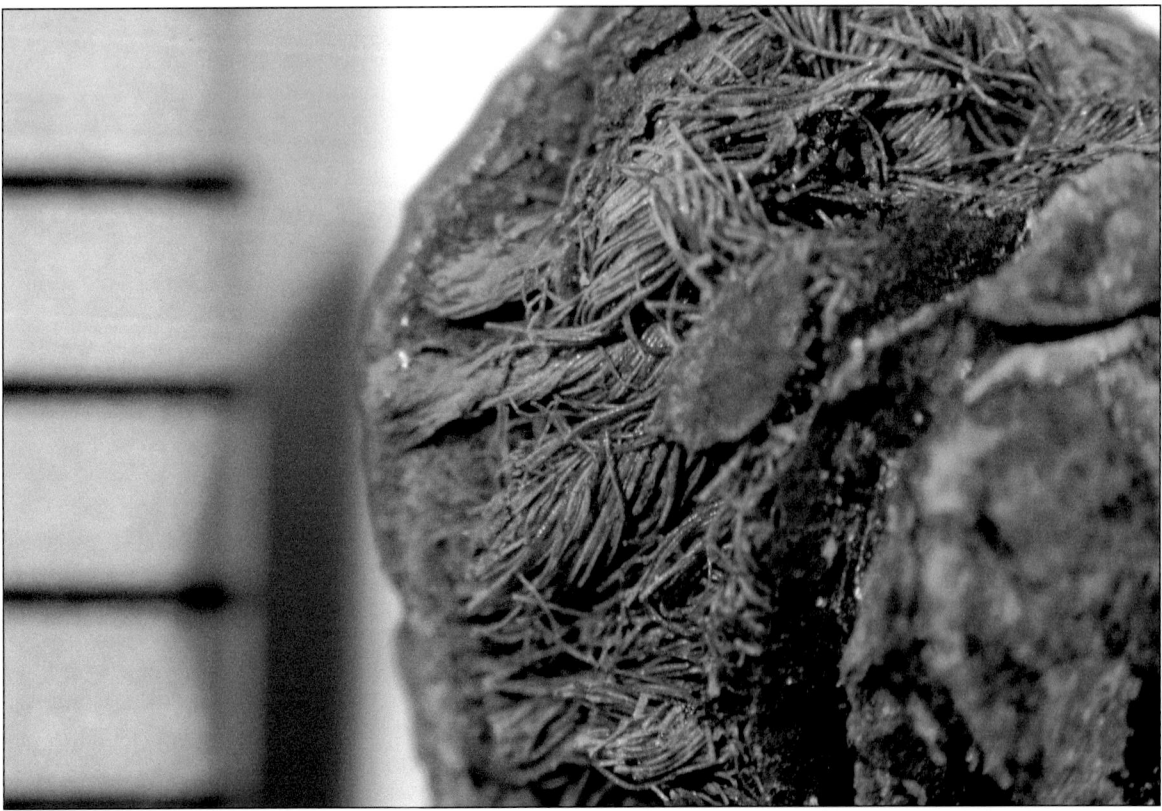

Fig. 26.4. Nydam, unidentified weave in z/s on a clasp (7093), scale in mm (Photo: S. Möller-Wiering).

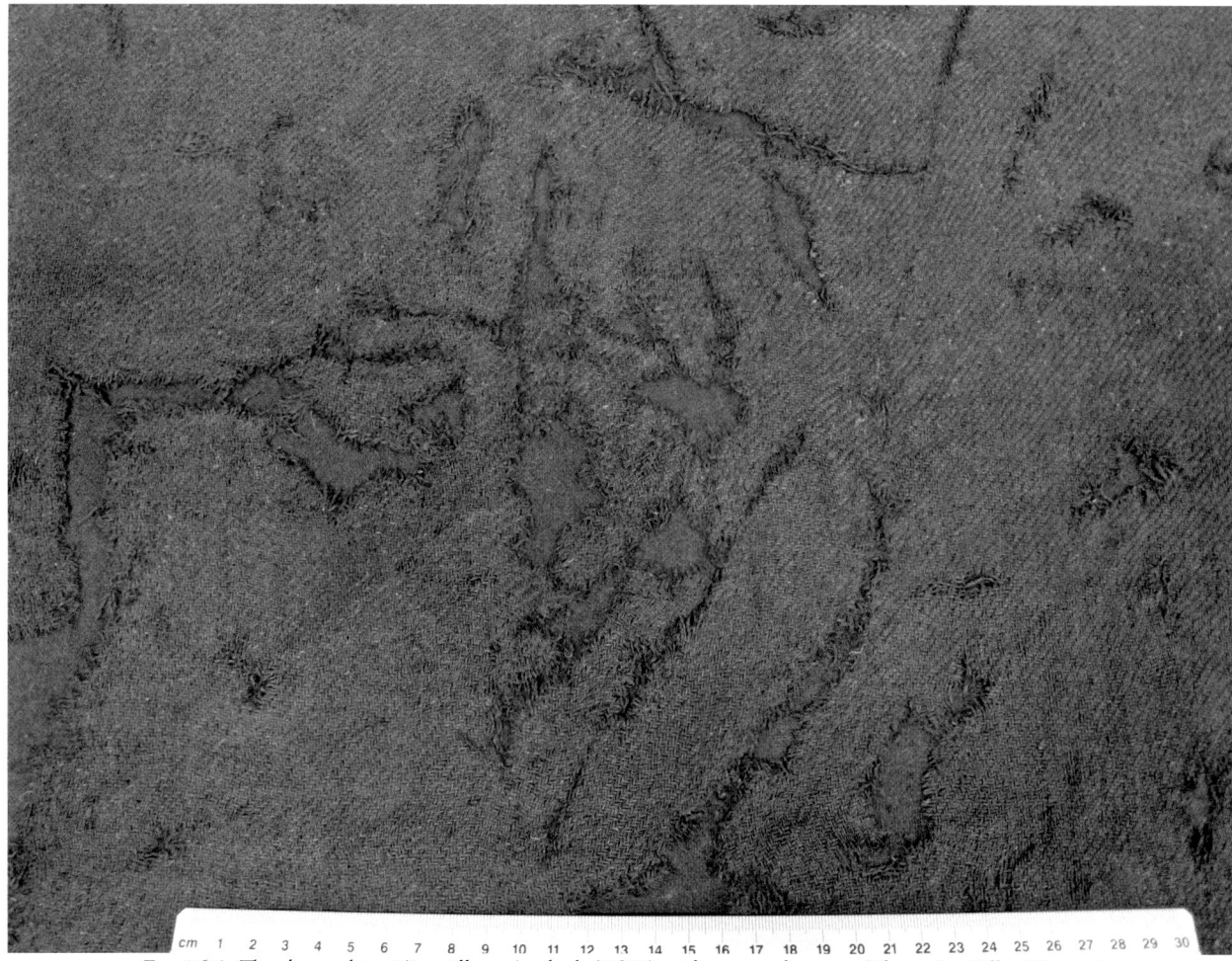
Fig. 26.5. Thorsberg, plain 2/2 twill in z/s, cloak (3688) with cuts, scale in cm (Photo: S. Möller-Wiering).

sometimes it is only their impression that can be analysed. Among the caulking material, there are 15 fabrics with a determinable spinning direction of both warp and weft. Only in two cases, are there s-spun yarns. All other textiles are made of z-spun threads in both warp and weft. However, in spite of this uniformity in spinning, the bindings and thread count are in a variety of qualities. Therefore, it may be said that the caulking material was not woven as a specific type of fabric for this purpose, but rather that it was reused material. In terms of the ritual, these textiles are in a group of their own, because they remained invisible and did not play any active role. The boat in its entirety was the sacrificial offering, the textiles were irrelevant here.

The third group of textiles from Nydam are connected to clasps or rather hooks and eyes. Most probably, they once belonged to the cuffs of tunics. About 200 single pieces and fragments were investigated, but only a smal amount of them still displayed traces of textiles. These remains are so tiny that it was usually impossible to determine all technical details. However, in 14 cases, the spinning direction of at least one thread system could be observed (Fig. 26.4). Out of these, at least 12 fabrics were partly or completely made of s-spun yarns; for the last two items, only z-spun yarns in one thread system could be determined, while the other system remained unclear. Although the numbers are low, they indicate a remarkable difference between caulking material and clasps.

Hypothetically, several reasons may account for this difference:

1. The boat could derive from another region with other spinning traditions than those observed for the clasps. However, in general z-spun as well as s-spun yarns were used in the entire central and northern Europe at that time (see Bender Jørgensen 1986).
2. Since the boat itself is most probably older than the textiles of the clasps, the chronology might account for the difference. Yet, even if it were some decades older, it belonged to the second half of the Roman Iron Age during which – as mentioned earlier – yarns in z and s were common everywhere.
3. Possibly, the function can provide an explanation. Since both textile variations – with and without s-spun yarn – are equally useful as caulking material, it might rather be the original function of the fabrics, *i.e.* their function before they were used in this boat. In other words: textiles of a particular function were seemingly preferred or rejected as caulking material.
4. Finally, other possible reasons must be taken into account, for example, the value or symbolic meaning of a particular type of weave.

Although this group of textiles, *i.e.* those on the clasps, is so badly preserved, they still bear further information about the ritual when the site maps are included. All registered clasps were found during the more recent excavation and apart from some displacement during the work due to water,[4] their position in the bog is well documented. Generally, the site maps reveal that the clasps were not spread all over the site but were concentrated in certain sections. On closer examination, it turned out that they were found in small assemblages and that in a few instances, the hooks and eyes even lay as pairs in rows. Since the latter is the way in which they were originally fastened to the clothes (Nockert 1991, 108–111), this proves that they were not cut off separately. Furthermore, the distribution of these small assemblages indicates that complete pieces of clothing were deposited. And finally, tiny remains of textiles – just fibres – on top of the plates of the clasps lead to the conclusion that the tunics might have been rolled up. Moreover, it should be emphasized that the silver clasps belonged to high status clothing.

Thorsberg

While the hitherto described material is widely unpublished as yet, the textiles from Thorsberg are rather well known (*e.g.* Schlabow 1976; Hald 1980, 70–73). As mentioned earlier, they were not preserved in combination with metals and are still in their organic state. Amongst others, the collection includes a complete tunic, two partially complete pairs of trousers and several more or less fragmentary cloaks. Only some general technical data and some aspects related to the rituals shall be mentioned here. About 30 different fabrics were determined. More than two thirds of these are 2/2 twills in z/s spinning, and within this group, there are more diamond twills than plain ones. This distribution thus resembles the material from Illerup Ådal. The remainder of only seven items encompasses six variants: plain twills in z/z, s/s and spin patterned, one broken twill in z/z spinning (which might be a weaving fault), and tabbies in z/z and s/s. Again, the preparation of the wool fibres and the weaves themselves meet the high quality standards represented in the non-textile artefacts.

Another parallel between Thorsberg and Illerup Ådal are cuts in the textiles as were also observed in Vimose. They are particularly obvious on one of the cloaks (no. 3688) (Fig. 26.5) and on the tunic (no. 3683). In the case of the tunic, these cuts hit the back of the piece, which might be an indication that this happened after the battle, in a similar manner to the damage on the weapons. One of the pairs of trousers (no. 3685) lacks their upper edge. It was cut off, which, again, leads to the interpretation that this was not due to the battle but was done deliberately at a later stage. Thus, the clothes were damaged just like the other artefacts. Moreover, as in Nydam, they were then rolled up – as was observed by Conrad Engelhardt during the excavation (*Diary* 1860, 52; 1863, 18–19) – and deposited as whole pieces.

Conclusions

The textiles from four large weapon deposits in Denmark and north Germany were analysed: Vimose, Illerup Ådal, Nydam and Thorsberg. The Roman Iron Age weapon deposits are interpreted as sacrifices of the local or regional population after armed conflicts with armies from abroad.

The textiles from all four sites are of good quality, and have nothing to do with sacking material. Most technical information was gathered from the Illerup Ådal and Thorsberg finds where the majority of textiles consists of 2/2 twills in z/s spinning, either diamond or plain woven. Technical differences between Vimose and the other sites may be due to chronology or lack of representativity. Moreover, the question of provenance still needs to be answered.

Textiles played a great role in the rituals after the battle when they were treated and damaged like the weapons. Some clothes were rolled up and then deposited as such. This was observed in Nydam and Thorsberg. Other pieces were used for carefully wrapping other sacrificial offerings, particularly weapons. This is known from Nydam, Vimose and Illerup Ådal. Due to preservation, not every praxis can be verified on every site. However, there are no contradictions. Instead, the results make up a picture of a common custom.

Notes

1. *E.g.* Bemmann and Bemmann interpreted this wrapping as a linen sack (1998, 2: 229).
2. The tablet-woven material from all four sites was examined by Lise Ræder Knudsen, Vejle. The results will be published in Möller-Wiering forthcoming.
3. Most of the caulking material as well as the clasps (see below) were originally analyzed by Ulla Mannering, Copenhagen, who kindly made the data available to the author.
4. Flemming Rieck, personal communication 03/08.

Bibliography

Bemmann, G. and Bemmann, J. (1998) *Der Opferplatz von Nydam. Bd 1: Text, Bd 2: Katalog und Tafeln.* Neumünster.

Bender Jørgensen, L. (1986) *Forhistoriske textiler i Skandinavien. Prehistoric Scandinavian Textiles.* Nordiske Fortidsminder, Serie B, 9. København.

Engelhardt, C. (1860) *Diary.* (Copy of unpublished manuscript).

Engelhardt, C. (1863) *Thorsbjerg Mosefund.* Kjöbenhavn (reprint 1969).

Engelhardt, C. (1865) *Nydam Mosefund.* Kjöbenhavn (reprint 1970).

Engelhardt, C. (1869) *Vimose Fundet.* Kjöbenhavn (reprint 1970).

Hald, M. (1980) *Ancient Danish Textiles from Bogs and Burials.* Copenhagen, Publications of the National Museum.

Ilkjær, J. (2003) Danske krigsbytteofringer. In *Sejrens triumf. Norden i skyggen af det romerske Imperium,* 44–46. København.

Jørgensen, E., and Petersen, P. V. (2003) Nydam mose – nye fund og iagttagelser. In *Sejrens triumf. Norden i skyggen af det romerske Imperium,* 258–284. København.

Lund Hansen, U. (2003) Våbenofferfundene gennem 150 år – forskning og tolkninger. *Sejrens triumf. Norden i skyggen af det romerske Imperium,* 84–89. København.

Möller-Wiering, S. (Forthcoming) *War and Worship: Textiles in Roman Iron Age Weapon Deposits*.

Nockert, M. (1991) *The Högom Find and other Migration Period Textiles and Costumes in Scandinavia*. Högom II, Archaeology and Environment 9. Umeå.

Petersen, P. V. (1995) *Nydam Offermose*. Aarhus, Dansk Historisk Håndbogsforlag.

Schlabow, K. (1976) *Textilfunde der Eisenzeit in Norddeutschland*. Neumünster.

27 Bewahren und Erfassen – Anmerkungen zum Umgang mit mineralisierten Strukturen auf Metallen in der Denkmalpflege

von Britt Nowak-Böck

Einleitung – Restaurierung Archäologie der Bayerischen Bodendenkmalpflege

Die archäologische Textilforschung hat in den letzten Jahrzehnten vielfach belegen können, dass aus den zunächst unscheinbaren mineralisierten Strukturen an Metallobjekten wie Textilien, Leder, Federn, Holz usw. weitreichende wissenschaftliche Erkenntnisse zu Bekleidung, Tradition und Handel zu gewinnen sind.

Die grundlegende Voraussetzung für eine wissenschaftliche Bearbeitung von organischen Resten, ganz gleich zu welchem Zeitpunkt sie stattfindet, ist eine nachhaltige Bewahrung und Bereitstellung der Befunde mit sämtlichen Informationsdetails. Die Arbeit in der praktischen Denkmalpflege zeigt, dass dies nur durch abgestimmte Arbeitsabläufe während der Grabung, der Zwischenlagerung, der Bearbeitung in den Restaurierungswerkstätten und dem Endverbleib der Funde möglich ist. Hierfür sind ganzheitliche Konzepte zum Umgang mit mineralisierten Strukturen auf Metallen erforderlich, die sich nicht nur auf exemplarisch ausgewählte Einzelobjekte oder Blockbergungen beziehen, sondern das hohe Fundaufkommen (z.B. bei frühmittelalterlichen Gräberfeldern) berücksichtigen.[1] Für die Restaurierung in der Denkmalpflege steht die materialgerechte Versorgung aller Funde im Vordergrund; eine konsequente, textiltechnologische Erfassung, Analyse und Auswertung sämtlicher Befunde im wissenschaftlichen sinne kann dagegen kaum geleistet werden und muss auf spezielle Projekte oder Fragestellungen beschränkt sein.

Grabung – Sensibilisierung für organische Materialien

In Bayern werden nahezu alle Grabungsaktivitäten von privaten Grabungsfirmen unter fachlicher Aufsicht des Bayerisches Landesamt für Denkmalpflege (BLfD) durchgeführt. Um eine fachgerechte Bergung und Versorgung der Funde zu gewährleisten, sind einheitliche und verbindliche Vorgaben für die Firmen, Grabungseinsätze der Restauratoren vor Ort, sowie konkrete Empfehlungen zur Handhabung von Objekten notwendig. Die von den Restauratoren formulierten *Empfehlungen zum Umgang mit organischen Resten während der Ausgrabung* verweisen detailliert auf wesentliche Aspekte bei der *Entdeckung, Dokumentation, Sicherung, Bergung* und *Verpackung* von Einzelfunden und In-situ-Blöcken (Ausführliche Empfehlungen des BLfD siehe unter *www.blfd.bayern.de* und weiterführende Literatur dort sowie Gasteiger 2007).

Der anschließende Weitertransport der Funde in die Werkstatt muss nach Absprache möglichst rasch organisiert werden. Für eine geeignete Zwischenlagerung sollten je nach Bedarf Kühl- und Tiefkühlmöglichkeiten, klimatisierte Räumlichkeiten und/oder präventive Verpackungssysteme bereitstehen.[2]

Abb. 27.1. Bergung einer frühmittelalterlichen Spatha mit Beschlägen der Aufhängung (Germering, Grab 101) (Bayerisches Landesamt für Denkmalpflege).

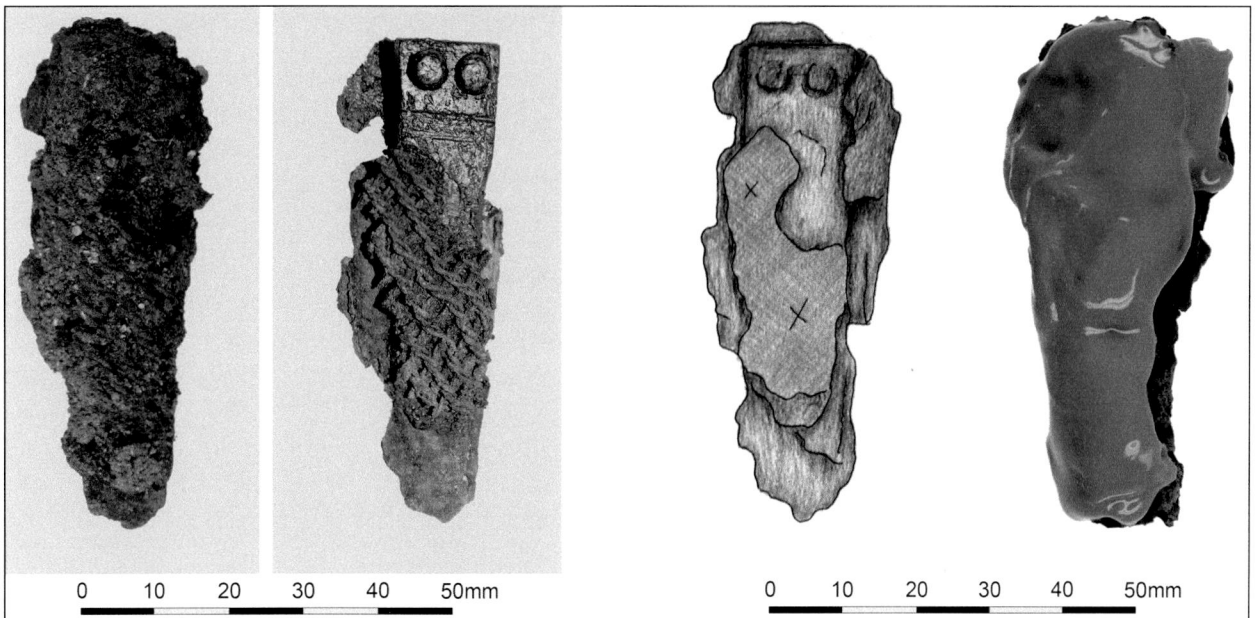

Abb. 27.2. a. Riemenzunge der Spathaaufhängung im Fundzustand; b. Riemenzunge mit mineralisierten Köpergewebenresten nach der Präparierung und Restaurierung; c. Zeichnerische Dokumentation der organischen Schichten; d. Beispiel einer Silikonkappe als Schutzabdeckung für organische Reste für eine weitere Freilegung der Metalloberfläche (Bayerisches Landesamt für Denkmalpflege).

Abb. 27.3. Detailzeichnung eines Köpergewebes 2/2 mit auffälliger Optik auf einer Riemenzunge der Spathaaufhängung (Germering, Grab 101) (Bayerisches Landesamt für Denkmalpflege).

Restaurierung – Bearbeitung von Fundstücken mit mineralisierten Strukturen

Anschließend an eine präventive Grundversorgung, datenbanktechnische Erfassung und Röntgenprospektion der Metallfunde und Blockbergungen muss im Hinblick auf die konservatorische Vorgehensweise die Qualität und Quantität der organischen Befunde ermittelt werden.[3] Im Folgenden werden verschiedene Aspekte der konservatorischen Bearbeitung thematisiert.

Präparierung – Freilegung

Die Untersuchung der organischen Reste erfolgt mittels Mikroskop und setzt meist Präparierungsarbeiten mit feinen Nadeln, Stäbchen und Pinseln voraus. Bei gut erfassbaren, einfachen Befunden ist die Reinigung von kleinen Untersuchungsfenstern ausreichend, um beispielsweise eine Gewebebindung oder den Verlauf einer organischen Schicht zu ermitteln. Das Präparieren von komplexen Befunden mit mehrschichtigem Aufbau ist wesentlich aufwändiger und kann mit dem gezielten Abtrag von organischen Schichten und dem Anlegen von mehreren Plana am Objekt verbunden sein. Eine Freilegung organischer Reste mittels Feinsandstrahlgerät erscheint im Vergleich zu grob und wenig kontrollierbar und ist somit nicht empfehlenswert. Eine chemische Freilegung von organischen Schichten wird nur in Ausnahmefällen (z.B. bei überlagernden Sinterschichten) und bei klar definierten Fragestellungen vorgenommen. (Zur chemischen Reinigung mittels Komplexon siehe u.a. Farke 1990, 154 und Farke 1992, 176–178).

Untersuchung – Substanzsicherung durch Dokumentation

Die Dokumentation umfasst in der Regel Übersichts- und Mikroskopfotos, Farbzeichnungen und ggf. Detailzeichnungen (z.B. mit Hilfe eines Mikroskop-Zeichentubus). Insbesondere für die Aufnahme von großen Fundkomplexen hat sich ein standardisierter Erfassungsbogen als Dokumentationsschema für sämtliche organische Schichten auf Metallen bewährt.[4] Die hausinterne Vorlage kann nach Einweisung von allen an Metallobjekten arbeitenden Restauratoren verwendet werden. Festzuhalten sind unter anderem detaillierte Angaben zu

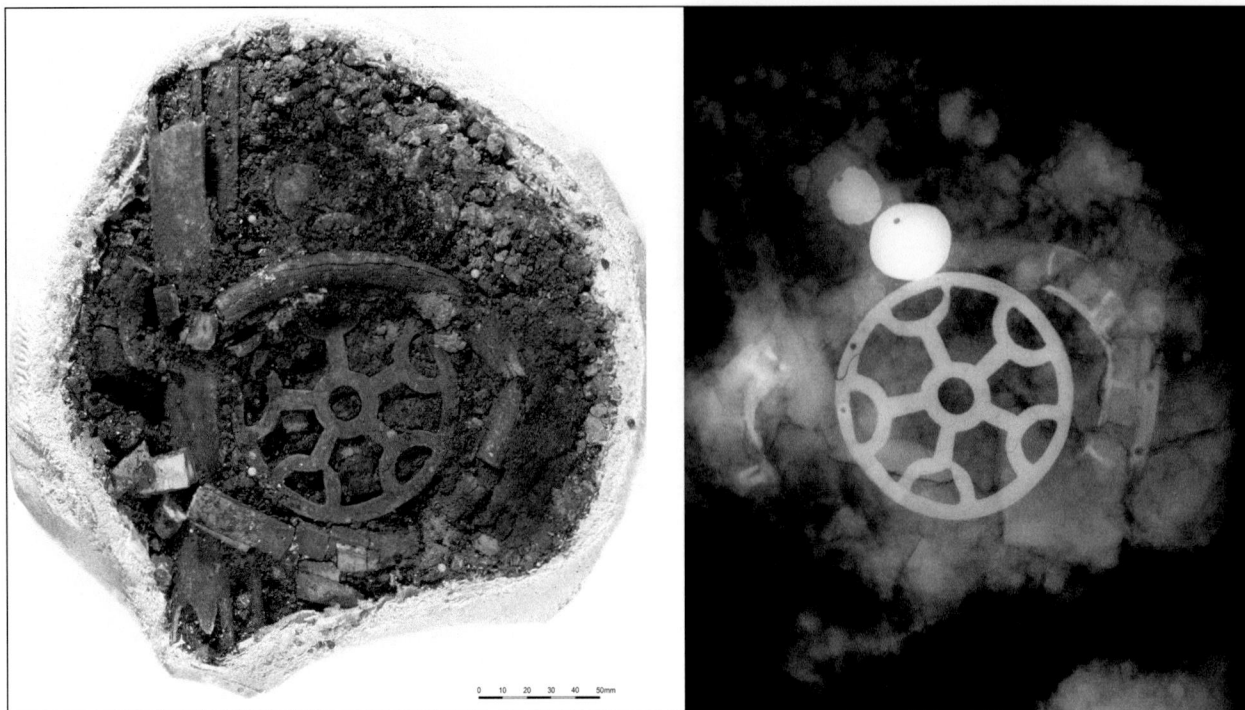

Abb. 27.4. Fundsituation und Röntgenaufnahme eines Gürtelgehänges mit Amulettscheibe und Umfassungsring mit dunklem, organischem Umfeld im Block geborgen (Lauchheim, Grab 407) (Bayerisches Landesamt für Denkmalpflege).

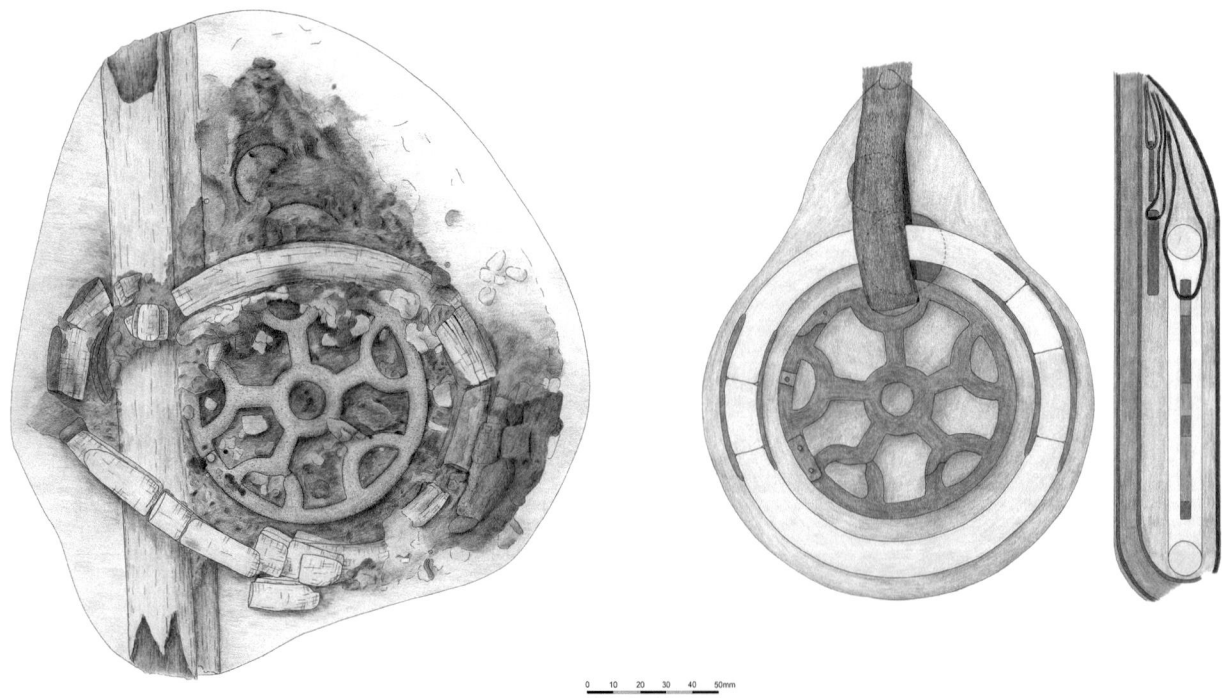

Abb. 27.5. Dokumentationszeichnung der Befundsituation (Planum 1, Maßstab 1:1) und Rekonstruktion der organischen Schichten zu einem beutelförmigen Behältnis und Aufhängeriemen (Lauchheim, Grab 407) (Bayerisches Landesamt für Denkmalpflege).

Fund bzw. Befund, Verlauf und stratigraphischer Abfolge der organischen Schichten, technische Analysedaten und eine erste Auswertung unter Einbeziehung der Lage und Ausrichtung sämtlicher Objekte im Befund.[5] Bei komplexen Befunden müssen die technischen Gewebeanalysen der einzelnen Schichten sowie die Auswertung bzw. Befundrekonstruktion von spezialisierten Textilfachkräften durchgeführt oder ergänzt werden.[6] Weiterführende Untersuchungen, z.B. zu Fasermaterialien, Besonderheiten durch webtechnische Verfahren, Überlegungen zur Herstellung, Verarbeitung und

Abb. 27.6. Oberflächenfreilegung an einer Spatha mittels Mikrofeinstrahltechnik (Bayerisches Landesamt für Denkmalpflege).

Funktion von Geweben und eine übergreifende Einordnung, steigen bereits weit in die archäologische Forschung bzw. Textiltechnologie ein. Auch hier sind speziell ausgebildete Fachkräfte gefragt.

Bei der Bearbeitung von Blockbergungen nach der mikrostratigraphischen Analysemethode wird der Befund durch das Anlegen von verschiedenen Plana nach und nach abgebaut. Die dabei angefertigte Dokumentation kann somit als „Befundsicherung" verstanden werden. Dies gilt insbesondere für die farbigen Planumszeichnungen, die wesentlich detaillierter und aussagekräftiger sind als entsprechende Fotoaufnahmen. Die Zeichnungen sind für die Erstellung einer späteren Rekonstruktion des Gesamtbefundes unerlässlich (Nowak 2002, 25–26; zur mikrostratigraphischen Analysemethode vgl. Hägg 1989, 431–439).

Freilegung der Metalle – Schutz der organischen Befunde

Bei einer weiteren (Teil-)Freilegung der Metallobjekte sind die mineralisierten Reste in besonderer Weise gefährdet. (Zur Thematik der Reinigung von archäologischen Metallobjekten mit organischen Resten s. Cronyn 1990; Fischer 2006). Für das sichere Hantieren ist eine temporäre Schutzabdeckung über den organischen Strukturen häufig hilfreich. Hierfür bewähren sich flüchtige Bindemittel[7] oder bei stabileren Strukturen extrem dehnbare Folien, ggf. kombiniert mit einer Silikon- oder Wachskappenabdeckung.

Wenn notwendig, kommen für eine endgültige Festigung der Organik in der Regel alterungsstabile Acrylharze, die auch als Schutzüberzüge für Metallflächen üblich sind, zum Einsatz (Farke 1986, 66–67; 1991, 27; Breuer 1993, 96–99; Fischer 1997, 65–69).

Da nicht immer davon auszugehen ist, dass ein und derselbe Restaurator alle Arbeitsschritte wie Erstversorgung, Dokumentation der organischen Befunde und Freilegung durchführt, sind klare Absprachen und ein Zugriff auf die datenbankgestützte Dokumentation und den Erfassungsbogen jederzeit notwendig.

Abb. 27.7. Natriumsulfit-Entsalzungsbad mit Eisenobjekten (Bayerisches Landesamt für Denkmalpflege).

Natriumsulfit-Entsalzung – Stabilisierung von Eisenfunden

Seit 2001 werden in den Restaurierungswerkstätten des BLfD Eisenfunde zur dauerhaften Stabilisierung mittels Natriumsulfit-Bädern nach üblichem Verfahren entsalzt. Die Behandlung zur Auswaschung der Salze besteht im Wesentlichen aus der Objektvorbehandlung, ca. 3–4 Natriumsulfit-Bädern, der Neutralisierung und der abschließenden Trocknung der Funde.[8]

Haften aber an den zu entsalzenden Fundstücken noch offensichtlich organische und nicht vollständig mineralisierte Reste an, so müssen diese mit alternativen Behandlungsmethoden konserviert werden, da alle organischen Materialien, insbesondere proteinische Werkstoffe, von warmer Lauge angegriffen werden.[9] Sind die erhaltenen Strukturen offensichtlich flächendeckend, mehrschichtig und komplex bzw. handelt es sich um sehr außergewöhnliche Befunde mit hohem Aussagewert, so werden sie ebenso von der Behandlung ausgeschlossen.

Bei einer Vielzahl an Funden hingegen sind die Strukturen vollständig mineralisiert oder nur noch als Abdrücke in der Korrosion erhalten und können relativ einfach erfasst und eingeordnet werden (z.B. Reste vom Holzgriff eines Messers). Es gilt zu überprüfen, ob eine Entsalzung für letztere Gruppe von Funden ohne Schädigung oder zumindest mit kalkulierbaren Risiken möglich ist. In der Fachliteratur wird diese Thematik bislang selten aufgegriffen; es werden andere Konservierungsverfahren, eine Probenentnahme bzw. eine komplette Abnahme vor der Behandlung oder die Abdeckung der Organik mit Schutzüberzügen während der Entsalzung empfohlen.[10]

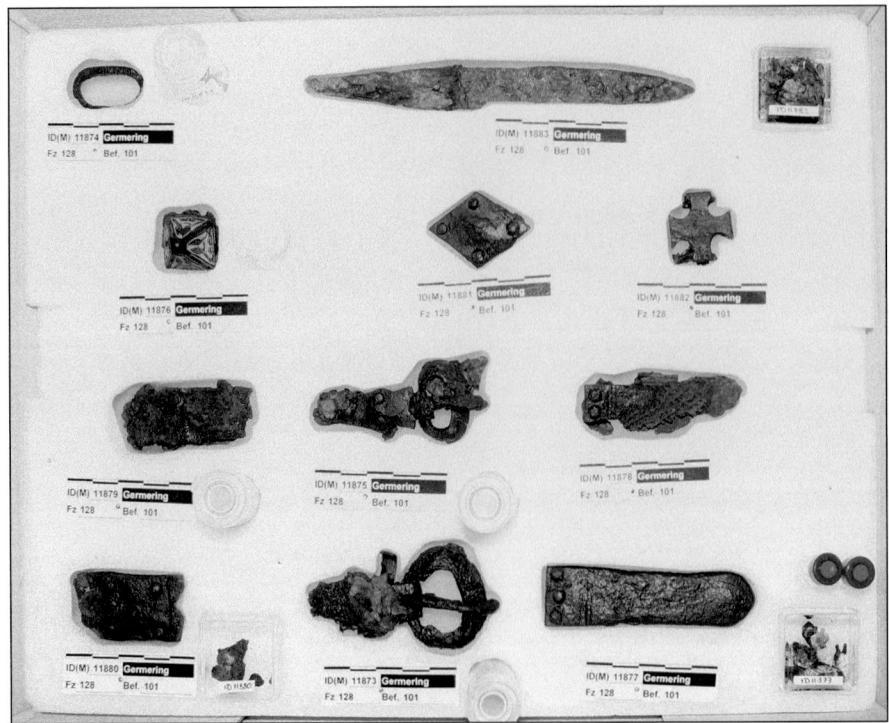

Abb. 27.8. Restaurierte Beschläge der Spathaaufhängung in chemisch inerter Plastazote-Verpackung (Germering Grab 101) (Bayerisches Landesamt für Denkmalpflege). (Restaurierung: Dipl.- Rest J. Wiesner).

Abb. 27.9. Zeichnerische Rekonstruktion der Spathaaufhängung zum Zeitpunkt der Niederlegung in Grab 101 (Zeichnung: H. Voß, Bayerisches Landesamt für Denkmalpflege).

Die bislang gesammelten Ergebnisse aus der Praxis zeigen, dass das konkrete Verhalten der organischen bzw. mineralisierten Reste während der einzelnen Vorgänge der Natriumsulfit-Entsalzung (Vorbehandlung, Entsalzung, Wässerung, Trocknung) noch nicht ausreichend einschätzbar ist.[11]

Beobachtungen und Erfahrungen sowie weitere Versuchsreihen (z.B. mit unterschiedlichen Abdeckmaterialien) bilden die Grundlage zur Verbesserung der Konservierungsmethode in Hinblick auf die organischen Befunde. Erste positive Ergebnisse in der Anwendung erbrachten temporäre Abdeckungen der organischen Strukturen mit flüchtigen Bindemitteln. Weitere Testläufe sollen die positiven Erfahrungen bestätigen.

Verpackung und Endlagerung – präventiver Schutz

Für die Aufbewahrung von restaurierten Metallobjekten mit organischen Resten kommen in der Regel säurefreie Kartons mit individuell angepassten, chemisch inerten PE-Schaumstoff-Auskleidungen zum Einsatz. Unbehandeltes Probematerial kann kühl und dunkel in Glasbehältnissen aufbewahrt werden. Die klimatischen Bedingungen sollten auf die Anforderungen der Metalle abgestimmt werden (max. 30 % relativer Luftfeuchtigkeit, unter 20°C), da bei stark abgebauten organischen Substanzen keine weiteren Veränderungen bei niedriger Luftfeuchtigkeit zu erwarten sind. Die Lagerung muss erschütterungs- und staubfrei sein. Die *Empfehlung zur Aufbewahrung archäologischer Funde* des BLfD richtet sich bei Angaben zur relativen Luftfeuchtigkeit, Temperatur, Licht und Luftqualität an allgemeingültige Standards (siehe unter *www.blfd.bayern.de*; auch: Fischer 1997, 76–78; Schichting 1994; Pearson 1987, 160).

Fazit – Weiterentwicklung von praktischen Konservierungsmethoden

Viele Aspekte zum konservatorischen Umgang mit mineralisierten Strukturen auf Metallen und eine fachliche Diskussion über verschiedene Arbeitsabläufe mit Fokus auf die organischen Befunde werden nur selten thematisiert. Der intensive Erfahrungsaustausch der Restauratoren zu einzelnen Arbeitsschritten von der Grabung bis zur Endlagerung ist somit für die Weiterentwicklung von Prozessen und üblichen Methoden dringend notwendig.

Anmerkungen

1. In dem Konzept der *Investigative Conservation* (erste Entwicklung in den 1980er Jahre vom English Heritage/ Ancient Monuments Laboratory) zur Versorgung und Auswertung großer Fundmengen unmittelbar nach ihrer Bergung wird in der Regel die Bearbeitung der organischen Auflagen auf Metallen zwar knapp thematisiert, aber es werden kaum restauratorische bzw. konzeptionelle Methoden zur praktischen Umsetzung aufgezeigt. So werden in unterschiedlichen Varianten eine datenbanktechnische Erfassung, Untersuchung, Präparierung, Materialanalysen mit modernem Untersuchungsspektrum und eine Sicherung gefordert, zumindest wenn neue wissenschaftliche Ergebnisse erzielt werden können (Niemeyer 1994, 288 und 290–291; Gebhard 1999, 179, 184–185; v. Freeden 2003, 34, 36 und 43; Gasteiger 2006, 380–381; Yoshida 2007). Eine fachgerechte Bearbeitung von Funden ist aber in der Praxis mitunter sehr zeitaufwändig und diffizil, wenn sie den maximalen Informationsgewinn für die Wissenschaft und die konservatorische Sicherung zum Ziel haben soll. Wesentlich komplexer und unkalkulierbarer ist die Aufnahme von In-situ-Blockbergungen mit organischen Befundsituationen (Nowak 2002).
2. Da das Bayerische Landesamt für Denkmalpflege (BLfD) keine sammelnde Behörde ist, unterhält es lediglich Zwischendepots für die temporäre Lagerung von Funden. Die Art der Aufbewahrung und die ggf. modifizierte Verpackung sind abhängig von Material und Zustand der Befunde und den gegebenen Möglichkeiten. Weiterführend zu Funden mit organischem Umfeld z.B. Newton und Logan 1992, 127–133; Biel und Klonk 1994, Kap. 24.2 und 24.3; Meier und Tegge 1996, 148–151; Gersbach 1998, 52 und 96; Höpfner 1999, 77–82.
3. Der Bearbeitungsaufwand ist in der Regel im Vorfeld schwer zu kalkulieren, da bei einer ersten Funddurchsicht häufig nicht sämtliche organischen Befunde und ihre Komplexität erkannt werden.
4. Es wurde ein hausinterner Erfassungsbogen in Anlehnung an Mitschke 2001 und die von Peek im Rahmen der VDR-Tagung Leipzig – denkmal 2004 „Schnittstellen in der Archäologie" vorgestellte Vorlage erarbeitet (auch Walton und Eastwood 1984; Farke 1986, 69).
5. Folgende Punkte werden erfasst: Angaben zum Befund (Befundnummer, Fundort, Grabnummer, Grabart und -bau, Störung ja/nein, Skeletterhaltung, -lage, Grabinventar, Bodentyp, Grabungsdaten), Angaben zum Fund (Werkstatt-, Fundzettelnummer, Objektansprache, Material, Datierung, Lage/Ausrichtung im Grab, Schauseite oben/unten, Röntgen/Foto/Zeichnung, Maßnahmen bei Erstversorgung/Lagerung und Restaurierung/Konservierung, Zustand Objekt und organische Schichten, Verlauf und stratigraphische Abfolge der Schichten) und alle technischen Analysedaten (Technische Zuordnung, Verarbeitungstechnik, Gewebeanalyse mit Angaben zu Fadenmaterial/Stärke/Farbe/Drehung/Dichte, Materialanalysen aller organischer Schichten).
6. Mitunter kann sich eine Gewebeanalyse sehr komplex gestalten und neue Untersuchungsmethoden wie z.B. die 3-D-Röntgencomputertomographie notwendig machen (Peek und Nowak-Böck 2007).
7. Flüchtige Bindemittel werden für die Festigung der Organik in der Praxis häufig verwendet; es besteht aber noch Forschungsbedarf, inwieweit der Einsatz wirklich bedenkenlos ist.
8. Zur Vorgehensweise am BLfD s. Gasteiger 2008. Eckdaten: a) Objektvorbehandlung: Dokumentation, ggf. Abdeckung/Sicherung, Verpackung; b) Natriumsulfit-Bäder: à 4 Wochen, bei ca. 40–60°C, ph-Wert: 14; c) Neutralisierung: in erwärmten Wasserbädern; d) Trocknung: unter Rotlichtlampe bei ca. 65°C.
9. Leider gibt es bislang keine objektive Untersuchungsmethode, die den Grad der Mineralisierung in einem Material aufzeigen kann.
10. Greiff und Bach sehen bei Substanzen, die nur noch als Strukturen in Korrosionsprodukten erhalten sind keine Gefährdung im Entsalzungsbad, empfehlen zur Sicherheit aber den Auftrag eines reversiblen Schutzüberzuges (Greiff und Bach 2000, 328). Schmidt-Ott und Oswald raten ebenfalls zur Abdeckung anhaftender organischer Materialien mit reversiblen Acrylschutzüberzügen (Paraloid B44 in Aceton 25%-ig, mehrfacher Auftrag) (Schmidt-Ott und Oswald 2006, 128). Generell stellt sich die Frage, ob eine oberflächliche Abdeckung überhaupt wirksam sein kann, da die Badlösung wohe das Objekt vollkommen durchdringt und somit von der Rückseite Zutritt zu den mineralisierten Resten hat. Scott und Seeley erwähnen die Problematik der basischen Natrium-Sulfit-Entsalzung von Objekten mit Organik und schlagen alternativ das *soxhet washing*-Verfahren vor (Scott und Seeley 1987, 73).
11. Durchaus positive Ergebnisse konnten an stabilen Strukturen z.B. von Leder, Textilien und Holz beobachtet werden. Lederoberflächen und Holzarten waren nach der Behandlung durch das Ablösen von Erdauflagen in der wässrigen Badlösung (gänzlich ohne Präparierungsarbeiten) sogar deutlicher zu bestimmen und blieben rein optisch offensichtlich unbeschadet. Negative Veränderungen und Verluste mussten bei einigen Lederschichten festgestellt werden, die nach der Behandlung in fein craquelierte Schollen zerbrochen waren. Auch zuvor stabile Holzreste zerfielen nach der Trocknung nahezu pulverförmig und konnten nicht mehr stabilisiert werden. Sehr nachteilig erschien auch die Veränderung von Schutzüberzügen zu unlöslichen Schichten und das "Herauskriechen" von Klebemitteln unter zuvor gefestigten Schollen mit mineralisierten Strukturen, deren aufwändige Beseitigung zu weiteren Beschädigungen führte. In Einzelfällen verloren mit Paraloid gefestigte organische Reste, die während des Entsalzungsprozesses über Wochen stabil waren, beim Wässerungsvorgang jeglichen inneren Zusammenhalt (hierzu auch Meissner 2007; Gasteiger 2008).

Literatur

Biel, J. und Klonk, D. (Hrsg.) (1994) *Handbuch der Grabungstechnik*. Stuttgart/Tübingen, Verband der Landesarchäologen in der Bundesrepublik Deutschland.

Breuer, H. (1993) *Die Restaurierung eines Gürtels aus dem "Fürstengrab" von Gommern*. Arbeitsgemeinschaft der Restauratoren und Römisch-Germanisches Zentralmuseum (Hrsg.), Arbeitsblätter *für Restauratoren* 26 (2) Gruppe 9, 96–99. Mainz.

Cronyn, J. M. (1990) *The Elements of Archaeological Conservation*. Cornwall.

Farke, H. (1986) *Archäologische Fasern, Geflechte, Gewebe – Bestimmung und Konservierung.* Restaurierung und Museumstechnik 7. Weimar.

Farke, H. (1990) Freilegung und Identifizierung von mineralisierten Geweberesten auf archäologischen Artefakten. Arbeitsgemeinschaft der Restauratoren und Römisch-Germanisches Zentralmuseum (Hrsg.), *Arbeitsblätter für Restauratoren Gruppe* 10, 152–159. Mainz.

Farke, H. (1991) Textilfunde aus dem 6. bis 7. Jahrhundert in Thüringen – Präparation, Bestimmung und Rekonstruktion. *Restauro* 1991 (1), Nr. 97.

Farke, H. (1992) Einsatz von Komplexon in der Textilkonservierung. Arbeitsgemeinschaft der Restauratoren und Römisch-Germanisches Zentralmuseum (Hrsg.), *Arbeitsblätter für Restauratoren Gruppe* 10, 176–178. Mainz.

Fischer, A. (1997) *Reste von organischen Materialien an Bodenfunden aus Metall – Identifizierung und Erhaltung für die archäologische Forschung.* Diplomarbeit 1994, Schriftenreihe des Instituts für Museumskunde an der Staatlichen Akademie der Bildenden Künste, Band 13, K.-W. Bachmann (Hrsg.). Stuttgart.

Fischer, A. (2006) *Zerstörung oder Informationsgewinn? Konzepte und Ziele beim Reinigen von archäologischen Metallfunden.* Verband der Restauratoren (Hrsg.), Oberflächenreinigung – Materialien und Methoden, Schriftenreihe 2.

Freeden, U. von (2003) Das frühmittelalterliche Gräberfeld von Tauberbischofsheim-Dittigheim. Erste Ergebnisse und Probleme seiner Publikation. *Berichte der Römisch-Germanischen Kommission* 84, 5–48.

Gasteiger, S. (2006) "Investigative Conservation" – ein zukunftsfähiges Restaurierungskonzept für Kleinfunde in der Archäologie und Bodendenkmalpflege. *Bericht der Bayerischen Bodendenkmalpflege* 45/46 (2004/05), 378–383.

Gasteiger, S. (2007) Empfehlungen zum Umgang mit archäologischen Funden. In Landesstelle für die nichtstaatlichen Museen in Bayern beim BLfD (Hrsg.), *Archäologische Funde im Museum – Erfassen, Restaurieren, Präsentieren.* MuseumsBausteine, Band 12. München.

Gasteiger, S. (2008) Nachhaltige Konservierung archäologischer Eisenfunde durch Natrium-Sulfit-Entsalzung. *Jahrbuch der Bayerischen Denkmalpflege*, 2006/07, Band 60/61 (im Druck).

Gebhard, R. (1999) "Investigative Conservation". Konzepte zur Eisen- und Bronzerestaurierung von frühmittelalterlichen Grabfunden. In Prähistorische Staatssammlung (Hrsg.), *Dedicatio. Hermann Dannheimer zum 70. Geburtstag*, 179–190. Kallmünz, Verlag Michael Lassleben.

Gersbach, E. (1998) *Ausgrabung heute. Methoden und Techniken der Feldgrabung.* Darmstadt, Wissenschaftliche Buchgesellschaft.

Greiff, S. und Bach, D. (2000) Eisenkorrosion und Nariumsulfitentsalzung: Theorie und Praxis. Arbeitsgemeinschaft der Restauratoren und Römisch-Germanisches Zentralmuseum (Hrsg.), *Arbeitsblätter für Restauratoren*, Gruppe 1, 319–339. Mainz.

Hägg, I. (1989) Historische Textilforschung auf neuen Wegen. *Archäologisches Korrespondenzblatt* 19 (4), 431–439.

Höpfner, M. (1999) Passive Konservierung großer Mengen archäologischer Eisenfunde. Arbeitsgemeinschaft der Restauratoren und Römisch-Germanisches Zentralmuseum (Hrsg.), *Arbeitsblätter für Restauratoren* Gruppe 21, 77–82. Mainz.

Meier, M. und Tegge, C. (1996) *Verpackung in Stickstoffgas – eine neue Methode zur Lagerung von archäologischen Funden!.* Berichte zur Denkmalpflege in Niedersachsen (16), 148–151.

Meissner, I. (2007) *Sicherungsmaßnahmen an archäologischen Eisenfunden vor der Natriumsulfit-Entsalzung.* Technische Universität München, Unveröffentlichte Seminararbeit 2006/2007, München.

Mitschke, S. (2001) *Zur Erfassung und Auswertung archäologischer Textilien an korrodiertem Metall.* Vorgeschichtliches Seminar der Philipps-Universität Marburg (Hrsg.), *Kleine Schriften* 51, Marburg.

Newton, C. L. und Logan, J. A. (1992) On-site Conservation with the Canadian Conservation Institute. In R. Payton (ed.), *Retrieval of Objects from Archaeological Sites*, 127–132. Denbigh, Archetype Publications Ltd.

Niemeyer, B. (1994) "Investigative Conservation" – das Restaurierungskonzept des Ancient Monuments Laboratory/English Heritage zur Untersuchung großer Fundkomplexe. Arbeitsgemeinschaft der Restauratoren und Römisch-Germanisches Zentralmuseum (Hrsg.), *Arbeitsblätter für Restauratoren* Gruppe 1, 287–292. Mainz.

Nowak, B. (2002) *Zur Bearbeitung von Blockbergungen mit organischen Resten aus archäologischen Ausgrabungen.* Staatliche Akademie der Bildenden Künste Stuttgart, unveröffentlichte Diplomarbeit, Stuttgart.

Pearson, C. (1987) *Conservation of Marine Archaeological Objects.* Oxford, Butterworth-Heinemann.

Peek, Ch. und Nowak-Böck, B. (2007) 3D-Computertomographie – Neue Möglichkeiten zur Untersuchung archäologischer Textilien. In A. Rast-Eicher und R. Windler (Hrsg.), *Archäologische Textilfunde – Archaeological Textiles. Report from the 9th NESAT Symposium, 18.–21.Mai 2005 in Braunwald*, 79–85. Näfels.

Schichting, C. (1994) *Working with Polyethylene Foam and Fluted Plastic Sheet.* Canadian Conservation Institute (ed.), Technical Bulletin 14, Canada.

Schmidt-Ott, K. und Oswald, N. (2006) Neues zur Eisenentsalzung mit alkalischem Sulfit. In Verband der Restauratoren (Hrsg.), *Beiträge zur Erhaltung von Kunst und Kulturgut*, 126–134. Bonn.

Scott, D. A., und Seeley, N. J. (1987) The Washing of Fragile Iron Artefacts. *Studies in Conservation* 32, 73–76.

Walton, P., und Eastwood, G. (1984) *A Brief Guide to the Cataloguing of Archaeological Textiles.* York, Worldplex Wordprocessing Bureau.

Yoshida, M. (2007) *Fundkomplexe in der Bodendenkmalpflege – Das Konzept Investigative Conservation am Beispiel eines frühmittelalterlichen Gräberfeldes.* Technische Universität München, unveröffentlichte Diplomarbeit, München.

28 Medieval Textiles from Trondheim: An Analysis of Function

by Ruth Iren Øien

Knowledge of medieval clothing has until recently primarily been obtained through finds of complete textile garments. However, through excavations in the major medieval towns in the last decades, quite a few small and large textile fragments have surfaced. In Trondheim, more than 1100 finds, all of which can be dated to AD 1050–1537, were excavated during 1969–1993. None of the textiles can be directly linked to human remains, but it may be assumed that the items had been passed down through generations, *i.e.* as wages for servants, second-hand usage, torn/cut for use as insulation, cloth, diapers, and toilet paper (Turnau 1994). Thus, most pieces of fabric are woven in one piece, while others have traces of folds, seams, buttonholes, or fragmentary features from the construction of plaiting and gores. The locations of these finds in Trondheim are Søndre gate, Erling Skakkes gate, Kjøpmannsgata 22–26, Televerkstomten, Erkebispegården, and Kjøpmannsgata/Bryggegaten (Fig. 28.1). All of these sites lie in the heart of medieval Trondheim and provide a good insight into the use of textiles by its population.

Which of these textiles have been used for clothing, and can these small fragments shed light on their construction? To answer these questions, Marianne Vedeler's concept was used: this relies on context, textile quality, seam, and form (Vedeler 2004). In this study, I discuss two groups of textiles: pleated fabrics and the remains of long gores indicating a rectangular pattern, as examples of how textile fragments can contribute to knowledge of medieval clothing.

Pleated Textiles

No complete piece of clothing has been found with this type of pleating conserved. However, finds of smaller and larger fragments found in excavations undertaken from 1960 to the present in an array of medieval towns in western Scandinavia provide us with clues. In Trondheim, I identified 93 fragments or groups of fragments, which could be linked to this type of textile. All these textiles are dated to AD 1050–1325, with the vast majority dated to AD 1150–1300. Each pleat is sewn down in the back of the fabric at the bottom of each fold; a vertical seam has been added using running stitch. The front edge of the fold has been folded or compressed, so that the fold has kept its shape. The stitches are very even and precise. Joints of two pieces of fabric are always made at the folds' back edge, either by carpet or overcast stitch.

In Trondheim, the weave is predominantly by 2/1 twill (90.4%). The remaining pieces are in 2/1 twill with reverse, and 2/2 twill. The pleated textiles in Bergen have a greater variety in weave during the earliest period, as there are equal amounts of 2/1 twill and 2/2 twill, but in the later period, the use of 2/2 twill decreases and 2/1 twill increases (Vedeler 2007, 92). Textile quality in Trondheim varies from medium coarse to fine, although most of the textiles have more than 15 threads per cm in warp and weft. The medium coarse fabric has a more equal number of threads in warp and weft. The weft threads rarely exceed 20 threads per cm, but the warp usually has a higher thread count than the weft (Fig. 28.2). The fabrics have a very even appearance and shine despite the difference in thread counts in warp and weft. The yarn seems to be made of long soft fibres where the undercoat is separated. The threads are tightly spun. The coarsest textiles in this group have lost some of their lustre. It is not possible to establish a correlation between thread count and the direction the threads are spun. This group of textiles is more frequently z/z spun than the other textiles in Trondheim.

Scholars have long agreed that pleated textiles are remains of medieval clothing (Nockert 1984, 191–192; Kjellberg and Hoffmann 1991, 92; Vedeler 2002, 222–223). What they looked like, whether they were worn by men or women, and what social rank the people wearing them had, is unknown. Textiles of this type found in Trondheim may give us more information regarding their appearance. The pleats are divided into two types: parallel pleats and *stråleplisse*. The parallel-pleated fragments have an even width along the entire fold. *Stråleplisse* are constructed from textile gores. Each pleat increases in width from top to bottom. The top can be as narrow as 0.5 cm, while the bottom rarely exceeds 3.5 cm. The use of *stråleplisse* tells us that the clothing has been cut for increased width.

In Trondheim, the pleated textiles are sewn together from single panels. The width of these panels varies from 8 to 16

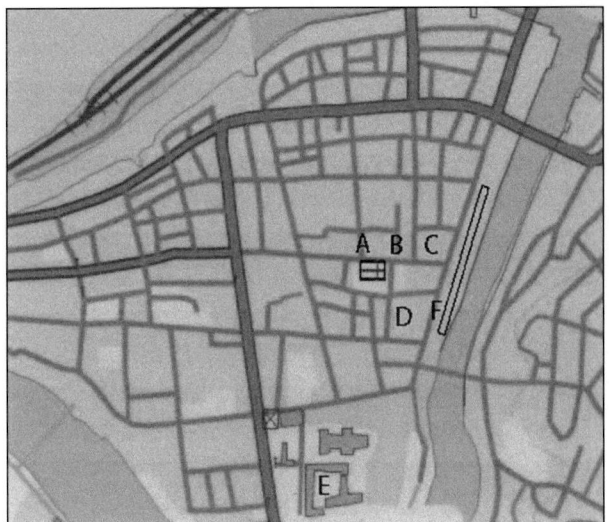

Fig. 28.1. Site locations (from Øien 2008).

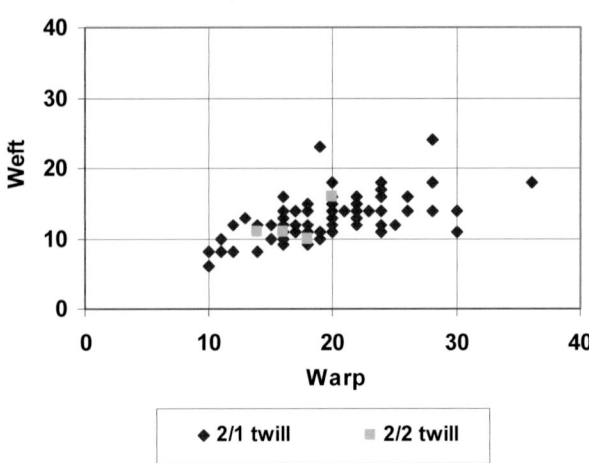

Fig. 28.2. Warp and weft in pleated textiles (from Øien 2008).

cm, where the following feature is often repeated: every panel is created of three pleats in the back before it is fastened to the next panel at the bottom of the fourth fold. The width of the panels in Tønsberg appears to have a width similar to those from Trondheim, while a fragment from Bergen was 61 cm wide with a folding edge at the bottom. This fragment comprises a whole section of preserved panels measuring 40 cm (Vedeler 2007, 97–98). All the finds from Trondheim seem to be pleated. Moreover, there are no other known finds of pleated textiles fastened to even unpleated fabric. I therefore agree with Vedeler who claims that: "This suggests that not only the middle part of the skirt has been pleated, but the whole front or back piece, possibly both" (Vedeler 2007, 98).

None of the fragments from Trondheim have kept their folded edges. The longest fragments are 56, 58 and 67.5 cm in length. There are preserved *stråleplisser* with intact folded edges from Bergen amongst other locations. These are 76 and 81 cm in length. *Stråleplisse* with intact folded edges on top and bottom are considered to come from skirts (Nockert 1984; Gjøl Hagen1992; Vedeler 2002). However, the material from Trondheim may provide us with insight into how the upper part of the dress may have looked. One of these fragments has a horizontal seam and is made up of two different kinds of fabric, one of which is coarser than the other. The coarser one is 2/2 twill while the finer is 2/1 twill. Today, the surface of the coarse fabric appears brighter compared to the other. The use of these two textile qualities yields a noticeable visual difference. Whether or not this difference would have been enhanced by using two different colours, remains to be seen until a dye analysis is performed.

Two other fragments from Trondheim provide more information on how a possible upper part of the pleated dress looked. One of these fragments has a horizontal seam, and is also made up of two different materials (Fig. 28.3). These are not sewn in on each side of the seam, but inserted as an additional panel in the fragment's upper part. One can also see remains of this fabric, which is sewn together using panel seams at the bottom. The two textiles are 2/1 and 2/2 twill. The 2/1 twill is noticeably finer than the other.

What is certain is that the fragment from Trondheim has been sewn together of two different textiles, which produces a visual effect. As noted before, whether this effect would be enhanced by using two different colours cannot be determined without a dye analysis. The horizontal seam is sewn together using running stitch and the textile is folded to each side, then sewn together using overcast stitching. On the front at the underside of the seam joint, down towards the wider part of the pleats one can vaguely see a seam in running stitches. This seam is not visible on the reverse side, but may be a marker of the horizontal joint. At the top of the pleats' width, there is about 1 cm between the seams, with a width of 0.5 cm from top to bottom.

There is no preserved finishing edge at the top of the pleats. Folds narrower than 0.5 cm from top to bottom are not known, indicating that the fragment's upper part could not have been much wider before it was sewn together with another piece of cloth or finished with an edge. This means that the pleats on this piece of fabric increase evenly by 2 cm, signifying that the spacing of the seams at the top is 1 cm, while it is 3 cm at the horizontal seam in the back. The distance from the top to the horizontal seam is 30 cm. On the underside of the horizontal seam, it appears as if the rate of increase declines. At the bottom edge, the fragment is ripped and torn – from seam to seam the folds do not measure more than 4–4.3 cm, which indicates that the increase is 1–1.3 cm in 20 cm. The rate of increase is lower. The part on the outer side of the horizontal seam has a greater rate of increase in fold width than what is known from other *stråleplisse* in Trondheim. It is also quite clear that the seamstress has been very accurate in shaping the folds to fit each other in the horizontal seam. The folds are fastened directly above each other with approximately the same width. The panel seam is also made in the same place on the two

Fig. 28.4. Reverse side of fragment N 23777 FA 342 (NTNU, Vitenskapsmuset, Trondheim).

Fig. 28.3. N 31326 FL 179 pleated textile with horizontal seam (Drawing: Nina Lundberg).

pieces. The fact that these fragments are sewn together with such precision emphasizes that the pleats' form and symmetry were important to the appearance of the garment.

Vedeler indicates in her interpretation of the trapezoid-shaped fragments with overcast stitches and hemming at each end that, the lower part of the cloth has been joined to an upper part, a bodice (Vedeler 2007, 98).

Can fragment number N 31326 FL 179 be the remains of such a bodice? It is not possible to tie the fragment to a specific body part from its length, and if the horizontal seam were disregarded, the fragment – as with other bigger pieces thought to be similar – has a skirt with a trapezoid-shaped cut. A detail on the fragment from Trondheim makes another placing plausible; on top of the fragment's back, a seam which creates a narrowing can be seen. This is at the most 1.5 cm wide, and in the downward direction the width gradually narrows down to nothing, about 20 cm from the top. Here the fragment has been shaped, in addition to the trapezoid-shaped cut. If one places the fragment under an arm, this horizontal seam is positioned at the underside of the

ribs of a person of 165 cm in height. This results in a fairly high-positioned bodice. A sculpture on Queen Katarina's tombstone in Gudhem Monastery in Västergötland, Sweden, dated AD 1100–1200 (Nockert 1984, 195) shows a section near the waist that may be a sewn horizontal joint or a belt that can indicate such a high-positioned bodice.

Fragment N 23777 FA 342 consists of parallel pleats and is 35 cm in length. It is made up of two panels, and there are also remains of a third which is barely connected to the rest today. The width today is approximately 25 cm. In one end, the edges of the pleats are cut on the bias and a triangular form appears from the top of each pleat. Each bias is about 2.5–3 cm in length. The cut appears to be shaped intentionally. This shape was also found in two additional fragments. Beneath the triangular edge on fragment N 23777 FA 342, there were small preserved horizontal seam holes. If one looks at the top of the pleats, there is some fabric at the top of each pleat. I think a triangular edge had been made to reduce the amount of fabric in the seam in order to avoid the fabric bunching together, and thus obtain good-looking seams, while at the same time, increasing the flexibility of the fabric. A similar edge can also be seen on the centre gores at the back and front of the Moselund garment, which has been ^{14}C dated to AD 1050–1155 (Hald 1950, 56–58; Østergård 2004, 135–139).

Beneath the triangular edge is an area of tiny pleats made with tunnel stitch, which goes vertically in the same direction as the folds. The seam is sewn is such a way that, today, it is barely visible on the front side. The seam goes on the side toward the top of each fold and the folds are drawn closer together. It appears the method is similar to that observed by Else Østergård in garments D10590 and D6473 from Herjolfsnæs (Østergård 2004, 99–100). These items have later been examined and one can spot in places a horizontally supporting seam on the fabric's reverse (Vedeler 2007, 134–138). This is not always the case in Trondheim, although the pleats here also seem to have a somewhat different function. A vertical seam of this kind was also found on fragment N 96960 FU 347. This piece is also a possible pleat. Two panel seams can be seen. However, the fabric beneath the gathered area is so fragmented that it is hard to see the pleating. Vertical wear marks can be spotted,

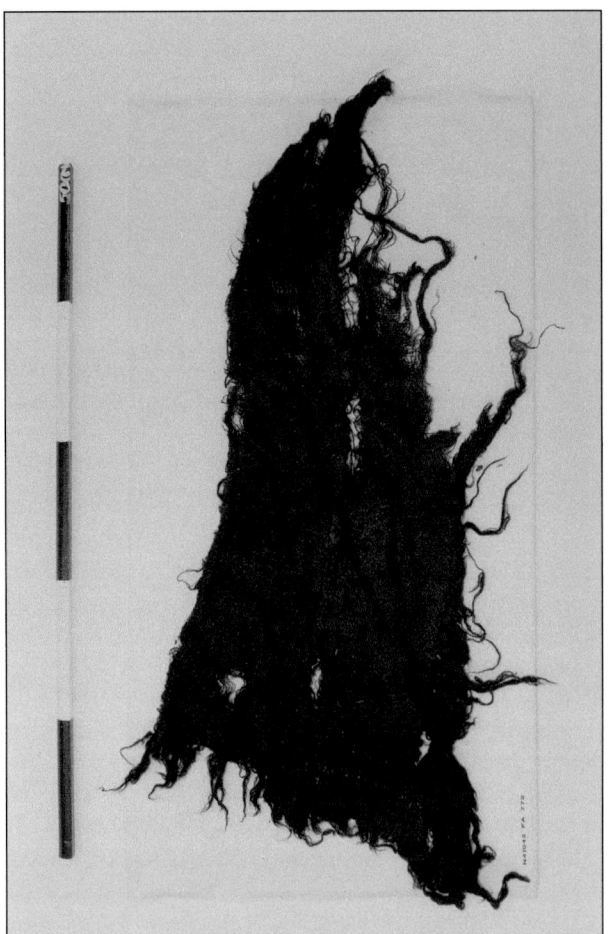

Fig. 28.5. N 39611 FA 772 fragment with gores (NTNU, Vitenskapsmuset, Trondheim).

Could fragment N 23777 FA 342 with a vertical seam be a part of the neckline, and how has it been covered – by a strip of fabric, ribbon or braid? Today, the fragment does not leave any clues as to the presence of a cut arc at the top of the fragment. The intact piece of the top is so narrow as to indicate any arc would have been hard anyway. Based on the fabric lengths of 35 cm, the pieces must have belonged to a piece of clothing at least 35 cm in length. It is not possible to determine if this is in the front or back of the bodice. In two fragments from Bergen, the thread does not follow the fabric's warp direction. One of the fragments has a slightly curved hem. Vedeler believes that this arc can come from the top of a sleeve which may have been connected to a bodice (Vedeler 2007, 99). If fragment N 23777 FA 342 belongs to a sleeve, it must have been designed differently from that on the sculpture on Queen Katarina's grave. The pleated material from Trondheim can yield knowledge of this type of garment, but regrettably not to such an extent that its appearance as a whole is revealed.

Two Fragments with Long Gores

Two fragments stand apart from the rest of the material in that their size and shape show them to have a cut connected by gores. N 39611 FA 772 has clear straight gores, while N 41042 FF 1316 has gores that are more oblong, possibly gores cut to shape. The material is 2/2 twill, the coarsest type of the 'medium coarse' group. The direction of spin is z/s. Both thread systems cover each other well and the fabric has a soft almost furry appearance.

The two fragments from Trondheim are sewn of materials of a textile quality similar to garments with rectangular shape and gores, and garments with gores in a rectangular pattern, from Bocksten, Guddal, Skjoldehamn, Kragelund and Moselund. These garments are woven in 2/2 twill and 2/1 twill. They all have coarse to medium coarse weft with 1–10 threads per cm (Hald 1950; 1980; Nockert 1997; Østergård 2004; Vedeler 2007).

The seams on the two fragments are made with running stitch, and then folded to one side and hemmed down by using overcast stitches. The exception is found on fragment N 39611 FA 772, where one of the seams at the outer edge on one of the gores is hemmed in the front with overcast stitches. Both fragments are dated to the second half of the 11th century AD. This makes it natural to connect these finds to the earliest medieval garments with a rectangular gore cut, such as Kragelund, Moselund, Skjoldehamn and Bocksten (Gjessing 1938; Nockert 1997; Hald 1980, 60, 70). However, the cut of N 41042 FF 1316 resembles garments with oblong gores from Herjolfsnæs, for instance item no. 38, D 10580; no. 41 D 10583; no. 43, D 10505.1 (Nørlund 1924, 102–114; Østergård 2004, 160–175). According to Vedeler, the garments cut with this type of gore may have been in use from the 12th century AD to the first half of the 14th century AD (Vedeler 2007, 113).

Fragment N 41042 FA 772 is made of two preserved gores. The piece is 56 cm in length with one fragmented edge at one end. The wedges are 1.5 cm wide and gradually

and a preserved seam can partly be observed. In addition, the vertical seam widens quickly and also obtains a thick folded edge at the top. Whether this piece of fabric has been parallel pleated or *stråleplisert* is impossible to discern. Above the gathered area a folded edge may be seen. Some fabric remains in the folded edge open a possibility that this fold may have been connected to another fabric. The use of long parallel pleated textiles suggests they were placed in an area with no need for increased width. When this need was present, it was achieved through the use of a vertical seam at the top of the folds which can be found in the above mentioned fragments. Pieces of cloth not requiring increased width can be a front or back of a garment or sleeves.

Fragment 23777 FA 342 has traces of stitch holes at one end, straight above the vertical seam. Could this fragment be placed on a neckline as the one seen in a representation on Queen Katarina's grave in Gudhem monastery in Sweden? On the sculpture, the bodice and the shoulder areas seem to have vertical folds that descend from the neckline. The lower arm is marked by the folds turning horizontal. This mark-up of a neckline can also be found partly preserved on a garment from grave 31 at Uvdal Stave Church. The sleeves are cut so that they descend from the neckline. This garment is dated later, AD 1300 (Vedeler 2007, 120–125).

increase to 12 cm at the bottom. To one side, an additional seam may be seen, while on the other side, a fold along the entire edge is seen. This supposed seam also represents a joint. Gores of this kind that increase in number from top to bottom are found on garments with rectangular cuts, where the wedges have been placed to enhance the width in the front, back, at the sides, and down. One can also find wedges like this on the garment from Skjoldehamn, and those from Bocksten, Kragelund and Moselund (Gjessing 1938; Hald 1980, 60, 70; Nockert 1997). Fragment N 41042 FA 772 from Trondheim has a length of at least 67 cm, but the folding edge at the bottom is not preserved. The length of the gores from the Moselund and Skjoldehamn garments are shorter than the fragments found in Trondheim. They may therefore be remains of gores from a garment. Whether these are centre back, centre front, or side gores cannot be determined. Placing these is difficult, as most of the garments from this period are asymmetrical and seem to be cut based on the availability of fabric.

The final fragment, N 41042 FF 1316 has a length of 72 cm at the top, width of 7 cm, and 34 cm at the bottom. Two complete gores are preserved. There are textile remains on both sides of the gores, but whether or not this represents a new gore or a textile without a special cut cannot be determined. The gores are narrow at the top and more oblong than those found in garments of a rectangular cut. At the top of the fragment, the edge is hemmed down. This hem is thick and it looks as if another fabric has been attached to it, possibly a sleeve. The textile is sewn together using coarse vertical overcast stitching, 25 cm from the top. This results in one of the gores being concealed. If one assumes that this fragment has been fastened to a sleeve, this seam would have come around the end of the ribs. This may be an attempt to fit a garment to an individual of different shape than its original owner. The fragment's length and possible placing beneath the arm makes it possible that it may have been made for a woman. As most women would have often been pregnant, they would have needed the possibility of adjusting the size of their garments.

Oblong gores of great length are also found in Nørlund's type Ia, b, c, and f (See Østergård 2004, 149–151). Type Ia has long gores from the armpits down. These are relatively wide at the top and have a greater give than that found in the fragments in Trondheim. In types Ib, c, and f, the side gores consist of 2–4 gores. The fragment from Trondheim can be determined with certainty to have two gores. The side gores in garments NR38 and NR41 from Herjolfsnæs have gores that are narrow at the top and increase toward the waist, where they increase greatly in width. This is not the case with the fragment from Trondheim. If one looks at the side gore in NR43, this looks more similar. The gore increases evenly in its entire length, which is also the case with the fragment from Trondheim. The latter is dated earlier than garments of similar cut. Garments with a rectangular cut from the earliest period most likely belonged to men. Can this fragment originate from a woman's garment, where gores have been placed into the sides in order to be able to adjust its size?

Conclusions

Studying textile fragments and ascertaining their function is similar to an intricate jigsaw puzzle. In order to be able to analyze its function, a textile fragment has to contain certain informative features. The benefit of context in such an analysis is limited, as the fragments are found separately from the body. The other three concepts of Marianne Vedeler mentioned at the outset can contribute to increase knowledge of medieval clothing to a much larger extent. The quality of a textile fragment may indicate its usage, but can alone only reveal whether or not a fragment was at all suitable as clothing. Seams, together with quality, show single groups of textiles that can be linked to certain pieces of clothing, as shown here with pleated textiles made of medium coarse and fine textiles. There are also other examples. Stocking fragments and mittens appear mainly to be made of coarse quality fabrics (Øien 2007a, 96–103; 2007b). Tabby weave rarely appear together with seams, and in the cases where seams are preserved, it is in the group of fine textiles. Moreover, if we look at the fully preserved dresses from the Middle Ages, only a small number of these are in tabby weave (Øien 2007a, 69–70). If the seam is not preserved in the fragments, the same often applies to the shape. A fragment's shape is the conceptual 'hook' best able to yield information about medieval clothing. If the shape of a garment is preserved, it may help to establish its function. The less shape there is, the more important the textile quality, seam and context become. Fragments with adequate shape can be placed in certain groups of clothing. The individual pieces of cloth within a group can be defined only to a slight extent.

Seam and textile quality are the two of Vedeler's conceptual 'hooks' with the most development potential, thus they can contribute to a larger extent to an analysis of function. This especially applies to the connection between seams done with hemming and seam joints, and to link these to textile quality and weave. This aspect may provide further knowledge of the technological choices made with regard to the making of the cloth. In addition, it may give information about the fragments' position on a finished piece of clothing.

Acknowledgement

I would like to thank Thomas Sletvik profoundly for helping me with the translating.

Bibliography

Gjessing, G. (1938) Skjoldehamndrakten. En senmiddelaldersk nordnorsk mannsdrakt. *Viking* 2, 27–81.

Gjøl Hagen, K. (1992) Solplissé – En reminisens av middelalderens draktutvikling? En komparativ studie i plisserte stoffer fra Birka, Vangsnes, Middelalderens Trondheim, Uvdal og Setersdal. *Varia* 25. Oslo, Universitetets Oldsaksamling.

Hald, M. (1950) *Olddanske tekstiler.* Nordiske fortidsminder V, Det kgl. Nordiske Oldskriftsselskab. København.

Hald, M. (1980) *Ancient Danish textiles from bogs and burials.* København.

Holck, P. (1988) Myrfunnet fra Skjoldehamn – mannlig same eller norrøn kvinne? *Viking* 51, 109–116.

Kjellberg, A., and Hoffmann, M. (1991) Tekstiler. *De arkeologiske utgravningene i gamlebyen, Oslo.* Bind 8. Dagliglivets gjenstander – del II. 13–81. Øvre Erivik.

Nockert, M. (1984) Medeltida dräkt i bild och verklighet. In *Den ljusa medeltiden: Studier tillägnade Aron Andersson. The Museum of National Antiquities, Stockholm studies 4,* 191–196. Stockholm.

Nockert, M. (1997) *Bockstensmannen och hans dräkt.* Varbergs museum årsbok 1997. Omarbeidet årgang fra 1985.

Nørlund, P. (1924) Buried Norsemen at Herjolfsnes. *Meddelelser om Grønland 67,* 1–270. København.

Øien, R. I. (2007a) *Fra tekstilfragment til drakt. – analyse og funksjonsbestemmelse av vevde tekstiler fra middelalderens Trondheim.* (Unpublished) Hovedfagsoppgave at NTNU- Vitenskapsmuseet. Trondheim.

Øien, R. I. (2007b) Middelalderens votter. *Spor 2,* 24–26.

Østergård, E. (2004) *Woven into the Earth: Textiles from Norse Greenland.* Aarhus, Aarhus University Press.

Schjølberg, E. (1998) 12th Century Twills from Bergen, Norway. *Textiles in European Archaeology,* 209–213. GOTARC Series A, vol. 1.

Strömberg, E., Hoffmann, M., and Geijer, A. (1979) *Nordisk textilteknisk terminologi, förindustriell vävnadsproduktion.* Oslo.

Turnau, I. (1994) *European Occupational dress from the fourteenth to the eighteenth century.* Warsaw, Institute of Archaeology and Ethnology, Polish Academy of Sciences. The Library of Polish Ethnography.

Vedeler, M. (2002) *Plisserte tekstiler fra middelalderske draktplagg. Universitetets kulturhistoriske museers skrifter nr. 1. UKM en mangfoldig forskningsinstitusjon.* Oslo.

Vedeler, M. (2004) Er dette rester etter klær? – Problemer knyttet til funksjonsbestemmelse av arkeologiske tekstiler. *Collegium Medieval. Tverrfaglig tidsskrift for middelalderforskning* 17, 75–86.

Vedeler, M (2007) *Klær og formspråk i norsk middelalder.* Acta Humaniora nr 280. Oslo, Det humanisktiske fakultet, Kulturhistorisk Museum Universitetet i Oslo.

29 Curry-Comb or Toothed Weft Beater? The Serrated Iron Tools from the Roman Province of Pannonia

by Judit Pásztókai-Szeöke

Introduction

Functional identification of tools from an archaeological context is never an easy task for scholars of past technologies. Some basic implements are very long-lived, but the formal similarities of these ancient objects to present-day ethnographic items can result in the misinterpretation of their function. One of the aims of my PhD project is to specify the function of a certain iron object known from the Roman province of Pannonia, the territory of modern Hungary. By focusing on its archaeological context and human handling it is hoped that a more unambiguous functional identification may be possible.

General Description

The tools in question are elongated rectangular iron blades 16–30 cm long, serrated with short teeth (1–2 per cm) on one of their longer sides and with an iron prong for the handle at a right or an obtuse angle in the middle of the other. Usually, at each corner of this latter side, they have one wavy prong bent back to the blade with a ring hanging from it (Fig. 29.1). To date, at least 27 iron objects of the same type are known among the archaeological finds from Pannonia (Burger 1979, Abb. 44; Müller 1982; Henning 1987; Péterfi 1993, table VI.56.3 and table XIV.102.8; unpublished item from Tác-Margittelep grave 216).

Dating

In Pannonia, the earliest of these items may have already been in use during the 2nd–3rd centuries AD, but from the 4th century onwards, there is no doubt about their use. Parallels are also well known from outside the province, with some evidence from the 2nd century AD, and the bulk dated to the 3rd–7th centuries AD (Müller 1982; Henning 1987; Gaitzsch 2005). Some morphologically identical items are also acknowledged from even later archaeological contexts in medieval Russia (Nikitin 1971; Kirpichnikov 1973 and 1986).

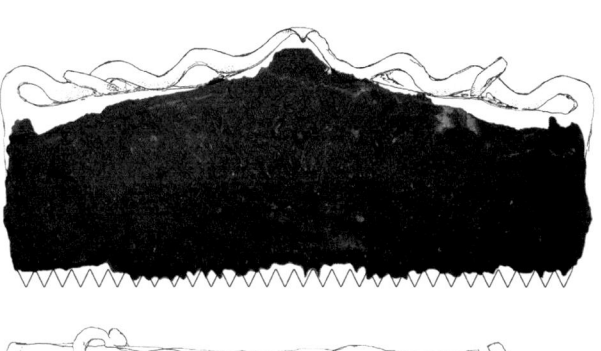

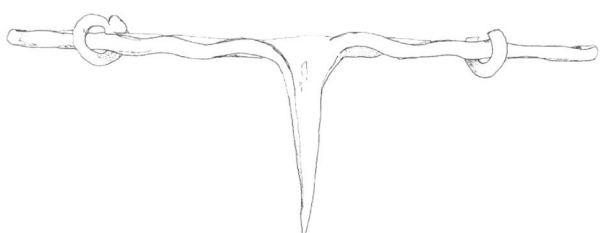

Fig. 29.1. Roman iron tool (weft beater). (Drawing: Marianne Bloch Hansen after Gaitzsch 2005, Taf.53. STR 2 and a photograph of a Roman item from Tác-Margittelep grave 216).

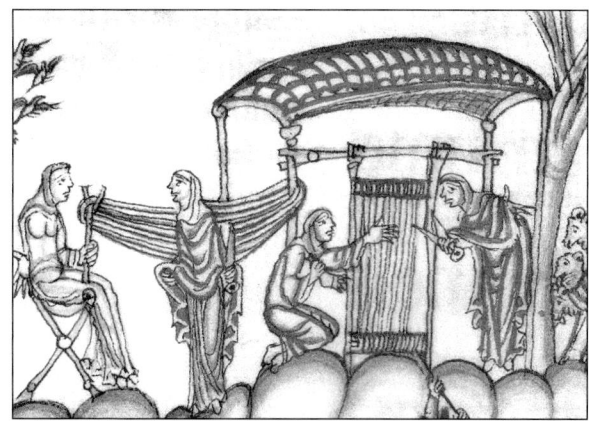

Fig. 29.2. Depiction of a vertical two-beam loom and the use of the toothed weft beater in the 12th century Eadwine Psalter (After Walton Rogers 1997, Fig. 817).

Function

Contrary to the generally accepted identification of these tools as scrapers for hides or curry-combs used for horse grooming, it is more likely that they were used as textile tools (toothed weft beaters) applied for beating the weft thread into its place while working on a two-beam vertical loom (Fig. 29.2). This interpretation is based on the archaeological context of the items from Pannonia, where they were found together with other textile tools in burials (Burger 1979, grave 67; Sági 1981, grave 111; Péterfi 1993, grave 56 and 102 and the unpublished grave 216 in Tác-Margittelep) and in association with textile producing activity in a settlement site as well (Thomas 1955). It is also supported by ethnographic parallels from Hungary, where a similar object was in use for weaving sacks, tent coverings of different animal hair, mainly goat hair (which was removed during the skin-tanning process called *tobak* in Hungary), and rugs made on a two-beam vertical loom even up to the 1930s (Ébner 1931; Morton 1936; Crowfoot 1936–1937; 1941; Domonkos 1954; 2000; Szolnoky 1954; Báldy Bellovics 1974; Lukács 2007, 69).

Future Perspectives

In the near future, in collaboration with two anthropologists, Mónika Merczi (Balassa Bálint Museum, Esztergom) and Zsolt Bernert (Hungarian Natural History Museum, Budapest), I will focus on the human handling of this particular iron object, on the basis of the above mentioned burial evidence. The aim of this research is to elucidate whether any activity markers, of horse grooming or weft beating, can be found in the skeletal material. The mechanical load, *e.g.* repetitive motions integrated into a technical gesture of occupational activities applied to living bone, influences the structure of bone tissue in terms of both its morphology and density. Since objects are involved in skills and actions performed by the human body, and similar objects used for different technical gestures require different skills and actions, this leads to different kinds of effects on the body. It is our hope that a study of this type would advance our knowledge of the relationship between people and their objects in the past.

Bibliography

Báldy Bellovsics, F. (1974) A torba, a bácskai buvenyácok tarisznyája. *Cumania* 2, 159–164.

Burger, A. Sz. (1979) *Das spätrömische Gräberfeld von Somogyszil*. Budapest.

Crowfoot, G. M. (1936–1937) Of the Warp-Weighted Loom. *Annual of the British School at Athens* 37, 36–47.

Crowfoot, G. M. (1941) The Vertical Loom in Palestine and Syria. *Palestine Exploration Quarterly* 73, 141–151.

Domonkos, O. (1954) Egy tiszántúli szőrtarisznyás műhely. *Néprajzi Értesítő* 37, 192–214.

Domonkos, O. (2000) Tarisznya- és pokrócszövés. In *Szőttes textíliák mindennapi életünkben. A Hevesen 1999. augusztus 13–15-én tartott konferencia előadásai*, 119–26. Heves.

Ébner, S. (1931) A szőrtarisznyás mesterség Dunántúlon. *Néprajzi Értesítő* 23, 165–169.

Gaitzsch, W. (2005) *Eisenfunde aus Pergamon. Geräte, Werkzeuge und Waffen*. Berlin.

Henning, J. (1987) *Südosteuropa zwischen Antike und Mittelalter. Archäologische Beiträge zur Landwirtschaft des 1. Jahrtausends u.Z.* Berlin.

Kirpichnikov, A. N. (1973) Снаряжение всадника и верхового коня на Руси IX–XII вв. *Arheologija SSSR Issue E1–36*. Leningrad.

Kirpichnikov, A. N. (1986) Russische Waffen des 9–15. Jahrhunderts. Waffen- und Kostümkunde. *Zeitschrift der Gesellschaft für historische Waffen- und Kostümkunde* 28:2, 85–129.

Lukács, L. (2007) *A tisztes ipar emlékei. Céhek, céhemlékek, az iparosok hagyományai Fejér megyében és Székesfehérváron*. Székesfehérvár.

Morton, H. V. (1936) *In the Steps of St. Paul*. London.

Müller, R. (1982) A mezőgazdasági vaseszközök fejlődése Magyarországon a későváskortól a törökkor végéig. *Zalai Gyűjtemény* 19, 1–2.

Nikitin, A. V. (1971) Работа русского кузнеца XVI–XVIII вв. *Arheologija SSSR. Issue E1–34*. Moscow.

Péterfi, Zs. (1993) Bátaszék-Kövesd pusztai későrómai temető. *A Wosinsky Mór Múzeum Évkönyve* 18, 47–128.

Sági, K. (1981) *Das römische Gräberfeld von Keszthely-Dobogó*. Fontes Archaeologici Hungariae 1981. Budapest.

Szolnoky, L. (1954) A bodrogközi függőleges szövőszék rekonstrukciója. *Néprajzi Értesítő* 36, 195–99.

Thomas, E. B. (1955) Die römerzeitliche Villa von Tác-Fövenypuszta. *Acta Archaeologica Hungarica* 6, 79–147.

Walton Rogers, P. (1997) *Textile Production at 16–22 Coppergate*, 17–59. York.

30 Textiles from the 3rd–12th century AD Cremation Graves Found in Lithuania

by Elvyra Pečeliūnaitė-Bazienė

The textile finds that belong to the Lithuanian museum collections have not been investigated for many years, and thus represent relatively pioneer territory for researchers. The textile material from cremation graves of Pryšmančiai barrow (western Lithuania) was briefly mentioned in the pre-war period by V. Nagevičius (1935, 12). Twelve fabric fragments that were found were given to the laboratory of Vytautas Magnus University in Kaunas. However, the research results were very superficial and non-comprehensive. Generally speaking, in Lithuania, the first serious textile research was carried out only in 1995 by S. Urbanavičienė. In the course of it, fabrics of the 15th–17th centuries AD found in Bečiai cemetery were researched and described (Senvaitienė *et al.* 1995, 104–116). Detailed analysis of textiles from both cremation and inhumation graves were performed by the present author in collaboration with the P. Gudynas Restoration Centre of Museum Treasures, where not only technological textile research, but also morphological, chemical analyses and the determination of colorants by the means of thin layer chromatography were carried out.

The period chosen was based on my doctoral dissertation that focused on all the textiles, amounting to approximately a 1000 pieces from the 1st–12th centuries AD found in the territory of Lithuania (Pečeliūnaitė-Bazienė 2007). The earliest fragments of woven textile on Lithuanian territory are dated to the 1st century AD. The bulk of those dated to the 3rd–12th centuries AD belong to inhumation graves, and only a very small part of textile fragments was recovered from cremation graves. However, these fabrics reflect general textile tendencies of the various periods in all the Lithuanian material, and allow us to look from a different perspective at conventional interpretations of textile functions in cremation burials. In this article, I would like to provide an overview of the textiles found in the cremation graves and to compare this material with inhumation graves, and also to attempt to find out if they were merely wrapping material of burnt burial items, special burial clothes or simply pieces of textiles.

Brief Overview of Burial Customs in the 1st–12th Centuries AD in Lithuania

In the 1st–4th centuries AD or the so-called Roman Period, inhumation graves dominated in all Lithuanian territory. In the 3rd–4th centuries AD, the cremation ritual appeared in the south of Lithuania and in the 4th century reached the eastern part. In the Middle Iron Age (5th–8th centuries AD), burial traditions changed and cremation became more popular. Around the 5th–6th centuries, cremation moved into central Lithuania and in the 6th century the tradition of cremation was established among the customs of the Highlanders; from the 7th–8th centuries AD, the Lamatians and Scalvians, too, increasingly used cremation (Tautavičius 1996, 285). This burial tradition became known to the Curonians in western Lithuania from the second half of the 8th century AD. In the Late Iron Age (9th–12th centuries AD), the cremation of the deceased was already widespread in Lithuanian territory. However, cremation did not replace inhumation entirely; both burial customs existed at the same time.

Textiles

Textiles from the Roman Period cremation graves are mentioned in the reports of archaeological excavations, however, regrettably no material is preserved. Among textile material of the 1st–4th centuries AD found in inhumation graves, the dominant weaves were z/z tabby and z/s 2/2 twill, mostly wool. At the same time, 2/2 diamond and 2/2 herringbone twills were found, but only in a small quantity. In neighbouring Latvia, one wool fabric of tabby, with plied S2z threads in both systems, was found (Zariņa 1990, 107). Fabrics of tabby that were found in Poland's territory most often were wool and the z/z spin direction dominated there (Maik 1977, 97, 99; 1988, 30–31). All these fabrics were found in inhumation graves.

During the Middle Iron Age (5th–8th centuries AD), the number of cremation burials increased; however, there are not many fabric fragments found in them either. A few textile

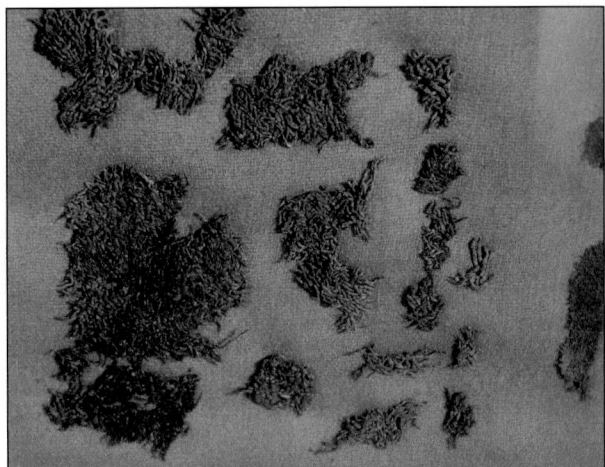

Fig. 30.1. Vidgiriai, grave 24, 2/2 diamond twill (Photo: E. Pečeliūnaitė-Bazienė).

Fig. 30.2. Viešvilė, grave 19, Tabby (Photo: E. Pečeliūnaitė-Bazienė).

Fig. 30.3. Genčai, grave 4, 2/2 twill (Pappilanmaki type) (Photo: E. Pečeliūnaitė-Bazienė).

fragments were found in the Vidgiriai barrow, grave 24 that was attributed to the Scalvians (south-western Lithuania) (Fig. 30.1). The grave is dated to the end of the 5th–6th century AD. The fabrics were woven in 2/2 diamond twill in z/s spinning, with a thread count of 12/10 threads per cm. In the cremation graves of Sudota (eastern Lithuania) barrow cemetery, a few textile fragments dated to the 5th–8th centuries AD were found. One of them was 2/2 twill in z/z-spinning, the rest were z/z-spun tabby. The thread count of the fabrics is around 10/10 threads per cm.

In the inhumation graves of the 5th–8th centuries AD, 2/2 twill in z/s spinning dominates, but from the 7th century onwards the number of fabrics of z/s 2/2 diamond twill increases; the latter is more often found in the western part of Lithuania. Tabby is very rarely found in the 5th–8th centuries AD, which is in contrast to the increasing usage of this weave in contemporary Scandinavia. In Latvia, tabbies account for almost 50% of all the textiles found. During the Migration Period (4th–8th centuries AD), 2/2 diamond twill was not often found in Northern Europe, except for the Northern German settlements of Hessens and Elisenhof, where textiles of 2/2 diamond twill comprise the major share (Hundt 1981, 24; Tidow 1990, 412). When this weave appeared again during the Viking period, it was most often found in high quality imported fabrics (Hoffmann 1974, 264). In Latvia, among all twills examined, just one fragment of 2/2 diamond twill was found in Liudvigova, dated to the 8th century AD and is thought to be imported (Zariņa 1990, 107–108). Meanwhile, the largest share of the 2/2 diamond twill fabrics that were found on Lithuanian territory in the 5th–8th centuries AD are not of a high quality (the thread count is around 12/10) and seem to be local products.

The vast majority of the fabrics found in cremation graves is dated to the 9th–11th centuries AD and are found in the region of western Lithuania. The cremation custom had begun to spread in this territory at that time and certain burial traditions determined the preservation of these fabrics. In most cases, cremated bodies were buried in wooden coffins, which also contained burial goods that were not burnt.

Textiles of tabby weave were found in the cremation graves of Viešvilė cemetery Žvirbliai and in the Kurklių Šilas barrows, which are dated to the 10th–11th centuries AD (Budvydas 2004; Iwanowska 2006, 63–64) (Fig. 30.2). The bulk was in z/z-spinning, with a few in z/s-spinning. Thread counts are about 10–12/12 threads per cm. Fabrics of vegetal fibre were found only in the Žvirbliai and Kurklių Šilas barrows. Tabby is in the minority among the textiles from inhumation graves of this period as well, while in neighbouring Latvia, and in Scandinavia, tabbies are found more often.

The vast majority of the fabrics among the textiles of the 9th–11th centuries AD comprise 2/2 twills. They were found in cremation graves of Genčai, Lazdininkai, Palanga, Viešvilė and Gintališkės (western, south-western Lithuania) (Fig. 30.3). The thread count is generally around 10/8 threads per cm. Similarly, both z/z-spun and z/s-spun fabrics were found. Only one fragment in s/s-spinning was recovered. Textile fragments found in Girkaliai, Genčai I and Palanga cremation graves belong to the so called *Pappilanmaki* type, according L. Bender-Jorgensen, and are of 2/2 twill, where warp is mostly 2-ply (S 2z) and weft slightly z-twisted (Bender-Jørgensen 1992, 96, 140) (Fig. 30.4). The graves are dated to the 9th–12th centuries AD. The thread count is around 8/10 threads per cm. Similar fabrics in quite a large

Fig. 30.4. Genčai, grave 20, 2/2 twill (Pappilanmaki type) (Photo: E. Pečeliūnaitė-Bazienė).

Fig. 30.5. Genčai, grave 16, 2/2 herringbone twill (Photo: E. Pečeliūnaitė-Bazienė).

Fig. 30.6. Viešvilė, grave 19, tabby buttonhole stitch, tubular border (Photo: E. Pečeliūnaitė-Bazienė).

amount were found in inhumation graves and almost all of them were found in western Lithuania. Many textiles with 2-plied threads were found in continental Finland where fabrics of 2/2 twill with 2-plied warp dated to AD 550/600–1070 formed the majority. Summing up this material of Northern Europe, Lise Bender-Jørgensen concludes that these fabrics were produced in Finland (Bender-Jorgensen 1992, 96, 140). Such textiles with 2-plied warp thread were also found in Latvia (Zariņa 1970, tab. 26; *Libiesi* 2001, 25). Describing the material from Novgorod, A. Nahlik indicates that 16.3% of textiles have 2-plied thread in one system (Nahlik 1964, 35). In the Finno-Ugric barrow Zalahtovye, Russia, textiles that were classified as of this type were found as well (Khvoshchinskaja 1992, 130). The textiles of this type found in western Lithuania were woven locally, but probably not by local weavers. It is known that, in the Palanga settlement, foreigners who came from the western coast of the Baltic Sea, *i.e.* from the southern part of the Jutland peninsula and its surrounding territories, used to live together with local Curonians (Žulkus 1997, 254, 290). There is a possibility that inhabitants of Finnish origin, who used to live there, wove this type of fabrics.

In the cremation graves of Genčai, Lazdininkai and Palanga fragments of 2/2 herringbone twill were found; they are dated to the 10th–12th centuries AD (Tautavičius 1962; Patkauskas 1976; Merkevičius 1984) (Fig. 30.5). The vast majority is z/z-spun, and the thread count is around 12/10 threads per cm. Generally, in Lithuanian territory, herringbone twill has been found from the 2nd–3rd centuries AD, although comparatively rarely. In the other territories, 2/2 herringbone twill was more popular. For instance, in Birka, such fabrics formed 35% of all textile finds (Nahlik 1963, 256). A similar case occurs in Wolin, Ukraine (9th–11th centuries AD), where such fabrics form 32% (Nahlik 1963, 256). On Russian territory, textiles in herringbone twill were found in the Finno-Ugric cemeteries of Liadinskij, Kriukov-Kuzhnovskij and Maksimovskij, and they form 40% of all fabrics found there (Yefimova 1966, 130). As in Lithuanian territory, these fabrics, too, are usually in z/z-spinning. In Poland, z/s-spun textiles were found and in Latvia both z/s and z/z-spun herringbone twills were found (Maik 1988, 128).

Just a few textile fragments of 2/2 diamond twill have been preserved in the material of cremation graves of the 9th–12th centuries AD. All the textiles found are in z/s-spinning. The thread count varies from 7/8 to 12/14 threads per cm. Comparing it to the material from inhumation graves, it can be observed that, in cremation graves, fabrics of diamond twill were not common, while it is one of the most popular weaves in the inhumation graves, except for the eastern part where very little of the textile material survived.

The only three-shed twill in z/z-spinning with a 20/10 thread count was found in Palanga, grave 262, which dates back to the 11th century. The textile was found in a rich merchant's grave and there is a high probability that this fragment was not of local origin. The earliest textiles of three-shed twill date back to the 3rd century AD in Lithuania, but were most probably imported. Very few fabrics of this type were found in inhumation graves of the 9th–12th centuries

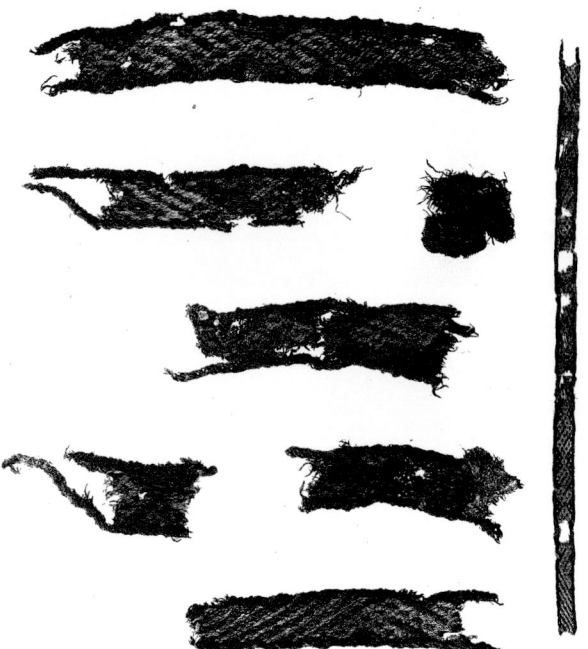

Fig. 30.7. Paragaudis, grave 59, tablet-woven patterned band (Photo: Archive of the P. Gudynas Restoration Center).

AD; all of them were of z/z spin. Among them, were found the so-called three-shed *Rippenkörper* textiles; all of them were of z/s spin direction. In Poland, three-shed twill was found in Opole, in the layers of the 9th–the beginning of the 12th century (Maik 1991, 14, 24, 39), and in Gdansk where they are dated to the 13th century (Maik 1988, 68); all of them were in z/z spinning. In eastern Latvia, one finds textiles of three-shed twill, which are dated the 12th–13th centuries AD. The dominant spin direction here is z/z, although z/s spin was found, too (Zariņa 1990, 109–110).

Only a few fragments of fabrics with preserved borders have been found. Thus, the borders were preserved in the textiles of tabby weave found in Viešvilė (Fig. 30.6). The fabric had a tubular selvedge, and additionally it was hemmed using two z-spun threads with buttonhole stitch. Tubular selvedges in the textile material were rather popular since the 8th century AD in all the territory of Lithuania.

Only one patterned tablet-woven band fragment was found in Paragaudis, grave 59, which is dated to the 10th century AD (Valatka 2004, 241). It is a colourful band, with edges of reddish ochre, and the weave in the middle of the band in grey and brown threads. The band is made of wool, but regrettably, its colourants could not be determined. The pattern of the band is made by rhythmically repeating stylized motif of the letter 'S' (Fig. 30.7). This pattern is also known as the 'serpent' pattern (Tumėnas 2002, 56, table 19, 33). The symbol of the serpent is primarily related to the culture of the western Balts; its popularity decreases eastwards (Tumėnas 2002, 5). In the inhumation graves, many of the tablet-woven bands have been preserved, although most of them were not patterned and had merely lengthwise coloured strips.

Dyes

Research was undertaken into the dyes of all textile fragments from cremation graves. It was determined that fabric fragments of blue colour dated to the 5th–11th centuries AD (Lazdininkai, Palanga, Viešvilė) were mostly dyed using indigotin. Blue was a very popular colour throughout various periods; the earliest textile fragment in which the indigotin was determined was dated to the end of the 1st century AD. Fabrics dyed blue were found in various cemeteries and barrows from different regions, in both cremation and inhumation graves. Both men and women wore blue garments. Although in Lithuanian territory, woad (*Isatis tinctoria*) is an adventive (Gudžinskas 1997, 233, 249) and very rare plant (*Lietuvos Flora* 1961, 527), usage of woad in northern Europe confirms the assumption that fabrics could be dyed blue by using this plant. Another possibility is that other local plants could be used, from which the indigo colourant was obtained, albeit in very low concentrations, and thus not economically viable. For example Devil's-Bit Scabious (*Succisa pratensis Moench*) is common all over Lithuania (*Lietuvos Flora* 1976, 538), as is Water Smartweed (*Polygonum amphibium*) (*Lietuvos Flora* 1961, 193). The other dyes have not been identified.

The Find Spots of Fabrics in the Graves

In western Lithuania's cremation graves, where fabric fragments were found, burial items such as weaponry and working-tools had not been burned, but placed in the grave in the same positions in which the deceased would have carried them when alive. Regrettably, not in all cases were exact textile find places indicated in the excavation reports. Hence, it is difficult to pinpoint the exact context of textile finds. However, in some cases, information is available. A hollowed-out oak log coffin was found in Gintališkė, grave 52, which dates to the 11th–12th century AD. It had a fur at the bottom, on which textiles and cremated bones with burnt remains of brass artefacts had been placed (Valatka 1971, 1). Fabrics of different weaves with burnt bones and a fur under the burial items in a coffin were also found in Genčai, grave 156 (Merkevičius 1984, 40). In the eastern part of Lithuania, in Kurklių Šilo barrow, textiles were found under burial goods on the barque (Butėnas 2000, 5). In Žvirbliai, textiles were found above burial goods. In Paragaudis (Samogitia), 10th century AD grave 59, a woman's body was found buried in a hollowed-out log coffin with cremated baby bones and burial goods placed by her side. These items were wrapped into two different 2/2 diamond twill textiles. This bundle was wrapped twice with a patterned tablet-woven band and pierced by two pins with chains, which were also wound around the bundle (Valatka 2004, 241).

For many years, it was thought by archaeologists that textile fragments found in cremation graves were a simple material, in which burial items and burnt bones were wrapped or covered. However, research indicates that very often the cremation graves of western and south-western Lithuania contained several textiles of different weaves and different colours, just as observed in inhumation graves. For instance, in Genčai, grave 16, textiles in 2/2 diamond twill and herringbone twill, were found, while 2/2 plain twills of different thread quality and fur were identified in grave 228. In Palanga, grave 296, three-shed twill and herringbone twill were found. In Viešvilė, four different fabrics were found. In Paragaudis, not merely the tablet-woven band and the textiles of a wrapped child, but also other textiles of different weaves have been preserved. These finds enable us to make the assumption that at least in western and south-western Lithuania, the textiles placed in the graves were ready-made clothes. Textiles found in the cremation graves of the different sexes are comparable to the textiles that originate from inhumation graves and can be related to certain garments. The grave material indicates that the graves of cremated individuals included not merely such burial goods as weaponry, working tools or jewellery, but also clothes placed in the order in which the deceased would have worn them in life (*e.g.* in Palanga). Textile fragments of various weaves and colours, with hemmed borders and remains of holes from brooches, indicate different garments. Textiles with metal spirals on them are also known from cremation graves (Genčai, grave 226, 230). In contrast, in eastern Lithuania a limited variety of textiles with a predominance of tabbies that are mostly found under and above burial items allow us to assume that, in this region, cremated bodies and burial items were wrapped into some fabrics or were covered by them.

Conclusions

The textiles of cremated graves do not differ markedly from material from inhumation graves of the 3rd–12th century AD, except for a limited amount of diamond twill in the late Iron Age. Almost in all Lithuanian territory, except for the areas inhabited by Semigallians and Lithuanians, in the 3rd–12th centuries AD, the z/s (s/z) spin combination dominated. This was determined by local traditions that have continued since the Roman Period and this peculiarity distinguishes our fabrics from our neighbours. Furthermore, 2/2 plain twill and 2/2 diamond twill have dominated since the 4th century AD. Some of these textiles had starting borders, woven and hemmed borders.

The detailed investigations of textiles and peculiarities of certain burial customs enable us to assume that these textiles could be clothes placed in the graves as burial items, at least in Western Lithuania. Clothes may have been placed in the order they were worn in life.

The investigations of fabrics showed that the textiles found in the cremation graves varied in weave and were often coloured blue (the identified colorant being indigotin). Finally, the fragments of tablet-woven bands found in the cremation graves were used for the wrapping of the burnt individuals or burial items.

Bibliography

Bender-Jørgensen, L. (1992) *North European Textiles until AD 1000*. Århus, Århus University Press.

Budvydas, U. (2004) Viešvilės kapinyno (Jurbarko raj., Viešvilės miestelyje) archeologinių tyrimų 2004 m. ataskaita. In *Lietuvos istorijos instituto rankraštynas*, F. 1, b. 4678. Vilnius.

Butėnas, E. (2000) Kurklių šilo pilkapyno, Anykščių raj., 2000 m. archeologinių tyrinėjimų ataskaita. In *Lietuvos istorijos instituto rankraštynas*, F. 1, b. 3627.

Gudžinskas, Z. (1997) Conspectus of Alien Plant Species of Lithuania. 3. Brassicaceae. In *Botanica Lituanica*, 215–249, 3(3). Vilnius.

Hoffmann, M. (1974) *The warp-weighted loom: Studies in the History and Technology of an Ancient Implement*. Oslo.

Hundt, H. J. (1981) *Die Textil-und Schnurreste aus der Frühgeschichtlichen Wurt Elisenhof. Studien zur Küstenarchäologie Schleswig-Holsteins. Serie A. Elisenhof: Die Ergebnisse der Ausgrabung der Frühgeschichtlichen Marschenseidlung beim Elisenhof in Eiderstedt 1957/58 und 1961/64*. Band 4. Frankfurt am Main/Bern.

Iwanowska, G. (2006) Cmentarzysko kurhanowe w Żwirblach pod Wilnem. Wprowadzenie w problematykę. Katalog. In A. Bitner-Wróblewska (ed.), *Warszawa Państwowe Muzeum Archeologiczne w Warszawie, Ośrodek Ochrony Dziedzictwa Kulturowego*. Stowarzyszenie Naukowe Archeologów Polskich, Oddział w Warszawie.

Khvoschchinskaja, N. (1992) New Finds of Medieval Textiles in the North of Novgorod Land. In L. Bender-Jørgensen and E. Munksgaard (eds), *Archaeological Textiles in Northern Europe. Report from the 4th NESAT Symposium 1.–5. May 1990 in Copenhagen*, 128–133. Copenhagen, Kongelige Danske Kunstakademi.

Libieši senatnē (2001). Rīga.

Lietuvos flora (1961) Volume 3. Vilnius.

Lietuvos flora (1976) Volume 5. Vilnius.

Maik, J. (1977) Tkaniny z okresu Rzymskiego z terenu Polski. In *Pomorania Antiqua*, Ossolineum, Vol.VII, 77–145. Łódź.

Maik, J. (1988) Wyroby włókiennicze na Pomorzu z okresu Rzymskiego i ze Średniowiecza. In *Acta Archaeologica Lodziensia* 34, 206. Łódź.

Maik, J. (1991) *Tekstylia wczesnośredniowieczne z wykopalisk w Opolu*. Warszawa and Łódź.

Merkevičius, A. (1984) Genčų km. I kapinyno Kurmaičių apyl., Kretingos raj. 1987 m. tyrinėjimų ataskaita. In *Lietuvos istorijos instituto rankraštynas*, F. 1, b. 1137. Vilnius.

Nagevičius, V. (1935) Mūsų pajūrio medžiaginė kultūra VIII–XIII amžiuje. In *Senovė*, Vol. 1, 131. Kaunas.

Nahlik, A. (1963) Ткани Новгорода. Опыт технологического анализа In *Матерьялы и исследования по археологии СССР* 123, 228–313. Moscow.

Nahlik, A. (1964) *Tkaniny wełniane importowane i miejscowe Nowogrodu Wielkiego X–XV w*. Wrocław.

Patkauskas, S. (1976) Lazdininkų senkapio (Kretingos r.) 1976 m. archeologinių tyrinėjimų ataskaita. I dalis. In *Lietuvos istorijos instituto rankraštynas*, F. 1, b. 453. Vilnius.

Pečeliūnaitė-Bazienė E. (2007) *Iron Age Textiles (I–XII c.) in Lithuania*. Doctoral dissertation. Vilnius University.

Senvaitienė, J., Vedrickienė L., Čeplinskaitė V., and Urbanavičienė S. (1995) Bečių kapinyno audinių konservavimas ir tyrimas. *Lietuvos archeologija* 11, 104–116.

Tautavičius, A. (1962) Palangoje, Komjaunimo g. esančio senkapio, archeologinių kasinėjimų, vykdytų 1962 m., ataskaita. In *Lietuvos istorijos instituto rankraštynas*, F. 1, b. 183. Vilnius.

Tautavičius, A. (1996) *Vidurinis geležies amžius Lietuvoje (V–IX a.)*. Vilnius.

Tidow, K. (1990) Frügeschichtliche Wollgewebe aus Norddeutschland – ihre Verbreitung und Herstellung. In *Experimentelle Archaeologie in Deutschland*, 410–417. Archäologische Mitt. aus Nordwestdeutschland Beih. 4. Oldenburg.

Tumėnas, V. (2002) *Lietuvių tradicinių rinktinių juostų ornamentas: tipologija ir semantika*. Lietuvos etnologija 9. Vilnius.

Valatka, R. (1971) Gintališkės kapinyno 1971 m. tyrinėjimai. In *Lietuvos istorijos instituto rankraštynas*, F. 1, b. 435. Vilnius.

Valatka, R. (2004) Paragaudžio juosta. In *Žemaičių žemės tyrinėjimai*, 240–242. Knyga I. Archeologija. Vilnius. Regionų kultūrinių iniciatyvų centras.

Yefimova, L. V. (1966) Ткани из фино-угрских могильников I т.н.э. In *Краткие сообщения о докладах и полевых исследованиях института археологии*, 107, 127–134. Moscow.

Zariņa, A. (1970) *Seno latgaļu apģērbs 7.–13.gs.* Rīga.

Zariņa, A. (1990) Herstellungsmethoden der in Gräberfeldern des 3.–13. Jh. im Gebiet Lettlands gefundenen Gewebe. In P. Walton and J. P. Wild (eds), *NESAT III: Textile Symposium in York, 6–9 May 1987,* 107–111. London, Archetype Publications.

Žulkus, V. (1997) Palangos viduramžių gyvenvietės. In *Acta Historica Universitatis Klaipedensis* VI. Klaipėda. Klaipėdos universiteto Vakarų Lietuvos ir Prūsijos istorijos centras.

31 Patterned Tablet-Woven Band – In Search of the 11th Century Textile Professional

by Silja Penna-Haverinen

In 1991, an impressive female grave (No. 27) was excavated in the Kirkkomäki burial ground in Turku, southwestern Finland. This grave and its textiles have previously been discussed in NESAT 9 by Heini Kirjavainen and Jaana Riikonen (Kirjavainen and Riikonen 2007). The grave is dated to the latter part of the 11th century AD. Numerous items of various clothing were found, among them, several skilfully made tablet-woven bands. One of them was used for attaching two sets of bronze bear-tooth pendants to the deceased's waist. This band has been part of experimental research since 2004.

My first task was to create a pattern reconstruction based on the best preserved piece of the textile while studying arts and crafts at the Southwest Finland Institute for Art, Craft and Design in Mynämäki (Penna-Haverinen 2005a). After that, I received the opportunity to continue my research at the archaeological experimentation workshop of Turku Provincial Museum. I began working with a textile reconstruction of the band, while at the same time studying its pattern and weaving techniques (Penna-Haverinen 2005b; 2006). Today my research covers all of the tablet-woven bands of the Kirkkomäki burial ground (Penna-Haverinen 2008). My experiments with the pattern and the textile reconstruction have had a major impact on these studies thus far – the experimental work with a single band has been beneficial for the material as a whole.

The Original Textile

There were several tablet-woven bands placed on the waist area of the deceased, which were very difficult to separate. Thus far, only one piece of this so-called bear-tooth band has been identified, while the rest of the few dozen pieces seem to belong to two, possibly three other bands. The identified piece is 7 cm long and 16–17 mm wide. The fibre analysis (Kirjavainen 2004) indicates that the 2-plied and s-twisted warp threads are made of wool from the Finnish stock of sheep. As the weft has almost completely vanished, it must have been some sort of plant fibre, possibly flax, hemp or nettle.

The band has been woven with 24 tablets. The ornamental pattern consists of 18 tablets and there are about 14 tablets per cm. The selvedge areas consist of three tablets each, of these only two are visible on the right side of the band and the third one can be seen only on the reverse side. There are four threads in each tablet: two dark-coloured as the background and two light-coloured as the pattern (Fig. 31.1). Two colours, red and blue, can be distinguished from the dark ground using a microscope. They form vertical stripes of three and six tablets on the background, while the light-coloured warp threads form a diagonal pattern above them. It seems likely that the pattern varied throughout the band. The pattern was recreated using a threading, where half the tablets are Z-twined and the other half S-twined. The tablets were turned individually.

Fig. 31.1 A schematic drawing of the pattern (Drawing: S. Penna-Haverinen).

sheep contains two types of fibres, long and hairlike coating wool and short and soft underwool. I combed all the fibres together without any sorting. When pulling the wool out of the combs, four different sorts of mixed fibres could be detected: (1) mixed wool containing many long, clean and healthy fibres with a small amount of short fibres, (2) mixed wool containing mostly short, clean and healthy fibres and some long fibres, (3) mixed dirty wool containing short and long fibres, dead hair and tangles and finally (4) noil. The first type produced good looking, durable and hard but relatively hairy yarn, whereas the third type produced lumpy, uneven and bad quality yarn. I found the second type the most suitable for producing good quality yarn for tablet weaving, since it turned out to be durable but less hairy. The healthy fibres gave a beautiful, shiny look to the yarn and, as it turned out, the softer fibres were more easily dyed than the long, hair-like fibres.

Four different low-whorl spindles were made for producing yarn. The shape of the whorls and their weight was selected by experimenting with different weights and using finds from Viking Age Birka as their models. A spindle weighing 25 g was considered most suitable. A heavier spindle, which weighed 45 g, was used for plying. The spindles are a bit heavy, but they worked well in the experiment.

Since the main questions concerned the production and usage of wool warp threads, the weft yarn was produced elsewhere. It was spun by a colleague, Toive Lehtinen, who used beautiful silvery flax of a fine quality, which was ideal to be hidden as weft of the band. The spinning was made with a spinning wheel and the yarn was left unplied.

Dyeing with Natural Colours

The blue colour was achieved using woad, but using modern procedures – there have been previous experiments on dyeing in the pre-modern manner in the archaeological experimentation workshop of the Turku Provincial Museum (Hannusas and Raitio 1997). I had two different yarns to dye, one spun of gray wool and one of white. Both of them came out as a very intense and beautiful blue.

As there were no traces of alizarin left on the yarns, it was suggested that, the red colour was achieved with some type of tannin. The modern Finnish dyeing recipes recommend the bark of rowan, willow or alder to obtain reddish brown colours. I tried dyeing with these using alum as a mordant, but the resulting colours were mostly different shades of brownish yellow. The biggest problem with dyeing was obtaining the intensity and proper shade of the colours. The problem presumably was caused by the long, hair like fibres. After several failed experiments, the best option turned out to be woad. Before dyeing, the dyed yarns were soaked in lime. It helped to intensify the colour. The resulting colour was more of a brown tone, but it came closer to the red colour than any other attempt.

Almost all of the spun yarn was used in the attempt to obtain the exact red colour. I already had the two base colours, red and blue, but not the colour for the pattern. Some bright yellow yarn left over from earlier experiments came in useful, as the bright yellow was more suitable for pattern colour than the brownish ones.

The yarns of the textile reconstruction, therefore, were light blue and reddish brown as the background colours and a bright yellow as the pattern colour. Although they differed from the original, I preferred to use natural dyes instead of modern chemical dyes.

Strengthening the Yarn for Weaving

One of my greatest concerns was the quality of the yarn. I made two experimental warps to examine possible problems. The first one was to test the undyed yarn without any treatment. It was spun from the combed wool of the first category, so it was very durable but hairy. To my surprise, the fine yarn was amazingly durable and there were no problems whatsoever.

The second experiment was to try out some natural treatments to strengthen the warp threads. I experimented with four different substances: (1) untreated cow milk, (2) skimmed cow milk, (3) egg white and (4) thin 'glue' from soaked flax seeds. Part of the warp was left without any treatment as reference. The first three treatments were all a great help for poor quality yarns, but there was no noticeable difference between untreated warp and the warp treated with flax glue. It is possible, that the glue was not thick enough. When it came to strengthening only, the egg white seemed to be the very best, but unfortunately it changed the colours of the warp permanently. When weaving, there was little difference between untreated and skimmed milk. The only and finally decisive difference occurred when preparing the warp, since the skimmed milk seemed easier to apply.

It was no surprise, then, to notice, that it would be best if no strengthening treatments were needed. All of the substances applied made the warp threads more hairy and although they made it easier to weave, the result did not necessarily appear optimal. I had to make a decision either to try weaving the textile reconstruction without any strengthening or to ensure that the fine warp threads would last the lengthy weaving process. I decided to opt for the strengthening, to be on the safe side.

Tablet Weaving

As there had been some experiments on using different materials for tablets (Grömer 2005, 87–88), this parameter was not tested in the experiment. I selected cardboard tablets for weaving, primarily because of their light weight. I wove with a weaver-tensioned warp, the cloth end of the warp attached to my waist, which is the most common way of weaving today. No major problems arose during weaving.

The only source of dissatisfaction was the proportions of the pattern. This was not a new problem, since it had also occurred when making the pattern reconstruction. Therefore, new experimental warps were made. All the experiments on how to produce the right proportioned pattern were made with warps composed of manufactured wool yarns. I made some experiments on different thicknesses of warp and weft yarns to find out how great an impact they had on the pattern proportions. It seemed that only a great difference in thickness had any noticeable effect. I also tried weaving with varying tension in the warp, different means and combinations of tightening the warp and weft, as well as many alternative ways of weaving positions. The most suitable for me seemed to be the combination of weaving with a stable and moderately tensed warp, relatively heavy tightening of the shed and loose tightening of the weft. To achieve this, a very simple tablet weaving loom and a beater were used. The results were promising and the pattern began to resemble the original.

Finally, to all these different aspects, which had impact on the proportions of the pattern, I had to add the undefinable factor of the weaver's personal touch. Regardless of the various weaving aids, the result was still dependent on my personal input.

Conclusions

The results of the pattern reconstruction and the textile reconstruction (Fig. 31.5) were compromises between the authentic execution and the skills of the researcher. This is hardly surprising, as the original tablet-woven band was made by an Iron Age professional, while the experiment was carried out by a mere student of the craft. Many of the crucial weaving techniques had to be rediscovered through experiments. Another, similar tablet-woven band was also studied to gain inspiration for the finish of the pattern reconstruction. The new information obtained by analyses needed to be adjusted to the personal style of the weaver, and the reconstruction had to be remodelled slightly from the original. The same can be said about the textile reconstruction, since the available materials and the personal skills were the major contributors to the final result.

The reconstruction of the band was not the only result gained from the experiments. New textile tools were produced with Iron Age tools as their model. The properties of Åland sheep wool and the use of different types of fibres were studied. Other tablet-woven bands from the Kirkkomäki burial ground were studied, as well, with some corresponding bands from nearby burial grounds. The whole process was also reported, and the knowledge from these experiments and studies has played a key role in the analysis of the Late Iron Age tablet-woven bands in general.

The experiments have also raised new questions regarding professionalism and the identification of these professionals. The production of this type of tablet-woven bands is most certainly a specialized craft. It takes time to learn and to create the detailed techniques of pattern weaving. It is possible to produce the same ornamental patterns with different techniques, resulting in different looks. And although the same pattern is executed with the same techniques, the result may still vary from weaver to weaver. The personal touch of a weaver must always be considered.

The rich variety of ornamentation of bands such as the Kirkkomäki band suggests that their weaver can produce any pattern of their choice. At that level of craftsmanship, it is possible to create one's own individual style of ornamental pattern composition and solve its technical construction. This individual style is similar to a signature or handwriting. It includes the repertoire of ornaments and the style of their composition, the detailed techniques of pattern construction and the personal logic in their use, which is designed to compliment the personal look of the weaver's products. It is assumed that, by studying these personal styles of tablet weaving, it may be possible to identify certain professionals. While the motifs of ornamentation seem very uniform in the large area along the Baltic Sea, the technical solutions differ from place to place. It might be possible, then, first to find the varying techniques used in pattern construction and then to identify the different levels of regional production, starting with a single craftsperson and local production, and finally expanding to a tradition within a larger area.

During the late 11th and early 12th centuries AD, these complicated and lavishly decorated textiles were luxury items, and they must have been quite expensive. Although there is no denying the professional quality of the bands, the social position of the craftspeople can be discussed however. Were the weavers independent producers, were they mobile or did they have some sort of distributor? Or were they members of a certain social group, who shared not only the luxury of leisure time for weaving, but who produced goods as gifts, as well?

Experimental archaeology can be the starting point in the search for answers to questions like these. It may not be possible to answer them solely by using this method, but it is still a highly necessary basic research, which can help to design further experimental research strategies and define the geographical distribution of different forms and styles of production.

Fig. 31.5. Two pattern reconstructions and a textile reconstruction (Photo: S. Penna-Haverinen).

Comment: Experimental Archaeology as a Process

Although my research began as an expedition to find an 11th century AD professional, at the same time it became a personal quest for the 21st century professional. In the beginning, I was a novice to both textile and experimental archaeology and I had very little experience in tablet weaving. During the process, I had to return constantly to the source, to the original textile, to compare the new skills I had learned and discoveries I had made. The more I learned, the more I discovered. Experimental archaeology is, indeed, vitally dependent on the researchers' own skills as craftspeople and as scholars. At first, it was frustrating to realize the limitations of my own abilities, but with time I saw this as the main strength of experimental archaeology. It became both a personal experience for me and a scientific experiment in archaeology (See Andersson Strand in this volume). I can only describe it as a multifaceted and continuous process of observing, analyzing and learning, not only using the original object, but also the reproduced objects. Therefore, it is important to include the process in discussions of the results of experiments in archaeology. The true results should arise from a common pool of accumulated experience, which can only be achieved by bringing the different opinions and outcomes together.

Acknowledgements

I would like to thank my teachers and mentors Jaana Riikonen and Hannele Köngäs for their constant support and help. I also wish to thank Markku Ikäheimo from the archaeological experimental workshop of Turku Provincial Museum for making my research possible.

Bibliography

Asplund, H., and Riikonen, J. (2007) Kirkkomäki. *Arkeologisia kaivauksia Turussa 1990-luvulla. Turun maakuntamuseon Raportteja* 20, 9–44. Tampere, Turun maakuntamuseo.

Collingwood, P. (1996) *The Techniques of Tablet Weaving*. 2nd ed. Arkansas City, Robin & Russ Handweavers, Inc.

Grömer, K. (2005) Tablet-woven Ribbons from the prehistoric Salt-mines at Hallstatt, Austria – results of some experiments. In P. Bichler, K. Grömer, R. Hoffman-de Kejzer, A. Kern, and H. Reschreiter (eds), *Hallstatt Textiles: Technical Analysis, Scientific Investigation and Experiment on Iron Age Textiles. BAR Int. Ser.* 1351, 81–90. Oxford, Archaeopress.

Hannusas, S., and Raitio, S. (1997) Morsinkovärjäys – historiaa ja kokeiluja. *Turun maakuntamuseo. Monisteita* 12. Saarijärvi, Turun maakuntamuseo.

Hansen, E. (1990) *Brickvævning – Historie, teknik, farver, mønstre*. Skive, Hovedland.

Kaukonen, T. (1965) Suomen kansanomaiset nauhat. Kansatieteellinen tutkimus. *Tietolipas* 41. Helsinki, Suomalaisen Kirjallisuuden Seura.

Kirjavainen, H. (2004) *Raportti kahdesta lautanauhan (nr. 237) loimilangasta Kirkkomäen haudasta nr. 27*. Turun yliopiston arkeologian oppiaineen arkisto.

Kirjavainen, H., and Riikonen, J. (2007) Some Finnish Archaeological Twill Weaves from the 11th to the 15th century. In A. Rast-Eicher and R. Windler (eds), *Archäologische textilfunde – Archaeological textiles. NESAT IX*, 134–140. Näfels, ArcheoTex.

Mäntylä, S. (2006) Rikalanmäen ruumiskalmisto – näkökulmia myöhäisrautakauden yhteisöön. In S. Mäntylä (ed.) *Miekka – Menneisyys – Maisema*, 36–67. Somero, Halikon kunta.

Pälsi, S. (1928) Puvustoaineksia Maskun Humikkalan kalmistosta. *Suomen Museo* 1928, 71–79. Helsinki, Suomen muinaismuistoyhdistys.

Penna-Haverinen, S. (2005a) *Löydöstä lautanauhaksi*. Artesaanin opinnäytetyö. Lounais-Suomen käsi- ja taideteollisuusoppilaitos, Mynämäki.

Penna-Haverinen, S. (2005b). *Työraportti. Kaarinan Kirkkomäen haudan 27 karhunhammasriipussarjoja kannattaneen kuviollisen lautanauhan, ns. karhunhammaslautanauhan, rekonstruktiokokeiluja*. Turun maakuntamuseon arkisto.

Penna-Haverinen, S. (2006) *Työraportti. Huomioita lautanauhan kudonnasta ja käsialasta*. Turun maakuntamuseon arkisto.

Penna-Haverinen, S. (2008) *Kaarinan Kirkkomäen myöhäisrautakautiset lautanauhat*. Proseminaariesitelmä. Turun yliopiston arkeologian oppiaineen arkisto.

Sarkki, S. (1979) *Suomen ristiretkiaikaiset nauhat. Helsingin yliopiston arkeologian laitoksen moniste* no 18. Helsinki, Helsingin yliopisto.

Stolte, H. (1990) Versuch der Musternachbildung eines Brettchengewebes. Teilstück der Manipel von Sankt Ulrich. In *Experimentelle Archäologie in Deutschland. Archäologische Mitteilungen aus Nordwestdeutschland. Beiheft* 4, 439–449. Oldenburg, Isensee.

32 Social and Economic Aspects of Textile Consumption in Medieval Tartu, Estonia

by Riina Rammo

Introduction

The aim of this paper is to characterize textiles and some aspects of textile consumption in medieval Tartu (Fig. 32.1), one of the Hanseatic towns in Estonia. On the basis of available material, the following questions are examined: what types of textiles, imported or local, were preferred in the domestic sphere, and by whom? As background, it should be noted that in Estonia, the Middle Ages begin after Christianization in the beginning of the 13th century and last until the middle of the 16th century AD. Thus, the truly medieval towns of Estonia were only established during the 13th century and their further development was influenced by the colonisation of the Germans who became the dominating upper classes. After the conquest of mainland Estonia by the Germans in 1224, in Tartu[1] the centre of the bishopric of Tartu – a political entity in medieval Livonia[2] – was established. By the 14th century, a Hanseatic town with *c.* 5000 inhabitants developed near the castle (for further details, see Tarvel 1980, 27, 50; Russow *et al.* 2006, 159–161). Regrettably, written sources on Tartu are scanty; the medieval archives of the town were not preserved.

The archaeological textile material in Estonia has attracted relatively little attention thus far. Research has primarily concentrated on material from the prehistoric period and from rural cemeteries of the Middle Ages (*e.g.* Peets 1987; 1992; Laul 1990; 1996; Laul and Valk 2007). As far as the medieval urban material is concerned, only two brief articles can be noted (Peets 2000; Rammo 2007).

The Find Context

It is estimated that over 2000 textile fragments were recovered from medieval Tartu, primarily from cesspits. The latter offer suitable preservation conditions for organic materials due to the anaerobic environment that prevails due to the waterlogged conditions caused by the high level of ground water. This paper is based on one of the largest collections of textile finds, which was recovered from excavations at Lossi (*i.e.* Castle) Street, in the central area of the medieval town (Fig. 32.2). Six wooden cesspits with a rich material were discovered here, in the dense layers of dung and garbage. Besides textiles, also numerous potsherds, wood and leather objects, imported glass beaker fragments, archaeobotanical and -zoological remains were found. Three of these cesspits, dated from the end of the 13th to the beginning of the 15th century, contained abundant textile fragments (Table 32.1) (Mäesalu 1990).

On the basis of the homogeneous finds, it has been assumed that these cesspits are closed find complexes, which were used during a certain period (Mäesalu 1990, 451), and that the finds could be associated with one particular group of people, for example, people belonging to one or more households.[3] The cesspits, which are situated next to each other and are assumed to have followed one another chronologically, offer a good opportunity to study the textiles consumed, used and discarded during the 14th century by the inhabitants in the central area of the medieval town.

Textiles

Altogether 1510 fragments[4] of textiles have been recorded, and 1114 of them were selected for further analyses, thus excluding tiny and degraded fragments. First, the finds were studied according to the basic textile research methods – the weave, thread count, twist direction, yarn diameter, colour and traces of finishing process and earlier usage were documented.

Due to the preservation conditions, the find material consists mostly of woollen pieces (99.2%), but some fragments of silk (0.3%) and linen (0.5%) have also been identified. The technical description of this material is similar to the finds from other contemporary medieval North European towns. The most common weave is 2/1 twill followed by tabbies; also the 2/2 twills are consistently present but only by 12% on average (Fig. 32.3). This kind of weave distribution is similar to some other textile assemblages in the Baltic Sea region, *e.g.* from Gdańsk, Elbląg, Schleswig and Lund (Maik 1998, 216–217). Exceptional finds are pieces of one four-shed lozenge twill, two six-shed twills and two repp weave fragments. The most frequent twist direction combination

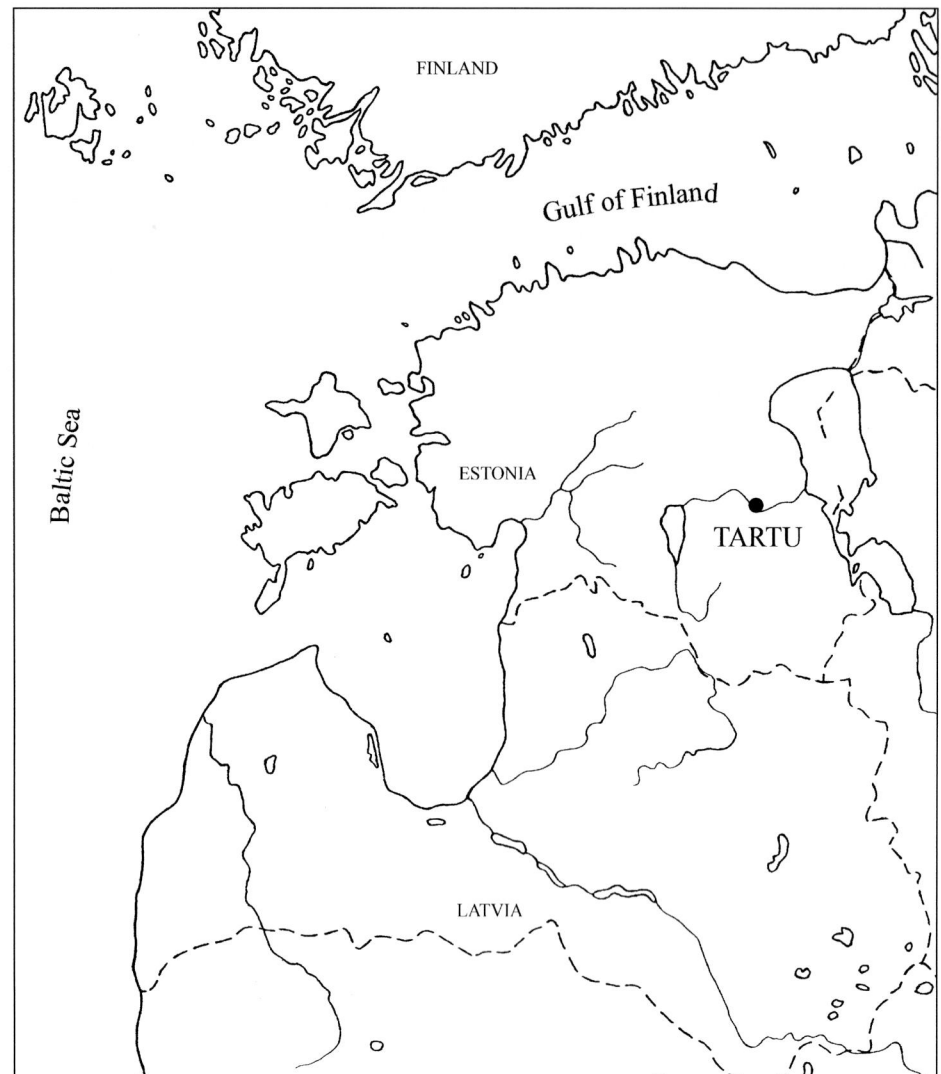

Fig. 32.1. Location of Tartu.

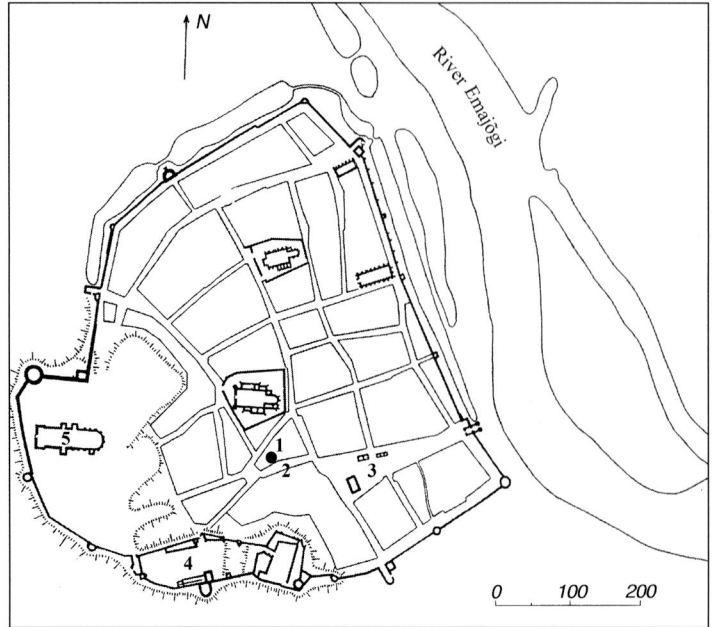

Fig. 32.2. Medieval Tartu: 1 – location of cesspits; 2 – Lossi Street; 3 – marketplace; 4 – episcopal castle; 5 – Dome Church (after Alttoa 1998, fig. 1).

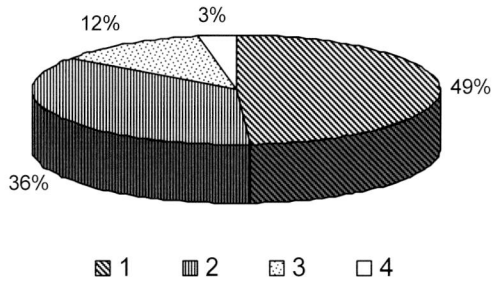

Fig. 32.3. Percentages of different types of weaves: 1 – 2/1 twill; 2 – tabby; 3 – 2/2 twill; 4 – others.

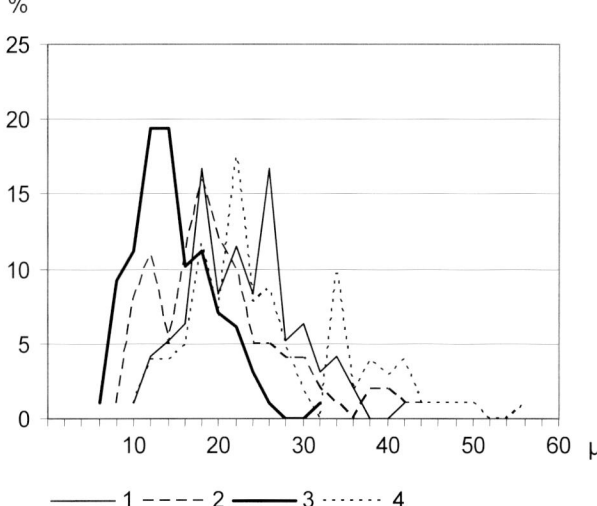

Fig. 32.4. Wool fibre measurements typical in group of imported textiles: 1 – ordinary tabby; 2 – broadcloth (1/1); 3 – fine worsted (2/2); 4 – striped cloth.

is z-spin in warp and s-spin in weft (85%), especially in 2/1 twills and tabbies; z/z spinning (15%) is characteristic primarily to the four-shed twills.

Today, the fragments are in different shades of brown. Although it is sometimes possible to distinguish various red hues (19%) and traces of dark blue or violet (1%) with the naked eye, it is not possible to say anything certain about the colours used. Fulling is typical for medieval textiles and more than half have been treated in this way; also further finishing processes (*i.e.* teaseling and shearing) have been commonly used.

Fragments with seams, stitches and buttonholes, as well as frequent traces of wear and cut edges (7%) indicate their usage as clothing items and household textiles (furnishing, bedclothes), thus they belong to the private sphere of daily life. They can be, for example, offcuts which are remainders from the making, or rather remaking, of clothing items. At least partly, the final use of these pieces was as toilet paper and sanitary towels (*e.g.* Walton 1989, 297).

It is a question of how representative this find assemblage is. This material certainly does not represent the entire range of textiles used by inhabitants of the medieval town. Besides bias caused by preservation conditions, it should be taken into consideration that the usage of cesspits spread here with the urbanisation process in the Middle Ages and probably relates to a rather higher social class. The other finds from these cesspits, *e.g.* glass beaker fragments, and the location in the central area of the town near the marketplace also support this assumption. Besides, the presence of these textiles in cesspits is related to the various habits and cultural behaviour patterns of medieval people in a very private sphere. If the fragments were used as toilet paper and/or sanitary towels, then perhaps not all textiles used by these inhabitants were

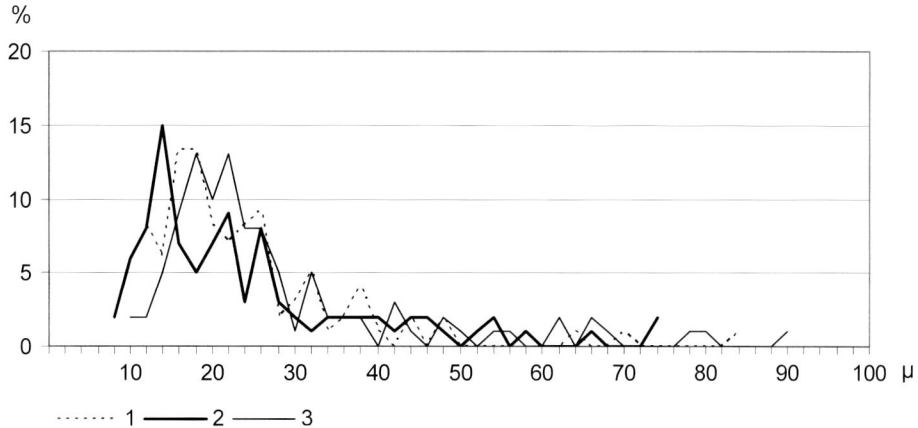

Fig. 32.5. Wool fibre measurements typical to group of others: 1 – coarse 2/2 twill (z/z, probably local); 2 – coarse 2/1 twill (z/s); 3 – coarse 2/1 twill (z/z, probably local).

	Datings	Number of textile fragments
1. cesspit	the end of the 13th – the beginning of the 14th century	596
3. cesspit	the 14th century	517
2. cesspit	the end of the 14th – the beginning of the 15th century	397
Altogether		1510

Table. 32.1. Datings and distribution of textile finds in cesspits.

discarded in this way. Perhaps, for these purposes, some special kinds of textiles were preferred, such as soft wool cloth or cheaper materials, while luxury products like silk items were reused in some other way.

Imported or Local?

One of the main objectives was to distinguish imports from local cloths, in order to gain a better understanding of the pattern of consumption. The foreign origin of textiles is indicated by the use of non-local materials (here the type of wool used) and technology, which can be clearly distinguished from those used in traditional local handicraft (Maik 1990, 120).

Therefore, the wool fibre analyses were included[5] (for the methods see: Nahlik 1963, 229–242; Maik 1998, 218–221; Peets 2000, 109–110; Ryder 2000, 4–6; Kirjavainen 2005, 134–138). On the basis of the earlier studies (Peets 1992, 8–34) and the reference group, it can be assumed that the local medieval sheep breed belonged to the northern short-tailed group, and that its wool used in textiles varied from the hairy to the generalised medium according to Michael Ryder's model (2000, 4–6). Probably, it was not possible to obtain fine, semi-fine and medium wool from this sheep's fleece and such samples may have originated from imported fabrics.

Furthermore, the technical data has been taken into account, and fragments under scrutiny have been compared with one of the pieces known with certainty to be local products. Of the latter, there is primarily the material from contemporary rural cemeteries which represents local traditions with prehistoric origins. The characteristic traits for these textiles made of wool are 2/2 twill weave and z-spun yarns in both systems; 2/1 twill and tabby were not so common and also these have been made wholly of z-spun yarns (Peets 1992, 65, 71, table 28). On the basis of these criteria, a test to distinguish the imported types of cloth was made.

Imports

The term 'imports' denotes here fabrics that reached Tartu from western European production centres via long-distance trade. These textiles constitute approximately 75% of the entire material from the cesspits in Lossi Street. It is also noteworthy that all fragments with red and other visible colours belong to these groups. Mainly semi-fine and fine types of wool are present, but also generalised medium occurred (Fig. 32.4). Among imported textiles, some special small groups can be named: the few silk finds (5 fragments); fine[6] worsteds in different twill weaves (35 fragments) – 2/2, 2/1 and also 3/3 twills; and fabrics made of wool and linen (12 fragments) (for parallels see *e.g.* Crowfoot *et al.* 2001, 36–39, 128). The striped cloths (28 fragments) of tabby weaves with weft-faced bands, well-known from the written sources, were spread and traded all over Europe and clearly indicate long-distance trade (Crowfoot *et al.* 2001, 52–55 and references cited there; Tidow 1988, 203; Wincott Heckett 1994, 149; Walton 2002, 2882–2883; Hammarlund and Vestergaard Pedersen 2007, 215).

The vast majority of the obviously imported textiles constitute quite a homogeneous group primarily of medium grade. Most common are tabbies and 2/1 twills[7] with z-spun yarn in warp and s-spun yarn in weft, which are felted and additionally finished (*i.e.* teaselled and sheared) in different grades (48 % of all finds, *cf.* with group 1 in Hammarlund *et al.* 2008, 79–80). This group probably represents broadcloth – well-known from written sources and the main cloth type traded during the Middle Ages. Not so common are the so-called ordinary plain 2/1 twills and tabbies (19%) similar to this broadcloth group according to the technical data, but without clear marks of fulling and finishing.

Fragments of the broadcloth type dominate and similar pieces of striped cloth and fine worsteds have also been found among other contemporary textile collections recovered from cesspits in Tartu.[8] In terms of technology and appearance, these textiles differ distinctively from textile finds from rural cemeteries and represent cloth types clearly related to the newly founded urban settlements.

Local Textiles

The remainder of the material (25%) comprises different types of coarser twills, made primarily of hairy medium wool resembling local sheep's fleece (Fig. 32.5). The question of the origin of coarser wool types, characteristic of the widespread northern short-tailed breeds, is more problematic than that of the finer ones. As a matter of fact, textiles made from such wool types could have been produced either locally or abroad – elsewhere in Northern Europe. Only few of these twill types can in fact be pointed out as being local products: for example, fragments of four-shed twills (87 fragments) resembling textiles found in rural cemeteries and one group of three-shed twills (31 fragments) – both made wholly of z-spun yarns.

However, these fragments are mostly 2/1 twills with z/s spin combination. Thus, the binding and yarn twist direction differ clearly from traditional local textiles. These new technical traits characteristic of the urban fabrics witness the spread of new textile production technology in Estonia during the 13th–14th centuries (*e.g.* horizontal looms). Nevertheless, the possibility that these textiles arrived here partly through trade contacts cannot be excluded either. In terms of local production, it should be noted that the fragments of tools found in Tartu are not numerous and also raw materials and other remains related to textile manufacture are almost absent. As far as may be ascertained today, textile production did not play an important role in the economy of medieval Tartu.

Textile Consumption in Medieval Tartu

At the outset of my research, I hoped to enquire further into local handicraft and textile production; however, during the work on the material, questions on the pattern of consumption and the consumers themselves arose. At least from the end of the 13th century to the beginning of the 15th century, according to the material that has been preserved in the cesspits, some of the inhabitants of Tartu clearly preferred imported cloth types for their clothing and household items. The demand for, and usage of, these kinds of cloth types occurred here in the newly developed towns and belonged only to an urban context. The reasons for this kind of consumption pattern can be various – cultural, economic and social – and are interlaced with questions of trade, social status and identity.

In the broadest sense, the demand for, and preference for, imports by consumers can be associated with the Hanseatic urban culture common to the Northern European towns in the Middle Ages, which signifies shared language, cultural values, identity and lifestyle expressed also through a similar material culture, such as art, architecture, fashion, or everyday objects. In the archaeological material, for example, the extensive consumption of imported ceramics in the Baltic areas (also in Estonian towns) has been interpreted through the similar dining practices in the Hanseatic region (Gaimster 2005, 412–413, 418; Russow 2006, 206). Obviously, consumers in this cultural sphere had perceptions about fashion and furnishing, hence about qualities of fabric, too, according to the accepted norms and habits. The uniformity of textile finds in the medieval Baltic area and the role of Hanseatic trade there have been emphasized earlier (*e.g.* Maik 1998, 216; 2005, 91).

In Tartu, the economic precondition for the similar consumption pattern was the existence of lively Hanseatic trade: in the 14th century, the main income of Tartu originated from transit commerce between western Europe and Russia and, according to written sources, cloth produced in western Europe was one of the primary trading products (*e.g.* Kivimäe 1998, 9, 14; Zeller 2002, 46). Besides the fact that written sources testify that the people of Tartu were middlemen in that trade, the archaeological data indicates that they also consumed a great amount of these textiles themselves. Thus, Tartu was also an important destination of this trade.

Who were these consumers? First, the value of the textiles should be estimated. Were these textiles luxury items meant only for the highest ranks of society[9] or were they a widely spread mass commodity?[10] It seems that these textiles can be classified as middle grade cloths or even rather cheap among imported types. The broadcloth fragments found in Tartu correspond well to the finds from London described as 'good middle quality' according to the written sources (Crowfoot *et al.* 2001, 44); and striped cloths, for example, have been regarded as one of the cheapest of the Flemish imports (Chorley 1987, 359). At the same time, the wool used in the latter group is clearly the coarsest type (generalised medium) among the imported textiles (Fig. 32.4:4). Furthermore, the mass occurrence of imported textiles makes it necessary to emphasize the fact that imports do not automatically signify luxury items, but rather that these cloths were probably a more widespread commodity, embracing quite a wide range of consumers, even if the latter all belonged to the wealthy sectors of the community. Thus, these cloths perhaps regularly featured in the daily life of the middle class burghers, *e.g.* artisans, and, even servants of wealthier households.

The choices made by consumers were not simply a reflection of their wealth and access to goods, but also the possibility to construct and express their various identities, aspirations and taste at a more symbolic level. Is it possible, on the basis of archaeological finds to determine, for example, the ethnic, cultural or social identity of these consumers at all? The consumption of uniform goods, characterizing Hanseatic communities in a broader sense, has been interpreted as self expression and the cultural maintenance of identity of the mainly German-speaking population in contrast to the locals (Gaimster 2005, 418–420). Jerzy Maik (2005, 91) has pointed out differences between those areas inhabited primarily by German settlers and those occupied by Slavs, characterizing the textiles and textile production of the latter as traditional. In an Estonian context, perhaps, this contrast can be related to the two main categories dividing the society – the *Deutsch* and the *Undeutsch*. It should be noted that, in the Late Middle Ages, these categories had rather a social than ethnic meaning. Obviously there were differences, for example, in the fashions of these two groups. Thus perhaps the consumption of imported textiles, too, can be related to the identity of the German-speaking *Deutsch*? In consideration of the material from Lossi Street and other cesspits in Tartu, the owner of the property in the centre of medieval town was most likely a burgher and thus belonged to the *Deutsch*. Nevertheless, the questions about different identities need further thorough and extensive study and the answers will probably remain inconclusive.

Conclusions

The beginning of the Middle Ages in Estonia brought about the development of towns, typical of medieval Europe. Alongside these, Hanseastic urban culture was introduced here mainly by German colonists and merchants, who brought their own habits of consumption. This process is observable on the basis of textiles recovered from cesspits in Tartu. Three-fourths of these fragments represent new cloth types, imported from Western Europe, that are characteristic only to the urban context. In view of the numerous occurrences of imported cloths, these were common, widespread commodities rather than luxury items. The preference for these kinds of textiles by consumers can be related to the questions of their social status and identity.

This kind of interpretation represents the preliminary results obtained on the basis of an examination of the material found in Tartu and it needs to be completed by further studies of different types of evidence, in particular the comparative study with other collections in Estonia and the Baltic Sea region and, especially, the material from other kinds of habitations, *i.e.* not merely the wealthy trading centers.

Notes

1 Known as Dorpat in the German-speaking countries.
2 Medieval Livonia comprised the present territories of Estonia and Latvia.
3 The cesspits were situated usually in the backyards and thus belonged to the private sphere. Nevertheless, written sources

indicate that sometimes a lavatory was shared by neighbouring households (Kamber and Keller 1996, 11).
4   The finds without known context (194 fragments) are not included in this research.
5   Samples of each cloth type (see above; altogether 33 fragments, both warp and weft yarns) and additionally a reference group of contemporary textiles certainly of local origin were analyzed.
6   The division of material according to the thread count (fine, medium, coarse) here follows the system used by Crowfoot *et al.* (2001, 44).
7   Exceptionally also some 2/2 twills (z/z and z/s).
8   Besides material from Lossi Street, there are 4 other larger assemblages of textiles from cesspits mainly dated to the 13th–15th century. They have not been studied yet.
9   The German-speaking upper class consisting mainly of merchants, clergy and craftspeople belonging to guilds.
10 There have been similar discussion and differing opinions among historians on this question of trade between Livonia (especially Tallinn) and Russia (*cf e.g.* Mickwits 1938, 58–59; Sass 1955, 83–84, Harder-Gersdorff 2002, 136 with references).

Bibliography

Alttoa, K. (1998) Das *Russische Ende* im mittelalterlichen Dorpat (Tartu). In A. Must, A. Rahi, L. Morits, and T. Mazur (eds), *Steinbrücke 1, Estnische Historische Zeitschrift*, 31–42. Tartu, Ajalookirjanduse Sihtasutus Kleio.

Chorley, P. (1987) The cloth exports of Flanders and northern France during the thirteenth century: a luxury trade? *Economic History Review*, 2nd ser. XL, 3, 349–379.

Crowfoot, E., Pritchard, F., and Staniland, K. (2001) *Textiles and clothing, c.1150–c.1450. Medieval finds from excavations in London*, 4. (New edition). Woodbridge, The Boydell Press.

Gaimster, D. (2005) A parallel history: the archaeology of Hanseatic urban culture in the Baltic *c*. 1200–1600. *World Archaeology 37, 3*, 408–423.

Hammarlund, L., and Vestergaard Pedersen, K. (2007) Textile appearance and visual impression – Craft knowledge applied to archaeological textiles. In A. Rast-Eicher and R. Windler (eds), *NESAT IX. Archäologische Textilfunde – Archaeological Textiles, Braunwald, 18.–21. Mai 2005*, 213–219. Ennenda.

Hammarlund, L., Kirjavainen, H., Vestergård Pedersen, K., and Vedeler, M. (2008) Visual Textiles: A Study of Appearance and Visual Impression in Archaeological Textiles. In R. Netherton and G. R. Owen-Crocker (eds), *Medieval Clothing and Textiles, 4*, 69–87. Woodbridge, The Boydell Press.

Harder-Gersdorff, E. (2002) Hansische Handelsgüter auf dem Grossmarkt Novgorod (13.–17. Jh.): Grundstrukturen und Forschungsfragen. In N. Angermann and K. Friedland (eds), *Novgorod, Markt und Kontor der Hanse*, 133–153. Quellen und Darstellungen zur hansischen Geschichte, Neue Folge, Band LIII. Köln, Weimar, Wien, Böhlau Verlag.

Kamber, P., and Keller, C. (1996) Latrinen und Abfallbeseitigung. In *Fundgruben – Stille Örtchen ausgeschöpft, Ausstellung vom 1. Juni bis 30. September 1996 in Historisches Museum Basel, Barfüsserkirche*, 10–23. Basel, Historisches Museum Basel.

Kirjavainen, H. (2005) The fleece types of Late Medieval textiles and raw wool finds from the Åbo Akademi site. In S. Mäntylä (ed.), *Rituals and Relations. Studies on the society and material culture of the Baltic Finns*, 131–146. Helsinki, Academia Scientiarum Fennica.

Kivimäe, J. (1998) Die Rolle von Dorpat (Tartu) im hansisch-russischen Handel im Spätmittelalter. In A. Must, A. Rahi, L. Morits, and T. Mazur (eds), *Steinbrücke 1, Estnische Historische Zeitschrift*, 9–17. Tartu, Ajalookirjanduse Sihtasutus Kleio.

Laul, S. (1990) Einige gemeinsame Züge in den vorgeschichtlichen Trachten der Ostseefinnen. In A. Viires (ed.), *Finno-Ugric Studies in Archaeology, Anthropology and Ethnography*, 29–43. Tallinn, Estonian Academy of Sciences, Institut of History.

Laul, S. (1996) Über die frühgeschichtlichen Elemente in den estnischen Volkstrachten. In *Historia Fenno-ugrica I: 1. Congress Primus historiae fenno-ugricae*, 733–753. Oulu, Societas Historiae Fenno-Ugricae.

Laul, S., and Valk, H. (2007) *Siksälä: a community at the frontiers*. Tallinn, Tartu, Tartu Ülikool.

Mäesalu, A. (1990) Sechs Holzkonstruktionen in Tartu (Lossi-Strasse). In *Eesti Teaduste Akadeemia Toimetised. Ühiskonnateadused, 39 (4)*, 446–452. Tallinn, Teaduste Akadeemia Kirjastus.

Maik, J. (1990) Medieval English and Flemish textiles found in Gdańsk. In P. Walton and J. P. Wild (eds), *Textiles in Northern Archaeology, NESAT III Textile Symposium in York 6–9 May 1987*, 119–128. London, Archetype Publications.

Maik, J. (1998) Westeuropäische Wollgewebe im mittelalterlichen Elbląg (Elbing). In L. Bender Jørgensen and C. Rinaldo (eds), *Textiles in European Archaeology. Report from the 6th NESAT Symposium, 7–11th May 1996 in Borås*, 215–231. GOTARC Series A, Vol. 1. Göteborg, Göteborg University.

Maik, J. (2005) Stand und Notwendigkeit der Forschungen über die mittelalterliche Wollweberei auf dem südlichen Ostseegebiet. In F. Pritchard and J. P. Wild (eds), *Northern Archaeological Textiles. NESAT VII. Textile Symposium in Edinburgh, 5th–7th May 1999*, 84–92. Oxford, Oxbow Books.

Mickwitz, G. (1938) *Aus Revaler Handelsbüchern: zur Technik des Ostseehandels in der ersten Hälfte des 16. Jahrhunderts.* Societas Scientiarum Fennica. Commentationes humanarum litterarum. IX, 8. Helsingfors, Akademische Buchhandlung.

Nahlik, A. (1963) = Нахлик, А. (1963) *Ткани Новгорода*. Жилища древнего Новгорода. Материалы и исследования по археологии СССР, 123. Moscow, Academy of Sciences of USSR.

Peets, J. (1987) Totenhandschuhe im Bestattungsbrauchtum der Esten und anderer Ostseefinnen. *Fennoscandia archaeologica IV*, 105–116.

Peets, J. (1992) *Eesti arheoloogilised tekstiilid kalmetest ja peitleidudest III–XVI saj. (Materjal, töövahendid, tehnoloogia).* MA thesis University of Tartu (unpublished).

Peets, J. (2000) Textile fragment from a church door – fieldwork and laboratory study. In Ü. Tamla (ed.), *Archaeological Field Works in Estonia 1999*, 108–112. Tallinn, Muinsuskaitseamet.

Rammo, R. (2007) Tekstiilitööd: vahendid, tehnikad, meetodid (Textile work: tools, techniques, methods). In A. Haak (ed.), *Pudemeid keskaegsest käsitööst Tartus. Näituse kataloog "Manu et mente – käe ja mõistusega"* (Exhibition Catalogue), 45–52. Tartu, Tartu Linnamuuseum.

Russow, E. (2006) *Importkeraamika Lääne-Eesti linnades 13.–17. sajandil*. Tallinn, Tallinna Ülikooli Ajaloo Instituut.

Russow, E., Valk, H., Haak, A., Pärn, A., and Mäesalu, A. (2006) Medieval archaeology of the European context: towns, churches, monasteries and castles. In V. Lang and M. Laneman (eds), *Archaeological Research in Estonia 1865–2005*, 159–192. Estonian Archaeology, 1. Tartu, Tartu University Press.

Ryder, M. (2000) Issues in Conserving Archaeological Textiles. *ATN* 31, 2–8.

Sass, K. H. (1955) *Hansischer Einfuhrhandel in Reval um 1430.* Wissenschaftliche Beiträge zur Geschichte und Landeskunde Ost-Mitteleuropas, 19. Marburg, Lahn, Johann Gottfried Herder-Institut.

Tarvel, E. (1980) Hansalinnana XIII sajandist Liivi sõjani. In R. Pullat (ed.), *Tartu ajalugu*, 27–60. Tallinn, Eesti raamat.

Tidow, K. (1988) Neue Funde von mittelalterlichen Wollgeweben aus Nord-Deutschland. In L. Bender Jørgensen, B. Magnus and E. Munksgaard (eds), *Archaeological Textiles: Report from the 2nd NESAT Symposium 1.–4.V.1984.*, 197–210. Arkaeologiske Skrifter 2. København: Arkaeologisk Institut.

Walton, P. (1989) *Textiles, Cordage and Raw Fibre from 16–22 Coppergate*. The Archaeology of York, The Small Finds 17/5. London, Council for British Archaeology.

Walton, P. (2002) Dress, dress accessories and personal ornament. Textile and yarn. In P. Ottoway and N. Rogers (eds), *Craft, Industry and Everyday Life: Finds from Medieval York.* 2880–2886. The Archaeology of York. The Small Finds 17/15. London, Council for British Archaeology.

Wincott Heckett, E. (1994) Medieval textiles from Waterford City. In G. Jaacks and K. Tidow (eds), *Archäologische Textilfunde – Archaeological Textiles: Textilsymposium Neumünster 4.–7.5. 1993. NESAT V*, 148–156. Neumünster, Textilmuseum Neumünster.

Zeller, A. (2002) *Der Handel deutscher Kaufleute im mittelalterlichen Novgorod.* Hamburger Beiträge zur Geschichte der Deutschen im europäischen Osten, 9. Hamburg.

33 Garments for a Queen

by Antoinette Rast-Eicher

The Merovingian burials from the Basilica of St. Denis, Paris were first excavated by Edouard Salin in 1953–54 and 1957 and then by Michel Fleury in 1957–1959, and by other excavators until 1980. Publications, especially of sarcophagus 49 – that of Arnegundis – took place in 1961, 1962 and 1979; the last large-scale publication of the material was in 1998 (Fleury and France-Lanord 1998).

New methods in archaeology developed in the recent decades led to a fresh study of the entire material, including anthropological, metal and stone analysis – and the textiles. This project, set up by Patrick Périn, director of the Musée d'Archéologie Nationale (MAN) in St. Germain-en-Laye, in France began in 2000, the textile part in 2006 (Périn *et al.* 2007). With regard to the textiles, Thomas Calligaro (Centre de Recherches et de Restauration des Musées de France, C2RMF) analysed the metal of the gold threads, Witold Nowik (Laboratoire de Champs-sur-Marne) is carrying out the colour analyses, and Sophie Desrosiers (École des hautes études en sciences sociales, EHESS) works on the comparisons of the silks. My task was to write the full catalogue of all the textiles, first sorting the organic material, taking samples, and doing the fibre analyses. The catalogue comprises textiles from 30 sarcophagi, most of them dated to the 6th century AD. An exhibition of the new results in the different fields is planned for 2009, followed later by a publication.

The graves were situated under the nave of the church. Sarcophagus 49 was opened in 1959 by Michel Fleury and attributed to Queen Arnegundis because the skeleton had a finger ring with her name and title engraved on it (*ARNEGUNDIS REGINE*). Merovingian kings and queens have been buried in the Basilica of St. Denis since Dagobert I († AD 639) – including Queen Arnegundis who was the wife of Clotaire I (511–561) and mother of King Chilperic († 584) (Périn *et al.* 2007, 182).

The new investigations showed that the queen was older than had been thought previously. According to the dental enamel analysis, she died at the age of 60 (+/-3), maybe during the dysentery epidemic in AD 580, a mere few years before her son, King Chilperic. Furthermore, she had poliomyelitis during her childhood, which left her with serious problems in her right leg.

After the excavation in 1959, the organic material was sent in blocks taken out of the sarcophagus to Albert France-Lanord (Nancy), a former engineer, who undertook the investigation of the textiles. He obviously did not analyse organic remains on the spot in the church, but received parts of the dried remnants (Fig. 33.1). Regrettably, his documentation (radiographies, photographs, drawings and notes) has not been found yet, so the new study of the textiles had to start with the published information (see titles in Périn *et al.* 2007, 204–206). There are also some textile pieces – such as the sleeves of garments – but they are now in much smaller fragments than those documented in the early 1960s.

In 2003, the textiles from Queen Arnegundis' grave were found in two different storage places in Paris, one of them being Michel Fleury's former office, left in aluminium, wooden or cardboard boxes of different sizes. These were marked by terms like '*vertèbres*' or '*os longs*', which refer to the places on the skeletons the organic material comes from. We assumed that these boxes had not been changed or re-opened since France-Lanord's study. Samples of the most important fragments have unfortunately been glued on a wooden board, and the gold embroidery was reconstructed after the radiographies on wax strips. However, compared to other excavations, the analyses of the Merovingian graves from the Basilica of St. Denis carried out in the late 1950s–early 1960s by Albert France-Lanord were certainly exemplary for that time.

The new investigation was begun by sorting through all the boxes, separating the materials, and examining the bones with textile remains – without documentation, they are the only certain location providing a stratigraphy of textiles. On closer examination, the preservation of the fine textiles was often so poor, that fibre determination was especially difficult. In certain cases, only the textile structure is visible, while the fibres are more or less deteriorated. Light microscopy was attempted, but could not be used, and scanning electron microscopy (SEM) became the only valid method of fibre analysis. Fibres preserved in such church sarcophagi have originally been wet, then partly decomposed and dried out, so that they become – compared to metal-replaced textiles – the

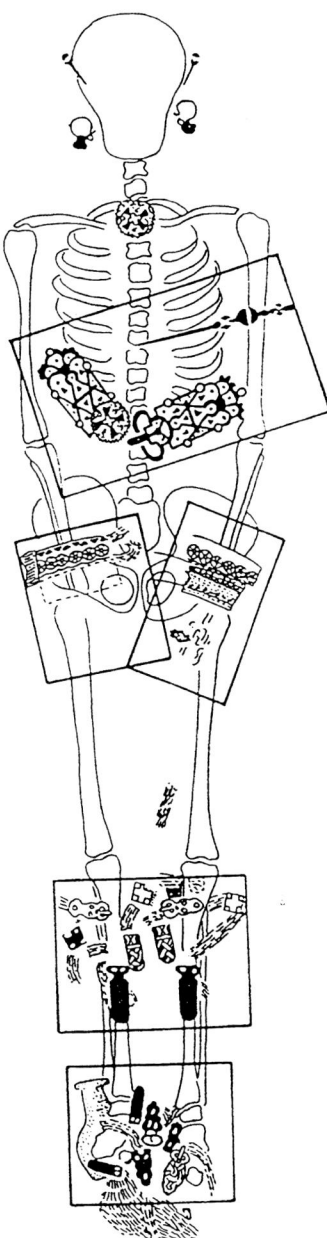

Fig. 33.1. Arnegundis' grave with placement of the blocks (After Fleury and France-Lanord 1998, 17).

Fig. 33.2. Samite S (Photo: A. Rast-Eicher).

a large girdle, and a mantle with decorated sleeves. Together with the published information, the new textile analysis is arriving at a slightly different interpretation of the textiles. From the top down, they are as follows:

1) Tabby cloth of flax/hemp found as the outer layer, and also on the glass, which was placed near her feet. The textile has small tablet-woven borders and was probably a shroud.

2) Wool textile, which is probably fulled. It is the second layer under the shroud and may be from something like a cover. The fibre analysis has shown a very fine wool; for such fibres, differentiation between sheep and goat is very difficult and may not be established in this case.

3) Samite 2/1 S, warp proportion 1:1, red/yellow (Fig. 33.2). According to Fleury and France-Lanord (1998), it is probably a veil, found on the head and shoulders. In 2006, tiny fragments, as well as a top layer on a vertebra and on a leather fragment from the girdle were found, which confirmed its position or use as an upper layer. Yet, as there were also gold threads on the head from a small decoration, there may have been some other fine textile used as a veil or a braid with gold thread was just a hair band.

4) A mantle with decorated sleeves and a large brocaded tablet-woven band is the key piece, but also the most difficult one, as parts of the textile are badly preserved. The main textile is the so-called '*violet*' piece, a textile with one system made of animal fibre, the other system is of plant fibre, but mostly not preserved.

The brocaded tablet-woven band was interpreted by France-Lanord as another garment (his so-called '*tunique de soie*'), and not as a decoration. The main textile ('*violet*'), on which the band is placed, is the textile he interpreted as a tunic ('*robe*'). Now we have reinterpreted the textiles as merely belonging to one garment made of a main textile ('*violet*') and the tablet-woven band as decoration.

The samite 2/1 Z has been woven with a warp proportion of 2:1. This samite was used as the end section of each sleeve. France-Lanord did not differentiate between samite S and samite Z, they were both called '*satin de soie*'. He merely mentions that it is similar to that of the veil on the head (France-Lanord 1979, 85). A fragment of the seam where

most difficult of materials to work with in my experience. Furthermore, fibres and the textiles shrink considerably, so that sizes are not reliable. Another problem in fibre analysis is the possible fibre contamination from an upper layer if the textiles are in very poor condition and the fibres are brittle. Therefore, it was important to undertake a careful examination of each sample to see if the fibres from one weaving system really crossed those of the other system.

There are several reconstructions of Arnegundis' garments (for example, France-Lanord 1979, 76; Fleury and France-Lanord 1998, II–132f). In archaeological studies, these reconstructions very often stand for the typical early medieval female garment. In the latest reconstruction (Fleury and France-Lanord 1998, II–133), she wears a short tunic held by

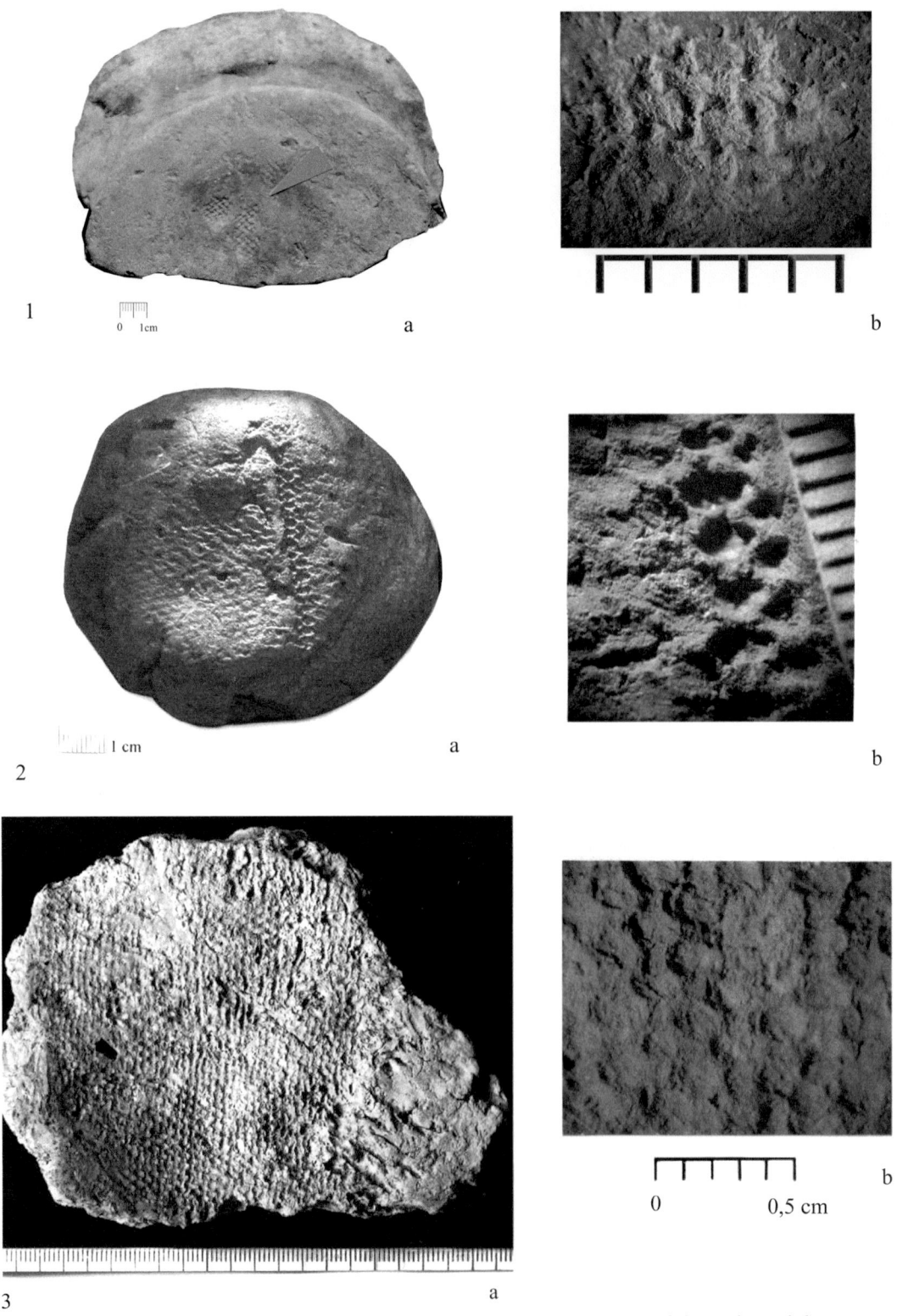

Fig. 34.1. Imprints of tabby textiles. 1. a–b. Pottery base, Szentes-Kiss Boldizsár farm site of the Early Neolithic Körös Culture, József Koszta Museum, Inv. No. 71.207.1; 2. a–b. Lump of clay, Dévaványa-Sártó site of the Middle Neolithic Szakálhát Culture, Hungarian National Museum, Inv. No. 60.35.28; 3. a–b. Inner side of pottery wall, Hódmezővásárhely Kökénydomb site from the Late Neolithic Tisza Culture, Inv. No. 43.43.30 (Photos E. Richter).

Diagonal Twill

Adding one binding point to the tabby structure in one direction and shifting the points diagonally in each following row produces the simplest form of diagonal twill (1/2). Thus, the pattern is created by regularly repeated strip-floats when a strip passes over two others and then goes under the next. The addition of more binding points to the pattern allows longer strip-floats to be produced. In the Hungarian Neolithic, 2/2 twill is known from the Late Neolithic Hódmezővásárhely-Kökénydomb sites (Fig. 34.2.2) (Csalog 1956) and a 3/2 twill from the Middle Neolithic Tiszaföldvár-Téglagyár (Richter 2003, fig. 1).

Chevron (herring-bone) Twill

In the creation of chevron patterns, the diagonal twill breaks along an axis at a 45° angle. The repetition of chevrons produces a zig-zag (herringbone) pattern. There is a 2/2 chevron twill from the Lengyel Culture (Fig. 34.2.3) (Bíró and Regenye 2002; Richter 2003, fig. 3) and two chevron twills from the Tisza Culture, which may have been a part of larger items. The one from Öcsöd-Kováshalom has a big pattern of chevron (Kalicz *et al.* 1990, Kat. No. 481; Richter 2005, fig. 4), whilst the other from Berettyóújfalu-Herpály has a displacement tip along its reflection line (Kalicz *et al.* 1990 Kat. No. 482, Richter 2005, fig 6.1).

Lozenge Twill

Using the next reflection line, the lines of chevron reverse at an angle of 45° along the line. A lozenge form develops as a result of this operation. We have only one known example of this structure from the Tisza Culture, Kökénydomb site (Fig. 34.3.1) (Csalog 1966, fig.1; Richter 2003, fig. 2).

Other Twill Variants

The addition of more binding points and/or inserted twill breaks creates further twill variants. They develop not necessarily from the lozenge, but from the simpler twills as well. The twill variants have been found among the Middle and Late Neolithic imprints from Hungary (Richter 2005). In Figure 34.3, an example can be seen from the Tiszaföldvár-Téglagyár site (Fig. 34.3.2), Middle Neolithic, Szakálhát Culture. As the imprints are fragments of larger structures, the complete patterns remain unknown.

On the basis of the mat imprint presented above, the following development of binding structures can be postulated: *tabby → diagonal twill → chevron twill → lozenge twill and other complex twill variants*. In the course of my professional work, I investigated and presented the above mentioned collection of archaeological samples in detail. Photographs and drawings of additional Neolithic imprints, with twill binding can be found in the archaeological literature (Dombay 1960, Taf. XCIII.I; Banner 1960, Taf. XXV. 26; Füzes 1990).

Plaited binding structures in the weaving

As indicated above, the tabby, the common twill bindings (diagonal, chevron, lozenge) and further twill variants were already developed in the plaiting technology of the Neolithic. Only the tabby was known in the weaving. Although the tabby structure is very simple, its adoption for weaving is a result of important technological innovations.

Prerequisites and Innovations

In plaited binding structures, the flexible, relatively wide strips can easily be selected manually, whereas in weaving, a tool is required to pick every second flexible warp. Such a tool may take the form of a long needle or a heddle bar, to which every second warp thread is fastened (Endrei 1952). The heddle bar can elevate and pull back the warps simultaneously along the whole width of the textile being produced, so this latter invention permitted warp threads to be moved at the same time in a directed movement, without being touched by human hand, making the weaving a mechanical work process (Endrei 1952). The question then arises of when this invention occurred? In the Eastern Gravettian Culture of the Upper Palaeolithic, twill binding (2/2) was used in matting according to archaeological evidence and one of the textile techniques used was tabby binding (Soffer and Adovasio 2002). This signifies that plaiting and weaving separated from each other at an extremely early stage, but it has not been possible to ascertain when weaving became widespread. At the beginning of the Neolithic, very sophisticated tabby textiles were made in the Near East, and, from the Early Neolithic, there are clear traces of warp-weighted looms in Hungary and South East Europe (Broudy 1979; Makkay 2001; Richter 2005).

There is another prerequisite which is needed to create tabby woven textiles. The threads lead us back to the Upper Palaeolithic once more, when the textiles were mostly made by the twining method (Soffer *et al.* 1998), which survived in the Neolithic – and continues to this day (Sentence 2002). In this method paired wefts are twisted around the fixed warps. The warps can be selected manually, therefore this work is almost the same as the basketry. However, a structural support such as a frame or loom is needed, to hold the flexible warps straight. According to ethnological comparisons, the simplest method has warps hanging down from the upper bar (beam) of a wooden upright frame (Broudy 1979, fig. 1–6). It is not necessary to use weights for thread tension, but using them can make the weaving process easier. In tabby weaving, warps should be held taut as in the case of twining. Thus, if the textile has been made on a loom, weights were unavoidable. We can tell that the innovations were important in the making of tabby textiles, which led to the origins of the warp-weighted loom that still exists in Scandinavia (Hoffmann 1974).

Archaeological Evidence of Loom Weaving

The use of a loom, especially the warp-weighted loom in the Neolithic in Hungary is indicated by the heaps of clay loom weights found on the house floors, and bone pin beaters. The shape of the loom weights is mainly conical. They were discovered in heaps and along the length of the house floors at the following sites dated to the Neolithic period: Tiszajenő-Szárazérpart (Selmeczi 1969), Battonya-Vidpart (Szénászky 1979), Dévaványa-Simasziget (Kalicz and Makkay 1977), Szarvas, Site 21 (Makkay 2001), Hódmezővásárhely-Kökénydomb (Banner and Korek 1949), Szegvár-Tűzköves (Korek 1972), Aszód (Kalicz 1985), Hódmezővásárhely-Gorzsa (Horváth 1982). Pin beaters were found in Early

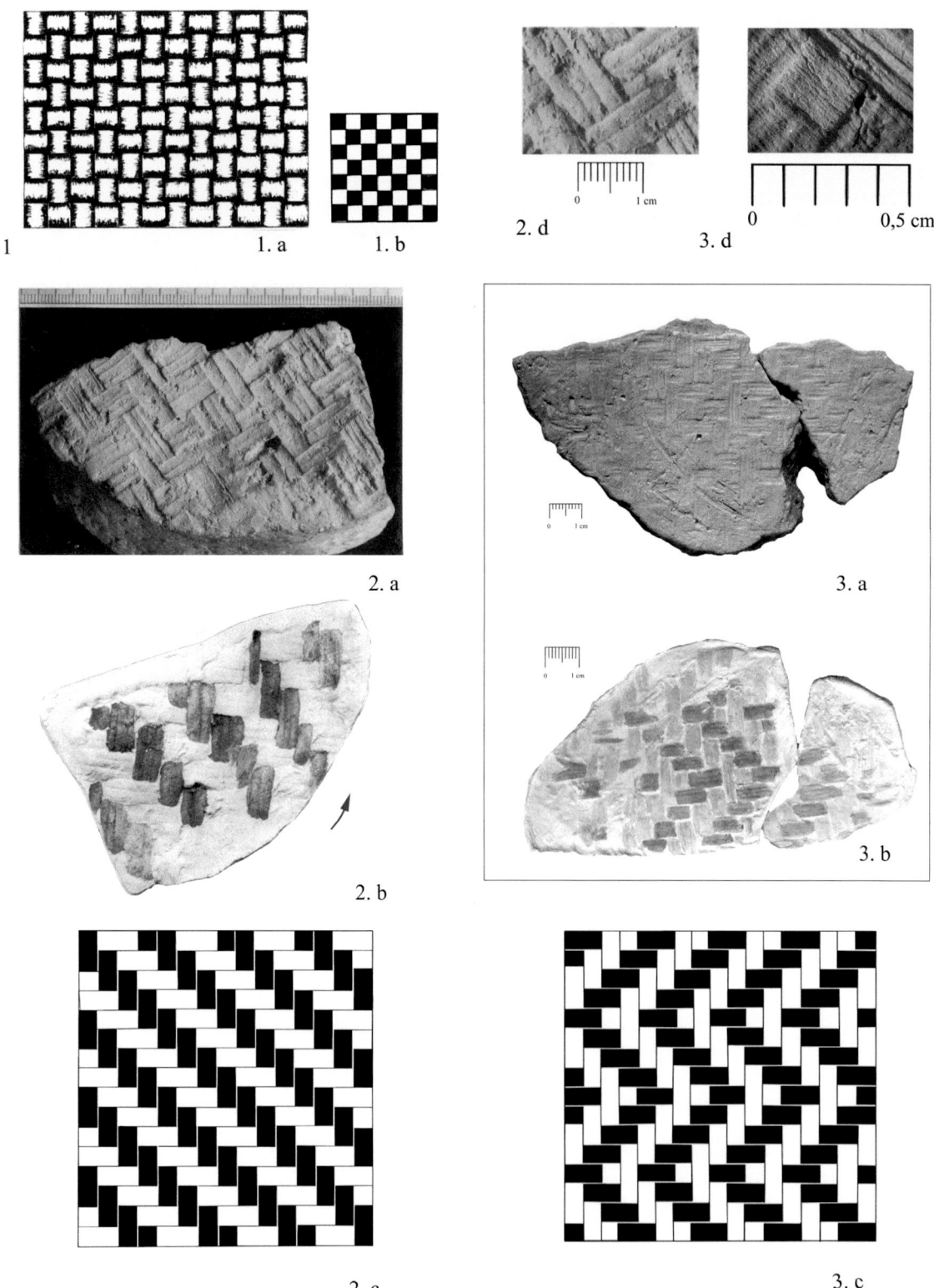

Fig. 34.2. 1.a. Drawing of tabby plaited mat, Bazsi site of the Middle Neolithic Transdanubian Linear Pottery Culture (After Füzes 1990, fig. 16); 1.b. Binding structure; 2.a. Imprint of plaited 2/2 diagonal twill on pottery base, Hódmezővásárhely-Kökénydomb site of the Tisza Culture, János Tornyai Museum, Inv. No. 61.11.30; 2.b. Plasticine positive; 2.c. Binding structure; 2.d. Detail; 3.a. Imprint of plaited chevron twill on pottery base, Kup-Egyes site of the Late Neolithic Lengyel Culture, Hungarian National Museum; 3.b. Plasticine positive; 3.c. Binding structure; 3.d. Microscopical photo of a detail (Photos E. Richter).

Neolithic sites Endrőd 39, 119, and Szarvas 23 (Makkay 2001, fig 18).

The imprints of tabby weave textiles suggest that they were made using a heddle bar. The production of more complicated tabby from Endrőd 39 (Makkay 2001) indicates that more heddle bars might have been added to looms

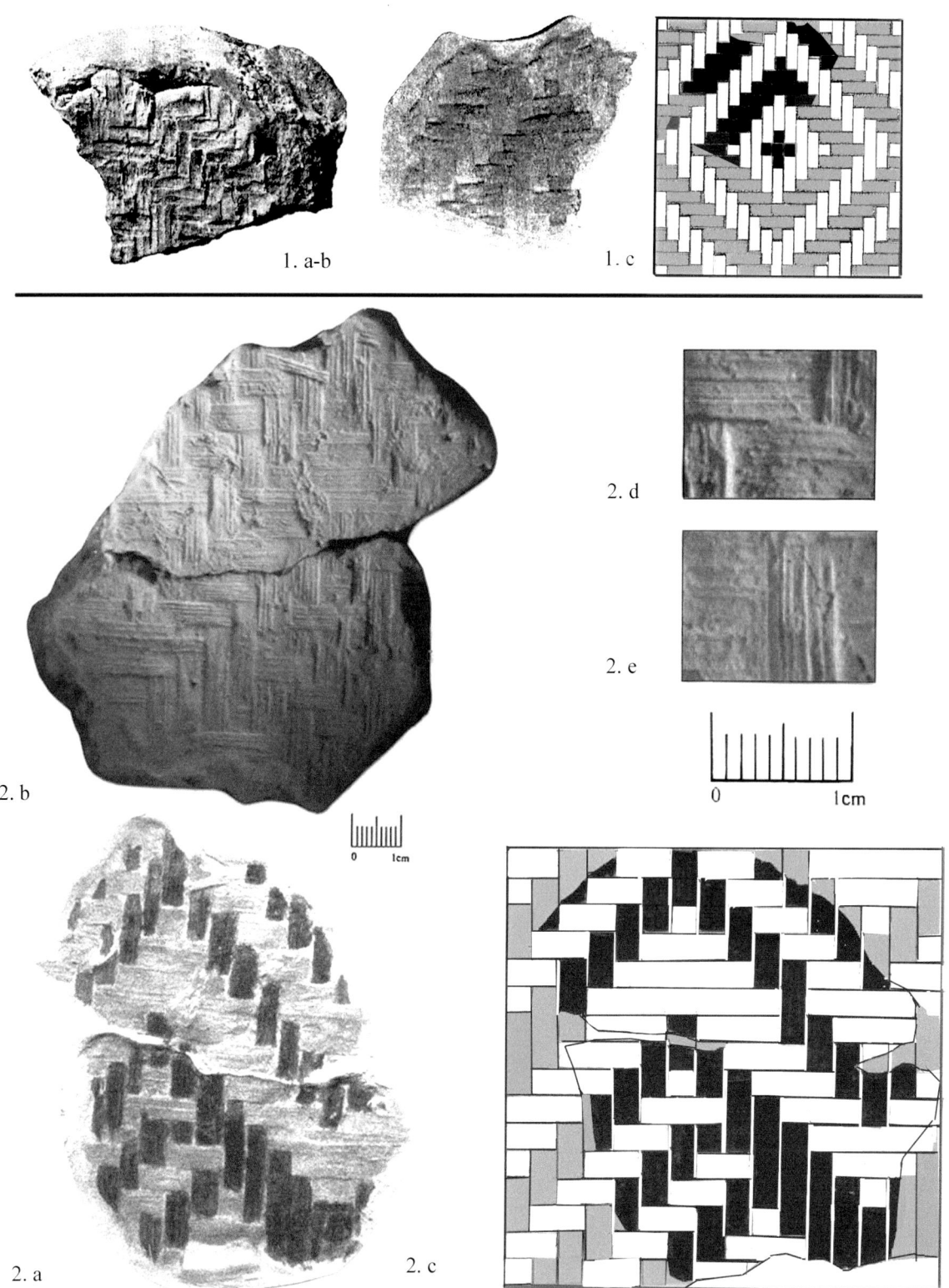

Fig. 34.3. 1.a. Plaited lozenge (3/3) on pottery base, Hódmezővásárhely-Kökénydomb site, Tisza Culture, János Tornyai Museum (After Csalog 1966, fig.1); 1.b. Gypsum positive; 1.c. Binding structure; 2.a. Plaited twill variant on the inner side of a pottery wall, Tiszaföldvár-Téglagyár site of the Szakálhát Culture, János Damjanich Museum, Inv. No.86.4.1707; 2.b. Plasticine positive; 2.c. Binding structure; 2.d–e. Microscopic photos (Photos: E. Richter).

during the course of the Neolithic in the area constituting Hungary today. A further stage in this weaving innovation occurred when the originally plaited twill binding structures appeared in textile crafts. The first twill textile is known from Alishar (Türkey) during the Copper Age, although they only became widespread during the Iron Age (Hallstatt C–D) in Europe (Good 1995, 331–335). In Hungary, published evidence for twill textiles from the Copper Age does not exist up to now.

Conclusions

Based on plaited binding structures of the Neolithic in Hungary, we can postulate the development row of twill binding at this early period. The twills are derivative versions of simple tabby:

tabby → diagonal twill → chevron twill → lozenge twill → other complex twill variants

However, the textile imprints have the structures of tabby binding (and, furthermore, only one tabby variant). The appearance and spread of plaited binding structures in loom weaving is the result of the invention of the heddle bar system. The major limiting factor is the nature of threads/strips used in each technology. While textile making requires mechanisation of processes, basketry cannot be mechanised.

Bibliography

Adovasio, J. M. (1977) *Basketry technology*. Chicago.
Banner, J. (1960) The Neolithic Settlement on the Kremenyák Hill at Csóka (Čoka). *ActaArchHung* 12, 1–58.
Banner, J., and Korek, J. (1949) A negyedik és őtődik ásatás a hódmezővsárhelyi Kökénydombon. – Les campagnes IV et V des fouilles pratiquées au Kökénydomb de Hódmezővásárhely. *Archaeológiai Értesítő* 76, 9–25.
Barna, J. (2004) Adatok a késő nelitikus viselet megismeréséhez a lengyeli kultúra újabb leletei alapján. – Some data to Late Neolithic costume according to new finds of the Lengyel culture. *Zalai Múzeum* 13, 178–201.
Bíró, K. T. (1992) Adatok a korai baltakészítés technológiájához – Data on the technology of early axe production. *Pápai Múzeumi Értesítő* 3–4, 33–79.
Bíró, K T., and Regenye, J. (2002) *Kup-Egyes. Újkőkori település ásatásának eredményeiből*. Kup.
Broudy, E. (1979) *The Book of looms. A History of the Handloom from ancient Times to the present*. Hannover and London.
Csalog, J. (1956) Újkőkori gyékénylenyomat Kökénydombról. – Imprints of matting on neolithic wares from Kökénydomb. *Archaeológiai Értesítő* 83, 183–185.
Csalog, J. (1966) A legújabb kökénydombi fonatlenyomat tanulságai. – Die Larne des neuesten Geflechtabdruckes von Kökénydomb. *Móra Ferenc Múzeum Értesítője* 1964/65, 17–45.
Csalog, Zs. (1962) Magyar népi fonástechnikák típusai. – Typen der Flechtarbeitstechnik des ungarischen volkes. *Ethnographia* 73, 302–323.
Dombay, J. (1960) Die Siedlung und das Gräberfeld in Zengővárkony. Beitrage zur Kultur des Aeneolithikums in Ungarn. *ArchHung* 37, Budapest.
Endrei, W. Gy. (1952) *A fonás és szövés története*. Budapest.
Füzes, M. (1990) A földmívelés kezdeti szakaszának (neolitikum és rézkor) növényleletei Magyarországon. – Die Pflanzenfunden in Ungarn der anfänglichen Entwicklungsphase des Ackerbaues (Neolithikum und Kupferzeit) *Tapolcai Városi Múzeum Évkönyve* 1, 139–298.
Good, I. (1995) Notes on a Bronze Age Textile Fragment from Hami, Xinjang. *The Journal of Indo-European Studies* 23, 313–345.
Hoffmann, M. (1974) *The warp weighted loom*. Oslo.
Horváth, F. (1982) A gorzsai halom késő neolit rétegei (The Late Neolithic Stratum of the Gorzsa Tell). *Archaeológiai Értesítő* 109, 201–222.
Horváth, F. (2005) Gorzsa. Előzetes eredmények az újkőkori tell 1978 és 1996 közötti feltárásából. – Gorzsa. Preliminary results of the excavation of the Neolithic tell between 1978–1996. In L. Bende, and G. Lőrinczy (eds), *Hétköznapok Vénuszai*, 51–83. Hódmezővásárhely.
Kalicz, N. (1985) *Kőkori falu Aszódon*. Múzeumi Füzetek 32. Aszód.
Kalicz, N. (1998) *Figürlische Kunst und Bemalte Keramik aus dem Neolithikum Westungarns*. Archaeolngua Ser. Minor. 10.
Kalicz, N., and Koós, J. (1997) Mezőkövesd-Mocsolyás. Újkőkori telep és temetkezések a Kr.e VI. évezredből. – Mezőkövesd-Mocsolyás. Neolithic settlement and graves from the 6th millennium B.C. In P. Raczky, T. Kovács and A. Anders (eds), *Utak a múltba. Az M3-as autópály régészeti leletmentései.– Paths into the Past*. Budapest 1997, 28–32.
Kalicz, N., Ljamic-Valovic, N., Meiner-Arendt, W., and Raczky, P. (1990) Katalog der ausgestellten Funde. In W. Meiner-Arendt (eds), *Alltag und Religion. Jungsteiunzeit in Ost-Ungarn*, 141–153. Frankfurt.
Kalicz, N., and Makkay, J. (1977) *Die Linienbandkeramik in der Grossen Ungarischen Tiefebene*. Studia Archaeologica 7, Budapest.
Korek, J. (1961) Neolitikus telep és sírok Dévaványán. – Eine neolitische Siedlung und neolitische Gräber in Déványa. *Folia Archaeologica* 13 (1961) 9–26.
Korek, J. (1972) *A tiszai kultúra (The Tisza culture)*. Budapest.
Makkay, J. (2001) Textile impressions and related finds of the Early Neolithic Körös culture in Hungary. With an Appendix: The ritual spinning. *Tractata Minuscula* 27. Budapest.
Marton, E. (2001) New approaches to the spinning and weaving of Neolithic-Aeneolithic people in the Carpathian-Basin (The "Shrewd Princess and looms). In J. Regenye, (ed.), *Sites and Stones. Lengyel Culture in Western Hungary and beyond*. Veszprém.
Nagy, K., Kralovánszky, M., Mátéfy, Gy., and Járó, M. (1993) *Textiltechnikák. – Textile techniques*. Budapest.
Richter, É. (2003) A tiszai kultúra szőttes jellegű díszítésvilágának kapcsolata a szövés és gyékényfonás motívumaival. *Ősrégészeti levelek* 5, 98–106.
Richter, É. (2005) *Textil és négyzetrendszeres fonatlenyomatok az Alföld neolitikumából. – Woven and Plaited Fabrics in the Neolithic of the Great Hungarian Plain*. In G. Lőrinczy and L. Bende (eds.), Hétköznapok vénuszai. Hódmezővásárhely.
Selmeczi, L. (1969) Das Wohnhaus der Körös Grouppe von Tiszajenő. – Neuere Ausgaben zu den Haustypen des Frühneolithikums. *Móra Ferenc Múzeum Évkönyve*, 19–22.
Sentence, B. (2001) *Basketry*. London.
Soffer, O., and Adovasio, J. M. (2002) Textiles and Upper Paleolithic lives. A focus on the perishable and the invisible. In: A. Jiři Švoboda and L. Sedláčková (eds), *The Gravettian along the Danube. Proceedings of the Mikulov Conference 20–21. Nov. 2002. The Dolni Vestonice Studies*, Vol. 11. Brno.
Soffer, O., Adovasio, M., Klíma B., and Svoboda, J. (1998) Perishable Technologies and the Genesis of the Eastern Gravettian. *Anthropologie* 36, 1–2, 43–68.
Szénászky, J. (1979) A korai szakálháti csoport települése Battonyán. – The settlement of the early Szakálhát group at Battonya. *Archaeologiai Értesítő* 106, 67–77.

35 The Neolithic Mats of the Eastern Baltic Littoral

by Virginija Rimkutė

Non-loom woven textiles from the eastern coast of the Baltic Sea have thus far been rarely analysed. One reason for this is the poor preservation conditions for organic materials with only two of the Neolithic site complexes – Šventoji 1A, 1B, 2B, 23 (peat-bog settlements) and Nida, Lithuania (Fig. 35.1), possessing major collections of plaited and twined mats in the Eastern Baltic coast.

Based on pollen data and the analysis of preserved mats, it is presumed that, not only the materials detected in archaeological finds, but also various other plants that grew in the area at the time could have been used for matting. Although the impressions on the potsherds could only be of vegetal fibre, the range of actual possible materials may be rather wide. Thus, it was necessary to experiment with several characteristics of plants, both those that grew during the Neolithic, and those that have been introduced to the area later.

Original Finds

On the Eastern Baltic coast, the Neolithic period is dated to *c*. 4800–1600 BC; the above mentioned sites date to the Middle Neolithic, *i.e. c*. 3700–2500 BC (Rimantienė 1989, 175–176; 2005, 41–42). Sites Šventoji 1B, 2B and Nida belong to the Narva culture (Girininkas 2005, 119), while Šventoji 1A belongs to the Baltic Coast Culture (Rimantienė 2005, 51). Fragments of mats were found in the sites of Šventoji 1A, 1B, 2B and 23 (Figs 35.2–35.4). The impressions

Fig. 35.2. A fragment of lime bark mat, found in Šventoji 1A (Photo: K. Vainoras; after Rimantienė 2005 252, fig. 115).

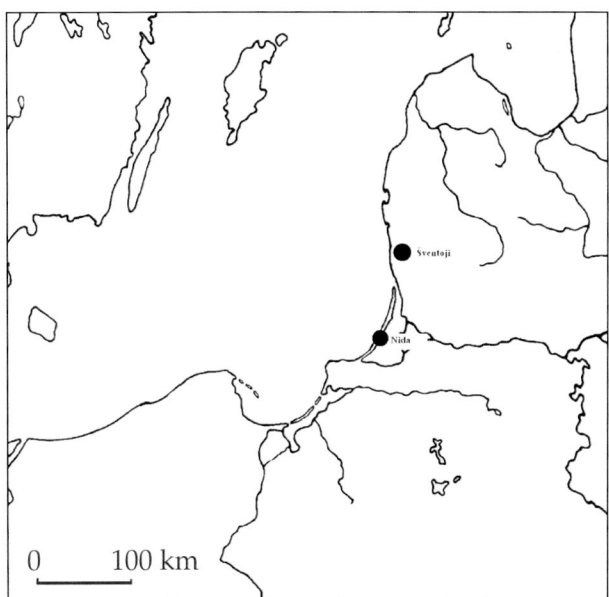

Fig. 35.1. Location of Šventoji and Nida sites on the eastern coast of the Baltic Sea (© V. Rimkutė).

Fig. 35.3. A fragment of lime bast mat or cloth, found in Šventoji 2B (Photo: K. Vainoras; after Rimantienė 2005, 326, fig. 197).

Fig. 35.4. A fragment of lime bast cloth, found in Šventoji 2B (Photo: K. Vainoras; after Rimantienė 2005, 326, fig. 195).

Fig. 35.5. A potsherd with an impression of twined cloth, found in Šventoji 2B (Photo: K. Vainoras; after Rimantienė 2005, 326, fig. 196).

of mats were found on potsherds at the Nida site and in some sites at Šventoji (Fig. 35.5). The pioneer scholar in this area is the archaeologist Rimutė Rimantienė, who directed the excavations of Neolithic Šventoji and Nida, and whose hypotheses on the question of matting, their function, as well as her identification of bindings is worthy of special mention (Rimantienė 1989, 80–82; 2005, 51, 198, 324).

Climate and Plants

Because of the alternating water level of the Littorina Sea, which existed around 7500–4000 BC, along the entire length of the eastern coast, shallow bays and sometimes even separate basins were common during this period (Kabailienė 2006, 120). Thus, lakeside plants would have grown. The relief, geological structure and the formation of the coast of Southern Latvia are identical to that of Northern Lithuania (Kabailienė 2006, 122), thus the variety of plants would have been similar too. During this period, mixed (broad-leaved) and coniferous woods flourished, while in the fields *Poaceae* and *Cyperaceae* species grew (Kabailienė 2006, 124). In the middle Neolithic in Šventoji, *Cerealia* appeared with *Urtica* as a companion plant (Kabailienė 2006, 122). In scorched earths, various bushes, *e.g. Salicaceae* and grasses (*Poaceae*) grew. The weather was warm and humid (Kabailienė 2006, 444).

Thus, a rather wide range of plants possessing certain qualities for matting and cordage were available. Furthermore, climatic conditions indicate a relatively long collecting season.

Experiences and Experiments

The materials identified in the original findings were selected for the experiments: bark and bast of common lime and broad-leaved lime (*Tilia cordata*, *Tilia platyphylos*). Then, according to other recovered fragments of basketry and pollen data, the bark and twigs of willows (*Salicaceae*, *e.g.* common osier *Salix viminalis*), bulrush, also known as cattail (*Typha latifolia*, *Typha angustifolia*) were selected. As for the grasses and other *Poaceae*, several of those growing presently that are suitable for weaving were chosen: orchard grass (*Dactylis glomerata*) and bents (genus *Agrostis*).

Preparation of the Material

Lime bark and other vegetal materials can only be gathered in certain seasons. For lime bark, as for other barks, the best time is spring–early summer. However, grasses and bents are at their best in full summer, while the leaves of the cattails are longest in late summer–early autumn. It is better to dry all materials before use. Optimally, they should be dried in the shade, out of direct sunlight. It takes longer, but the flexibility, so necessary in weaving, remains.

The qualities significantly depend on the season. For instance, stems of orchard grass (*Dactylis glomerata*) are soft and flexible until the beginning of June, but from mid-June they become straw-like, thus stiffer and not suitable as weft for twining. However, the greatest length of the stem is reached from the middle of June. Therefore, it may be supposed that, stems of such grasses on the Neolithic Eastern Baltic littoral could be used as weft in twining until the maturing of the stem; afterwards, stems were suitable for warp only.

However, there is a part of *Poaceae* which is always suitable for twisting and twining: the leaves. In late autumn–early spring these leaves do not match the criteria for cordage or matting. Thus, it may be presumed that, materials and their combinations in cordage and matting were alternated according to season.

As far as the tools are concerned, very few are needed. The most important 'tools' are fingers. Simple flint flakes for cutting, a spinning-hook for spinning, and some bone or wooden awls may be sufficient (Fig. 35.6).

Matting

The bindings identified by R. Rimantienė are closed simple twining, open simple twining, diagonal twining, 1/2, 2/1 and 2/2 twills. Some technological aspects of these bindings are discussed here.

Twined Mats

Rimantienė claims twined mats to be warp-twined or even tablet-woven (with cordage as warp and other material as weft). However, weft-twined mats are more credible. For this a wide range of materials can be used (Fig. 35.7).

Freshly collected thin lime bark, torn into strips, was tried for twinning with stems, collected in the late summer of 2007. The mat from Šventoji 1A was made from narrow lime bark

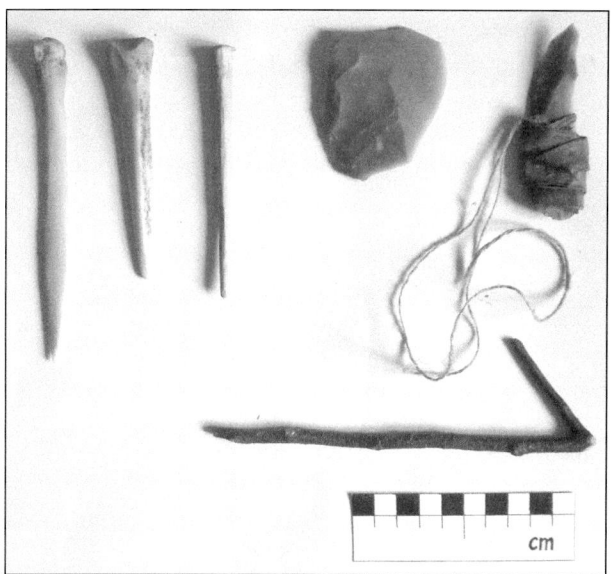

Fig. 35.6. Tools, needed for the preparation of raw material and for working with it: flint flakes, bone or wooden awls, and spinning-hook (©Photo: V. Rimkutė).

Fig. 35.8. A mat of retted lime bast; 2/2 twill (©Photo: V. Rimkutė).

Fig. 35.7. Various reconstructed mats at the outset (©Photo: V. Rimkutė).

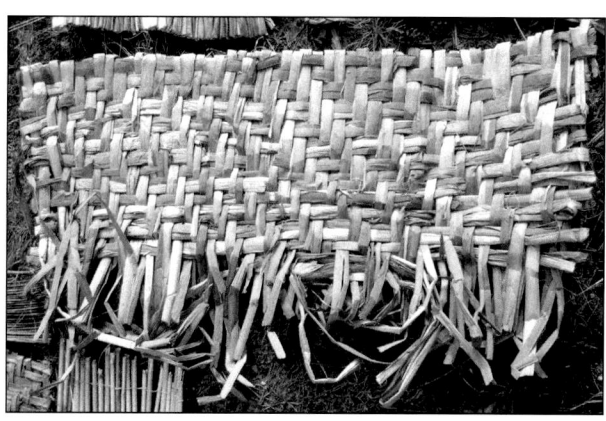

Fig. 35.9. A mat of cattail leaves in 2/2 twill at the outset (©Photo: V. Rimkutė).

strips in 2/2 twill. During the process of reconstructing this mat, it was discovered that when lime bark is fresh, it is very slippery, *i.e.* hard to work with. The twists slide out rather easily, thus a big piece of a bark strip is used for interlacing. Assuming that a strip of young lime branch bark is about 0.5–1 m long, interlacing appears to be a very frequent and inefficient process. Thus it is likely that the elaborate piece of matting from the Šventoji 1A site was probably made from the retted remnants of cordage, rather than from fresh bark, disregarding the fact that, the weave was 2/2 twill (See Fig. 35.2).

Twining gives a very specific and convenient feature: the mat can be rolled up easily and carried anywhere. This also works for reed stems and osier twigs, twined with lime bast cordage. Cattail leaves proved to be a suitable material for twining soft smooth mats, either short or long, wide or narrow, according to need.

Twills

The cordage weft allows twill mats to be rolled up. In 2005, this was attempted with common osier bark stripes and lime bast cordage. However, the bark of the *Salix viminalis* appeared to be rather brittle, and the ends of warps become loose while being used. Therefore, this way of making mats would be efficient with a thicker material. In the same year, retted lime bark was tried out in order to reconstruct the above mentioned 2/2 twill mat of Šventoji 1A. The result is rather

Fig. 35.10. View of the plaited mat of cattail leaves in twill 2/2 after heavy rain and drying (©Photo: V. Rimkutė).

rough and stiff, but quite durable: after several years of usage, no significant wear-marks were noticed (Fig. 35.8).

The 2/2 twill worked out rather well, *e.g.* with *Typha latifolia* (Fig. 35.9). However, as cattail leaves are soft, it was found difficult to fix them as weft. Then, an attempt was made to plait cattail leaves diagonally (Fig. 35.10). This way of matting was found to be quite convenient and productive.

The absence of wicker ware/tabby binding in the Eastern Baltic littoral was surprising, as this type of binding is much simpler than twills. However, reasons for this cannot be determined at present.

Conclusions

A substantial range of plants was found to be suitable for matting. However, each of these plants has certain qualities, which influence the matting process. Thus, some plants and their parts are better for twining, while others are more suitable for plaiting. Most of the mats made during the experiment are rather small, but functional: their qualities are adequate for covering and transportation. Making a large mat for sleeping or as roof-shelter requires a greater quantity of material and longer time, necessitating more experiments. In the future, other plants and their various combinations will be subject to experimentation: *e.g.* common club-rush (*Schoenoplectus lacustris*), hemp (*Cannabis sativa sativa*), rye (*Secale cereale*), common stinging nettle (*Urtica dioica*), common flax (*Linum usitatissimum*), Scotch pine (*Pinus sylvestris*), and soft rush (*Juncus effusus*). Some other types of bindings and weaves will also be tried in the framework of archaeological experimentation.

Acknowledgements

I am grateful to the National Museum of Lithuania for kind permission to publish Figures 35.2–35.5.

Bibliography

Girininkas, A. (2005) Neolitas. In A. Girininkas, A. Dubonis, Z. Kiaupa, J. Kiaupienė, Č. Laurinavičius, R. Miknys, G. Sliesoriūnas, and G. Zabiela (eds), *Akmens amžius ir ankstyvasis metalų laikotarpis. Lietuvos istorija, Vol. 1*, 103–196. Vilnius, Baltos lankos.

Kabailienė, M. (2006) *Gamtinės aplinkos raida Lietuvoje per 14 000 metų*. Vilnius, Vilniaus universiteto leidykla.

Rimantienė, R. (1989) *Nida: senųjų baltų gyvenvietė*. Vilnius, Mokslas.

Rimantienė, R. (2005) *Akmens amžiaus žvejai prie Pajūrio lagūnos: Šventosios ir Būtingės tyrinėjimai*. Vilnius, Lietuvos Nacionalinis Muziejus.

36 The Impact of Dyes and Natural Pigmentation of Wool on the Preservation of Archaeological Textiles

by Maj G. Ringgaard and Annemette Bruselius Scharff

Introduction

This article presents an experiment in progress investigating how mordants, dyes and natural pigmentation affect the deterioration of textiles buried in wet and waterlogged peat respectively, thus imitating the aerobic and anaerobic conditions found in a bog. The aim of the project is to document and analyse the changes that occur during burial in soil in wet conditions, and to investigate how the different factors such as textile fibres, mordants, dyestuffs and natural pigmentation affect the rates of deterioration. Some textiles may experience accelerated breakdown due to either fibre type or components within the fibres. The result of the experiments will help us to interpret what the archaeological brown rags may have looked like when they were in use. Furthermore, it can also improve our knowledge of the kind of textiles that can be found in a waterlogged archaeological excavation.

During the past 40 years, knowledge of the identification of fibres and dyes in archaeological textiles has improved greatly (Jørgensen and Walton 1986; Walton 1988; Vanden Berghe 2008). Much information on the use of dyes and fibres in ancient times has been gathered and, furthermore, burial experiments have added to our knowledge of the deterioration processes of textiles in different soils (Needles et al. 1986; Needles 1987; Janaway 2002; Peacock 2004).

However, the subject is vast. In previous works, the focus has been on the preservation of the dyestuff itself: if, and to what extent, the dyes fade or change colour, and only little has been written on how the dyes affect the preservation of the fibres. In this project, we have expanded the range of materials tested to gain further knowledge of how the dyes, mordants and natural pigmentation change in the soil and affect the degree of deterioration of wool and silk.

Many well-preserved prehistoric textiles have been recovered in the northern European countries due to the preservative effects of the waterlogged or wet soil in bogs, but the textiles have lost their original colours and are preserved in different shades of brown (Munksgård 1974; Hald 1980). By performing these experiments, we hope to find, if not the key, then at least an aid to recreate the colours of the past. Furthermore, we hope to deepen our knowledge of the selective deterioration that occurs in soil and, thereby, to better understand why some fibres are rarely or seldom found.

It is well known that protein fibres are better preserved in waterlogged soil compared to cellulose fibres, when we are dealing with slightly acidic conditions. Thus, hemp and linen are seldom found in wet soil where wool and silk textiles are preserved. However, textile finds show that protein fibres also may vary in degree of deterioration depending on the colour of the yarn. Fragments from Lønne Hede (1st century AD) clearly show this phenomenon. The striped fragment in Figure 36.1 is made of the same type of wool, but while the white wool has nearly disappeared, the brown wool seems very well preserved (Fig 36.1).

These observations show that the coloured yarns must contain something that prevents their deterioration. This preservative could, for example, be metal salts (mordants), dyestuffs or natural pigmentation, all of which may inhibit deterioration. An analysis to clarify this phenomenon is being conducted by Scharff and results will be published later. Here, we take the reverse approach to the problem. We start with

Fig. 36.1. Textile from Lønne Hede, Jutland, dated to the 1st century AD. The wool in the bright bands is nearly disintegrated, while the darker bands of natural pigmented fibres are better preserved (Photo: M. G. Ringgaard and A. B. Scharff).

Fabric	Ronald-say wool				Green-land wool				Hand spun wool 2/2 twill							Hand Spun wool			TF silk			Test fabric (TF) wool							
Dye/Mordant/pigmentation	I	II	III	IV	V	II	III	–	O	N	III	=	I+II	I+III	–	O	=	I+II	–	O	I+II	–	O	N	III	=	I+II	I+III	I+V	–	O
0 Undyed	×		×	×		×	×		×	×	×	×			×	×			×			×	×	×	×					×	×
1 Madder						×											×		×											×	
2 Cochenille											×					×			×	×										×	×
3 Brazilin											×						×		×											×	
4 Indigo	×	×					×		×										I+V ×	×										I+V ×	
5 Weld										×		×	×	×					×					×		×	×		×		
6 Walnut																×															×
7 Oakwood																×															×
8 Gallnut						×										×			×							×					×
9 Other									•																						

Table 36.1. Treatments of fabrics.

known dyes, try to imitate archaeological deterioration and see if we are still able to identify dyes or mordants. Several silk and wool textiles, with and without pigmentation, were treated with mordants and dyestuff before burial in wet and waterlogged peat. Here we present the first results after 8 months of burial and focus on the visible colour change and the macroscopic and microscopic observations of natural pigmented wool plus wool dyed with madder and indigo.

Materials and Methods

The experiment was carried out with 5 different types of wool (3 non-pigmented and 2 both pigmented and non-pigmented), plus a natural silk fabric. The treatments of the fabrics are summarized in Table 36.1, where the mordants are given Roman numerals (I: 18% Alum ($K_2O_4Al_2(SO_4)_3$), II: 5% Iron ($FeSO_4 7H_2O$), III: 10% Copper ($CuSO_4 5H_2O$), IV: 10% Cream of Tartar ($KC_4H_5O_6$), V: 2% Tin ($SnCl_2 2H_2O$). Some dyed samples had smaller pieces of undyed white wool attached to investigate if dyes or mordants are able to migrate from one fabric to another (Table 36.1).

Sixty different combinations of fabrics, mordants and dyes have been prepared for burial in wet and waterlogged peat respectively. The textile samples were placed horizontally, 2 layers in each box: peat (8 cm), samples (Layer 2), peat (8 cm), samples (Layer 1), peat 12 cm. The boxes, imitating anaerobic conditions, were watered with tap water ensuring saturation of peat and fabrics. The pH in the boxes was slightly acidic, pH 6. These boxes were covered with lids. Boxes imitating aerobic conditions were watered monthly and without lids.

The boxes were placed in a greenhouse at stable temperature and humidity. A greenhouse was chosen because this gave the opportunity for a slightly raised temperature accelerating the processes. Textiles are excavated after 8, 16 and 24 months (Fig. 36.2).

Methods used for the determination of the degree of deterioration include visual colour changes documented by digital photography, specification of colours using

Fig. 36.2. The boxes with textiles buried in peat imitating the condition of a slightly acidic bog are placed in a greenhouse at stable temperature and humidity (Photo: M. G. Ringgaard and A. B. Scharff)

the CIELAB and Munsell colour systems; observation of fabric surfaces in a stereo microscope; observation of fibre morphology, colour and pigmentation in transmitted light microscopy, and observation of fibres in SEM; quantitative and qualitative element analysis using SEM-EDX and IPC-MS; and the testing of dyestuff remains with HPLC.

Preliminary Results

The first 2 boxes have been excavated after 8 months. The textile samples were gently rinsed in deionised water and freeze dried (two month in a normal chest freezer). The samples from waterlogged conditions are shown in Figures 36.3–36.4.

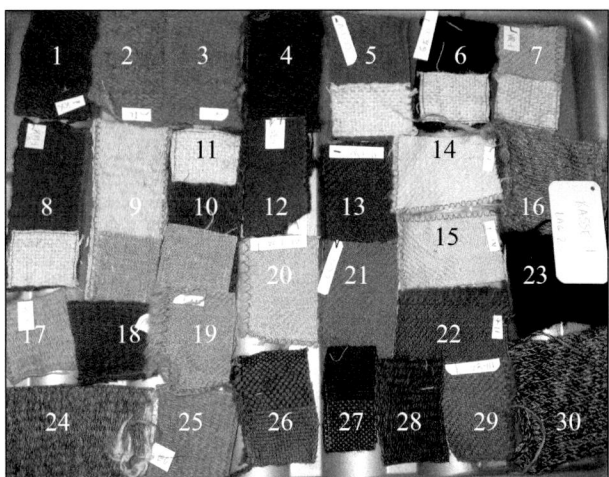

Fig. 36.3. Textile samples, white or natural brown wool, with different dyes and mordants before burial (Photo: M. G. Ringgaard and A. B. Scharff). Reproduced as a colour plate on page 305.

Fig. 36.4. Samples after 8 months burial under waterlogged conditions in closed boxes (Photo: M. G. Ringgaard and A. B. Scharff). Reproduced as a colour plate on page 305.

1 Lichen *Ochrolechia tartara*	6 Gallnut +(Fe)	11 Ronaldsay wool	16 Greenland wool Pigmented	21 Cochenille+(Al)	26 Indigo
2 Lichen *Pamelia omphalodes*	7 Weld +(Al) *Reseda luteola*	12 Brazil wood +(Al) *Caesalpinia sappan*	17 Gallnut	22 Iron (Fe)	27 Indigo
3 Lichen *Cetraria islandica*	8 Madder +(Al) *Rubia tinctoria*	13 Weld +(Al)+ (Fe)	18 Walnut	23 Gallnut +(Fe)	28 Indigo
4 Fungus *Dermocybe sanguinia*	9 Greenland wool	14 Alum (Al)	19 Copper (Cu)	24 Greenland wool pigmented	29 Weld +(Cu)
5 Cochenille +(Al)+(Sn) *Dactylopius coccus*	10 Indigo *Indigofera tinctoria*	15 Cream of Tartar (IV)	20 Weld +(Al)+IV	25 Oak wood	30 Ronalsdsay Pigmented

Table 36.2. Legend to the samples shown in Figs. 36.3–36.4.

Fig. 36.5. SEM image of natural white wool (test fabric) after 8 months burial (Photo: M. Taube and M. Ringgaard).

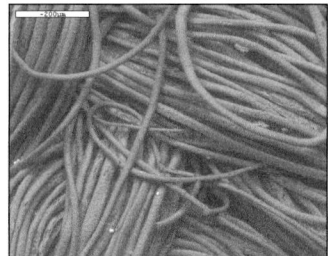

Fig. 36.6. SEM image of alum mordanted madder dyed wool (test fabric) after 8 months burial (Photo: M. Taube and M. Ringgaard).

The general picture is that nearly all samples have changed considerably in colour, except for wool and silk dyed with madder, and the lichen and fungus dyed wool which only show few changes (Figs 36.3–36.4).

The white samples have turned beige. Yellow samples dyed with weld have lost their bright colour and turned pale greyish yellow. Red samples dyed with cochineal and Brazil wood still have the red hue, but are considerably paler.

Indigo samples are all pale blue to yellow-beige. The greatest colour changes are observed in silk samples and the woollen test fabric samples (thin fabrics), while the thicker wool fabrics were changed to pale blue. Another remarkable observation was the difference between open and closed boxes. The blue colours were better preserved in the boxes without lids compared to samples in boxes with lids. It looks as if the anoxic conditions change indigotin into its leuco state. The mordanted samples have changed colour, too: the black and dark colours with iron mordant have turned much brighter, while samples with copper mordant have turned from pistachio green to dark olive (Table 36.2).

Until now, only some of the samples from the waterlogged burials have been analysed under the microscope. Most of the analysed wool fibres display signs of biodegradation, of pitting and erosion of the surface. Scale patterns are less visible than on the non-degraded reference samples, in some places the scales are completely eroded.

An exception is alum mordanted wool samples dyed with madder, which only display little changes, both in the waterlogged (closed box) and wet (open box) samples. The samples of the same white wool fabric, both the untreated and the alum mordanted, look very much deteriorated when observed in SEM (Figs 36.5–36.6).

It is not possible to detect any difference between naturally brown and white North Ronaldsay wool when observed under a transmitted light microscope. Although some archaeological textiles indicate that pigmentation protects fibres from degradation, this could not be seen in these experiments. There were no visible differences in the preservation of the natural brown and the white wool.

Conclusions

The colour change observed in the indigo-dyed samples corresponds well to the observations of samples from the Lønne Hede excavation and from Scottish bog finds. Here, indigotin was identified in textiles which appeared visibly uncoloured (Surowiec et al. 2006, 213). This indicates a reduction of the indigo in anaerobic conditions, changing indigotin into its uncoloured leuco state.

In some archaeological excavations (18th century garbage dump at Churchillparken, Copenhagen) reddish textiles were found still coloured and well preserved among several deteriorated brownish textile fragments.[1] Furthermore, Elizabeth Peacock has conducted experiments with dyed and undyed fulled wool fabric, buried for 8 years in two kinds of bogs (one raised and one fen). Over the years, samples have been excavated, and the results indicate that dyed samples are better preserved than the undyed. Her experiments showed a stepwise deterioration, with madder dyed samples best preserved, then indigo, weld and finally uncoloured samples (Peacock, personal communication 2007). In our experiments, SEM pictures of the alum mordanted wool textile dyed with madder showed only little degradation, while the fibres in the undyed fabric showed severe signs of degradation. Fibres from the sample with only alum but no dye showed similar degradation as the undyed sample. These results clearly indicate the protective, preservative effect of madder.

We found that, in some of the textiles, the dyes did not penetrate the inner core of the yarns. This was not visible before the burial. Degradation of the outer layers of fibres exposed the poorly dyed core of the treads. This fibre deterioration was most prominent at the edges of the textile. These parts, damaged frayed areas and edges, are those where samples for dye analyses usually are taken from archaeological textiles, suggesting that there is a great risk of a blank test result even if the textile had been dyed in the past. The darker colour of the textiles with copper mordants could be due to the formation of black copper sulphides.

Future Perspectives

The results of the experiments can contribute to our knowledge of ancient textiles, for instance, by demonstrating that brown textiles from an archaeological context may originally have been coloured. They also enhance our understanding of which textiles would survive, and thereby more likely to be found in archaeological excavations, compared to those textiles that experience an accelerated breakdown due to either fibre type or the components within the fibres.

The hues of the textile after excavation from the water-logged soil gives no indication of the original colour of the textile, except in the case of the reds: after burial, many of the undyed textiles had similar colours as the dyed textiles. The nuance of brown seems to be more dependent on the mordant used, than the original colour of the textile.

We envisage further investigations to detect mordants by SEM-EDX and ICP-MS, and to determine if mordants migrate, or if there is an ion exchange between wool and soil. Moreover, dyes will be detected by HPLC (and TLC) UV/VIS spectrometry.

Acknowledgements

The authors wish to thank Theodor Bøsterli at the Faculty of Life Sciences, University of Copenhagen for housing the boxes with samples in their greenhouses; Michelle Taube, National Museum of Denmark, for the SEM pictures; and Elizabeth Peacock for sharing her expertise in experimental burials of textiles.

Note

1 Maj Ringgaard has investigated this phenomenon in the course of her research for her PhD thesis.

Bibliography

Hald, M. (1980) *Ancient Danish Textiles from Bogs and Burials*, Publication of the National Museum, Archaeological-Historical Series vol. XXI, Copenhagen.

Janaway, R. C. (2002) Degradation of Clothing and Other Dress Materials Associated with Buried Bodies of Archaeological and Forensic Interest. In W. D. Haglund and M. H. Sorg (eds), *Advances in Forensic Taphonomy: Method, Theory, and Archaeological Perspectives*, 379–402. Boca Raton, CRC Press.

Jørgensen, L. B., and Walton, P. (1986) Dyes and Fleece Types in Prehistoric Textiles from Scandinavia and Germany. *Journal of Danish Archaeology* 5, 177–188.

Munksgård, E. (1974) *Oldtidsdragter*. København, Nationalmuseet.

Needles, H., Cassman, V., and Collins, M. J. (1986) Mordanted, Natural-Dyed Wool and Silk Fabrics. Light and Burial-induced changes in the color and tensile properties. In *Historic textiles and paper materials: conservation and characterization, Advances in Chemistry, Series 212.* ACS, 199–210.

Needles, H. L. (1987) Burial-induced color changes in unmordanted and mordanted alizarin-dyed cotton and wool fabrics. In *AIC Preprints presented at the 15th meeting, Vancouver, British Columbia, Canada, May 20–24*, 78–84. Washington D.C.

Peacock, E. E. (2004) Moseforsøg – Two generations of Bog Burial Studies. Interim Textile Results. In J. Maik (ed.), *Priceless Invention of Humanity – Textiles! NESAT VIII*, 185–194. Acta Archeologica Lodziensia 50/1, Łódź.

Surowiec, I., Quye, A., and Trojanowicz, M. (2006) Liquid chromatography determination of natural dyes in extracts from historical Scottish textiles excavated from peat bogs. *Journal of Chromatography* A 1112, 209–217.

Walton, P. (1988) Dyes and Wools in Iron Age Textiles from Norway and Denmark. *Journal of Danish Archaeology* 7, 144–158.

37 Wear on Magdalenian Bone Tools: A New Methodology for Studying Evidence of Fiber Industries

by Elisabeth Ann Stone

Introduction

The role of textiles within contemporary, historic, and late prehistoric communities is widely acknowledged to be socially, culturally, and economically important (Schneider 1987; Croes 1997). In nearly every known context, vegetal resources are manipulated into forms that include cordage, baskets, nets, mats, and knitted, knotted and woven fabrics. These various objects serve a number of utilitarian functions as well as forming rich, malleable technologies in which gender, social class, occupation and many other social identities are expressed. However, despite their modern near-ubiquity and late prehistoric importance, little is known about the early exploitation of vegetal fibers or the development of such perishable technologies. This is unfortunate, because the simple presence or absence of such an enormous class of raw materials has vast implications for our understanding of technology, ideology, social relations, economy and environment. When did this technology come into being? Generally, it has been assumed that Paleolithic populations did not use many vegetal resources. Reconstructions of even Late Upper Paleolithic (LUP) life often reflect a vision of a harsh world of big-game hunters living off their kill, in terms of both subsistence and raw materials for a variety of objects.

The goal of my PhD research is to identify ways in which the manufacture of objects from plant-derived materials – cordage, basketry, mats, or even woven fabrics – can be identified in contexts where these objects are not preserved in today's archaeological record. Although the identification of specific forms is, at present, far beyond the scope of the research design, the basic question of whether or not bone and antler technologies were used to prepare and manipulate vegetal fibers would be of immense help in further identifying areas where early fiber technologies existed. As yet, it is still unknown if even vegetal cordage or matting was produced in the LUP of Western Europe.

Perishables in the Upper Paleolithic

Perishable materials comprise up to 95% of material culture in ethnographically documented forager groups (Osgood 1940; Damas 1984; Owen 1993; 2005). Archaeological research at wet sites and other sites of extraordinary preservation indicates that the remains found in most archaeological contexts may be far from representative of the complete array of items used by prehistoric populations (Tuross and Fogel 1994; Croes 1997; Nadel *et al.* 1994; 2004). However, the role of plant materials in the Upper Paleolithic (UP) has still been difficult to identify and interpret. The preservation of organic materials in the archaeological record is generally poor and standard archaeological methods are not tailored to the recovery of these materials. The UP archaeological record is dominated by stone and bone artifacts, facilitating research on subsistence, mobility, human cognitive evolution and adaptation to environmental conditions. Ethnographically, plant use is less well documented than are hunting and other predominantly male activities, as archaeology and ethnography have historically been dominated by male researchers who often focused on men's work. These factors have limited our understanding of female roles (Kehoe 1990; Conkey 1991; Hurcombe 1994; Owen 1999; 2005). However, Owen's (2005) examination of modern foragers in a range of ecozones demonstrates that plants play an important role in economics and subsistence, even in areas where plant life is not abundant (Lee and DeVore 1968; Hawkes *et al.* 1997; Lupo and Schmitt 2002).

Finally, there are some very specific reasons to believe that vegetal fibers may have begun to play a role in the material culture of Western Europe during the LUP. Castro Curel (1990) suggests that certain depictions on engraved plaquettes from Late Magdalenian levels at Parpalló (eastern Spain) are images of vegetal fibers and that images previously identified as snakes are actually twined cordage. Lines crossed at right angles she identifies as woven textiles. One image of a spiral with diagonal lines extending between the sections of the spiral she identifies as coiled basketry. Bahn has argued that many of the enigmatic 'signs' in UP art are images of plants or organic artifacts, including textiles (Tyldesley and Bahn 1983; Bahn 2001). Straus (1992) has suggested that some grid symbols in Cantabrian cave art in Northern Spain could represent nets used in fishing or game drives. Soffer and her

Fig. 37.1. Two Magdalenian bone needles found within centimeters of each other in the same level at El Mirón (© Photo: E. Stone).

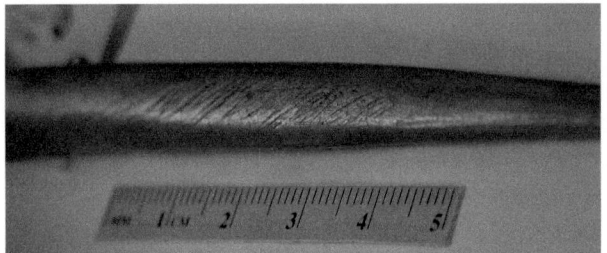

Fig. 37.2. Zuni Weaving Needle; AMNH 50.1/8789 (Courtesy of the Division of Anthropology, American Museum of Natural History; Photo: E. Stone).

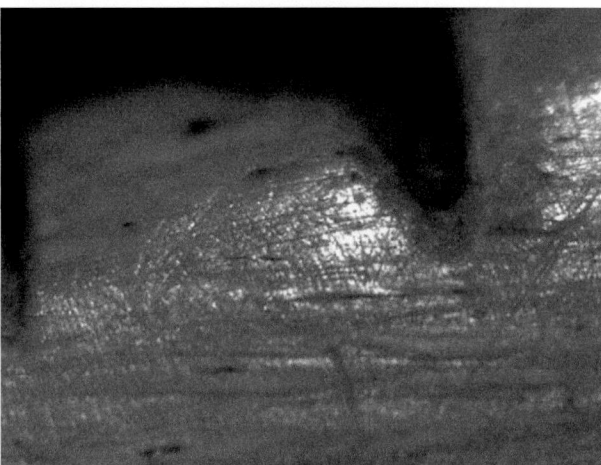

Fig. 37.3. Hawikuh Weaving Batten NMAI 066490 (Courtesy of the National Museum of the American Indian, Smithsonian Institution; Photo: E. Stone).

Fig. 37.4. Cree bone hide-scraper NMAI 253335 (Courtesy, National Museum of the American Indian, Smithsonian Institution; Photo: E. Stone).

colleagues (Soffer *et al.* 2000a) have studied female figurines from Eastern Europe and argue that they are dressed in apparel that has been twined or knit. The oldest known preserved plant-fiber technologies come from the Israeli site of Ohalo II, where cordage and bedding dated to 21,000 BC were preserved in a waterlogged context (Nadel *et al.* 1994; 2004). Rope preserved as an apparent natural cast in the French cave of Lascaux is dated to *c.* 20,000 BC and may have been made from plant materials (Glory 1959; Leroi-Gourhan and Allain 1979), while the Pavlovian sites of Moravia (Czech Republic) have yielded clay fragments imprinted with cordage, knots and woven fabrics from levels dated to 30,000 BC (Grigor'ev 1993; Adovasio *et al.* 1996; 2001; Soffer *et al.* 2000b). Kehoe (1990; 1999) has argued that many UP osseous artifacts interpreted as hunting implements were actually used in the production of textiles. Owen's research (1993; 1994; 1999; 2005) on the division of labor in the Magdalenian of Germany and her studies of bone needles and awls also suggest that plant materials played a significant role in LUP material culture. Soffer (2004) identified wear on osseous *sagaies* ('points') from throughout Ice Age Europe, which suggests that these objects were really used in the production of cordage and weaving. In this preliminary study, she focused on certain antler tools generally interpreted as projectile points and on other functionally ambiguous artifacts, such as modified ribs. Wear identification was based on comparison with ethnographic battens and other weaving tools.

Vegetal Resources in the Late Upper Paleolithic of Northern Spain

Although we often envision Paleolithic people dressed in furs and hides as they produce stone and bone instruments, there is extensive evidence that the extraction of vegetal fibers and the manufacture of textiles and basketry were, at a minimum, possible. Bone and lithic technologies from the LUP are extremely complex and sophisticated; the production of perishable technologies was undoubtedly also highly developed. This period is also known for a dramatic increase in the number and variability of osseous tool forms (Fig. 37.1), the function of many of which remains unknown. Despite the stereotype of the LUP climate as a cold, dry, windy heath-steppe-tundra, the necessary resources for the extraction of vegetal resources were available. The Magdalenian began under cold conditions but sedimentation and pollen spectra indicate that, while still situated in the Ice Age, the period was marked by generally rising temperatures and humidity, punctuated by several cold shifts (Butzer 1971; González Morales and Straus 2005; Straus 2005). Pollen spectra from Late Upper Paleolithic sites in Northern Spain indicate the

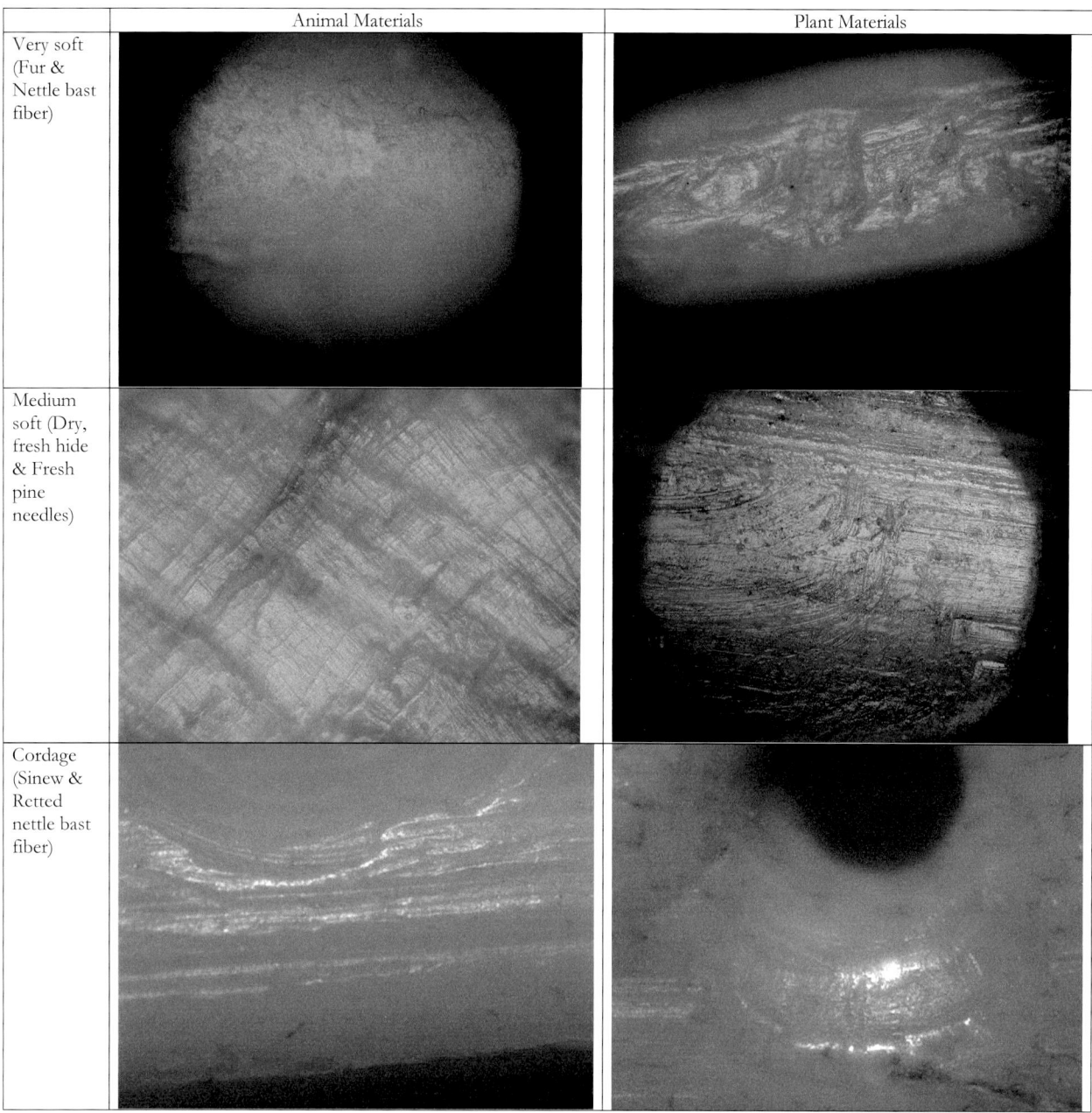

Fig. 37.5. Use-wear from experiments with distinct soft materials on bone (© Photo: E. Stone).

local availability of fiber sources such as nettle and willow that could have been incorporated into the local material culture (*e.g.* Boyer-Klein 1981; Freeman and Gonzalez Echegaray 1995). Bast fibers from plants such as nettle, flax and hemp, some of the earliest documented fibers used in the production of woven cloth and cordage (Hald 1942; Barber 1991), are not physically linked to a spine or point that could be used as a natural needle and would be sewn with the aid of a tool. Possible sources of vegetal fibers in the LUP of Northern Spain include:

Urtica: nettle – bast fibers from this plant can be extracted directly or by retting and can be twisted into cordage;

Pinus: fir – traditional uses documented ethnographically include the use of twigs and roots for basketry production, boiled roots used as cords;

Salix: willow – traditional uses documented ethnographically include the production of twine and fine threads used for clothing production from bark fibers, basketry production, dye;

Populus: aspen, cottonwood, poplar – traditional uses documented ethnographically include basket production, use of bark to make a fiber for rope and clothing production, as well as dye;

Quercus: oak – traditional uses documented ethnographically include as a dye, a mordant, a tanning chemical for hides and a basketry material;

Corylus: hazelnut – traditional uses documented ethnographically include basket production, rope production from twigs, and dye.

Museum	Origin Area	Eyed Sewing Needles	Eyed Snowshoe Needles	Eyed Mat Needles	Eyed Fish Needles; Large Needles Use Unk.	Completely Worked Awls	Articular Awls	Basketry Awls	Weaving Awls	Bone Points	Pins, Bodkins	Wound Plugs, Pegs	Hide Scrapers	Worked Rib Tools	Total
AMNH	Arctic		5		8	14	5			11	1	1			45
	Calif. Coast						9	1							10
	NE N. Am.		2			1	6					5			14
	Pacific NW			1	1	7	27			4		12			52
	N. Am. Plains	1	4	10			2		1						18
	SW N. Am.					2	9	1	8						20
	Total AMNH	1	11	11	9	24	58	2	9	15	1	18			159
NMAI	Arctic	4			6		4			6	11			5	36
	Calif. Coast			1	1	2	2			2	1	13		1	23
	NE N. Am.	8	66	21	4	8	10			2			2	1	122
	Pacific NW			6			4			1					11
	N. Am. Plains	2		22	1	4	2			1		4	1		37
	SW N. Am.						13								13
	S. Am.	1					6	4	2						13
	Total NMAI	15	66	50	12	14	37	8	2	12	12	17	3	7	255
NMNH	Arctic	21	7		1		2				1		4		36
	Calif. Coast						1	1							2
	NE N. Am.	10		1											11
	Pacific NW			1			1	2							4
	N. Am. Plains			5											5
	SW N. Am.	1				5	2	3	3						14
	S. Am.	2				1	1	1							5
	Oceania			3	1		2	4	1						11
	Siberia				2										2
	Total NMNH	34	7	10	4	6	9	11	4	1		4			90

*types drawn from museum catalogs; **AMNH = American Museum of Natural History; NMNH = Smithsonian National Museum of Natural History; NMAI = Smithsonian Museum of the American Indian; Am. = America; Calif. = California

Table 37.1. Ethnographic Collections.

Ethnographic Collections

In ethnographic contexts bone and antler tools are widely used in the production of perishable technologies: hide-processing, basketry, weaving, and sewing (Fig. 37.2). Among the groups that created and used the osseous implements included in my own study are Comanche, Cree, Eskimo, Fox, Haida, Inuit, Kwakiutl, Menomini, Navajo, Pomo, various Pueblos, Salish, Seri, Tlingit, Winnebago, Yupík, Yurok (North America); Fuegian, Guaymi, Urubu (South America); Australian Aborigine, Chimbu, Micronesian (Oceania); and Siberian (Eurasia). This is by no means an exhaustive list of either my research or the peoples using such bone and antler objects, but rather shows the ubiquity of these tools in contemporary and late historic contexts. Tools employed in the production of perishable objects – basketry awls, hide-piercing awls, shuttles, net gauges, matting needles and weaving battens – can take forms similar to those of LUP artifacts (Densmore 1929; Kidder 1932; Osgood 1940; Campana 1989; Burnham 1992; Kehoe 1999; Soffer 2004). Research was done in four major ethnographic collections housed in the US: the American Museum of Natural History, New York, the Burke Museum, Seattle, and two Smithsonian Institutions, the National Museum of Natural History and the American Museum of the American Indian, both in Washington, D.C. All bone tools available and appropriate to the study were examined, allowing me to begin to identify the general differences between wear from vegetal and animal fibers (see Table 37.1 and Figs 37.3–37.4). Most studies of UP material use arctic populations as the ethnographic reference for comparison and analogy, due to climatic similarity (*e.g.* Binford 1978; 1980). However, given the climatic variation within the Magdalenian, the period between the Last Glacial

Resource Group	Resource	State	Form
Vegetal	Nettle	Retted 4 weeks Retted 8 weeks Stripped	Woven fabric Twisted Twined Knotted
	Pine needles	Fresh	Bundled Twisted
	Roots (pine, spruce)	Boiled Fresh	Flat Twisted
	Saplings	Green Soaked	
	Bark (willow, spruce, birch, oak)	On wood Fresh Soaked	Twisted Plaited
Animal	Skin	Fresh Dry Tanned	Flat Thong Knotted thong
	Fur		Un-worked (on hide) Braided
	Sinew	Dried and softened	Plain Twisted Twined Knotted

Table 37.2. Experiments.

Maximum and the onset of the Holocene, 18,000–12,000 years ago (these and all dates in calibrated radio-carbon years), other populations provide equally good, if not better, models for the potential technologies employed in the LUP. These collections, rather than providing analogs for archaeological forms will serve to illustrate the variability of uses of bone tools in the manufacture of fiber industries, and the range of uses of tools with similar forms to archaeological artifacts, and will provide samples of general wear from plant and animal fiber on bone surfaces. The use-wear on these objects was produced in a dynamic, day-to-day context by people who actually used these objects, which reflects not only the mechanical properties of the interacting materials, but also personal variation, developed motor skills and long-term, authentic utilization. The physical and mechanical properties of bone and fibers are not greatly affected by variation in environment, and considering a wider range of possibilities in the ethnographic record may allow the recognition of unexpected patterns on archaeological artifacts (Owen 1993).

Osseous tools for basketry, weaving, hide-working and sewing display a wide range of variation in the ethnographic record. In some cases, that variation reflects functional differences: needles for sewing sinew in clothing, rush roofing and hide snow-shoes vary dramatically in size and form. In other cases, a wide range of forms may serve the same purpose *even within the same human population*: bone needles created by Arctic populations for sewing hide and gut with sinew display a considerable amount of variation in form, shape and size, despite their overall similarity in function. Other forms may be utilized for one of multiple functions, despite being formally identical, while others are multi-purpose tools such as many awls. Finally, there are 'ideal' cases where a certain form is closely tied to one, and only one, function: snow shoe needles in northern North America are remarkably similar across regions, cultures and time.

Experimentation

As noted by Hurcombe (1994; 2008) and Owen (1999; 2005), identifying plant processing can be difficult because of the wide range of techniques and uses of plant materials and the lack of detail in ethnographic information on organic materials. As a complement to the study of ethnographic materials, experimentation was used to produce a collection of worn surfaces where variables including time used, direction of work, specific gestures used, and the state of raw materials could be controlled. This allows the association of particular wear patterns with defined parameters. Activities were only replicated in rare occasions (the use of a bone weaving batten, for example); instead, experiments focused on isolating the effects of different forms of contact between bone and soft, fibrous objects (see Table 37.2).

My experiments show that different soft materials create distinct wear patterns (Fig. 37.5). For example, perforations on elk long-bone abraded for one hour with plied cordage made from retted nettle fiber can be distinguished from those abraded with sinew. Nettle cordage produces a developed, non-invasive polish. Striations are wide and deep and run parallel or sub-parallel to the direction of work. Wear from sinew is invasive, with a less developed polish visible on high and low points of the original surface topology. Few striations are visible, and those that are present are finer than those left by nettle cordage. Other researchers have had similar results. LeMoine's (1997) experimentation with braided sinew on bone tools produced a smooth polish with separate, distinct, long striations. Buc (2005) identified an invasive polish with fine, occasionally intercrossed striations from working hide with bone tools. Bone tools used on plant fibers, however, had non-invasive polish with approximately parallel striations. Campana's (1989) experimentation with Neolithic bone and antler points led him to suggest that finely tipped tools were used on fine or loose materials, such as woven fabrics, while tools with thick tips were probably used on leather or other strong materials.

Use-Wear on Needles, Awls, and other Osseous Artifacts

I have focused my research on two major groups of artifacts: bone needles and awls. One of the primary reasons is that these tools are both abundant and nearly ubiquitous in archaeological sites in the north of Spain. Bone preservation is often good in Magdalenian sites situated in karstic caves, and while this surely does not provide a complete picture of the range of sites and activities during the epoch, it does provide an easily identifiable sample of objects with high potential for the preservation of wear. Eyed needles first appear in Western Europe, c. 20,000 years ago. Sites in Moravia have yielded needles 8000 years older, along with clay fragments imprinted with cordage, knots and woven fabrics. The new tools suggest a change in the production sequence of soft materials which is likely to be linked to other changes. Ethnographic evidence shows that sinew can be sewn easily and effectively without the aid of a needle, using only a fine awl. Sinew that has been moistened and dried near heat is hard and can be guided through a perforation with the same ease as a bone needle (Amato in press). However, most types of vegetal cordage would be sewn with the aid of a tool, including bast fibers from plants such as nettle, flax and hemp – possibly some of the earliest fibers used in the production of woven cloth and cordage (Barber 1991). Stettler (1998) noted a decrease in the minimum size of needle perforations between the Solutrean (during the cold snap known as the Last Glacial Maximum, 21,000–18,000 years ago) and the Magdalenian (18,000–12,000). Macroscopic analysis of the collection used for this project indicates that the range of variation in Magdalenian bone needles (Fig. 37.1) fluctuated during the period; width – a variable with clear functional implications – varied little at the beginning of the Magdalenian, followed by a period of extreme variability that then stabilized near the end of this period.

Using the standards provided by ethnographic collections and experimentation, a study of archaeological assemblages of osseous tools from Northern Spain is underway. Macroscopic observations have been made on artifacts from Mirón, Entrefoces, Castillo, Altamira, Pila, Rascaño, and Juyo; microscopic observations have been completed for the latter four sites. Preliminary observations indicate that wear preserved on many of these artifacts is indicative of their use on soft surfaces. Many of the non-invasive polishes that may indicate vegetal resources are present but the more detailed comparisons of polish variations and the characterization of striations are yet to be completed.

Implications

An understanding of the role of osseous technology in the preparation of perishable artifacts will contribute to a broader understanding of the entire suite of technologies and activities employed by Ice Age occupants of Western Europe. The dominant image of LUP individuals revolves around the hunt and all that goes with it: men, large trophy animals, hide working, danger and weapons. Although these elements did contribute to the lives of LUP populations, that resulting impression is greatly impoverished by its simplicity. Ethnographically, women and men generally do different tasks (*e.g.* Lee and DeVore 1968; Burnham 1992; Kaplan *et al.* 2000; Hawkes and Bliege Bird 2002). Waguespack (2006) contends that as dependence on hunting increases, women's activities shift in emphasis from subsistence to manufacture. In LUP Europe, where hunting of large game was a primary subsistence source, women should have designated a significant amount of their time to manufacturing, among other things, perishable goods. If plant materials were regularly used to produce fiber technologies during the LUP of Western Europe, a holistic, more balanced understanding of this period is impossible without studies aimed at both direct and indirect evidence of such activities.

Acknowledgements

This research is supported by the National Science Foundation through a Graduate Research Fellowship and a Doctoral Dissertation Improvement Grant, as well as a grant from the Office of Graduate Studies of the University of New Mexico. Permission to study collections from the following institutions is greatly appreciated: American Museum of Natural History, Burke Museum, El Mirón Prehistoric Project, Maison Méditerranéenne des Sciences de l'Homme, Museo de Altamira, Museo de Cantabria, Museo de Oviedo, National Museum of the American Indian, National Museum of Natural History, Universidad de Cantabria.

Bibliography

Adovasio, J. M., Soffer, O., and Klima, B. (1996) Upper palaeolithic fibre technology: Interlaced woven finds from Pavlov I, Czech Republic, c. 26,000 years ago. *Antiquity* 70(269), 526–534.

Adovasio, J. M., Soffer, O., Hyland, D. C., Illingworth, J. S., Klima, B., Svoboda, J. (2001) Perishable Industries from Dolní Vestonice I: New Insights into the Nature and Origin of the Gravettian. *Archaeology, ethnology, and anthropology of Eurasia* 2:6, 48–65.

Amato, P. (in press) Sewing with or without a needle in the Upper Palaeolithic? In I. Sidera *et al.* (eds), *Proceedings of the 6th Meeting of the ICAZ Worked Bone Research Group in Paris, 27th August–1st*

September 2007. Colloques de la Maison René-Ginouvès. Paris, De Boccard editions.

Bahn, P. G. (2001) Palaeolithic weaving – a contribution from Chauvet. *Antiquity* 75, 271–272.

Barber, E. W. (1991) *Prehistoric Textiles: The Development of Cloth in the Neolithic and Bronze Ages with Special Reference to the Aegean*. Princeton, Princeton University Press.

Binford, L. R. (1978) *Nunamiut Ethnoarchaeology*. Academic Press.

Binford, L. R. (1980) Willow Smoke and Dogs' Tails: Hunter-Gatherer Settlement Systems and Archaeological Site Formation. *American Antiquity* 45:1, 4–20.

Boyer-Klein, A. (1981) Análisis palinológico del Rascaño. In J. G. Echegaray and I. B. Maestu (eds), *El Paleolítico Superior de la Cueva del Rascaño (Santander)*, 214–220. Monografías del Centro de Investigación y Museo de Altamira. Vol. 3. Santander, Ministerio de Cultura, Dirrección General de Bellas Artes, Archivos y Bibliotecas.

Boyer-Klein, A., and Leroi-Gourhan, A. (1985) Análisis palinológico de la Cueva del Juyo. In I. Barandiarán, *et al.* (eds), *Excavaciones en la Cueva del Juyo*, 55–61. Madrid, Ministerio de Cultura.

Buc, N. (2005) *Análisis de microdesgaste en tecnología ósea. El caso de punzones y alisadores en el noreste de la provincia de Buenos Aires (humedal del Paraná inferior)*. Universidad de Buenos Aires.

Burnham, D. K. (1992) *To Please the Caribou: Painted Caribou-Skin Coats Worn by the Naskapi, Montagnais, and Cree Hunters of the Quebec-Labrador Peninsula*. Seattle, University of Washington Press.

Butzer, K. W. (1971) *Environment and archeology; an ecological approach to prehistory*. Chicago, Aldine.

Campana, D. V. (1989) *Natufian and Protoneolithic Bone Tools: The Manufacture and Use of Bone Implements in the Zagros and the Levant*. BAR Int. Ser. 494. Oxford.

Castro Curel, Z. (1990) Información gráfica en plaquetas del Parpalló: consideraciones sobre inicios de tecnologías vegetales. *Cypsela* VIII, 15–20.

Conkey, M. W. (1991) Contexts of Action, Contexts for Power: Material Culture and Gender in the Magdalenian. In J. M. Gero, and M. W. Conkey (eds), *Engendering Archaeology: Women and Prehistory*, 57–92. Oxford, Basil Blackwell.

Croes, D. R. (1997) The north-central cultural dichotomy on the Northwest coast of North America: its evolution as suggested by wet-site basketry and wooden fish-hooks. *Antiquity* 71, 594–615.

Damas, D., ed. (1984) *Arctic*. Washington, D.C., Smithsonian Institution.

Densmore, F. (1929) *Chippewa Customs*. Smithsonian Institution Bureau of American Ethnology Bulletins 86. Washington, D.C., Government Printing Office.

Freeman, L. G., and Gonzalez Echegaray, J. (1995) The Magdalenian Site of El Juyo (Cantabria, Spain): Artistic Documents in Context. *Bollettino del Centro Camuno di Studi Preistorici* 28, 25–42.

Glory, A. (1959) Débris de corde Paléolithique à la Grotte de Lascaux (Dordogne). *Memoires de la Société préhistorique française* 5, 135–169.

González Morales, M. R., and Straus, L. G. (2005) The Magdalenian sequence of El Mirón Cave (Cantabria, Spain): an approach to the problems of definition of the Lower Magdalenian in Cantabrian Spain. In V. Dujardin (ed.), *Industrie osseuse et parures du Solutéen au Magdalénien en Europe*, 209–219. Mémoire de la Société préhistorique française XXXIX. Paris, Société préhistorique française.

Grigor'ev, G. P. (1993) The Kostenki-Avdeevo Archaeological Culture and the Willendorf-Pavlov-Kostenki-Avdeevo Cultural Unity. In O. Soffer and N. D. Praslov (eds), *From Kostenki to Clovis: Upper Paleolithic–Paleo-Indian Adaptations*, 51–65. New York City, Plenum Press.

Hald, M. (1942) The Nettle as a Culture Plant. *Folk – Liv. Acta Ethnologica et Folkloristica Europaeae* VI, 28–49.

Hawkes, K., and Bliege Bird, R. (2002) Showing off, handicap signaling, and the evolution of men's work. *Evolutionary Anthropology* 11:2, 58–67.

Hoffecker, J. F. (2005) Innovation and technological knowledge in the Upper Paleolithic of Northern Eurasia. *Evolutionary Anthropology* 14:5, 186–198.

Hurcombe, L. (1994) Plant-Working and Craft Activities as a Potential Source of Wear Variation. *Helinium* XXXIV:2, 201–209.

Hurcombe, L. (2008) Organics from inorganics: using experimental archaeology as a research tool for studying perishable material culture. *World Archaeology* 40(1), 83–115.

Kehoe, A. B. (1990) Points and Lines. In S. M. Nelson and A. B. Kehoe (eds), *Powers of Observation: alternative views in archeology*, 23–37. Archeological Papers of the American Anthropological Association. Vol. 2. Washington, D.C., American Anthropological Association.

Kehoe, A. B. (1999) Warping Prehistory: Direct Data and Ethnographic Analogies for Fiber Manufactures. *Urgeschichtliche Materialhefte* 14, 31–41.

Kidder, A. V. (1932) *The Artifacts of Pecos*. Foundations of Archaeology. Clinton Corners, Percheron Press.

Lee, R. B., and DeVore, I. (1968) *Man the Hunter*. Aldine.

LeMoine, G. M. (1997) *Use Wear Analysis on Bone and Antler Tools of the Mackenzie Inuit*. BAR Int. Ser. 679. Oxford, Hadrian Books.

Leroi-Gourhan, A. (1986) The Palynology of La Riera Cave. In G. A. Clark and L. G. Straus (eds), *La Riera Cave: Stone Age Hunter-Gatherer Adaptations in Northern Spain*, 59–64. Anthropological Research Papers. Vol. 39. Tempe, Arizona State University.

Leroi-Gourhan, A., and Allain, J. (1979) *Lascaux Inconnu*. Paris, France, Centre National de la Recherche Scientifique.

McCorriston, J. (1997) The Fiber Revolution: Textile Extensification, Alienation, and Social Stratification in Ancient Mesopotamia. *Current Anthropology* 38:4, 517–549.

Nadel, D., Danin, A., Werker, E., Schick, T., Kislev, M. E., and Stewart, K. (1994) 19,000-Year-Old Twisted Fibers from Ohalo-II. *Current Anthropology* 35:4, 451–457.

Nadel, D., Weiss, E., Simchoni, O., Tsatskin, A., Danin, A., and Kislev, M. (2004) Stone Age hut in Israel yields world's oldest evidence of bedding. *Proceedings of The National Academy of Sciences Of The United States Of America* 101:17, 6821–6826.

Osgood, C. (1940) *Ingalik material culture*. New Haven, Yale University Press.

Owen, L. (1993) Materials worked by hunter and gatherer groups of northern North America: implications for use-wear analysis. In P. C. Anderson, S. Beyries, M. Otte and H. Plisson (eds), *Traces et fonction: les gestes retrouvés*, 3–12. Études et Recherches Archéologiques de l'Université de Liège. Vol. 1. Liège, Centre de Recherches Archéologiques du CNRS.

Owen, L. (1994) Gender, Crafts, and the Reconstruction of Tool Use. *Helenium* XXXIV:2, 186–200.

Owen, L. (1999) Questioning Stereo-Typical Notions of Prehistoric Tool Functions – Ethno-Analogy, Experimentation and Functional Analysis. *Urgeschichtliche Materialhefte* 14, 17–30.

Owen, L. (2005) *Distorting the Past: Gender and the Division of Labor in the European Upper Paleolithic*. Tübingen, Kerns Verlag.

Schneider, J. (1987) The Anthropology of Cloth. *Annual Review of Anthropology* 16, 409–448.

Soffer, O. (2004) Recovering Perishable Technologies through Use Wear on Tools: Preliminary Evidence for Upper Paleolithic Weaving and Net Making. *Current Anthropology* 45, 407–413.

Soffer, O., Adovasio, J. M., and Hyland, D. C. (2000a) The "Venus" figurines – Textiles, Basketry, Gender, and Status in the Upper Paleolithic. *Current Anthropology* 41:4, 511–537.

Soffer, O., Adovasio, J. M., Illinsworth, J. S., Amirkhanov, H. A., Praslov, N. D., and Street, M. (2000b) Palaeolithic perishables made permanent. *Antiquity* 74, 286, 812–821.

Stettler, H. (1998) *Material Culture and Behavioral Change: Organic Artifacts from the Site of El Juyo and the Cantabrian Upper Paleolithic*. PhD diss., University of Chicago.

Stordeur-Yedid, D. (1979) *Les aiguilles à chas au Paléolithique.* XIII Supplément à Gallia Préhistoire. Paris, CNRS.

Straus, L. G. (1992) *Iberia before the Iberians*. Albuquerque, University of New Mexico.

Straus, L. G. (2005) The Upper Paleolithic of Cantabrian Spain. *Evolutionary Anthropology* 14:4, 145–158.

Tuross, N., and Fogel, M. L. (1994) Exceptional Molecular Preservation in the Fossil Record: The Archaeological, Conservation, and Scientific Challenge. In D. A. Scott and P. Meyers (eds), *Archaeometry of Pre-Columbian Sites and Artifacts: Proceedings of a Symposium organized by the UCLA Institute of Archaeology and the Getty Conservation Institute, Los Angeles, California, March 23–27, 1992*, 367–380. Los Angeles, Getty Conservation Institute.

Tyldesley, J. A., and Bahn, P. G. (1983) Use of Plants in the European Palaeolithic: A Review of the Evidence. *Quaternary Science Reviews* 2, 53–81.

Waguespack, N. M. (2006) The Organization of Male and Female Labor in Foraging Societies: Implications for Early Paleoindian Archaeology. *American Anthropologist* 107, 666–676.

38 A Bronze Age Plaited Starting Border

by Amica Sundström

In the Stockholm Museum of National Antiquities (SHM), there is a substantial collection of prehistoric textiles. These surviving textiles originate from a limited area of Sweden, the majority from Skåne and Halland in southern Sweden (Bender Jørgensen 1986, 232–233). This is where the geographic prerequisites for peat formation are in place, and it was here that the peat mounds, which were necessary for the creation of an environment conducive to the preservation of organic materials, were built.

The fragments are all very tiny, homogeneous and exclusively tabby. With few exceptions, the thread count is between 4 and 6 threads/cm in one thread system and 3–4 in the other. The spin directions vary between s/s, s/z and z/s. All the fragments are dark brown, no matter whether they come from an urn, an oak coffin or a stone cist.

In the Stockholm Museum of National Antiquities (SHM) collection, there is one sample that is slightly larger than the others. The textile fragment (inventory number 9822: 834: find 27 4344) was acquired by the museum through the purchase of Baron Claes Kurck's collection of ancient objects.

It is recorded in the museum catalogue as number 834 under the heading "Skåne diverse" (Skåne miscellaneous) and is described as "a piece of woollen cloth, coarse, plain weave, probably Bronze Age, a border with hem, approximately 14 square centimetres, with very uneven edges, site of find unknown" (see Fig. 38.1).

The fragment measures 15 × 15 cm. It is a tabby. On closer inspection, the hem was found to be a plaited starting border. Two threads are plaited together and every second pair of threads goes on forming the weave; the other pairs end in a loop where the plait changes to become the weave. The plait consequently consists of twice the number of threads compared to the weave. The plait must have been made before starting the weave as a pair of threads is connected with a second pair in a loop two threads away (see Figs 38.2 and 38.3). The woven piece has 4.5 s-spun threads/cm in the warp and 4 z-spun threads/cm in the weft.

A piece of the starting border has been reconstructed in order to understand the plaiting technique. The starting border was plaited by hand. As new pairs of threads were added, it became easier to plait. The actual weaving of the cloth continued with plaiting of the plain weave (see Fig. 38.4).

The textile find has been ^{14}C dated to 1390–1120 BC (Ångström Laboratory, Uppsala University, 2007). The thread count, colour and spin direction of the fragment are similar

Fig. 38.1. The fabric, 15 × 15 cm (Photo: Christer Ålin, SHM).

Fig. 38.2. Detail of the plaited starting border (Photo: Christer Ålin, SHM).

to those of the other Bronze Age textiles in the collection.

The fragment has an unclear provenance, as documentation of where it was found is missing. In the future, isotope tracing may contribute to the discussion of the context from which the fragment derives (see Frei in this volume). At present, I can think of three possible contexts:

1   the fragment has been produced and used in Skåne
2   the fragment has been imported, but used in Skåne
3   the fragment may have been bought abroad and added to the collection of ancient finds in modern times.

Isotope tracing might make it possible to exclude one of these possibilities, and would therefore give us a better basis for discussion. Likewise, fibre analysis would be of interest to find out if the fragment is made of the same type of fibre as the other Bronze Age fragments in the museum collection, or if it corresponds better with the finds elsewhere. Two textile finds with plaited edges are known from Austria. They originated from an excavation of salt mines in Hallstatt, Austria, and have been dated to the middle of the Bronze Age (1460–1245 BC). One of them must be a starting border, because both threads after the plaiting go down into the tabby weave (Grömer 2007, 95–96). The difference between this and the Swedish find is that here all the threads go down into the weave, whereas in the Swedish find, the pair of threads turns in a loop and only alternate threads go down to form the tabby weave.

There are no looms preserved dated to the Swedish Bronze Age. The different types of looms that we believe have been used during the Swedish Bronze Age are the two-beam loom and the warp-weighted loom. The textile fragments found in Sweden are far too small for any assumptions to be made as to the type of loom that might have been used to make them. From Early Iron Age Sweden, there is a find of one large textile, the oval cloak of Gerum (SHM 16719, find 266931), which has turning threads in all the sides of the cloak, in its largest part (Franzén and Lundwall 2006, 283). This implies that a two-beam loom could have been used.

The question is, if a plaited starting border implies that a two-beam loom has been used. Compared to other starting borders, this find is more similar to those made on a warp-weighted loom. In the small sample I have made to determine how it was plaited and if it might be possible to make the plait after the weaving, I continued to plait the tabby ground weave. To a weaver, it is easy to identify the gains if one, in some way, tightens up the warp threads and uses a weaving sword and a half-heddle stick when weaving. Such a construction does not necessarily have to be a loom. A few sticks stuck into the ground to fasten the starting boarder and one or two sticks (depending on the width of the weave) to fasten the warp, would be quite sufficient to weave a tabby. The fabric could also have been plaited by hand in its entirety. 'Time' and 'efficiency' are difficult parameters, as plaiting skills or the tradition of plaiting does not necessarily have to be affected by the fact that weaving is a more efficient method. Finally, the most time-consuming part of the process is making the yarn.

Questions about when looms were introduced, which types of looms they were, and from where the impulses and the knowledge of how to use them derived, cannot be answered merely based on one small fragment alone. The analysis of this fragment is rather to be seen as a find that contributes to further discussion. As knowledge is gathered, the questions will be further defined and the answers will hopefully become clearer.

Bibliography

Ångströms Laboratory, Tandem Laboratory, Uppsala universitet, Uppsala, Sweden, unpublished analytical report 2007.10.12.

Bender Jørgensen, L. (1986) *Forhistoriske textiler I Skandinavian. Prehistoric Scandinavian Textiles*. Nordiske Fortidsminder Ser. B: 9. Copenhagen, Det Kgl. Nordiske Oldskriftselskab.

Borg, G. Ch., Jonsson, L., Lagerlöf, A., Mattsson, E., Ullén, I., and Werner, G. (1994) *Konserveringstekniska studier. Nedbrytning av arkeologiskt material i jord: målsättning och bakgrund*, 37–39, 54. Stockholm.

Franzén, M.-L., and Lundwall, E. (2006) Nya upptäckter på Gerumsmanteln. *Fornvännen* 2006/101, 283.

Grömer, K. (2007) *Bronzezeitliche Gewebefunde aus Hallstatt – Ihr Kontext in der Textilkunde Mitteleuropas und die Entwicklung der Textiltechnologie zur Eisenzeit*. PhD Dissertation, Universität Wien.

Sundström, A. (2007) *En Sammanställning av textilsamlingen från brons- förromersk och romersk järnålder vid statens historiska museum*. Unpublished report.

Fig. 38.3. Drawing of the starting border (Drawing: A. Sundström).

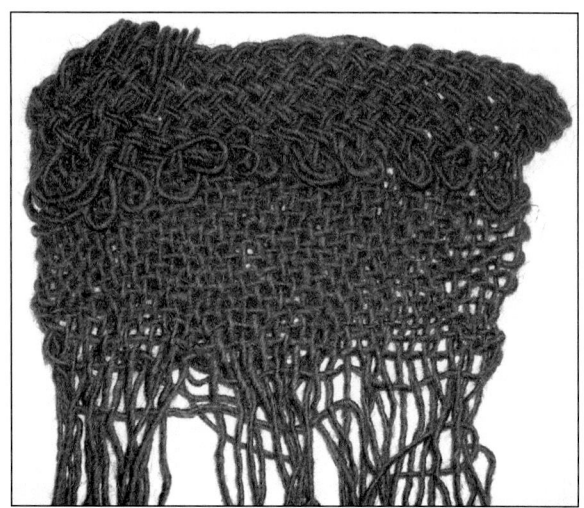

Fig. 38.4. The reconstruction of the starting border, made by Amica Sundström (Photo: A. Sundström).

39 Textile Craftsmanship in the Norwegian Migration Period

by Synnøve Thingnæs

The Norwegian Migration Period (*c.* AD 400–570) stands out from other periods, with its rich and distinctive textile material, such as the Snartemo and Evebø chieftain graves, which are similar to the Högom chieftain grave in Sweden (Dedekam 1926; Hougen 1935; Nockert 1991). These fine textiles are characterized by the elaborate tablet-woven bands with horsehair inlays, individually twined around the warp threads.

On the other hand, there are textiles in the same chieftain graves that represent ordinary textile production used for clothes and blankets. The aim of this paper is to show that textiles may have been produced within different production units (see Andersson 2003): the production of household textiles was linked to common knowledge, whereas the horsehair woven details may have been part of a more specialized knowledge controlled by specialists, in line with the theories of *e.g.* Mary Helms (1993; 1998) and Jan Apel (2001). The textiles have been analysed by using a theoretical perspective, connected to technological and social aspects of production (Thingnæs 2007). This has been done to show how textiles, and technology in general, can be used as a key to understand social practices, and how material culture can be a component in the production and reproduction of social systems.

The main focus of this paper is the theoretical framework connected to technology, followed by a brief presentation of the archaeological material, and the interpretation of the material from the perspective of technological choices connected to social values.

The Archaeological Material

I analysed textile remains from rich male graves, mainly from Southern Norway. The two chieftains' graves from Snartemo in Vest-Agder (Grave II and Grave V, both from *c.* AD 500) constituted the primary basis of my study (Thingnæs 2007). The analyses also included the material from Øvre Berge (6th century AD) and Vemestad (5th century AD), both male graves from Vest-Agder. The Snartemo material was divided into different categories of textile qualities, whereby I attempted to establish the types and quantities of fabrics in the material, as well as their use (tunics, cloaks, trousers and blankets). This was based on technical analyses, combined with stratigraphical information from the graves. The material has earlier been thoroughly published, by Hans Dedekam (1926) and Bjørn Hougen (1935), both of whom were concerned with the material's technological aspects. However, my aim was to use more recent theoretical approaches and my own background as a handweaver to re-examine this textile material.

Snartemo grave II contained 97 items; these have been divided into 13 types. With the exception of the tablet-woven textiles, all the wool fabrics are in 2/2 diagonal and diamond twill. Snartemo grave V contained 164 fragments which could be analysed. These pieces were divided into seven to eight different types, and excluding the tablet-woven bands, all were in 2/2 diagonal twill. My analyses and interpretations of the tablet-woven bands from Snartemo have been supplemented by Lise Ræder Knudsen's studies (1996; 2001) and Amica Sundström's reconstructions of the Högom horsehair band (1997).

Supplementary information about textile material was obtained through Lise Bender Jørgensen's publications and catalogues of Scandinavian and North European textiles (1986; 1992) and Margareta Nockert's publication of the textiles from Högom in Sweden (1991).

Theoretical Framework

Archaeologists have, at least for some decades, examined the technological side of archaeological material to understand various aspects of the past. Understanding the technology used to make a certain object can help us to obtain information about the social systems surrounding the object, for example, trade, raw material exchange and diffusion of techniques (Hodges 1964, 13–15). This is achieved through basic studies of technological sequences connected to a specific handicraft. In recent years, some archaeologists, such as Marcia-Anne Dobres (1999; 2000) and Jan Apel (2001), have demonstrated that these basic studies can also help us understand the social aspects of the craft.

This understanding of technology and archaeology is connected to the concept of *chaîne opératoire*, or the Operational Chain. It is the analysis of each step in the production of an artefact, and has its foundation in French stone tool investigation from the 1960s (Leroi-Gourhan 1964; Lemmonier 1993; Bender Jørgensen 2003a; Hurcombe in this volume). The main principle of the Operational Chain, as it is used today, is that every step in the production of an artefact is potentially subject to cultural pressure and symbolic discourse (Lemmonier 1993). Thus, technology has the social aspect incorporated into the analyses. This approach offers an alternative to analyses where the social meaning is discussed at a higher level, *after* the technological issues have been accounted for. In essence, we can speak of social structures as manifestations of material culture, in this case through technology.

Technology and Social Mechanisms

Different handicrafts, like textile craftsmanship, are in their essence technologies. The material included in this study is first and foremost visible traces of a piece of work, executed with a certain technology. Second, but not less important, the material remains are also traces of social and cultural activity, even if these activities are more difficult to trace. Thus, by using technological theory, combined with social theories, it is possible to understand the social mechanisms behind a handicraft.

Technology is closely linked to skills, and in this respect, the terms *knowledge* and *know-how*, and especially the relationship *between* the two terms, are useful. They can both be defined as types of memory, used in almost all practical actions. The degree of the type of memory involved varies, depending on the type of action performed. *Know-how* is an unconscious and intuitive type of memory that can be learnt through practical and bodily experience, in other words, learning by doing. The practical knowledge is learnt step by step, often through a system of apprenticeship. Knowledge is on the other hand, in this connection, a communicative type of memory, and can be achieved through language or observation, as, for example, through recipes (Apel 2001, 27).

Craft and technological knowledge can function as tools to renegotiate social identity and status. This includes possessing and displaying knowledge through controlling technology (Dobres 1999, 134–135). This control of knowledge can be achieved by displaying only parts of a skill, while keeping other parts secret. The control and display of knowledge can be connected to types of knowledge. The parts of a process based on communicative knowledge, and which demand a low degree of practical skills, can be surrounded by secrets and taboos. On the other hand, the parts of the handicraft which demand a high degree of practical skills can safely be on public display, without anyone being able to copy the process. This can again increase the prestige and power of the craftsperson (Apel 2001, 326). Three main factors can be said to control the spread of a technology: knowledge, know-how and the availability of raw material. The theoretical knowledge can, as mentioned earlier, be kept secret from the public. One can keep secret, for example, recipes, or knowledge of certain processes in a craft, such as smelting. The practical skills needed to perform a technique, can be controlled by introducing rules and apprentice systems, which for example, can be connected to sex and kinship, restricting the skills available to segments of the society. Thirdly, the raw materials can be controlled by monopolizing their sources (Apel 2001, 30).

Jan Apel uses this system as an example of how the flint dagger production in the Neolithic period was organized (2001). Another example from an ethnographical context is the dyeing of textiles with indigo on the Kodi Islands in East Indonesia (Hoskins 1989). The active ingredient is indigotin, which reacts in an oxidation process, and the blue colour does not appear before the fabric is in contact with air (Cardon 2003, 265, 285). Indigo dyeing on Kodi is connected to magic and is surrounded by many taboos and cosmological symbols. Only a few persons are initiated into the secrets of the dyeing process. The colouring consists of a complex of technological skills and magic, which is controlled in a matrilineal apprentice system. The final colouring process demands a high degree of practical skills, and occurs in public. Yet the recipe for the colour bath and the mixing process, and the location of the best raw material, are kept secret (Hoskins 1989, 152–153).

Textile Craftsmanship in the Norwegian Migration Period

When it comes to textile craftsmanship in the Scandinavian Iron Age, it is assumed that most people knew the principles of weaving, and that most farms were self-sufficient in most of the textiles needed. Within the analysis of this textile material, I have worked from the hypothesis that most of the fabrics, like the regular clothing fabrics, had been made by regular weaving skills, while certain details, such as some of the tablet-woven bands attached to the chieftains' costumes, could have been produced within a limited and specialized production unit (Thingnæs 2007). Some of these items are of such highly skilled standards, that they stand out as being unique. Certain steps in the clothing production process could also have been surrounded by secrets and magic.

Textile Technology and Chaîne Opératoire

The study of the Snartemo textiles has its basis in the Operational Chain. As mentioned earlier, each step in the production process and the subsequent use of the textiles were examined: raw material, spinning and spinning direction, weaving, colouring, sewing, wearing/use, repairs and final deposition, in this case, in the graves. By considering every step in the life of a textile, an opportunity is provided to create a nuanced impression of textile production. Different tasks can have been carried out by different persons, differentiated for example, on the basis of status or sex. Furthermore, each step can have been charged with certain symbolic meaning, both connected to the production process, and to the textile and its owner when finished and when in use.

Technology and Cosmology

Some of the textiles in the graves are not part of the chieftains' costume, but seem to have a function precisely as funerary textiles (Hougen 1935, 71–79; Nockert 1991, 35, 64). Here, it seems that technology can directly indicate a symbolic meaning. This is reflected in the spinning direction of the thread. The Norwegian Migration Period textiles consist of a large group of z-spun 2/2 diagonal twill fabrics. In the rest of Scandinavia, too, a major part of the contemporary textiles consists of z-spun twills (Bender Jørgensen 1986, 67–69). In several of the graves included in my analysis, as well as in additional material, the *only* textiles with alternating spinning direction are those used as a blanket under, or, to cover the dead. These textiles are also different in quality; their fabric is much coarser, while most of the others are of a finer quality. The blankets also differ when it comes to colour; the finer qualities are mostly dyed in red, yellow or blue colours, while the blankets are undyed, but in a pattern or shade using the natural pigments in the wool (Hougen 1935, 16, 18; Magnus 1983, 299; Walton 1990, 144–158; Nockert 1991, 35; Thingnæs 2007, 66–67, 74–75, 91–93). The diverging spinning direction could result from them being imported, but previous analyses of the wool fibres indicate that the wool is from local animal stocks, as are most of the other textiles in the graves (Walton 1990, 148–155). Here, one should be aware that both the provenance of wool fibre and the spinning direction is a topic for further discussion. However, I suggest that the alternating spinning direction can be a result of an intentional choice: due to various factors in textile craft, Iron Age society and Norse belief systems, the spinning direction can be connected to cosmology and rituals for the dead.

Norwegian, Swedish as well as Slavic folklore recounts shared beliefs on spinning and magic, especially taboos connected to spinning (Lysenko and Komarova 1992; Tin 2007). General magic in this archaic belief system also emphasizes magical effects by carrying out certain tasks, including spinning, in accordance with the apparent direction of the sun, in other words, in the direction of the sun's perceived movement across the sky. These beliefs are often tied to death and funeral rituals (Storaker 1923; Christiansen 1925; Moe 1925). The spinning has been connected to the transitional phases in life, such as birth, wedding and death, and is a symbol of a change in the cosmological realm (Lysenko and Komarova 1992; Tin 2007). Furthermore, in Norse society, the textile craftsmanship and especially spinning is connected to magic. Eldar Heide argues in his doctoral thesis (2006) that, in both Norse and Sami culture, spinning is part of the rituals for making or controlling the wind and casting spells, and that spinning is actually the equivalent of performing magic. It is possibly some of the same belief system that is reflected in a Migration Period textile find from Tegle in Jæren (Halvorsen 2008, see also Halvorsen in this volume). It consists of a whole warp with a tablet-woven starting border and yarns that have been laid in a bag and deposited in a bog, and has been interpreted by archaeologists as an offering (Hoffmann and Trætteberg 1960; Hoffmann 1991, 143). The unwoven warp can possibly represent a material in transformation and a transitional phase.

Certainly, some caution is necessary when employing historical and anthropological sources to develop theories about archaeological material; for example, the interpretations of Heide and other scholars (Helms 1993; Dobres 2000; Apel 2001; Heide 2006) can be criticized for attaching too much importance and mystique to normal and daily tasks. To spin a thread or to weave a cloth has probably been a completely mundane activity most of the time. Yet, a great many sources indicate that textile craftsmanship at least in the Late Iron Age and subsequently, is strongly tied to rituals, rules and magic. I believe that even if an action plays a large role in daily life, as spinning must have done, it can still be connected to a universal belief: craft is both an activity and at the same time a symbol of something larger and all-embracing.

Colouring and Secret Knowledge

Another interesting feature about the Migration Period textiles worth emphasizing is the use of the colour blue, derived from woad (*Isatis tinctoria* L.). Textiles coloured with woad appear in several of the chieftains' graves (Walton 1990). Woad has the same colouring agent as indigo, *indigotin*, and reacts in the same way when exposed to air. As mentioned earlier, the textile does not obtain its colour before the yarn or fabric is removed from the colour bath (Cardon 2003, 265, 285). Seeds from woad have been found in Scandinavia, both from the Early Roman Age in Denmark, and the Viking Age Oseberg grave in Norway, which indicates that the Scandinavians dyed with woad during the Iron Age (Hald 1950, 138; Næss 2002, 32; Vander Berghe in this volume). Some of the other colours used in the Migration Period graves for tunics and cloaks derive from rare and imported dyestuffs, probably according status to the owner merely by being rare and valuable (Walton 1990; Hedeager Krag 2001, 73–74). However, in the case of woad, it is also possible that it is the dyeing process itself that gives the value and symbolic meaning to the clothing. As with the earlier mentioned indigo colouring process in Indonesia, here, too, the process could have been surrounded by secretiveness and mystique, and the find of woad seeds in the later Oseberg grave can possibly be interpreted as a symbol of the control and knowledge of the dyeing process.

Tablet-Woven Horsehair Bands

The third aspect of the Migration Period textiles I would like to emphasize is the making of the tablet-woven bands, which were attached to tunics, cloaks and used as sword belts. The weaving of these bands is a type of production that clearly stands out, on equal terms with, for example, glass and metal work. The bands are made in different techniques, but especially the bands woven with inlays of unspun horsehair are characteristic of the Scandinavian Migration Period, and are intricate and time-consuming to make. A professional weaver, Amica Sundström has reconstructed parts of the horsehair bands from the Swedish Högom grave, and has estimated that it takes one hour to weave one millimetre, sometimes even longer if the pattern is complicated. It is also estimated that the whole assembly of bands from the Högom costume

40 Textilfunde aus Ausgrabungen in Heidelberg

nach Klaus Tidow

Im Rahmen eines Forschungsprogrammes, daß sich mit der Erfassung und wissenschaftlichen Bearbeitung aller hoch- und spätmittelalterlichen sowie frühneuzeitlichen Textilfunde aus Ausgrabungen in Baden-Württemberg befaßt, konnten mit der textiltechnischen Analyse der Textilfunde vom Kornmarkt in Heidelberg ein weiterer Fundkomplex abgeschlossen werden. Unter der Leitung von Dr. Johanna Banck-Burgess vom Landesamt für Denkmalpflege in Baden Württemberg, Abteilung Textilarchäologie, wurden in den letzten Jahren die textiltechnischen Daten von insgesamt 3.500 Textilien erfasst. Sie stammen aus einer Schlackenhalde in Wiesloch (11./12. Jh., siehe *ATN* 33 Seite 44), aus einem Münzhortfund in Ladenburg (12./13. Jh., siehe ATN 35, Seite 6–9), aus der Latrine des Augustiner-Eremiten-Klosters in Freiburg (13.–16. Jh., siehe *ATN* 39, 26) und aus den Latrinen vom Kornmarkt in Heidelberg. Die Untersuchungen umfaßten die Beschreibung des Erhaltungszustandes, die Bestimmung der Größe und Form der Textilien sowie die Ermittlung der textiltechnischen Daten wie Herstellungstechnik, Bindung, Garndrehung, Garnstärke, Einstellung, Ausrüstung und Bearbeitungsspuren (Nähte, Säume, Schnittkanten).

Die Textilfunde

Mit den über 2700 Textilfunden aus den Latrinen vom Kornmarkt in Heidelberg haben wir die bisher größte Sammlung von Textilien aus Ausgrabungen in Süddeutschland bearbeitet. Die Ausgrabungen fanden in den Jahren 1986 und 1987 statt. Erste Ergebnisse wurden 1992 im Rahmen einer Ausstellung vorgestellt und in einem Katalog veröffentlicht. Von den Textilfunden wurden allerdings nur 14 Fragmente gezeigt, darunter die Reste eines geknüpften Haarnetzes und eines Baretts aus Seidensamt (Sarri 1992, 120–124). Nach einer weiteren Begutachtung der Textilfunde hinsichtlich ihres Erhaltungszustandes und möglicherweise notwendigen Konservierungen, zeigte sich, daß die Erfassung der textiltechnischen Daten dringend erforderlich ist. Diese Aufgabe übernahm Dietlind Hachmeister, die im Auftrage des Landesamtes für Denkmalpflege in Esslingen zwischen Oktober 2005 und September 2007 den Fundkatalog erstellte.

Insgesamt wurden auf dem Kornmarkt in Heidelberg 10 Latrinen ausgegraben, von denen sieben Textilien enthielten. Die ältesten Funde können aus der frühen ersten Hälfte des 15. Jahrhunderts und die jüngsten aus dem Ende des 17. Jahrhunderts stammen. Sie müssen vor dem großen Brande mit der Zerstörung des Kornmarkts 1693 in die Latrinen gelangt sein (Wendt 1992, 63). Neben der textilkundlichen Bewertung hinsichtlich ihrer Herstellung, Herkunft und Verwendung werden auch Aussagen zur Sozialstruktur möglich sein, da die Latrinen dem ehemaligen Hospital oder Bürgerhäusern, von denen häufig die früheren Bewohner bekannt sind, zugeordnet werden können. So unterscheiden sich die Textilien aus der Latrine des Hospitals deutlich von den Textilien aus Bürgerhäusern. Die Gemeinschaftslatrine des Hospitals wurde bis zur Auflassung 1556 benutzt, während z.B. die Latrine eines Bürgerhaus, die ebenfalls zunächst zum Hospital gehörte, noch bis 1693 weiter gebraucht worden ist. So wurden in dieser Latrine vor allem hochwertige Gewebe wie Tafte, Samte und Damaste aus Seide und sehr feine Wollgewebe in Atlasbindung gefunden. In der Gemeinschaftslatrine des Hospitals fehlen solche Gewebe (nur ein Seidentaft wurde geborgen), während anderseits außer Wollgeweben in Tuch- und Köperbindungen verhältnismäßig viele Gewebe aus pflanzlichen Fasern (vermutlich aus Flachs/Lein) und Filze aus Wolle und anderen tierischen Fasern bestimmt werden konnten. Ob allerdings für diesen und auch für die anderen Latrinen eine differenzierte Datierung der Funde möglich ist, kann z.Z. nicht gesagt werden.

Breitgewebe

Unter den Heidelberger Textilfunden kommen Gewebe aus Schafwolle am häufigsten vor. Es sind ca. 2500 Gewebefragmente, davon 79,5 % Gewebe in Tuchbindung, 16,3 % in Köperbindung K 2/2 und 1,6 % in Köperbindung K 2/1 und 2,1 % in Atlasbindung 1/4. Von den Ableitungen der Tuch- und Köperbindungen konnten nur sechs Panama (P 2/2 2-fädig) und ein Spitzkaro (aus K 2/1) bestimmt werden. Dieses Gewebe besteht aus Zwirnen (z/S) in einem und einfachen Garnen (z) im anderen Fadensystem. Es gehört zu den mittelfeinen bis feinen Qualitäten. Da es sich um ein sehr kleines Fragment von 7 × 2 cm handelt, konnte allerdings der

Bindungsrapport nicht bestimmt werden. Sowohl unter den spätmittelalterlichen und als auch unter den frühneuzeitlichen Gewebefunden aus Nord- und Süddeutschland konnten wir bisher Spitzkarogewebe in dieser Bindung nicht nachweisen. Alle übrigen Gewebebindungen sind auch von anderen Fundstellen des 15.–17. Jh.s aus Deutschland bekannt, wenngleich es Unterschiede in den Zusammensetzungen in den einzelnen Fundkomplexen gibt.

Nur 117 Fragmente von mittelfeinen und feinen Leinengeweben konnten unter den Heidelberger Textilfunden bestimmt werden, darunter neun sehr kleine Fragmente (max. 1x1 cm) eines Spitzkarogewebes (aus K 2/2 oder K 3/3?), dessen Bindungsrapport wir ebenfalls nicht bestimmen konnten.

Unter den Seidengeweben finden wir die für das Spätmittelalter und die Frühneuzeit auch von anderen zeitgleichen Fundstellen bekannten Webtechniken und Gewebebindungen wieder, nämlich vor allem Tafte (34), Samte (10) und Damaste (3). Die textiltechnischen Daten ließen sich zwar für alle Gewebe ermitteln, doch konnten die Musterrapporte der Damaste aufgrund der geringen Größe nicht rekonstruiert werden.

Zu den Breitgeweben gehören noch neun Reste von drei verschiedenen mittelfeinen Mischgeweben aus Lein und Wolle, vermutlich ein Gewebe aus Baumwolle und Wolle, alle in Tuch-/Leinwandbindung, und der Rest eines Köpergewebes (K 2/1). Von diesen Geweben sind die Fäden eines Fadensystems vergangen, das andere besteht aus Wolle.

Bänder

Unter den Heidelberger Textilfunden befinden sich nur sechs Bänder, darunter drei aus Seide und drei aus Baumwolle. Die schmalen, einfachen Baumwollbänder (0,6 bzw. 1,5 cm) gehören zu zwei verschiedenen Geweben. Als Kette wurden jeweils Zwirne (z/S) und als Schuß einfache Garne (s) genommen. Die Bänder sind noch 40, 44 bzw. 46,5 cm lang. Ebenfalls sehr schmal ist ein Seidenband in Brettchenweberei, das mit 10 Vierlochbrettchen gewebt worden ist. Etwas breiter (etwa 4 cm) ist ein gemustertes Seidenband in Brettchenweberei. Das sechste Seidenband ist in Taftbindung gewebt und mit kleinen Plättchen bisher nicht bestimmten Materials verziert ist.

Nicht gewebte Textilien

Zu den nicht gewebten Textilien gehören die Reste eines geknüpften Haarnetzes aus Seide, 18 Gestricke aus Wolle (vermutlich von sechs verschiedenen Gestricken) von grober und mittelfeiner Qualität und 36 Filzfragmente aus Schafwolle oder anderen tierischen Fasern, die noch bestimmt werden müssen. Bemerkenswert ist, daß die Farbpalette der Filze verhältnismäßig groß ist, nämlich hell-, mittel-, dunkel- und schwarzbraun sowie rötlichbraun, rotbraun und hellrot. Farbstoffanalysen müssen zeigen, welche davon gefärbt sind und welche aus naturfarbigen Fasern bestehen.

Zusammenfassung

Mit den textiltechnischen Analysen an den Textilfunden vom Kornmarkt in Heidelberg haben wir unsere Untersuchungen an den hoch- und spätmittelalterlichen sowie frühneuzeitliche Textilien aus Ausgrabungen in Baden-Württemberg vorläufig abgeschlossen. Die Untersuchungen haben einmal mehr gezeigt, daß bei großen Fundkomplexen wie in Heidelberg eine Vielzahl von Textiltechniken, Gewebebindungen und –qualitäten nachgewiesen werden können. Aufgrund der Erhaltungsbedingungen in Latrinen bzw. Kloaken sind es vor allem Gewebe aus tierischen Fasern wie Wolle und Seide, aber auch solche aus Lein und Baumwolle kommen vor. Verhältnismäßig zahlreich sind auch Filze und Gestricke. Vereinzelt haben sich auch geknüpfte Netz erhalten. Solche Fundzusammensetzungen kennen wir auch aus Ausgrabungen in Norddeutschland, zum Beispiel in Lübeck. Der größte Fundkomplex mit rund 9000 Einzelfragmenten stammt aus einem aufgegeben Brunnen, der als Kloake benutzt wurde (Grabung Schrangen). Auch hier sind es überwiegend Wollgewebe, die uns einen Überblick über die im 15. und 16. Jh. bekannten Gewebequalitäten vermitteln. Seidengewebe kommen in verschiedenen Techniken (Breitgewebe und Bänder) vor, während Leinengewebe auch in diesem Fundkomplex unterrepräsentiert sind. Von den nicht gewebten Textilien sind es wiederum Gestricke und Filze aus Wolle bzw. anderen tierischen Fasern (Tidow 1992; Jaacks 1993).

Bezieht man die Textilfunde aus Gebäuden wie zum Beispiel aus Kempten im Allgäu (St. Mangplatz „Mühlberg-Ensemble") mit ein, wo überwiegend Leinengewebe in Leinwandbindung, aber auch in Köperbindung K 2/1 und Ableitungen des Köpers 3/1 und K 3/3 sowie Mischgewebe aus Leinen und Baumwolle (K 3/1-Kreuzköper und K 3/1-Spitzgrat) geborgen wurden, so haben wir nunmehr einen guten Überblick die Gewebe, Filze, Gestricke und Bänder, die im 15. bis 17. Jh. hergestellt und vorwiegend zur Kleidung verarbeitet wurden (Rast-Eicher und Tidow 2005).

Die sich nun anschließenden Untersuchungen befassen sich mit den früheren Verwendungen der Heidelberger Textilien und ihrer Herkunft sowie mit einem sozialgeschichtlichen Vergleich der in Heidelberg und Freiburg gefundenen Textilien des Spätmittelalters und der Frühneuzeit. Hierzu ist allerdings eine intensive Zusammenarbeit zwischen den Archäologen Kostümgeschichtlerinnen, Archivaren und Textilforschern/-innen erforderlich.

Literatur

Jaacks, G. (1933) Kostümgeschichtliche Untersuchungen an den Gewebefunden aus den Grabungen Hundestraße, Schrangen und Königstraße zu Lübeck. *Lübecker Schriften zur Archäologie und Kulturgeschichte* 23, 283–293.

Rast-Eicher, A., und Tidow, K. (2005) Die Textilien aus dem „Mühlberg-Ensemble". In *Depotfunde aus Gebäuden in Zentraleuropa. Bamberger Kolloquien zur Archäologie des Mittelalters und der Neuzeit* 1, 83–86. Berlin.

Sarri, K. (1992) Die Textilfunde. In: *Vor dem großen Brand*, 120–124. Stuttgart.

Tidow, K. (1992) Die spätmittelalterlichen und frühneuzeitlichen Wollgewebe und andere Textilfunde aus Lübeck. *Lübecker Schriften zur Archäologie und Kulturgeschichte* 22, 237–271.

Wendt, A. (1992) Entwässerung und Abfallbeseitigung. In: *Vor dem großen Brand*, 58–63. Stuttgart.

41 Textile Remains on a Roman Bronze Vessel from Řepov (Czech Republic)

by Kristýna Urbanová and Helena Březinová

Location of the Finds

In 1904, two large metallic vessels were found next to each other at the location known as 'Na Včelníku' on the eastern edge of the Municipality of Řepov (District of Mladá Boleslav). One of them was full of smaller items, including textiles and 'some bones' according to the workmen who discovered the find. Because of this reference and the nature of the find, it may be assumed that it was not a deposit, but a burial of an important individual (possibly a man), who was buried in the first half of the 2nd century BC (Roman Iron Age period B1). That it was an inhumation is clear from the fact that none of these objects were deformed by the heat of the funeral pyre. In 1909, the full set was given to the National Museum in Prague.

Description of the Items

The main items of this set are two large metallic vessels – a bronze situla and a bronze pan with the head of a maenad and hanging ringlets. The situla contained a strainer with a ladle (the latter with the stamps L COMPITVRICIN and ///· POLIBI), a shallow bronze bowl, two bronze forged drinking horns decorated with silver rivets, a finger ring, a razor, a pair of shears and textile remains.

The bronze situla is of the Östland type E 39 (Motyková-Šneiderová 1967, 40, Taf. I.–IV.; Sakař 1970, 40, fig. 22:1–3, 6–16; Waldhauser and Košnar 1997, 78, 137–139; Karasová 1998, 79, fig. 6b; Droberjar 2002, 284–285), 24.2 cm high, dated to the mid-2nd century BC, preserved in perfect condition and, according to the available archive materials, it is probable that it has not yet been subject to conservation procedures.

The sides of the situla are a true treasury of organic remains. Almost half of the outer surface is covered by corrosion and textile fragments or their imprints, cords and remnants of animal fur and plant material. The remains of a fur were also preserved inside the vessel. Several textile techniques are encountered here. Most of the vessel's surface is covered by a fine cloth in plain weave. Furthermore, on the outer side, two small imprints of twill weave and a fragment of a tablet-woven band may be seen. It is noteworthy that the organic remains have been preserved only on one half of the vessel – the second half is devoid of organic material. Thus, it cannot be ruled out that they may have been destroyed as a result of careless handling or in undocumented efforts to clean the surface of the vessel. It is also remarkable that no textiles or organic remains were found on the other items in the set, not even on the items inside the situla.

Technical Analysis of the Textile Fragments

The analysis of the textile remains was undertaken by the present authors using microscopic photography (microscope Olympus SZX 9) at the National Museum in Prague[1] and at

Fig. 41.1. View of bronze situla, on the side with textile rests (Photo: L. Káchová).

the Restoration Laboratory of the Institute of Archaeology of the Academy of Sciences of the Czech Republic, Prague. The technical analysis of the textile fragments included the structure, thread count, spin direction, twist angle, number of twists per cm, and thickness of the thread.

The structure was determined according to the typology

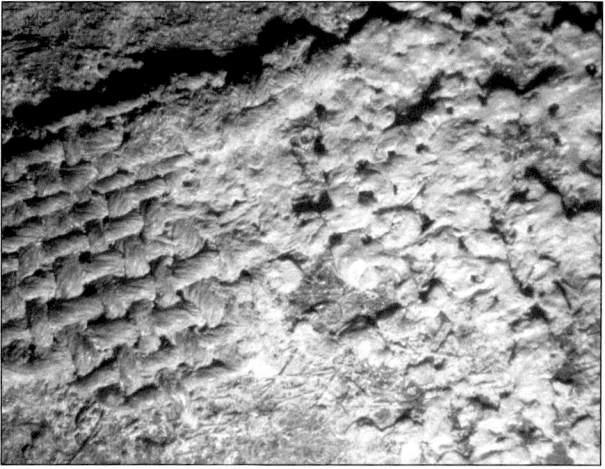

Fig. 41.4. Detail of plain weave under 10× magnification. One half is in its organic state, the other half has mineralised (Photo: M. Králík).

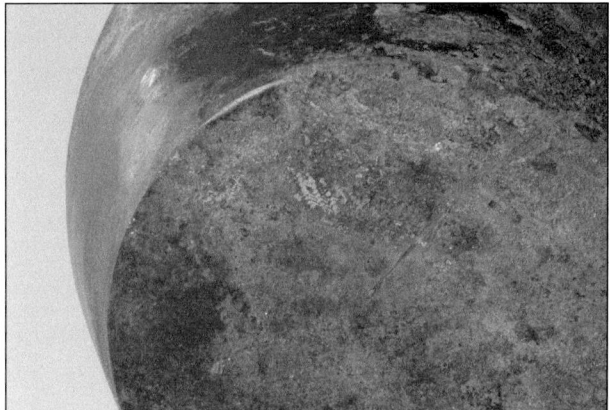

Fig. 41.2. View of lower side of bronze situla with remains of twill and tablet weave (Photo: L. Káchová.)

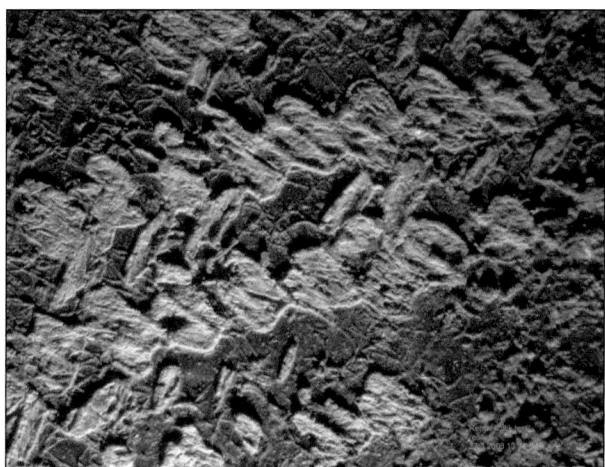

Fig. 41.5. Detail of twill weave under 8× magnification (Photo: M. Králík).

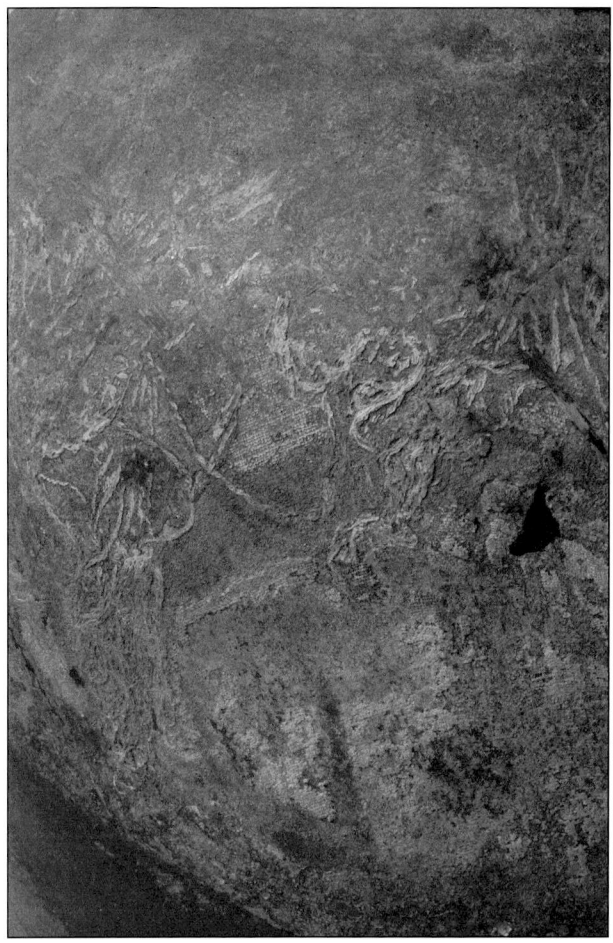

Fig. 41.3. Detail of twisted threads and remains of plain weave (Photo: L. Káchová).

Fig. 41.6. Detail of tablet weave under 8× magnification (Photo: M. Králík).

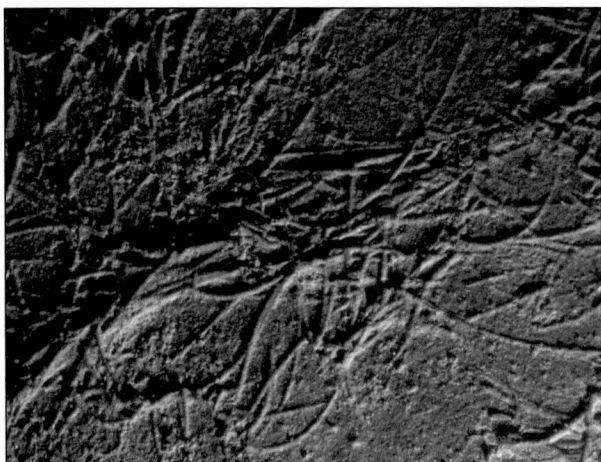

Fig. 41.7. Detail of animal fur under 6.3× magnification (Photo: M. Králík).

created by the international organisation C.I.E.T.A.[2] It was not always possible to determine the warp and weft direction – most of the textile fragments were too small. In these cases, the terms 'warp and weft' are replaced by 'first and second sets of threads' (Březinová 2007, 319).

The textile raw materials were determined by molecular spectroscopy at the Institute of Chemical Technology in Prague.[3] The analysis of the animal fur remains was done in co-operation with the Criminology Institute in Prague. A negative impression of the sample was lifted at the National Museum using a non-destructive method (the impression material used was Mikrosil), which was then analysed with an electron microscope.[4]

Cords

Numerous remains of cords are preserved on almost half of the situla's surface, on an area of 15 cm (height of situla) × 25 cm (width of situla); the largest portion of the remains is found on the shoulder of the vessel.

The cords are preserved in three varying forms: in their organic state, sometimes covered with a layer of encrustation; as remains closely adhering to the situla's surface; and as mineralized and metamorphosed cords. The cords are preserved in random clusters, without any interlinking or clear system. In certain places, the threads are intertwined (but not intentionally). Sometimes, individual threads that have unravelled have accidentally entangled together. The longest discernible cord is 13 cm long. The twist of the threads is S 2z; the number of ply per cm is 8. The thickness of the cords is 1 mm and the thickness of the individual threads is 0.5 mm. The current colour of the surviving cords is greenish-grey with light brownish-green shade at some points.

The cords survive on the same surface of the situla as the remains of a textile in plain weave. Both types overlap in various ways, but their relationship or the sequence of layers cannot be determined. Between the cords and cloth there are surviving remains of grass, which were probably the filling of the grave (together with soil).

The cords could be interpreted as a decorative fringe of a garment or another textile product, which was placed close to or around the situla as part of the funerary rites. The thickness of the individual threads of the cords and their twist direction indicate that, they could have been connected to the fine plain weave.

Fabric of Plain Weave

Small remains of a fabric of plain weave are preserved at various places on almost half of the situla's surface, on an area of 15 cm (height) × 25 cm (width). The fabric is preserved in three forms: as a fabric in its organic state, sometimes covered with a layer of corrosion; as only remains of the fabric threads that closely adhere to the situla's surface; and as mineralized and metamorphosed threads. It is clearly visible at several points that the fabric was folded in two layers.

It is impossible to determine the warp and weft of the fabric. In one direction it has 20 threads per cm, thread thickness 0.3–0.4 mm and thread twist z. In the other direction, it has 18 threads per cm, thread thickness 0.5 mm and thread twist z. The current colour of the textile remains is greenish-grey, with a light brownish-green colour at some points. Imprints of very minute and fine fibres 3–5 mm long could be traces of the surface finish of the plain weave as it is discernible in the immediate surroundings of this fabric.

Fabric of Twill Weave

A small fragment of a 2/1 twill weave is preserved on the outside of the base of the situla on an area of 10.3 cm (length) × 2.5 cm (width). The textile fragment is in very poor condition and the fabric structure is indiscernible. One direction has a thread count of 18–19 threads per cm, z twist and thread thickness 0.6 mm. The other has a thread count of 9 threads per cm, thread twist s and thread thickness 0.6 mm. The current colour of the fabric is light grey-green.

Tablet-woven Band

Close to the twill weave, a small fragment of a tablet-woven band was found. There is no contact between the two weaves, and their relationship cannot be determined. The condition is very poor, only the lower parts of the mineralized threads are visible at both edges and in the middle section of the band. The width of the band delimited by traces of the surviving edges is 1.9 cm, and the discernible length of the band is 8 cm. The number of tablets used cannot be determined with certainty; the width of the threads from one tablet is respectively 1.5 by 2 mm, giving an estimated number of about 10 tablets.

The tablet weave can be interpreted as a girdle or belt, which could be used independently or as an edge of a fabric (possibly the twill weave present close by). However, the small size of the fragment and its very poor state of preservation do not permit any conclusions about the relationship between the band and the twill fabric.

Raw Material

Microscopic samples were taken from remains of plain weave and from cords and analysed using infrared spectroscopy with the objective of ascertaining the raw materials used. For plain weave, plant fibre (probably linen) was used and for cords, a combination of plant and protein fibre (probably silk) was utilised.

Animal Fur

Both on the inner and the outer side of the situla, there are remains of animal fur. There are two significant clusters of hair (both approximately 8 cm in length, and 6 cm in width) preserved mainly under the upper edge of the situla, but they do not appear below the edge. We cannot determine the species of furry animal. The analysis made at the Institute of Criminology in Prague was not successful, because the fur was too degraded by bronze corrosion products.

Comparanda

A similar find with bronze vessels is known from the territory of modern-day Poland, at Leśnie, where a royal tomb of a female from the 2nd century AD was discovered in 1985. The grave contained a number of textile items, preserved also on a bronze vessel deposited at the head of the deceased (Maik 2005, 98). The cloth in this case is interpreted as remains of funeral attire or parts of clothing. A further example of textile remains on a metallic vessel is the recovery of a silver pan with silk fabric from an inhumation grave at Stráže, Slovakia, dated to the beginning of the 4th century AD (Ondrouch 1957, 167). Ondrouch suggested that the silk shroud indicates the high status of the deceased person.

As far as the wrapping of the items buried with the body is concerned, this find is not unique. Many items with traces of cloth coverings have also been found in other burials from the Roman era in Bohemia (*e.g.* Třebusice, 1st and 2nd century AD). It seems that some items that were not subjected to the funeral pyre were placed in the grave only upon the burial of the deceased. In the case of some smaller shallow graves, there are indications of the placement of cremated remains directly into organic containers, such as textile bags.

The current find is a rare example of the preservation of the imprints of fur, and until now such direct proof of the use of furs in inhumations has not been encountered in the Czech Republic. However, evidence of bones from wild furry animals (beaver, bear) has previously been found, for instance, in the imperial tomb in Muschau (see Peška and Tejral 2002, 493–494). Furthermore, bear claw finds from cremation cemeteries in the territory of the Czech Republic have been interpreted as evidence of the presence of bear skin, which the deceased was laid upon on the funeral pyre (see Schönfelder 1994, 217–227; Droberjar 2002, 189).

Conclusions

This find is one of the most beautiful burial assemblages from the Roman era that has attracted the attention of researchers since its discovery. However, this interest has only been focused on the study of the precious vessels and items, which are unique in Bohemia. Yet, these items also conceal another dimension of ancient daily life, which has been ignored to date. We could interpret the fabrics found adhering to the situla surface as remains of a garment or other textile product, which was placed close to the situla as part of the funeral paraphernalia. The find's primary importance lies in the fact that it shows a variation of weaving techniques. The textile remains on the bronze situla from Řepov document most of the known weaving techniques, which were used by the Germanic tribes in Europe (Schlabow 1976; Hald 1980). Here, plain weave, twill weave and bands in tablet weaving are present. To date, no evidence of the latter technique in the prehistoric period has been found in the territory of the Czech Republic. Hopefully there will be more similar finds in the future, which can help put together the picture of textile production in the territory of Bohemia and Moravia in the Roman era.

Notes

1. We would like to thank J. Vykouková for allowing us to use the photographing equipment, and M. Králík for providing us with a camera and his assistance during the photo documentation.
2. CIETA 1979: Tracés techniques (français - anglais). Lyon.
3. Performed by M. Novotná, Central Laboratories, Laboratory of molecular spectroscopy, University of Chemical Technology in Prague.
4. Performed by M. Eliášová, Criminology Institute in Prague.

Bibliography

Barber, E. J. W. (1991) *Prehistoric Textiles. The Development of Cloth in the Neolithic and Bronze Ages with Special Reference to the Aegean.* Princeton.

Bender-Jørgensen (1986) *Forhistoriske textiler i Skandinavien. Prehistoric Scandinavian Textiles.* Nordiske Fortidsminder Ser. B:9. Det Kgl. Nordiske Oldskriftselskab.

Březinová, H. (2007) *Analýza textilních fragmentů dochovaných na předmětech z hrobu 2349 z Tišic.* Archeologické výzkumy v jižních Čechách 20. České Budějovice.

Droberjar, E. (2002) *Encyklopedie římské a germánské archeologie v Čechách a na Moravě.* Praha.

Hald, M. (1980) *Ancient Danish Textiles from Bogs and Burials. A Comparative Study of Costume and Iron Age Textiles.* Copenhagen.

Karasová, Z. (1998) *Die römischen Bronzegefässe in Böhmen. Fontes Archaelogici Pragenses.* Volumen 22. Pragae.

Köhler, R. (1975) *Untersuchungen zu Grabkomplexen der älteren römischen Kaiserzeit in Böhmen unter Aspekten der religiösen und sozialen Gliederung.* Neumünster.

Maik, J. (2005) Tkaniny z grobu książęcego v Leśnie. In *Brusy i okolice w pradziejach na tle porównawczym,* 89–111. Brusy.

Mitschke, S. (2001) *Zur Erfassung und Auswertung archäologischer Textilien an korrodiertem Metall. Eine Studie zu ausgewählten Funden aus dem Gräberfeld von Eltville, Rheingau-Taunus-Kreis (5.–8. Jh. n. Chr).* Marburg.

Motyková-Šneiderová, K. (1967) *Weiterentwicklung und Ausklang der älteren römischen Kaiserzeit in Böhmen.* Pragae.

Ondrouch, V. (1957) *Richly decorated tombs from the Roman era in Slovakia, newer findings.* Bratislava.

Peška, J., and Tejral, J. (2002) *Königsgrab von Mušov.* Mainz.

Sakař, V. (1970) *Roman Imports in Bohemia.* Pragae.

Schönfelder, M. (1994) Bear-claws in Germanic graves, *Oxford Journal of Archaeology* 13/2, 217–227.

Schlabow, K. (1976) *Textilfunde der Eisenzeit in Norddeutschland.* Neumünster.

Waldhauser, J., and Košnar, L. (1997) *Archaeology of the Germans in the Jizera area and the Bohemian Paradise.* Prague, Mladá Boleslav.

42 Dyes: to be or not to be. An Investigation of Early Iron Age Dyes in Danish Peat Bog Textiles

by Ina Vanden Berghe, Margarita Gleba and Ulla Mannering

Introduction

Evidence of the use of dyeing in early Scandinavian textiles was presented in several publications during the 1980s. Dyes were identified using UV/Visible spectrophotometry after dye extraction, confirmed with paper and thin-layer chromatography (Taylor and Walton 1983b; Bender Jørgensen and Walton 1986; Walton 1986; 1988). The research programme 'Textile and Costume from Bronze and Early Iron Ages in Danish Collections' conducted by the Centre for Textile Research at the University of Copenhagen, Denmark, provided the opportunity for a new, large-scale, natural organic dye study on 45 textile finds from 26 sites in Denmark. To this objective, high performance liquid chromatography with photo diode array detection (HPLC-PDA), currently the most appropriate technique for natural organic dye analysis was used. The applied method has proven its efficacy for the detection of natural organic dyes in textiles in a wide historical context and geographical area (Wouters and Rosario Chirinos 1992; Wouters 1998; Vanden Berghe and Wouters 2004a; 2004b; Cardon *et al.* 2004).

The methodology and a complete list with the results in terms of detected dye components, for the series of Early Iron Age Danish bog textile samples (186 in total) is published elsewhere (Vanden Berghe, Gleba and Mannering 2009). Here, a general overview of the types of dyes found is given, followed by two specific examples presented in detail.

Methodology

The samples were analysed using high performance liquid chromatography. Each dye analysis requires a sample thread of about 5 mm long. The samples are examined under binocular microscope prior to analysis to avoid visible contamination. After extraction of the dyes from the fibres using acidic methanol followed by a second extraction in ethyl acetate, the analyte is redissolved in methanol/water and 20 µl is injected into the chromatographic system. Results obtained by using other extraction techniques as tested out in the first phase of the study are also included in the result in Table 42.1. More details regarding the analytical part of the work is published elsewhere (Vanden Berghe, Gleba and Mannering 2009).

Results

Dye Components/Dye Sources

As dye sources are composed of many different characteristic dye components and only few of them survived in these prehistoric samples, species identification of the dye sources is hardly possible. The dye results rather suggest a specific range of dye sources (Cardon 2007).

Traces of luteolin were found in the majority of the samples. Other, more rarely detected flavonoid dye components were rhamnetin and, in combination with luteolin, apigenin and quercetin. Alizarin and purpurin were detected in only one textile (see below). Indigotin and indirubin were found as well, frequently in combination with luteolin. Table 42.2 gives an overview of a selection of possible biological dye sources that can be related to the dye components found in a European context.

Apart from the known components, five unidentified peaks were signalised, also probably referring to colouring matter. Four of them have their maximum absorption around the range of 400 to 500 nm, hence suggesting a reddish colouring matter. Three of them could not be identified at all, while unknown 1 shows a high correlation with the reference of yellow wall lichen. For unknown 3, high similarity is found with Scandinavian orchil, with alkanet and with anthocyanidin glycosides. Based on both the analytical aspect and on the historical context of the samples, Scandinavian orchil is the most probable source.

Dyes in a Geographical and Historical Context

An overview of the identified dye sources for each analysed site is presented in Table 42.1. Only in two textiles, from Stidsholt and an unknown location, both dated between 400–200 BC, were no dyes detected. Evidence for woad

Site and C14 dating	Threads without dyes	Luteolin yellow	Yellow berries	Woad	madder/ ladies bedstraw /dyer's woodruff	Other
Rebild (360-160 BC)		lu		in		
Elling Mose (381 BC-AD 167)		lu		in		tannin
Huldremose I (210-41 BC)		lu	rht	in		Scand. Orchil?, tannin, unknown 5
Bredmose (370 BC-AD 10)	no dyes	lu/ap		in		unknown 2 and 4
Corselitze (AD 210-410)	no dyes	lu		in		
Grathe Hede (190 BC-AD 10)	no dyes	lu		in		
Krogens Mølle Mose (399-181 BC)	no dyes	lu		in		tannin
Søgårds Mose II (AD 130-340)	no dyes			in		
Skærsø (210 BC-AD 90)	no dyes				al/pu	
Fræer Mose (110 BC-AD 60)		lu/ap				
Rønbjerg I (230-50 BC)		lu/ap				
Ålestrup (520-380 BC)	no dyes	lu/ap				
Haraldskaer Mose (210-20 BC)	no dyes	lu/qu				tannin
Borremose I (365-157 BC)		lu				Yellow wall lichen?
Stokholm Mose (360-50 BC)		lu				
Thorup Mose I (240-50 BC)		lu				
Auning Mose (200 BC-AD 140)		lu				
Borremose II (416-95 BC)		lu				
Borremose III (311-209 BC)		lu				
Borremose V (370-180 BC)	no dyes	lu				
Huldremose II (210-30 BC)	no dyes	lu				tannin
Karlby Mose (170 BC-AD 140)	no dyes	lu				
Ømark (390-200 BC), painted textile	no dyes	lu				
Thorup Mose II (400-200 BC)	no dyes	lu				
Stidsholt (392-204 BC)	no dyes					
Unknown site (400-200 BC)	no dyes					

Table 42.1. Identified dyes in relation to the sites of excavation.

Detected dye components	Type of dye	Common name	Latin nomenclature
luteolin, apigenin, quercetin	flavonoid	weld saw-wort dyer's broom chamomile	*Reseda luteola* L. *Serratula tinctoria* L. *Genista tinctoria* L. *Anthemis* sp.
rhamnetin	flavonoid	berries (yellow berries)	*Rhamnus cartharticus* L. or other species
indigotin, indirubin	indigoid	woad	*Isatis tinctoria* L.
alizarin, purpurin	anthraquinone	madder ladies bedstraw dyer's woodruff	*Rubia* sp. *Galium verum* L. *Asperula tinctoria* L.

Table 42.2. Possible European biological dye sources related to the identified dye components.

dyeing is found in eight of the sites spread over a time span of the whole Early Iron Age (500 BC–AD 400). The oldest indigotin-dyed find is from Rebild, dated between the 4th and the 3rd century BC. Søgårds Mose II is the only site where woad is the solely detected dye source. On the other samples, woad was found together with other dyes like the luteolin yellow, yellow berries, a lichen dye, probably Scandinavian orchil and tannin. For all the textiles that do contain luteolin, it is quite remarkable that, it seems to have been used often in combination with woad dyeing for the creation of stripes and checked patterns through the use of different coloured warps and wefts. Skærsø is the only textile where a red dye from the type of madder, dyer's woodruff or ladies bedstraw was used. The presence of lichen dyes could

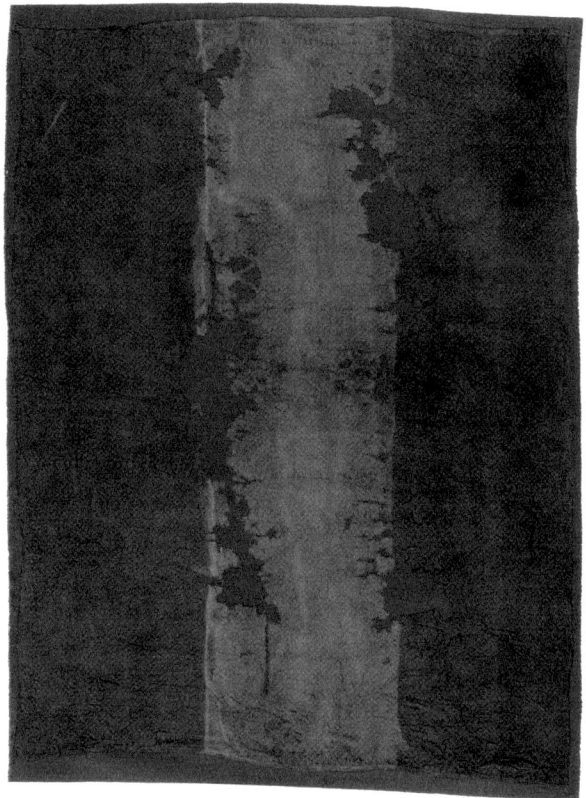

Fig. 42.1. Textile with tassels from Skærsø, Koldinghus Museet (Photo: Roberto Fortuna; National Museum of Denmark).

Fig. 42.3. Scarf from Huldremose, Inv. no. 3473, National Museum of Denmark (Photo: Roberto Fortuna; National Museum of Denmark).

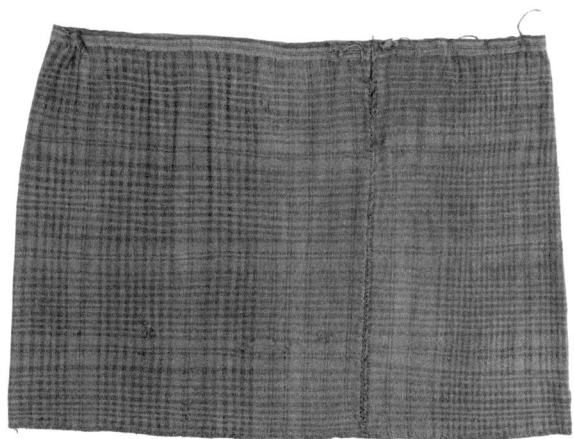

Fig. 42.2. Skirt from Huldremose, Inv. no. 3473, National Museum of Denmark (Photo: Roberto Fortuna; National Museum of Denmark).

be suggested based on the analyses of finds from Huldremose I and Borremose I.

The Textile from Skærsø

This textile was found during peat cutting in 1944. It is a large, dark brown, almost complete rectangular piece of 2/2 diamond twill with all four edges tablet-woven (Hald 1980, 66–67; Mannering and Gleba forthcoming). It is radiocarbon-dated to 210 BC–AD 90 (Østergaard 1996, recalibrated in 2007). Tassels were preserved on three corners, each consisting of three groups of tablet warp threads wound with a much thinner plied yarn of red colour. The textile was analysed for dyes in the 1990s and a madder-like dye was identified in the tassels but not in the textile itself (Walton 1988, 148; Nørgaard and Østergaard 1994, 18).

The new tests have identified alizarin in the warp thread of the ground weave, while on the weft, only a trace of alizarin was found. It is not possible to say with certainty that both weft and warp were dyed. It is rather likely that the weft was undyed or only slightly coloured compared to the warp. In the threads belonging to the tablet-woven border, no dyes were detected, while in the thread of the red-coloured tassel, both alizarin and purpurin were identified with the relative ratio of 90 alizarin over 10 purpurin at 255 nm.

This was the only textile from the whole analysed collection, in which a red dye source of the *Rubiaceae* family was identified, indicating the use of either ladies bedstraw (*Galium verum* L.), dyer's woodruff (*Asperula tinctoria* L.) or a madder type (*Rubia* species). The presence of alizarin as the only or major anthraquinone dye rather suggests the use of a madder-like type of dye instead of the more local *Galium* or *Asperula* species, which generally have purpurin as the major characteristic component (Cardon 2007, 127).

The Skærsø textile provides to date the earliest evidence for madder-like dye in Scandinavia, although it has been identified in later finds by previous studies: in two Danish archaeological textiles from the Migration period (AD 400–550) and in a tablet-woven band from a tunic from the same period, found in the grave-mound at Högom in Sweden (Hofenk de Graaff 2004, 125). However, in these textiles, purpurin was identified without alizarin.

Huldremose I

In 1879, a body of a woman was found during peat cutting. The find is radiocarbon-dated to 350–41 BC. The woman was dressed in two skin capes, a skirt and a scarf (Hald 1980, 47–54; Mannering and Gleba forthcoming). The skirt is a chequered wool piece in 2/2 twill, with one tubular tabby selvedge and one wide warp-faced tabby band with tubular tabby selvedge, which was used as a waistband. The scarf is a chequered wool rectangular cloth in 2/2 twill with tubular tabby selvedges.

Different dye components were identified for skirt and scarf.

Skirt 3473:

	woad	luteolin	unknown 5
Light warp	x	x	x
Dark warp	x	x	x
Light weft	x	x	x
Dark weft	-	x	x

Dye components (all at trace level) referring to both woad and a luteolin based yellow dye source were found in three out of the four analysed threads. In addition, an unidentified component (unknown 5) is systematically found in this textile.

Scarf 3474:

	yellow berries	Scand. orchil	unknown 5
Light warp	-	-	x
Medium warp	x	x	x
Dark warp	-	x	x
Light weft	-	x	x
Medium weft	x	x	-
Dark weft	-	x	-

No indication of a blue dye source was found in the scarf, while a different yellow dye source was used, found on the 'medium brown' coloured threads both in warp and weft. Apart from the light coloured warp, all threads were dyed with a lichen dye, probably Scandinavian orchil. Here, the same unknown component 5 is found as detected in the skirt.

Conclusions

The results of the natural organic dye analysis, using HPLC-DAD, clearly indicate that the majority of Danish Early Iron Age bog textiles were originally dyed. Eight different dye constituents, belonging to the three main groups of natural organic dyes were identified, significantly enlarging our knowledge of the use of biological dye sources in the Early Iron Age. However, dye components were only detected at trace level; hence it has to be considered that some dyes might have been missed. A detailed historical and lab-oriented study of local dye sources, also including less common ones, would be useful to improve the knowledge and identification of local dye sources.

Acknowledgements

We would like to thank Mrs. Dean, Mrs. Grierson and Dr. Schweppe for the references offered and Mrs. Marie-Christine Maquoi, for her very accurate lab work.

Bibliography

Bender Jørgensen, L., and Walton, P. (1986) Dyes and Fleece Types in Prehistoric Textiles from Scandinavia and Germany. *Journal of Danish Archaeology*, 5, 177–188.

Cardon, D. (2007): *Natural Dyes – sources, tradition, technology, science*. London, Archetype Publications.

Cardon, D., Wouters, J., Vanden Berghe, I., Richard, G., and Breniaux, R. (2004) Dyes analyses of selected textiles from Maximianon, Krokodilô and Didymoi (Egypt). In C. Alfaro, J. P. Wild, B. Costa (eds), *Purpurae vestes: Actas del I Symposium Internacional sobre Textiles y Tintes del Mediterráneo en época romana (Ibiza, 8 al 10 de noviembre, 2002)*, 145–154. Valencia.

Hald, M. (1980) *Ancient Danish Textiles from Bogs and Burials*. Copenhagen.

Hofenk de Graaff, J. (2004) *The Colourful Past, Origins, Chemistry and Identification of Natural Dyestuffs*. London, Archetype Publications.

Mannering, U., and Gleba, M. (forthcoming) *Designed for Life and Death*. Copenhagen.

Nørgaard, A., and Østergaard, E. (1994) A Reconstruction of a Blanket from the Migration Period. *Archaeological Textiles Newsletter* 18/19, 17–19

Østergaard, E. (1996) More on ^{14}C Dating from Denmark. *Archaeological Textiles Newsletter* 22, 22.

Surowiec, I., Quye, A., and Trojanowicz, M. (2006) Liquid chromatography determination of natural dyes in extracts from historical Scottish textiles excavated from peat bogs. *Journal of Chromatography A*, 1112, 209–217.

Taylor, G. W., and Walton, P. (1983a) Lichen purples. *Dyes on Historical and Archaeological Textiles 2*, 14–19.

Taylor, G. W., and Walton, P. (1983b) Detection and identification of dyes on Anglo-Scandinavian textiles. *Studies in Conservation* 20, 153–160.

Vanden Berghe, I., Gleba, M., and Mannering, U. (2009), Towards the identification of dyestuffs in Early Iron Age Scandinavian peat bog textiles, *Journal of Archaeological Science* 36, 1910–1921.

Vanden Berghe, I., and Wouters, J. (2004a) The Dyes in the Cangrande Textiles (original title: coloranti delle stoffe di Cangrande). In *Cangrande Della Scala, La morte e il corredo di un principe nel medioevo europeo*, 104–111. Venice, Marsilio Editori.

Vanden Berghe, I., Wouters, J. (2004b) Dye analysis of Ottoman Silks. In *The Ottoman Silk Textiles of the Royal Museums of Art and History in Brussels*, 49–60. Turnhout, Brepols.

Walton, P. (1986) Dyes in early Scandinavian textiles. *Bulletin of Dyes on Historical and Archaeological Textiles* 5, 38–43.

Walton, P. (1988) Dyes and Wools in Iron Age Textiles from Norway and Denmark. *Journal of Danish Archaeology* 7, 144–158.

Wouters, J. (1985) High Performance liquid chromatography of anthaquinones: Analysis of plant and insect extracts and dyed textiles. *Studies in Conservation* 30, 119–128.

Wouters, J. (1998) The dyes of early woven Indian silks. In K. Rhiboud (ed.), *Samit & Lampas. Indian motifs*, 145–152. AEDTA/Calico Museum.

Wouters, J., and Rosario-Chirinos, N. (1992) Dyestuff analysis of Pre-Columbian Peruvian textiles with high performance liquid chromatography and diode-array detection. *Journal of the American Institute for Conservation* 31, volume 2, 237–255.

Zhang, X., and Laursen, R. A. (2005) Development of Mild Extraction Methods for the Analysis of Natural Dyes in Textiles of Historical Interest Using LC-Diode Array Detector-MS. *Analytical Chemistry* 77, 2022–2025.

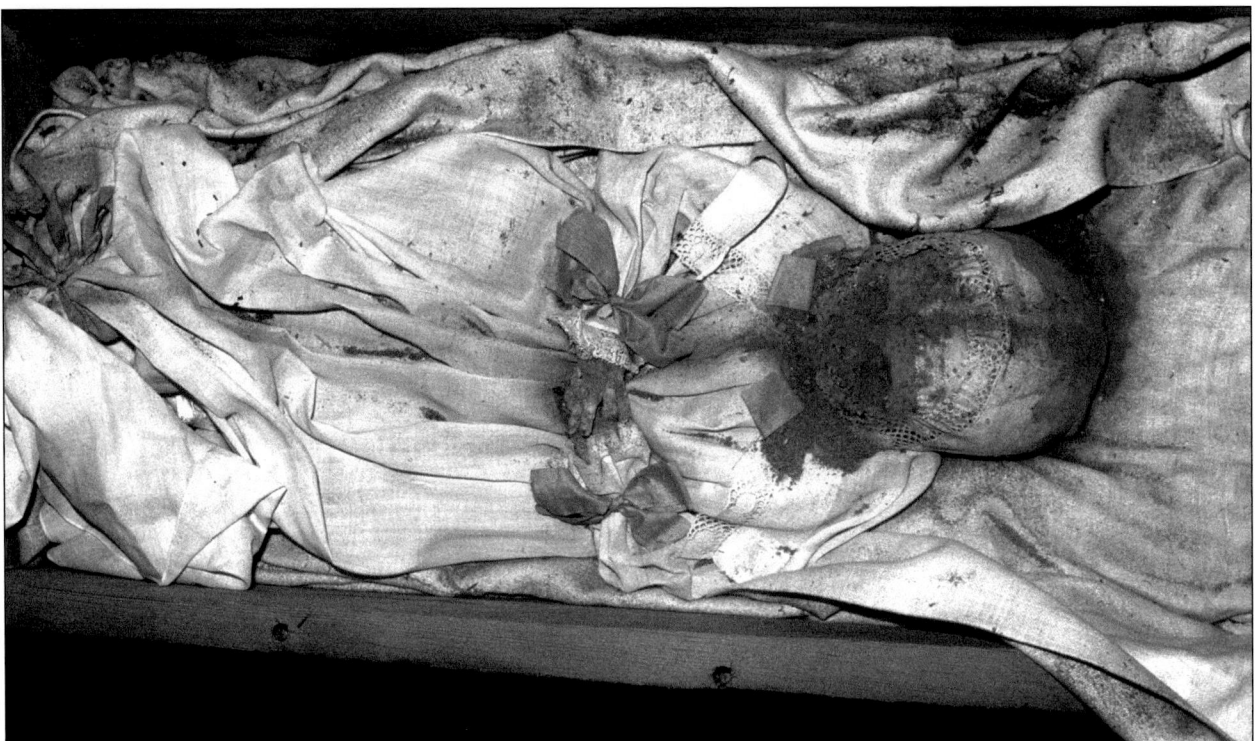

Fig. 43.3. Child burial from Ringebu Stave Church, Norway, mid-18th century (Photo: Bø museum). Reproduced as a colour plate on page 306.

found in the slightly later burial of a woman, probably dating to the first half of the 18th century (Fig. 43.2). The coffin has the form of a chest or trunk, rectangular with a slightly curved, nearly flat lid, made of two boards. The woman was dressed in a headcovering of black lace and probably an imitation burial shirt. Her body was covered with a simple shroud made of linen tabby. However, on top of the shroud was a smaller piece of silk in green and gold, decorated with a simple cross. Another smaller piece of blue brocade covered part of her legs. These smaller pieces of precious material make up for the otherwise modest impression that the burial gives. It is as if the mourners wished to follow a moderate line, yet give the grave a touch of splendour (Vedeler 2008b).

It seems as if simpler forms of burial dress were increasingly accepted in the upper social strata in the middle of the 18th century. By 1750, silk cloth and metal embroidery were no longer in common use as burial dress for the bourgeoisie, judging by the few examples observed in this study. Undyed linen seems to be the preferred material for both burial shirts and coverings. Yet, the use of ribbons made of silk, tied to wrists, feet or to the chest of the deceased was widespread. The ribbons embellished an otherwise simple burial dress. A fabulous example was observed in the Stave Church of Ringebu (Fig. 43.3), where a small child was found buried in a specialised burial shirt made of undyed linen. Her wrists, neck and feet were decorated with silk ribbons, still shiny and in a bright pink colour. Many similar ribbons have been found in numerous Norwegian parish churches, although in most cases their original colours have faded to dark brown.

The Material Image

Written sources indicate that large sums were spent on funerals in the late Middle Ages, and that both the Church and the king sought to reduce the expenditure on funerals after the Reformation (Gittings 1984, 19–38 letter no. 220). Yet, both archaeological and written sources from the 17th century suggest that the burial clothing traditions developed in a more ostentatious manner in the century following the Reformation. Why? A part of the explanation may lie in a new understanding of the funeral ritual.

In the late medieval funerary ritual, it was of uttermost importance to secure the salvation of the soul. According to preserved wills, people were anxious to spend a great deal of money on prayers for the dead, in order to secure their afterlife(DN b.XI., 192.[1]). Even though the location of the grave and the physical condition of the body were important issues, the primary goal was to save the soul. Seen from this perspective, burial dress and the other parts of the grave interior were probably of lesser importance.

From the Reformation onwards, the funeral ritual was almost stripped of eschatological purposes. The burial service was shortened, and prayers for the dead and the requiem mass were no longer part of the ritual. The funeral rites that in Late Medieval times could last for several days and were repeated after a month, were now increasingly restricted to a single day (Houlbrooke 1989, 34).

Clare Gittings has suggested that the late medieval doctrine of purgatory and the rituals which developed around it created a strong bond between the bereaved and the deceased. By organising prayers and rituals for the soul, the survivors could assist the deceased in the afterlife (Gittings 1984, 22–24, 40).

In the 17th century, personal wealth seems to be symbolised by silk and glittering metal embroidery used in burial furnishing. The archaeological material from Norway demonstrates that it was not uncommon to make the coffin resemble a bed, with pillow and flower decorations. Decorations made in gold and silver threads and other expensive materials were increasingly used for funeral preparations. In many aspects, this custom contradicts Luther's advice of moderation and temperance. A crucial point in Luther's *Thesis*, is that eternal life cannot be bought for money (Spaeth *et al.* 1915/2008). In this perspective, one would expect a more moderate and simple burial dress in 17th-century Norway. In fact, the Danish-Norwegian government passed a law to restrict the use of extravagant burial dress in this period. In 1663, King Frederik III issued a decree that forbade any use of silk and other extravagant materials in burial dress.[2] Nearly twenty years later, in 1683, the religious community was threatened by a penalty of 50 *riksdaler*[3] for any contravention of the decree. The king condemned the people's excessiveness and all the extravagant use of money on burial costume. Nobody was allowed to dress their dead in anything but cotton or home-made linen (Troels-Lund 1914).[4] This new attitude could be seen as an expression of the new religious beliefs. On the other hand, it is likely that the king's desire to restrict the extravagance of the upper class was based more on a wish to control and sustain the order of power and social status, than on a desire to make the burial costume correspond to a religious wish for moderation. Yet, how could a worldly or secular explanation be applied to a religious ritual?

The historian Edward Muir claims in his work, *Ritual in Early Modern Europe*, that the Lutheran Reformation in the 16th century divided Europe into two different ways of understanding rituals (Muir 2005). In most Catholic areas of Europe, the religious rituals in the church were still understood as divine acts. The material images, like sculptures of the saints could be understood as a medium through which God communicated with the congregation. In Protestant communities, a different understanding of the sense and meaning of rituals seemed to be developing. The iconoclasm of the 16th century raised issues concerning spiritual presence in material forms as such. According to Protestant doctrine, rituals are primarily means of understanding and structuring the actions of human beings. They divulge little if anything of the actions of God (Amundsen *et al.* 2006). An important observation here is that the iconoclastic movement questioned sight itself as a source of information of the divine (Muir 2005).

On this basis, it is possible to understand the changes in burial dress in the centuries succeeding the Reformation. The increasing use of precious materials and ostentatious display in burial dress could be seen simply as a possibility of displaying secular wealth and power. In this sense, the funeral ritual became even more secular than before. It is possible that it was more acceptable to display wealth through lavish burial in a world where rituals were perceived more as human actions and less as ways to communicate with God. The king's condemnation could then be understood as an attempt to control the upper social strata becoming more powerful every day. Nevertheless, the material from Norway suggests that burial dress did finally become more moderate in the 18th century.

In any case, there is no doubt that the authorities wished to force through the use of a simpler and less extravagant burial dress. The graves underneath Norwegian church floors that were examined demonstrate that changes did not occur immediately. Judging from the material found in the preserved graves, several decades passed from the time when the first royal decrees were issued, to actual change towards a simpler burial dress took place.

Notes

1 DN b. XI s.192. Letter no. 220.4 mai 1466, Sogn. *Gunnor Jonsdotter gaf setthe oc afhende sire Eric Nielsson prester ath Warfrv kirkiu j Tunsbergh jordh som heiter Ramfnebergh er ligghe j Heiterdaal j Skydesislo War -fru kirkia j fornempda stadh, segh til bønahaldz och salegagn oc syno foreldre som henne thet left hafde til ærwerdeligh æigho oc alzafrædes med lutnmlutumetter (...).*

2 Timme 1842: "Den dødes klæder skal være uden silketøj og bebræmmelse".

3 *Riksdaler* was the basic unit in the Norwegian monetary system from the 16th to the 19th century.

4 Troels-Lund 1914 B XIV:125 "Ingen maae til at klæde deres Liig eller andre Maader i Kisten bruge andet, end Cartun (bomull) eller hiemmegiort Lærred".

Bibliography

Amundsen, A. B., Hodne, B., and Ohrvik, A. (2006) *Ritualer: kulturhistoriske studier*. Oslo: Universitetsforlaget.

Gittings, C. (1984) *Death, burial and the individual in early modern England*. London, Routledge.

Gravjord, I. (2005) *"de sal: folches kisteklæder" i Bø gamle kyrkje: graver frå 1600- og 1700-åra*. Skien, Telemark museum.

Houlbrooke, R. (1989) Death, church and family in England between the late fifteenth and the early eighteenth centuries. In R. Houlbrooke (ed.), *Death, ritual, and bereavement*, 25–42. London, Routledge.

Hovin Stub, K. (1990). Tekstilmaterialet fra Heddal prestegård. Arkeologiske undersøkelser på Heddal prestegård. *Varia* 20.

Jantsch, A. K., and Ødegården, M. (In press) *Gravkjellerne under Vår frue kirke. Gravskikken blant byens borgerskap i perioden 1681–1805*.

Muir, E. (2005) *Ritual in Early Modern Europe*. Cambridge, Cambridge University Press.

Nøstberg, I. (1993) Kirkegårder og holdninger til de døde: en studie med utgangspunkt i den arkeologisk undersøkte kirkegården i Heddal, Telemark. Thesis University of Oslo.

Østergård, E. (2003) *Som syet til jorden: tekstilfund fra det norrøne Grønland*. Århus, Aarhus Universitetsforlag.

Pylkkänen, R. (1955) Gravdräkter från 1600-talet i Åbo domkyrkomuseum. *Åbo stads Historiska museum årsskrift 1953–1954*, Vol. 17–18, 3–40.

Timme, F. (1842) *Kongelige forordninger. Aabne Breve og andre trykte Anordninger for Norge udkomne i Tidsrommet 1648–1814, bind 2*. Christiania.

Troels-Lund, T. (1914) *Daglig liv i Norden i det sekstende aarhundrede*. København.

Vedeler, M. (2007a) *Klær og formspråk i norsk middelalder.* Oslo, Det humanistiske fakultet, Universitetet i Oslo.

Vedeler, M. (2007b) Medieval Clothing in Uvdal, Norway. In *Archäologische Textilfunde: NESAT 9: Nordeuropäisches Symposium für archäologische Textilien, Braunwald, 18.–20. Mai 2005,* 141–147. Ennenda.

Vedeler, M. (2008a) Gravskikk og gravdrakt i Ringebu. *Heimgrenda. Gravskikk og gravdrakt i Ringebu.* 29–34. Ringebu historielag.

Vedeler, M. (2008b) Paramenter og gravtekstiler fra Veøy gamle kirke. Møte mellom liturgi og folkelig sørgeskikk. In *Veøy Kirke. Romsdalsmuseets årbok 2008.* Molde, Romsdalsmuseet.

Spaeth, A. *et al.* (1915) Disputation of Doctor Martin Luther on the Power and Efficacy of Indulgences. 1517. Project Wittenberg http://www.iclnet.org/pub/resources/text/wittenberg/luther/web/ninetyfive.html(Accessed 14.04.08).

Diplomatarium Norvegicum. Dokumentasjonsprosjektet 2001. http://www.dokpro.uio.no/perl/middelalder/diplom_vise_tekst.prl?b=9939&s=n&str= (Accessed 14.04.08).

44 The Moment of Inertia: a Parameter for the Functional Classification of Worldwide Spindle-Whorls from all Periods

by André Verhecken

Introduction

Spindle-whorls are often found in relatively large numbers in archaeological excavations. Archaeologists studying assemblages of spindle-whorls have proceeded in different ways. Some researchers (*e.g.* Kuhn 1988, 101, 136) simply state the number of whorls found, and add the average values for *e.g.* whorl diameters and weights. However, these data are not very instructive (an average diameter value of *e.g.* 5 cm can be obtained from data sets spanning 4.5 to 5.5 cm, but also from 1 to 9 cm) and averages are of very little use when the extreme values or the standard deviation from statistics are not mentioned. Moreover, averages would give a distorted idea when more than one distribution is present in the material studied, as *e.g.* in Fig. 44.7. Other researchers (*e. g.* Bulleid and Gray 1948, tables I–XVI) have recorded the essential parameters of every spindle-whorl and presented them in tables, but without any further use of these data.

Much work has been done by those researchers (*e.g.* Kemp and Vogelsang-Eastwood 2001, 269–273, 282–286) who studied the individual spindle-whorls in great detail and generally recorded diameter, height and weight.[1] Their data, as well as those from the previously mentioned group, have been used for the calculations of the present study. Individual distribution histograms can be made for each of these data sets (*e.g.* Médard 2006, figs 52–54, 56 has published the distribution histograms of these data), but then the question

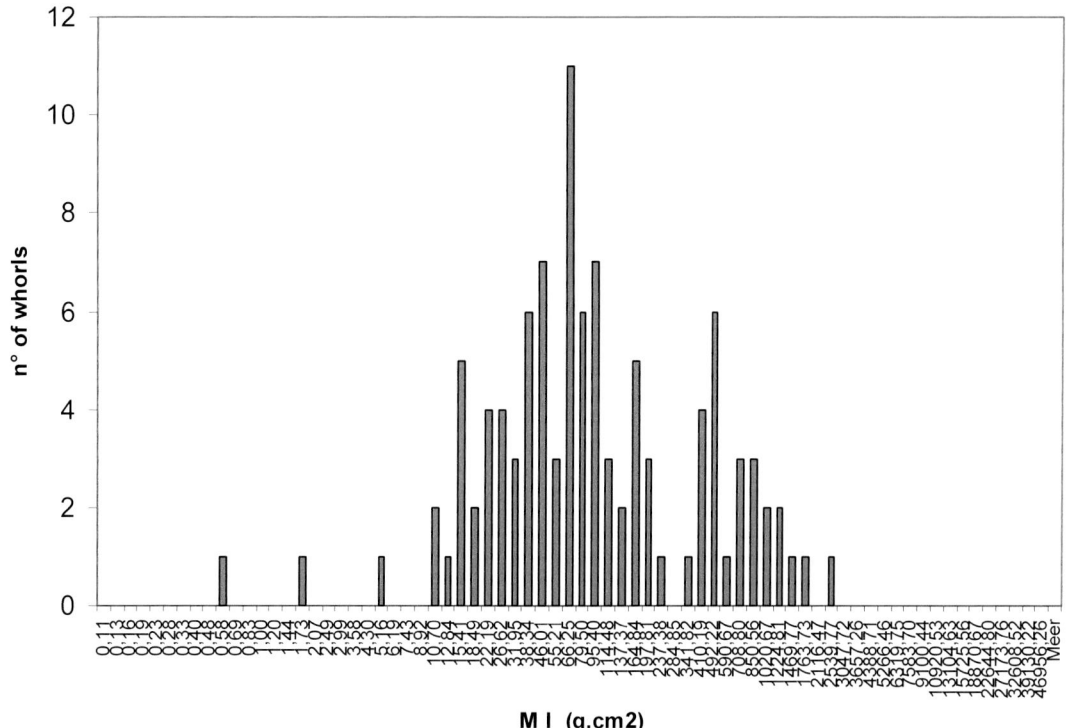

Fig. 44.1. MI histogram, Pharaonic Egypt, based on Kemp and Vogelsang-Eastwood 2001 (n= 103).

arises as to the practical interest of these histograms. Any given spindle-whorl is represented in a given bar in each histogram, but the relationship between the different histograms is not clear. In principle, it is possible that a spindle-whorl is represented by a mid-value bar in the weight histogram, a high-value bar in the diameter histogram and a low-value bar in the height histogram; or any other combination. Thus, these histograms, correct as they are, provide no easily interpretable information as to the main function of a spindle-whorl, which is its rotational characteristics.

It is clear that another approach is needed. In physics, the parameter relative to a rotational movement is the Moment of Inertia (MI), and this will be applied here to spindle-whorls. Using this approach, the essential rotational parameters of spindle-whorls can be 'condensed' into the single numerical value of the MI, rather than having to use form, diameter, mass, height and bore diameter as has mainly been done until now. This parameter greatly simplifies the comparison of whorls with one another. The MI histograms also allow a direct comparison of spindle-whorl assemblages from different geographical and/or historical backgrounds, such comparison being very difficult without the use of the MI.

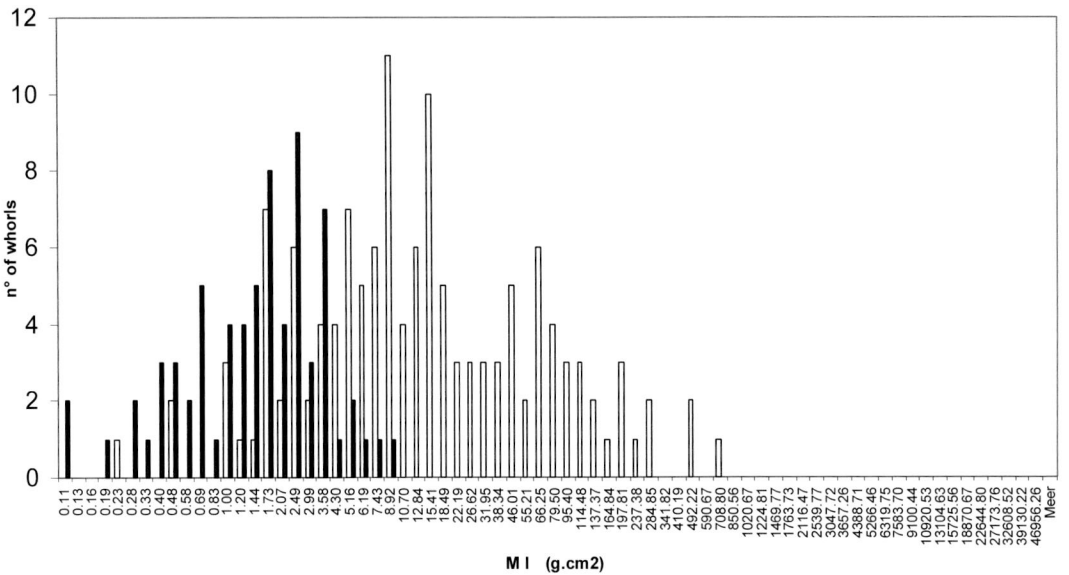

Fig. 44.2. MI histogram, Greece, Nichoria: spindle-whorls (n= 134, white bars) and 'spindle whorls or conuli' (n=70, black bars) based on Carington Smith 1992.

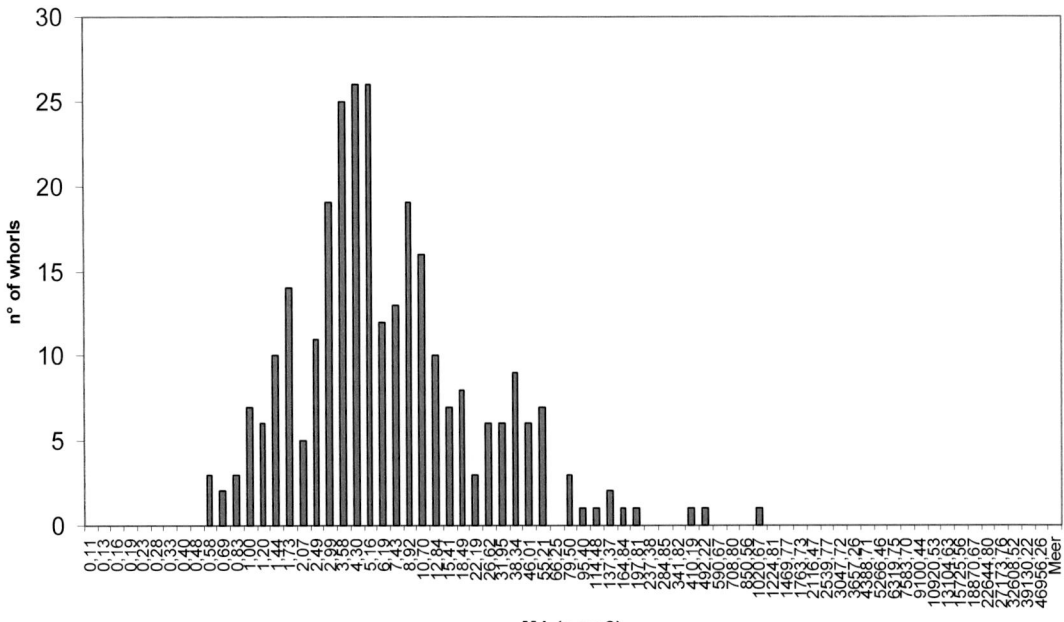

Fig. 44.3. MI histogram, Post-pharaonic Egypt, mainly 1st millennium D (n= 291), based on material in Musée du Louvre, Paris; Benaki Museum, Athens; Royal Museums of Art and History, Brussels; and material said to be from Egypt in the Katoen Natie collection, Antwerp.

The results based on the MI are sometimes rather similar to those already obtained before: *e.g.* the distribution width for Birka and Hedeby spindle-whorls (Figs 44.4–44.5, and Andersson, personal communication), or to give a further example, the several objects classified as *conuli* by Carington-Smith (1992, 685) are found here (Fig. 44.2) to have the same MI as local spindle-whorls.

During this study, it was found that the MI has already been applied to spindle-whorls in a few recently published papers (Loughran-Delahunt 1996; Heckman 1997; Wendling and Wendling 2001; Bocquet 2004; Médard 2006, 105–118). Yet, apart from mentioning the MI values, little or no further

D	H	m	MI		D	H	m	MI
2	16	50	25.1	a	2	8	25	12.5	b
3	3.1	21.9	24.8		3	3.5	25	28.1
4	1	12.6	25.1		4	2	25	50
5	0.4	8.4	25.2		5	1.3	25	78.1

Table 44.1. Combinations of D and H (both in cm) of cylindrical (unbored) spindle-whorls yielding a constant m (in g) with variable MI (in g.cm^2) for $\varrho = 1$ (table 44.1.b); and for a practically constant MI with a variable m (table 44.1.a).

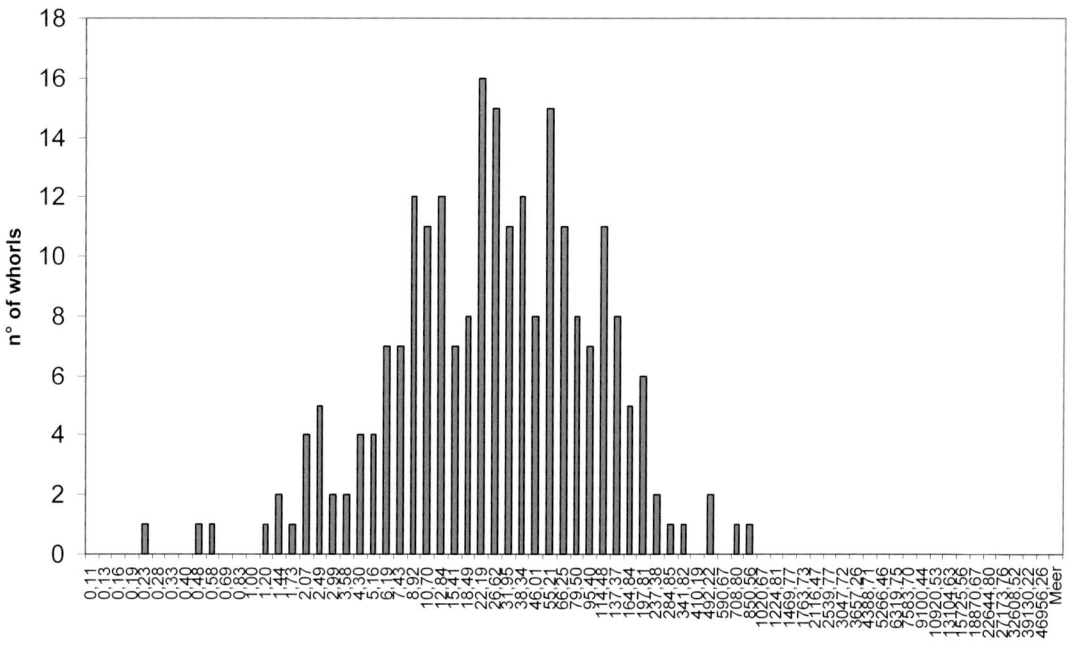

Fig. 44.4. MI histogram, Sweden, Birka, 8th–11th century AD (n= 232) based on Andersson 2003.

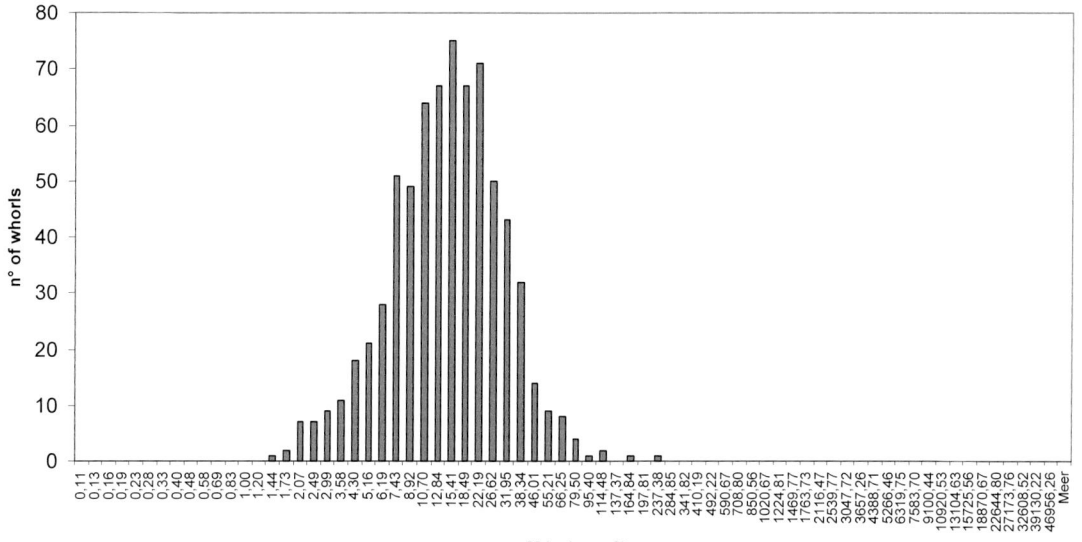

Fig. 44.5. MI histogram, Denmark, Hedeby, 8th–11th century AD (n= 712) based on Andersson 2003.

Whorl form	D (cm)	H (cm)	Bore d (cm)	MI whorl (g.cm²)	MI spindle-shaft (g.cm²)	Relative MI for spindle-shaft (%)
cylinder	4	2	0.7	83.4	0.57	0.7
cone	2	1	0.4	0.53	0.06	10.2

Table 44.2. Relative MI of cylindrical spindle-shaft in wood (ϱ= 0.7, L = 30 cm) as compared to that of the complete spindle with a ceramic (ϱ= 1.66) whorl, for a large cylindrical and a small 'conical' spindle-whorl.

	cylinder		tapered	
d	mass	MI	mass	MI
0.3	2.12	0.02	0.71	0.00
0.4	3.77	0.08	1.26	0.02
0.5	5.89	0.18	1.96	0.04
0.6	8.48	0.38	2.83	0.08
0.7	11.55	0.71	3.85	0.14
0.8	15.08	1.21	5.03	0.24
0.9	19.09	1.93	6.36	0.39
1	23.56	2.95	7.85	0.59
1.1	28.51	4.31	9.50	0.86
1.2	33.93	6.11	11.31	1.22
1.3	39.82	8.41	13.27	1.68
1.4	46.18	11.31	15.39	2.26
1.5	53.01	14.91	17.67	2.98

Table 44.3. Mass (in g) and MI (in g.cm²) of cylindrical and tapered spindle-shafts, calculated for ϱ = 1 and L = 30 cm; d = largest diameter of shaft, in cm.

use was made there of the information obtained (except for Médard 2006).

The present study concentrates on the physical characteristics of spindle-whorls such as form, dimensions and mass, needed for calculation of the MI. Decoration (painted, scratched or incised) is not accounted for here, as long as it does not interfere with these physical parameters. The shape is very important, therefore publications showing figures of spindle-whorls in top-view only (*e.g.* Crowfoot 1931, figs 42–44), without mentioning their shape, are useless in this respect. Ideally, a top- and front-view of each whorl should be given, together with a scale factor, and, if possible, with a cross-section. As a second choice, a front view provides most of the data needed, except the central bore diameter, but this is of less importance (see below).

In this study, a distinction is made between the *spindle-*

author	parameters	yarn	conclusion	MI
Wendling and Wendling 2001		'fine'		13
Wendling and Wendling 2001		'coarser'		32
Carington Smith 1992, 674	m<10 g, D<2 cm	-	suspect	3 *
Crewe 1998, 13	m>10 g, D>=2 cm, d >0.4 cm	flax, wool	acceptable	8.5 *
Mårtensson et al. 2006	m = 3.7 g	wool	proven	1 *

Table 44.4. Lowest values of spindle MI (in g.cm²) needed for spinning: published, or (*) calculated here.

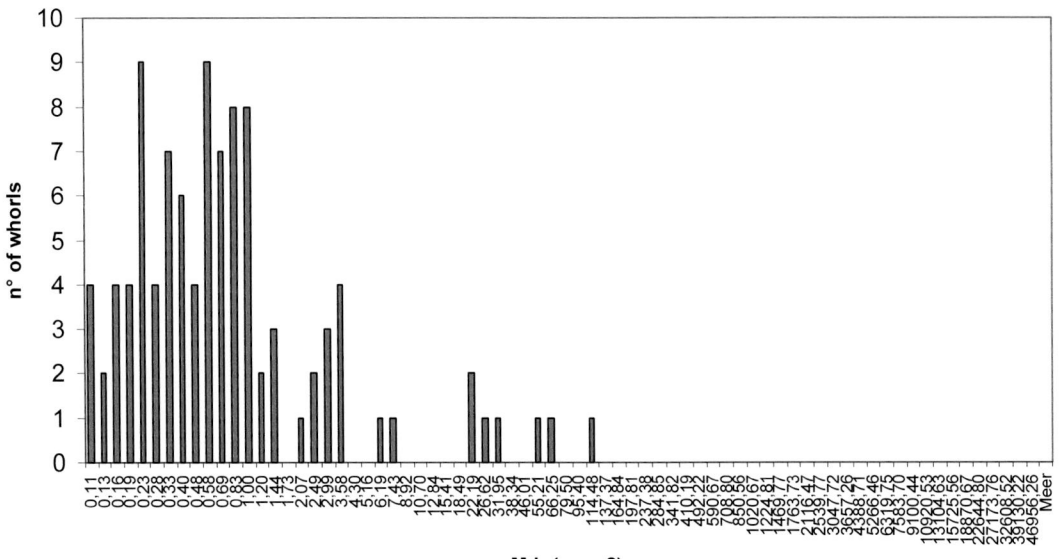

Fig. 44.6. MI histogram, Peru, pre-Columbian, based on Snethlage 1931 and material said to be from Peru in the Katoen Natie collection (n= 25).

whorl, the *spindle-shaft*, and the *spindle* being the combination of both. An arbitrarily chosen spindle-shaft length of 30 cm has been used for calculations of theoretical examples.

Abbreviations used are as follows: **MI** or **I** in formulas: moment of inertia; **D**: diameter of spindle-whorl; **d**: diameter of central bore; **H**: height of spindle-whorl, measured parallel to the rotational axis; **L**: length of spindle-shaft; **m**: mass (in commonly used language: weight); ϱ: density of the material; **RPM**: rotations per minute; **rad**: radian (circumference of a circle divided by 2 π); **s**: second.

Moment of Inertia

Inertia means that a body tends to remain in its actual state of rest or movement, unless it is influenced by an external force. A body without rotation will remain so unless activated by an external force; and a rotating body will keep the same rotation unless it is accelerated or slowed down by an external force. The *moment of inertia* (MI) of an object rotating around a given axis is a measure for indicating how difficult it is to change an object's rotational motion about that axis.

Imagine a point with mass m, rotating around an axis at a distance r. Then the MI is *defined* as

$$I_{point} = m\,r^2$$

Reg. n°	weight (g)	Frequ.1 (turns/6s)	RPM	MI (g.cm²)	Energy (joule)
523	259	59	767	1670	4480
30269/3	8	160	2080	9	178
24827	8	154	2002	4.2	77
Gr.95	20	161	2093	50	999

Table 44.5. MI and activation energy, calculated for some whorls published by Grömer 2005.

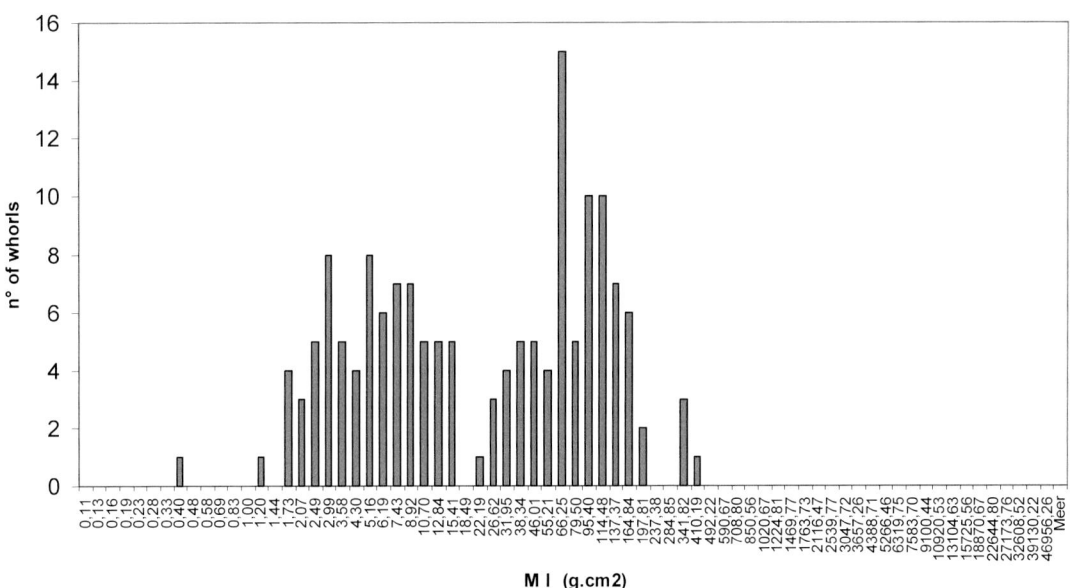

Fig. 44.7. MI histogram, Mexico, Xaltocan (n= 156), c. the Aztec period, based on Brumfiel (in press).

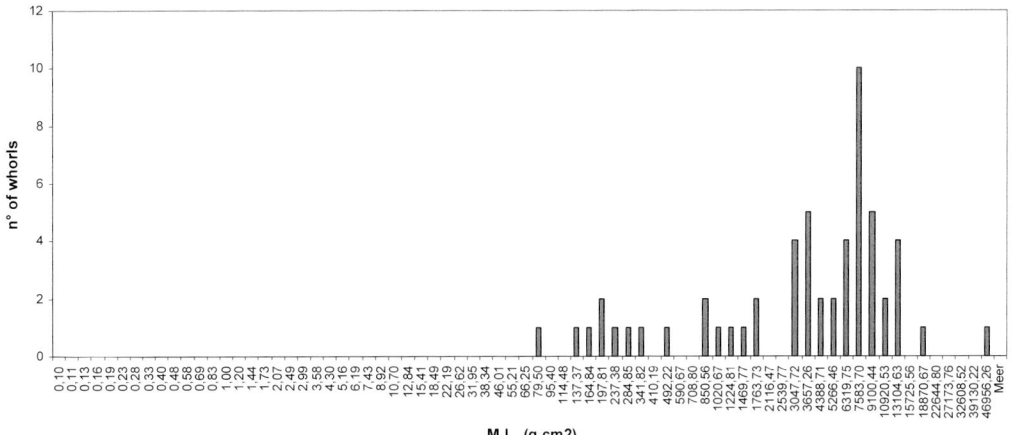

Fig. 44.8. MI histogram, American NW Coast Indian spindle-whorls, 19th–20th century AD (n= 56), based on Loughran-Delahunt 1966.

A real body consists of a very large number of infinitesimal mass points, many of them at different distances from the axis. All these values have to be summed up:

$$I_{body} = \Sigma\, m_i\, r_i^2$$

When the body can be described geometrically, the mathematical technique of *integration* allows drawing up a formula for most of the forms commonly used for spindle-whorls, and thus to calculate the MI of a given whorl. These formulas are rather simple for simple geometrical forms; for example:

(I) Sphere: $I = m\,D^2 / 10$ Cylinder: $I = m\,D^2 / 8$
Cone [2]: $I = 3\,m\,D^2 / 40$

If the mass is unknown but H and ϱ are, then adapted formulas can be used:

(II) Sphere: $I = \pi\,\varrho\,D^5 / 60$ Cylinder: $I = \pi\,\varrho\,D^4\,H / 32$
Cone: $I = \pi\,\varrho\,D^4\,H / 160$

This shows that two factors define the MI: the mass (in

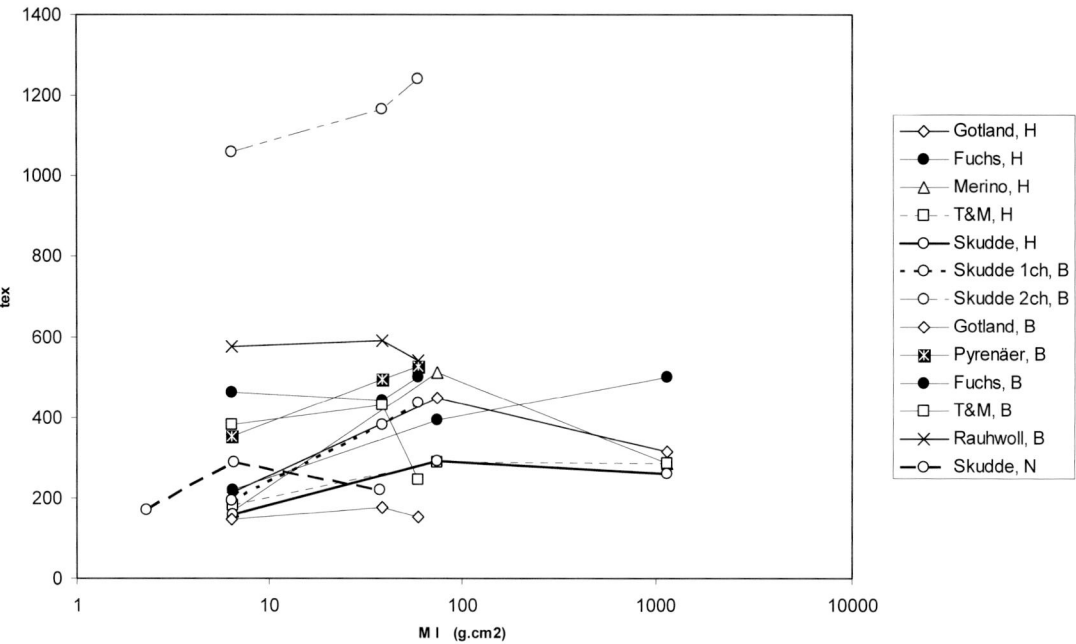

Fig. 44.9. Relation between MI of the spindle-whorls and the TEX of the yarns spun from different races and qualities of wool; number of tests: 33, spun in Museumsdorf Düppell, Berlin. Pyrenäer: hair of Pyrenees dog; T & M: Thones et Marthod sheep; H, B, S: initials of the spinning technicians (see Acknowledgements).

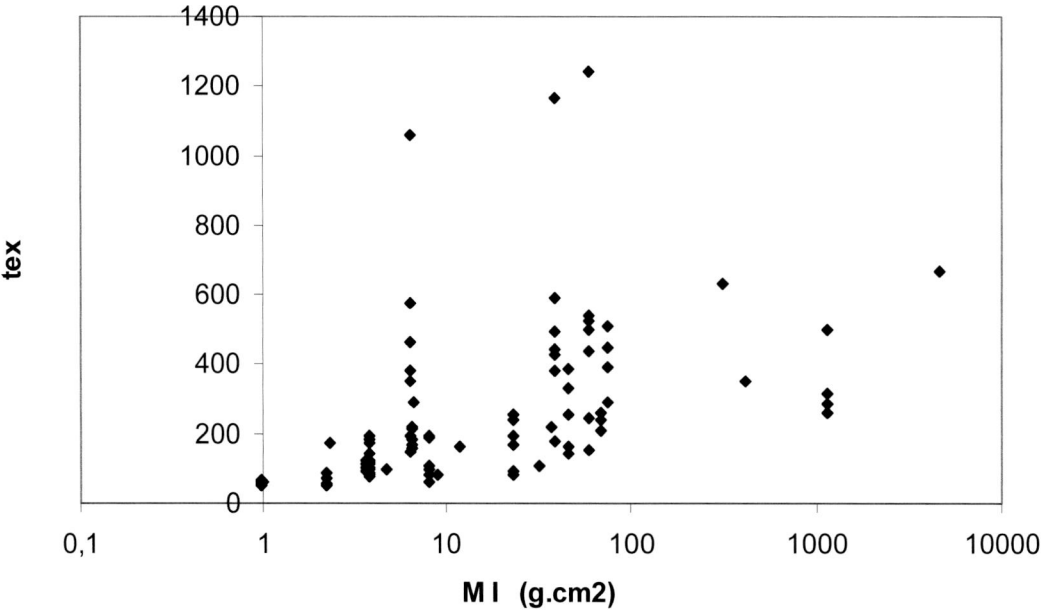

Fig. 44.10. Relation between MI of the spindle-whorls and the TEX of the wool yarns spun with them, based on data from Andersson 2003, CTR reports and spinning tests. Number of data: 111.

commonly used language: the weight), and the mass distribution with respect to the rotational axis (the shape).

For more complex forms, the relevant formulas can be much more complicated. Therefore, the examples given in the present paper are based on simple shapes only (cylinder, cone). However, the MI has also been calculated for the other archaeological spindle-whorls studied here. The formulas used for the calculations are given in Table 44.6.

The official unit of MI in the SI unit system is "kg m^2", but this value is far too large to be used for spindle-whorls. In order to obtain manageable figures, the MI is expressed here in "g cm^2" (there being 10 million g cm^2 in 1 kg m^2).[3]

Why use MI instead of Mass

The mass of a whorl is not uniquely responsible for its rotational properties; it only forms part of the formula for calculating the MI (although evidently, in suspended spinning, the mass can be a limiting factor for the fineness of the yarn, although for reasons independent of the MI).

The distinction between mass and MI is easily demonstrated by the examples in Table 44.1.

For hypothetical unbored cylindrical spindle-whorls with $\varrho = 1$ and *with the same* MI, the variations of mass and H in function of D are given in Table 44.1a; similarly, Table 44.1b gives for whorls *with the same mass* of 25 g, how MI varies in function of D and H.

This shows that D, H and m can vary together and yet give the same MI. There are, of course, practical limits to D and H, because of reasons of strength. For example, a disc made of wood or ceramics, thinner than, say, 2 mm may be too fragile for practical use as a spindle-whorl. Similarly, D, H and MI can vary together and yet give the same m.

Also, imagine two cylindrical whorls with the same height and *the same mass*, but made from different materials, *e.g.* wood ($\varrho_w = 0.7$) and some kind of stone (with $\varrho_s = 2.1$). It is easily calculated from formulas (II) that the wooden whorl, in order to have the same mass as the stone whorl, must have a diameter 1.73 times that of the stone whorl, and then its MI is 3 times that of the stone whorl. Thus, widely different masses can lead to the same MI, and equal masses can lead to different MI values. This clearly shows that the mass cannot be the unique relevant factor for describing rotational movements.

For any given form, the relation MI / m (the MI per mass unit) can be interesting; it could be called the *mass efficiency*. For a cylinder this is: m D^2 / 8 m = 0.125 D^2; a sphere gives 0.1 D^2, and a cone 0.075 D^2. This shows that a cylinder yields the highest mass efficiency.

Even for a simple form like a cylinder, an extreme variability in MI can be achieved for the same mass, by varying its dimensions. A large-diameter flat disc has the highest mass efficiency. This was recognised, albeit in other words, by Barber (1991, 303): "anything fancier than a simple round disk or blob with a centred hole is unnecessary". Thus, theoretically speaking, there was no need to develop the wealth of forms that have been used for spindle-whorls ever since the Neolithic. All kinds of fancy forms can be considered mere local artistic expressions, perhaps perpetuated by tradition, unless they lead to some practical advantage in the context of spinning. Examples are the "hollow whorls" (sometimes referred to as *scooped out*, or *excavated*) mentioned by Barber (1991, 391) with concave top, possibly useful for retaining the spun yarn wound on the spindle-shaft; or tiny elongated whorls useful as ready-made shuttles directly fit for weaving (Barber 1991, 305).

The use of the MI greatly simplifies the study of the rotational parameters of spindle-whorls: it embodies the data of diameter, height, mass, form and material in one single numerical value. Moreover, it obviates the need for a standardized method of recording and interpreting spindle-whorls.

The Accuracy of MI Calculation

There are two reasons why calculated MI values may not give an accurate figure for the real MI value of a spindle-whorl.

The most obvious reason is an insufficient agreement between the real form of the whorl, and the theoretical, geometrically perfect, form on which the calculation formula is based. Similarly, when the mass is unknown, and has to be calculated based on the calculated volume and the density, the density value found in the literature may not correspond exactly with that from the spindle-whorl material (densities of natural materials like wood, bone, antler, stone, may be rather variable).

The second reason is the *error propagation* in the calculation. Every measurement has an inherent 'error', depending on the measuring instrument used. A pair of callipers can measure a length with an accuracy of 0.1 mm, but a measuring rod measures only to 0.5 or 1 cm accuracy; and different types of balances give mass measurements with an accuracy varying from *e.g.* 2 g to 0.1 mg. Also, dimensions or masses that were only estimated, not measured (*e.g.* as pointed out by Holstein 2003, 192) may produce huge errors. Error propagation means that these measuring errors for D, H and m are compounded during the calculation of the MI. There is no room to go into detail here, but errors of *e.g.* 20% and more of the calculated value can be obtained. Therefore, it makes no sense to distinguish between closely related MI values differing by less than that error.

The combination of these factors yields an 'error'. It will be clear from the foregoing that this does not mean that something was done incorrectly: it only points to an uncertainty regarding the result obtained.[4] The error can be assessed for each combination of measuring techniques used, but for simplicity in this paper an error of less than 20% of the measurement value is postulated.

Spindle-shaft, Spindle-whorl and Bore

To allow the mounting of a spindle-whorl onto a spindle-shaft, the whorl must have a central boring. This obviously leads to the loss of some MI, corresponding to the small cylinder of material removed from the whorl; the missing MI is easily calculated with the formulas (II) given above.

264   André Verhecken

form	volume	M I (density is known)	M I (mass is known)
sphere	$\frac{\pi}{12} \cdot D \cdot (2D^2 - 3d^2)$	$\frac{\pi \cdot \rho}{480} \cdot D \cdot (8D^4 - 15d^4)$	$\frac{m}{40} \cdot \frac{(8 \cdot D^4 - 15 \cdot d^4)}{(2 \cdot D^2 - 3 \cdot d^2)}$
spherical cap	$\frac{\pi}{24} \cdot H \cdot (3D^2 + 4H^2 - 6d^2)$	$\pi \cdot \rho \cdot \left[\frac{(8 \cdot R^5 - 15 \cdot R^4 \cdot h + 10 \cdot R^2 \cdot h^3 - 3 \cdot h^5)}{30} - d^4 \cdot \frac{H}{32} \right]$	$m \cdot \left[\frac{16 \cdot (8 \cdot R^5 - 15 \cdot R^4 \cdot h + 10 \cdot R^2 \cdot h^3 - 3 \cdot h^5) - 15d^4 \cdot H}{20H \cdot (3 \cdot D^2 + 4 \cdot H^2 - 6 \cdot d^2)} \right]$
spherical frustum	$\frac{\pi}{24} \cdot H \cdot (3 \cdot D^2 + 3 \cdot D1^2 + 4 \cdot H^2 - 6 \cdot d^2)$	$\frac{\pi \cdot \rho}{480} \cdot \left[240R^4 \cdot (q-p) - 160 \cdot R^2 \cdot (q^3 - p^3) - 48(p^5 - q^5) - 15d^4 \cdot H \right]$	$m \cdot \left[\frac{16 \cdot (8 \cdot R^5 - 15 \cdot R^4 \cdot h + 10 \cdot R^2 \cdot h^3 - 3 \cdot h^5) - 15d^4 \cdot H}{20 \cdot H \cdot (3 \cdot D^2 + 4 \cdot H^2 - 6 \cdot d^2)} \right]$
paraboloid cap	$\frac{\pi}{8} \cdot H \cdot (D^2 - 2 \cdot d^2)$	$\frac{\pi \cdot \rho}{96} \cdot H \cdot (D^4 - 3 \cdot d^4)$	$\frac{m}{12} \cdot \frac{(D^4 - 3 \cdot d^4)}{(D^2 - 2 \cdot d^2)}$
paraboloid frustum	$\frac{\pi}{8} \cdot H \cdot (D^2 + D1^2 - 2 \cdot d^2)$	$\frac{\pi \cdot \rho}{96} \cdot H \cdot [D^2 \cdot (D^2 + D1^2) + D1^4 - 3 \cdot d^4]$	$\frac{m}{12} \cdot \frac{[D^2 \cdot (D^2 + D1^2) + D1^4 - 3 \cdot d^4]}{(D^2 + D1^2 - 2 \cdot d^2)}$
ellipsoid	$\frac{\pi}{12} \cdot H \cdot (2 \cdot D^2 - 3 \cdot d^2)$	$\frac{\pi \cdot \rho}{480} \cdot H \cdot (8 \cdot D^4 - 15 \cdot d^4)$	$\frac{m}{40} \cdot \frac{(8 \cdot D^4 - 15 \cdot d^4)}{(2 \cdot D^2 - 3 \cdot d^2)}$
ellipsoid frustum	$\frac{\pi}{12} \cdot H \cdot (2 \cdot D^2 + D1^2 - 3 \cdot d^2)$	$\frac{\pi \cdot \rho}{480} \cdot H \cdot (8 \cdot D^4 + 4 \cdot D^2 \cdot D1^2 + 3 \cdot D1^4)$	$m \cdot \frac{(8D^4 + 4D^2 \cdot D1^2 + 3D1^4 - 15 \cdot d^4)}{40 \cdot (2 \cdot D^2 + D1^2 - 3 \cdot d^2)}$
hyperboloid frustum	$\frac{\pi}{4} \cdot H \cdot (D \cdot D1 - d^2)$	$\frac{\pi \cdot \rho}{96} \cdot H \cdot [D \cdot D1 \cdot (D^2 + D \cdot D1 + D1^3) - 3 \cdot d^4]$	$\frac{m \cdot [D \cdot D1 \cdot (D^2 + D \cdot D1 + D1^2) - 3d^4]}{24 \cdot (D \cdot D1 - d^2)}$
torus	$\frac{\pi^2}{4} \cdot H^2 \cdot \frac{(D-H)}{4}$	$\frac{\pi^2 \cdot \rho}{64} \cdot (D-H) \cdot H^2 \cdot [3 \cdot (D-H)^2 + 4 \cdot H^2]$	$\frac{m}{16} \cdot [3 \cdot (D-H)^2 + 4H^2]$
donut	$\frac{\pi}{4} \cdot H \cdot [D \cdot (D - 0.5 \cdot H) + 0.122 \cdot H^2 - d^2]$	$\frac{\pi \cdot \rho}{32} \cdot H \cdot (D^4 - D^3 \cdot H + 0.733 \cdot D^2 \cdot H^2 - 0.315 \cdot D \cdot H^3 + 0.058 \cdot H^4 - d^4)$	$\frac{m \cdot (D^4 - D^3 \cdot H + 0.733 D^2 \cdot H^2 - 0.315 D \cdot H^3 + 0.058 H^4 - d^4)}{8 \cdot [D \cdot (D - 0.506 H) + 0.122 H^2 - d^2]}$
conoid/general	$\frac{\pi}{12} \cdot \left[H \cdot (D^2 + D \cdot D2 + D2^2 - 3 \cdot d^2) + e \cdot [D1 \cdot (D+D1) - D2 \cdot (D+D2)] \right]$	$\frac{\pi \cdot \rho}{160} \cdot \left[D^5 \cdot \frac{[e \cdot (D1-D2) + H \cdot (D-D1)]}{(D-D1) \cdot (D-D2)} - \frac{e \cdot D1^5}{D-D1} - \frac{(H-e) \cdot D2^5}{D-D2} - 5 \cdot d^4 \cdot H \right]$	$3 \cdot m \cdot \frac{\left[D^5 \cdot \frac{[e \cdot (D1-D2) + H \cdot (D-D1)]}{(D-D1) \cdot (D-D2)} - \frac{e \cdot D1^5}{D-D1} - \frac{(H-e) \cdot D2^5}{D-D2} - 5 \cdot d^4 \cdot H \right]}{40 \cdot H \cdot [D^2 + D \cdot D2 + D2^2 - 3 \cdot d^2] + e \cdot [D1 \cdot (D+D1) - D2 \cdot (D+D2)]}$
trapezoid/general	$\frac{\pi}{12} \cdot \left[e \cdot (D^2 + D \cdot D1 + D1^2) + (H-e-b) \cdot (D^2 + D \cdot D2 + D2^2) + 3 \cdot (D^2 \cdot b - d^2 \cdot H) \right]$	$\frac{\pi \cdot \rho}{160} \cdot \left[e \cdot \frac{(D^5 - D1^5)}{D-D1} + (H-e-b) \cdot \frac{(D^5 - D2^5)}{D-D2} + 5 \cdot (b \cdot D^4 - d^4 \cdot H) \right]$	$\frac{3m \cdot \left[e \cdot \frac{(D^5 - D1^5)}{D-D1} + (H-e-b) \cdot \frac{(D^5 - D2^5)}{D-D2} + 5 \cdot (b \cdot D^4 - d^4 \cdot H) \right]}{40 \cdot [e \cdot (D^2 + D \cdot D1 + D1^2) + (H-e-b) \cdot (D^2 + D \cdot D2 + D2^2) + 3 \cdot (bD^2 - d^2 \cdot H)]}$
cylinder	$\frac{\pi}{4} \cdot H \cdot (D^2 - d^2)$	$\frac{\pi \cdot \rho}{32} \cdot H \cdot (D^4 - d^4)$	$\frac{m}{8} \cdot (D^2 + d^2)$

Help values for spherical cap: $R := \frac{(D^2 + 4H^2)}{8 \cdot H}$, $h := \frac{(D^2 - 4H^2)}{8 \cdot H}$

Help values for spherical frustum: $p := \frac{(D^2 - D1^2 - 4 \cdot H^2)}{8 \cdot H}$, $q := H + p$, $R := \sqrt{\frac{D1^2}{4} + q^2}$

Table 44.6. Formulas for calculating the volume of spindle whorls of commonly found forms, and their MI in case their density or mass is known. Forms and measurements abbreviations are explained in Figs. 44.11 and 44.12.

The spindle-shaft itself also has its contribution to mass and MI of the spindle combination. However, in general the MI of the spindle-shaft is negligible as compared to that of the whorl, at least for the usual shafts made of wood and with diameters less than about 0.8 cm. Yet, the relative fraction of a cylindrical shaft's MI to the MI of the complete spindle becomes more important for smaller whorls, as shown in Table 44.2 for ceramic whorls ($\varrho = 1.66$) with a cylindrical wooden spindle-shaft ($\varrho = 0.7$; L = 30 cm). The values obtained are negligible and certainly smaller than the uncertainty caused by the errors mentioned above.

Also here the measurement error must be considered. If the spindle-shaft diameter d is known with an accuracy of only 0.1 cm, this leads to large errors. For example: for a cylindrical spindle-shaft [d = 0.7 cm (error 0.1 cm); L = 30 cm (error 0.1 cm)] of wood ($\varrho = 0.8$ (error 0.1)], the relative error on the MI can be estimated to be 70%. Consequently, the absolute error for the spindle-shaft measurements should be less than 0.1 cm in order to obtain a spindle-shaft MI with a reasonable accuracy. In the above examples, a spindle-shaft length of 30 cm is admittedly unusual [although lengths between 27 and 37 cm were used in pharaonic Amarna (Kemp and Vogelsang-Eastwood 2001, 265), Coptic Antinoe (Rutschowscaya 1986, 45), and 20th century Morocco, Turkey and Pakistan (in the collection of F. Sorber, Wijnegem, Belgium; unpublished)]; but shorter lengths have even larger relative errors on the MI.

In most cases, and especially for larger spindle-whorls, the effect of the spindle-shaft MI on the total spindle MI can be neglected.

Normally, spindle-shafts are cylindrical, or they are tapered towards one or both ends. Table 44.3 gives the mass and MI for spindle-shafts (cylindrical, and tapered) for L = 30 cm and $\varrho = 1$. For other values of these parameters, the calculation is straightforward. *E.g.*, for a wooden spindle-shaft with $\varrho = 0.7$ or a bronze spindle-shaft with $\varrho = 8.8$, just multiply the values obtained from the table with the appropriate ϱ values. For different lengths, divide the value by 30 and multiply by the real length.

Which Object is a Spindle-whorl?

Sometimes, it is not easy to distinguish spindle-whorls from other perforated bodies such as small round loom-weights or mace heads, net sinkers, pump drill balances, beads, toy wheels, *conuli*, buttons, pinheads, or pendants. Therefore, it would be useful to have a criterion by which this distinction can be made.

A number of general statements can be found in the literature (Liu 1978, 90; Carington Smith 1992, 674): a spindle-whorl must have a bore placed as centrally as possible, and as vertically as possible (to avoid wobbling) (Barber 1991, 52). The bore should not be smaller than about 4 mm (Liu 1978, 97; Carington Smith 1992, 685), nor have the form of a strongly tapered hour-glass. However, Médard (2006 figs 69, 73) includes several whorls with such a boring, and mentions that 8.5 % of the whorls studied were fixed on the shaft by means of a lump of mastic (Médard 2006, 141, fig.

143). A large D combined with a very small d are unlikely to give a spindle-whorl (Crewe 1998, 14). It can be shown from the formulas (II) that D/d depends on the relationship between the MI's of the whorl and the shaft, and, in simple cases, it depends on the increase in MI.[5]

Crewe (1998, 13) also considers the diameter of the central bore, as compared to the 'size/weight' of the whorl, as an indication for deciding if the object can be a spindle-whorl, but no limit value was given. However, 'size/weight' is a rather vague expression. The weight depends on d, ϱ and the volume of the whorl, and the volume depends on its form and dimensions. So, it appears that what Crewe actually had in mind was the MI.

Carington Smith (1992, 685–686) distinguished spindle-whorls from *conuli*, and classified 87 objects from Nichoria, Greece as "conulus or whorl" (p. 706), primarily based on their weight and bore diameter. Figure 44.2 shows that the MI distributions of the two groups (after eliminating from the *conuli* 11 objects she thinks could be used as spindle-whorls) strongly overlap, thus creating some doubt about her classification.

Minimal MI needed for Spinning

A good criterion for deciding if a bored object can be used as a spindle-whorl might be to ascertain if the combined spindle reaches the minimal MI needed for spinning (a given yarn). This directly leads to the question of what MI is needed for spinning. The few data obtained in the literature, and MI values calculated from them,[6] are given in Table 44.4.

The most relevant data are those by Mårtensson *et al.* (2006, 4, 7), who reported spinning fine wool yarn with a spindle weighing 3.62 g, for which an MI of 1 g cm^2 and an MI increase factor of 373 were calculated here. If this value of 1 g cm^2 is indeed the minimum needed to allow spinning *wool* (as yet unproven), then Table 44.3 shows that *e.g.* a cylindrical shaft with d = 0.8 cm should be sufficient without a whorl. Is this realistic? If not, this may also have to do with the fact that not enough RPM can be given by finger-flicking, the shaft diameter being too high. A tapered shaft is better in this respect; a maximum diameter of 1.2 cm would be needed to obtain 1.2 g cm^2. The MI values for cylindrical and tapered spindle-shafts are given in Table 44.3; the calculation for different values of ϱ and L is straightforward.

From Figure 44.6 it can be deduced that the lowest MI needed for spinning cotton (in pre-Columbian Peru) is only about 0.1 g cm^2. Most probably this was done with the slender spindle with a tiny whorl in the horizontal position, a technique still in use in northern Peru and Ecuador in the 1980s (Meisch 2003–2004, 77).

As already remarked by Crowfoot (1931, 11), the spun yarn wound on the spindle also has a mass, and this works as an extra whorl. Thus, the MI of the complete spindle (shaft + whorl + yarn) changes continually during the spinning process, until the yarn is taken off the spindle. Obviously the spindle, both in its 'empty' and 'full' state, must allow the spinning of the intended yarn type.

It must be concluded that, there is no simple criterion

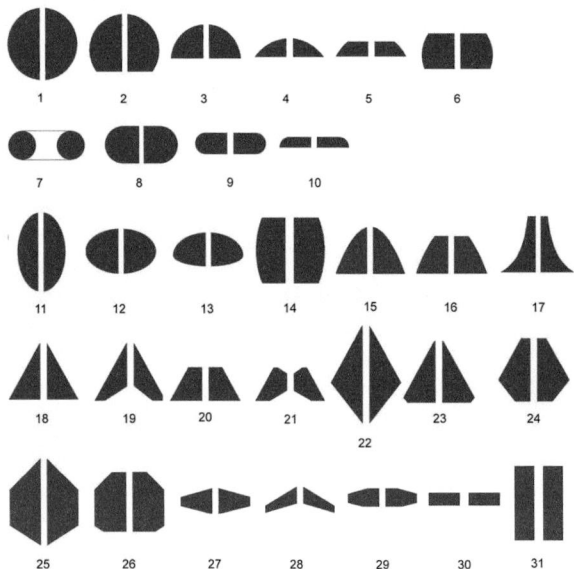

Fig. 44.11. Schematic representation of cross-section of spindle-whorl forms for which formulas are given in Table 44.6.
1: spherical; 2–4: spherical cap; 5–6: spherical frustum; 7: torus (ring); 9–10 "donut"; 11–12: ellipsoid; 14: ellipsoid frustum; 15: paraboloid cap; 16: paraboloid frustum; 17: hyperboloid frustum; 18–24: conoidgeneral; 25–29: trapezoidgeneral; 30–31: cylinder.

for deciding whether or not a given perforated object can be used as a spindle-whorl in general. The MI necessary for spinning depends on the fibre to be spun.

Rotation Speed of Spindle-whorls

Energy is needed to activate a spindle-whorl at rest, but in practice that amount of energy is unknown. In an attempt to compare spindle-whorls with one another, Médard (2006, 106) introduced the concept of a *rotation index*. This is the angular velocity (expressed in rad/s) of any spindle-whorl when activated with an identical, but unknown energy pulse: it depends only on the MI. [7] However, this is a purely theoretical concept, only valid if the activating energy is the same for the different spindle-whorls studied.

We know from physics that this energy, expressed in SI units, is:
(III) $E = I \omega^2 / 2$
where E is the kinetic energy in *joule* (kg m^2/s^2), I is the MI of the spindle-whorl (in kg m^2), and ω is the *angular velocity* (in rad/s). Converting this formula to more familiar units, MI in g.cm^2 and rotation speed in RPM, gives: [8]
(IV) $E = 4.56 \; 10^{-6} \; I \; (RPM)^2$

An approximation of the initial rotation speed of spindles was given by Lindner (1967, 56), repeated by Bohnsack (1981, 57). It was not obtained by spinning fibres but by twining two nylon threads. From the number of turns during the first 6 seconds after activation, Lindner deduced that 10 times that value equals the rotations per minute (RPM) or *frequency*, but this is not entirely correct.[9] Grömer (2005, 112) used a different concept of *frequency*: the number of turns obtained after 6 seconds of rotation.

The energy discussed here is that of the impulse given to the spindle at rest, resulting in a RPM that gradually decreases until the spindle stops. Therefore, the initial rotation speed during the first second of rotation is higher than that based on the value of 10 times the RPM after 6 s. The data given by Lindner (1967, 56) allow an approximation of the starting RPM of *e.g.* his one-piece biconical wool spindle from Tessin (Lindner 1967, 54, fig. 50, n° 5), with m = 23 g and MI = 11.3 g cm^2 (here calculated): 2730 RPM was found instead of his 2100 RPM. [10] His use of the data for 6 seconds led to an underestimation of the starting RPM by a factor 1.3.

Based on Grömer's (2005, 112) data, the MI of some of her spindle-whorls was calculated, and also the RPM and energy input (using formula IV) she gave to her spindles, assuming Lindner's underestimation factor.

Table 44.5 shows that the ratio between the energy inputs given to the spindles with the largest and the smallest MI was 4480/77 = 58 times. Consequently, the assumption of activation by a *constant* energy (necessary for the concept of the *rotation index*) can only be realistic when the whorls have almost the same MI (not the same mass: see rows 2 and 3 in Table 44.5).

Classification of Spindle-whorls based on their MI

Using formulas from the literature or obtained by integration (see above), the MI has been calculated for spindle-whorls for which the following is known:
- The *form*: spherical, cylindrical, ellipsoid, paraboloid, conical, and derived forms.
- All *dimensions* needed for a complete description of the whorl.
- The *mass*. When this is missing, but the *material* of the whorl is known, as a second choice (but with loss of accuracy for the MI) its *density* can be looked up, and this then allows the calculation of the mass.

For rather large numbers of whorls originating from the same site or period, the values thus obtained can be used for drawing up a histogram, showing which MI ranges have the highest occurrence in the material studied. Obviously, also small numbers of spindle-whorls can be used for drawing up a histogram, but its information value will be rather limited. In fact, such histograms only show that a given range of MI values is not void.

Histogram Classes

Before drawing up a histogram, the classes in which the MI values will be classified have to be defined. In the present application, a linear scale is not relevant because:
- The very wide span of the MI values obtained: from about 0.1 g cm^2 for pre-Columbian Peru (Fig. 44.6), till about 50.000 g cm^2 for 19th–20th century spindle-whorls from Indian tribes from the American North-West Coast (Fig. 44.8).
- An increase in MI from 1 to 20 g cm^2 (= × 20) may be very important, but the same difference between 1000 and 1020 g cm^2 (= × 1.02) most probably is

Carington Smith, J. (1992) Spinning and Weaving Equipment. In W. A. McDonald and N. C. Wilkie (eds), *Excavations at Nichoria in Southwest Greece.* Vol. 2, *Bronze Age occupations*, 674–711. Minneapolis.

Crewe, L. (1998) *Spindle Whorls. A study of form, function and decoration in prehistoric Bronze Age Cyprus.* Jonsered, Paul Aström Förlag.

Crewe, L. (2002) Spindle whorls and loom-weights. In G. W. Clarke *et al.* (eds), *Jebel Khalid on the Euphrates: report on excavations 1986–1996,* Vol. 1, 217–243. Mediterranean Archaeology Supplement 5. Sydney.

Crowfoot, G. M. (1931) *Methods of handspinning in Egypt and the Sudan.* Bankfield Museum Notes Ser. 2 n° 12. 1978 reprint of 1974 reprint. Carlton, Ruth Bean.

Dabney, M. K. (2000) Ceramic loomweights and spindle whorls. In J. W. Shaw and M. C. Shaw (eds), *Kommos IV. The Greek Sanctuary. Part 1* (Text), 352–357. Princeton, Princeton University Press.

Gleba, M. (2008) *Textile Production in Pre-Roman Italy.* Ancient Textiles Series Vol. 4. Oxford, Oxbow Books.

Grömer, K. (2005) Efficiency and technique – Experiments with originall spindle whorls. In P. Bilcher, K. Grömer *et al.* (eds) *Hallstatt textiles. Technical analysis, Scientific investigation and experiments on Iron Age Textiles.* Oxford, BAR Int. Ser. 1351.

Heckman, R. A. (1997) Worked sherds and other ceramic artifacts. In S. Whittlesey and B. Montgomery (eds), *The Lower verde Archaeological Project,* Vol 3. *Material culture and physical anthropology*, 87–93. Tuscon, SRO Press.

Holstein, D. (2003) *Der Kestenberg bei Möriken (AG). Auswertung der Ausgrabungen 1950–1953 in der bronze- und eisenzeitlichen Höhensiedlung.* Basel, Holstein.

Kemp, B. J., and Vogelsang-Eastwood, G. (2001) *The ancient textile industry at Amarna.* London, Egypt Exploration Society.

Kuhn, D. (1988) Textile technology: spinning and reeling. In J. Needham (ed.), *Science and Civilisation in China.* Vol. 5. Chemistry and Chemical technology, Part ix, 1–520. Cambridge, Cambridge University Press.

Lindner, A. (1967) *Spinnen und Weben, einst und jetzt.* Lüzern/Frankfurt am Main, C. J. Bücher.

Liu, R. K. (1978) Spindle whorls: Pt. I. Some comments and speculations. *The Bead Journal* 3, 87–103.

Loughran-Delahunt, I. (1996) *A functional analysis of Northwest Coast spindle whorls.* Master's thesis, Western Washington University.

Mårtensson, L, Andersson, E., Nosch, M.-L., and Batzer, A. (2006) Whorl or Bead? Technical Report, Experimental Archaeology, part 2:2, Copenhagen, Centre for Textile Research. Available at: http://ctr.hum.ku.dk/upload/application/pdf/f51d6748/Technical%20report%202-2,%20experimental%20arcaheology.PDF

Médard, F. (2006) *Les activités de filage au Néolithique sur le Plateau suisse.* Paris, CNRS Editions.

Meisch, L. A. (2003–2004) 'Introduction' and 'Cotton'. In Meisch, L. A., Miller, L. M. and Rowe, A. P. (2003–2004) Spinning in Highland Ecuador. *The Textile Museum Journal*, 42–43, 77–97.

Melton, N. D. (1999) Post-medieval spindle whorls in the Northern Isles: examples made with reworked potsherds. *Proceedings of the Society of Antiquaries of Scotland* 129, 841–846.

Murray, E. (1998) Spindle whorls from Laguna de On. In M. Masson and R. M. Rosenzweig (eds), *Belize Postclassic Project 1997: Laguna de On, Progresso Lagoon, Laguna Secca*, 157–162. Institute for Mesoamerican Studies, Occasional publication N° 2. Albany.

Nichols, D. L., McLaughlin, M. J., and Benton, M. (2000) Production intensification and regional specialization. Maguey fibers and textiles in the Aztec city-state of Otumba. *Ancient Mesoamerica* 11, 267–291.

Pritchard, F. (1991) Small Finds. In A. Vince (ed.), *Finds and environmental evidence.* Aspects of Saxo-Norman London: 2, 120–278. London & Middlesex Archaeological Society, special paper 12.

Rutschowscaya, M. H. (1986) *Catalogue des bois de l'Egypte copte.* Paris, Musée du Louvre.

Shamir, O. (1996) Loomweights and whorls. In D. A. Ariel and A. De Groot (eds), *Excavations at the City of David 1978–1985*, Vol. IV, Various reports. *Qedem* 35, 135–179.

Shamir, O. (1999) Spindle whorls. In A. Golani and E. C. M. Van den Brink (eds), *Salvage excavation at the early Bronze Age IA settlement at Azor*, 38. *'Atiqot* 38, 1–50.

Shamir, O. (2001) Spindle whorls. In A. Mazar and B. Panitz-Cohen (eds), *The finds from the first millennium BCE. Text*, 259–262. Timnah (Tel Batash) final reports II. *Qedem* 42. Jerusalem, Hebrew University.

Shamir, O. (2003) Spindle whorls. In A. Golani (ed.), Salvage Excavations at the Early Bronze Age Site of Qiryat Ata, 209–214. *Israel Antiquities Authority Reports* 18. Jerusalem.

Shamir, O. (2004a) The spindle whorls from Ashkelon, Afridar – Area E. *'Atiqot* 45, 97–100.

Shamir, O. (2004b) Loom weights of the Persian period from Horbat-Rogem, Horbat Mesura and Horbat Ha-Ro'a. In R. Cohen and R. Cohen-Amin (eds), *Ancient settlements of the Negev Highlands. Vol. 2, The Iron Age and the Persian period*, 19–27. *Israel Antiquities Authority Reports* 20. Jerusalem.

Shamir, O. (2004c) Spindle whorls. In Y. Hirschfeld (ed.), *Excavations at Tiberias, 1989–1994*, chapter 13. *Israel Antiquities Authority Reports 22*, Jerusalem.

Shamir, O. (2006) Objects associated with the weaving industry. In A. Mazar (ed.), *Excavations at Tel Beth-Shean 1989–1996, Vol. 1. From the late Bronze Age IIB to the medieval period*, 474–483. Jerusalem, The Israel Exploration Society.

Snethlage, E. H. (1931) Form und Ornamentik altperuanische Spindeln. *Beiträge zur Völkerkunde* 14 (2), 77–89. Berlin.

Vandemeulebroeke, I. (1986) *Textiel en textielvervaardiging in België, van prehistorie tot de vroege Middeleeuwen.* Master's thesis, University of Leuven.

Walton Rogers, P. (1997) *Textile production at 16–22 Coppergate. The Archaeology of York*, Vol. 17, *The Small Finds*, fasc. 11. Walmgate, Council for British Archaeology.

Warren, P. (1972) *Myrtos. An early Bronze Age settlement in Crete.* British School of Archaeology at Athens, Supplementary volume No. 7. Oxford, Thames and Hudson.

Wendling, E. and Wendling, J. (2001) *Fusaïoles, fuseaux et moment d'inertie.* (Text drafted 2001-07-07) (http://pagesperso-orange.fr/rouelles/momentdinertie.htm) (Accessed 2008-05-03).

45 Elite and Military Scandinavian Dress as Portrayed in the Lewis Chess Pieces

by Elizabeth Wincott Heckett

About ninety-three 'Chessmen' were found in 1831 on the Isle of Lewis in the Outer Hebrides, Scotland (Robinson 2004, 5–7). Since then some have been lost, but at present seventy-eight pieces are in safe keeping in the British Museum and the National Museums of Scotland (Robinson 2004, 11). There are six kings, five queens, thirteen bishops, fourteen knights, ten warders or rooks and nineteen pawns in the British Museum (BM78-159), and two kings, three queens, three bishops, one knight and two warders in the National Museums of Scotland (NMS19-29) (Figs 45.1–45.2).

Stratford (1997) and Robinson (2004, 13–14) place the manufacture of these chess sets in the second half of the 12th century AD, most probably in Trondheim, Norway. Other single pieces, carved in somewhat different styles have been found in Norway, Denmark, France, Greenland, Ireland, Scotland and Spain.[1] For example, a single late 12th-century queen was found in a bog in County Meath, Ireland; this piece is now in the National Museum of Ireland (Ó Floinn 2002, fig. 7:16, 268, 282) (Fig. 45.3).

Fig. 45.1. Chessmen set out for a game (©The Trustees of the British Museum).

Fig. 45.2. The eleven chessmen held in the National Museums of Scotland (©The Trustees of the National Museums of Scotland).

Fig. 45.3. Queen, late 12th century, found in County Meath, Ireland (Drawing by Joanna Heckett; Courtesy of the National Museum of Ireland).

Fig. 45.4. *Purple silk headcovering, early 11th century, Fishamble Street, Dublin (© The National Museum of Ireland).*

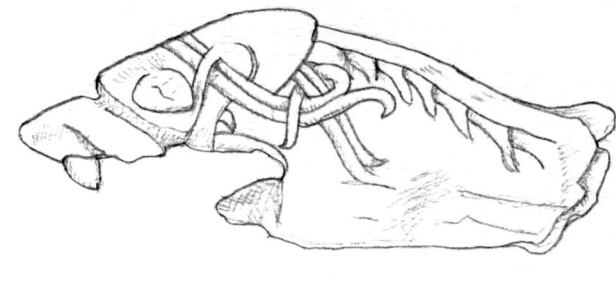

Fig. 45.5. *Animal head chair terminal, wood, 10th century, Fishamble Street, Dublin (Drawing by Joanna Heckett, Courtesy of the National Museum of Ireland).*

Since almost all the Lewis pieces have beautifully delineated clothing, armour and accoutrements, they afford an insight into elite, religious and military Scandinavian dress of the 12th century AD.

In total, eight kings, eight queens, sixteen bishops, fifteen knights, twelve warders and nineteen pawns have survived; almost all made of walrus ivory. This was very likely culled from the seas around Greenland by Norwegian settlers who traded ivory very profitably with northern Europe. The unusually large number of figures found, representing four different chess sets, provides an unrivalled opportunity to study and compare clothing and accessories.

It became apparent to the author, whilst analysing silk and wool headcoverings from Viking Age Dublin that comparisons can be made between specific archaeological finds and the Lewis Chessmen. For example, the queens wear shoulder length veils under their crowns, and a purple silk cloth-piece from Dublin has the right dimensions to be worn in this way (Wincott Heckett 2004, 4–6, 146–147) (Fig. 45.4). Again, Else Østergård's comprehensive work on the medieval clothes from Greenland has shown clearly that some of the full length, male gowns are closely related to those worn by the Lewis warders (Østergård 2004, 96, 128–129). This suggests strongly that the clothes displayed on the different characters are accurate contemporary representations.

Not only do the clothes appear to do this; the ceremonial chairs used by kings, queens and bishops are in some cases shown carved into familiar interlinked animal patterns, with the uprights of the back legs finishing as chair terminals that were carved into animal heads. Excavations in Fishamble Street/St. John's Lane, Dublin produced a 10th century piece of carved wood representing an animal head that was identified as being part of a chair (Fig. 45.5). The slots in one edge show that it originally stood upright at the side of the chair with the throat fastened to another member with a dowel or mortise. The carved head showed traces of gilding and is of high quality (Lang 1988, 15–16, 53–54). James Robinson suggests that a whole walrus tusk, used as a reliquary in the 14th century (British Museum M & ME 1959, 12–2, 1), was carved with similar scrollwork to some of the Lewis thrones and thus may have originally formed part of a chair (Robinson 2004, 30–37).

There are also differences from the dress shown on chesspieces from other regions. For example, the headcoverings of both the late 12th century Irish and Spanish queens differ from those of the Lewis queens (See endnote 1).

Subtle differences confirm the individuality of each piece; an example of this is seen in the representation of braids decorating the hems, neckbands and cuffs of dresses and tunics where different patterns are employed. Some bishops are shown wearing chasubles whereas others are wearing copes. The simple orphreys on the backs of the chasubles and copes are in different styles, as are the devices on the shields of the knights.

The Kings

Although it is clear that there is an accepted style within which the figures have been carved, and in the particular way they are clothed, it is informative to examine closely the clothes each set of figures is wearing. All the seated kings are wearing long, semi-circular cloaks fastened on the right shoulder and reaching to the ground. In one case, a brooch or clasp seems to be shown on the shoulder. Incised lines around the edges of the cloaks, some decorated with dots suggest that narrow bands (perhaps tablet-woven) were either sewn around the edges or were an integral part of the garment. A full robe or tunic can be seen under the cloak, again reaching the ground. These robes also have more or less ornamental edgings at the hem and in one case a beaded effect is created.

The closure of the cloaks at the right shoulder enables the kings to use the swords that each one carries in the right hand, underlining his role as military commander. This is in contrast to the ceremonial robes used by the Norman

King Roger II of Sicily at his court in Palermo at the end of the 12th century (an outstanding example is on view in the Schatzkammer in the Hofburg Castle in Vienna). Such cloaks are clasped at the centre front of the neck so that the wearer's arms have a very limited ability for action but can carry the kingly orb and sceptre. However the pattern for the chess kings' cloaks is similar to that of Roger II. With the Lewis kings, the semi-circular cloaks fall straight down to the ground on the right-hand side leaving the right knee displaying folds in the tunic. The folded tunic is also shown at the left knee.

It is not possible to see how the tunic is cut since most of it lies under the cloak. It may be that it is in the type of dalmatic that we know was made in late 12th century Sicily (an example is also on view in the Schatzkammer in the Hofburg Castle, Vienna). Certainly the kings' sleeves that show below their cloaks are narrow cut between the elbow and the wrist, as they are on the Sicilian dalmatic. Some kings may be wearing an undershirt as well as a tunic. For example, the king BM 78 seemingly has a separate cuff visible below the decorative edging on the tunic sleeve. The undershirt may well be related to such undergarments as the 13th century linen shirt of Louis the Good that is held in the Cathedral Treasury of Notre-Dame, Paris. Another fine example of a linen undergarment is the so-called Alb of St. Hugo (La Valsainte bei Charmey, Chartreuse, Switzerland), dated to the 12th century (Schmedding 1978, 302–303).

Shoes under the robes can be seen on some kings; so little is visible that their pattern is unclear. However, we do know, again from Palermo, how ceremonial silk footwear, a kind of loose boots, was cut. Alternatively, since the kings are armed for action, perhaps the shoes would have been more like a type of 12th and 13th century ankle boots found in Waterford City that were cut in a style widely used in Europe. These boots have also been found in Oslo, in layers pre-dating the 12th century, and Borgund, Norway in the 12th and 13th centuries (Class III, O'Rourke 1997, 704–706, fig. 18:3). Many shoes and boots of this pattern have central vamp stripes like those worn by the British Museum King (BM 78).

The regal hairstyles are very formal, with long, twisted locks arranged over the kings' shoulders and under a crown. In Irish iconography of the 11th century, clerics are shown with similar formally twisted locks on the *Breac Maodhóg* house-shaped shrine, traditionally in the keeping of the O'Farrellys of Drumlane, County Cavan (Wallace 2002, 219, fig. 6.30, 234, 254). One king's crown has a 'double band' at the base of its outer side while another has a band with a line of dots like those on the bishops' mitres.

The Queens

The queens are also wearing full length semi-circular cloaks but these are fastened at the centre front; the dresses seen below are pleated or fall in loose folds to the feet. Circular pleated folds in varying lengths edge the long sleeves of the dresses. It seems that the habit of pleating garments, as seen in the linen shirts found in the 11th century graves at Birka, Sweden may well have persisted into the late 12th century (Hägg 1983, 318, 343–349). Part of a narrow pleated wool skirt in 2/2 twill was found in Herjolfsnæs (Nørlund No. 59, Museum No. D6473), Greenland, dating between the later 12th and the 13th centuries (Østergård 2004, 189).

A feature of the way the queens are represented is the 'thinking position' where the chin rests in the right hand. Four examples have the right elbow cupped and supported in the left hand. In two cases, the left hand rests on the left knee and in another two cases the queens hold a ceremonial horn in the left hand. In European chess, the queen replaced the Arabic vizier as the counsellor of the king, reflecting the different, more powerful role played by aristocratic or royal women by the 12th century. It is suggested that the thoughtful pose represents contemplation, which could be understood as the queen planning strategies for the military campaign.

The horn may have been used for ceremonial drinking or as a container for money (Robinson 2004, 15–19, 44–45). One of the very few cherished survivals from the old Irish royal families is the Kavanagh 'Charter' horn. This is a 12th century elephant ivory horn with 15th-century metal mounts that was preserved by the Kavanagh family in their ancestral home at Borris, County Carlow. It is a symbol of the kingship of Leinster and a unique survival of Irish royal regalia (Ó Floinn 2002, 267, 279, fig. 7.13). Another ceremonial horn with obvious Scandinavian associations, the elephant ivory Horn of Ulph, that has been one of the treasures of York Minster, England since the 11th century, illustrates another function of the ceremonial horns of the period. Ulph Thoroldsson, a local chieftain, gifted his land to the Minster and sealed the bargain symbolically by filling this horn with wine which he swiftly drank, and then placed it on the high altar to seal his gift.

The Bishops

There are nine bishops in chasubles and six wearing copes, both of which are made from semi-circular pieces of cloth. The bishops' vestments seem to show signs of the hoods that were an integral part of the pattern of both the cope and the chasuble when they were first worn in the 5th and 6th centuries AD.[2] The folds of cloth carved at the neck of the pieces are seen clearly and certainly seem more substantial than a collar.

The chasubles have a cross piece of cloth fastening at the centre neck, and a decorated band down the front of the vestment that would hide the stitching together of the cloth that creates the bell shape. The English word 'chasuble' derives from the Latin *casula* or 'little house', since the bell chasuble with its hood enclosed its wearer and created a separate, enclosed space. In the early church, both copes and chasubles were used for ceremonial occasions such as the consecration of a church, although by the 13th century, the chasuble became the required vestment for conducting the Mass and ceased to be used at other ceremonies.

The dalmatic can be seen beneath both the cope and the chasuble; this garment derived from the secular tunic woven in one piece that was worn throughout the Roman Empire. Beneath the dalmatic is another garment known as the alb,

Fig. 45.6. Felted wool cloth, 13th century, Christchurch, Cork (Photo: E. Wincott Heckett).

usually made from linen. The alb was originally the *linea alba*, a white linen undershirt worn by upper class Romans and is also related to the linen undershirts of the elite in 12th and 13th centuries Europe. The examples of the Alb of St. Hugo and the undershirt of St. Louis of France have been discussed earlier.

The Knights

The knights are shown riding small horses, carrying a lance, shield and sword. They are wearing long-sleeved garments, with the skirts falling to each side of the horse. This suggests either a coat or a gown with slits at the front and back of its skirts. It may be of interest that the Kragelund gown worn by a male corpse from Viborg, Denmark had a divided gusset at the front which would have been practical for riding (Hald 1950, 33; 1980, 39; Østergård 2004, 124–125). There are no carved markings on the chests of the knights to suggest the front closures of a coat. The raised pommel of a saddle can be seen under one knight, and all have saddle cloths with differing trims hanging down on both sides of the horses. From 13th century Cork, a piece of orange-coloured felted cloth with a fringed edge was found that could have been part of such a saddle cloth (Wincott Heckett 1985, 83–87) (Fig. 45.6).

The dating of the chessmen suggests that the knights may be seen both as warriors and perhaps participants in the popular jousting tournaments, known as *melées*. At the end of the 12th century, jousting was not yet a sport for single encounters; indeed, large groups of knights faced each other for an opening lance charge, and then fought to unhorse opponents for ransoms and booty. The Lewis knights carry the pointed lances and swords that were widely used throughout northern Europe at the time. The differing emblems on the knights' shields would be important for the identification of opponents in the heat of a *melée* or battle. It is in the 12th century that these devices became part of the early heraldic tradition.

The Warders or Foot-soldiers

Nine foot-soldiers are wearing long gowns with a pleated gore in the front and three seem to be wearing chain armour. These last have been identified as *beserkers*, the fierce fighters who flung themselves into battle as if possessed. This is signalled by the fact that each one is biting his shield. All carry a sword, two are wearing pointed helmets decorated round the lower edge and the third very likely wears a mail hood.

At least three of the other foot-soldiers wear conical helmets with ear-flaps, perhaps another two have conical helmets without flaps, and one a modified conical helmet with a flattened top and a circle of decoration in the middle. It seems that four of the long wool gowns worn by the warders have the gore pleated outwards. One warder, BM 118, however wears a gown with an inverted pleat in reverse to the others. Many of the Herjolfsnæs gown types seem to have these inverted pleats at the front (Østergård 2004).

Conclusions

We have seen that the Lewis hoard consisted of several sets of chessmen and other gaming pieces. Found on a beach near Uig on the island of Lewis, it was presumably hidden from sight. Probably whoever concealed the hoard intended to reclaim his valuable possessions but some event prevented him from doing so. It seems likely that there were merchants trading from Norway around the Orkneys and Hebrides past the Isle of Man towards Ireland. In the 12th century, the three sons of the Duke of Norway each had kingdoms in the Orkney Islands and the Isle of Man. Further to the south, in Ireland, were the Hiberno-Norse town settlements of Dublin, Waterford, Wexford, Cork and Limerick. The island itself was divided into many Irish kingships, each with its own court and power structure. In AD 1169, Anglo-Norman knights under the direction of Henry II of England invaded Ireland and built up power structures and large landholdings as fiefdoms of their own. All these would be customers for the chess sets and gaming pieces brought by the merchants.

Chess as it was played in the 12th century had become a reflection of European feudal society. This is already shown in the poem written at the monastery of Einsiedeln, Switzerland probably at the end of the 10th century. The terms king, queen, count, knight, rook (*rochus*) glossed as margrave (a noble vassal who controlled a border area) are used in this poem, and not those used in earlier chess (*shatray*) in Arab lands as *shah* (king), *firz* (vizier), *al-fil* (elephant), *farus* (horse), *rukh* (chariot), *baidaq* (footman) that go back to ancient India (Eales 2007, 164–165). As we have seen, in medieval Europe royal women had achieved public duties and recognition and so replaced the eastern vizier or royal adviser and this is reflected in the Lewis chess pieces and other finds. The importance of the Christian Church is emphasized by the appearance of the bishops. In their turn, the knights are dressed and mounted for chivalric battle, reflecting the development of courtly practices. The first craftsmen of Trondheim to carve such figures for the pleasure of northern elite circles must have copied chess sets that were imports from the south of Europe. They would of course have been very familiar with these regal and aristocratic northern patrons and their way of life since they were already satisfying their needs for finely carved items in bone, walrus-ivory and wood. It is not surprising that these master craftsmen were so well fitted to represent with keen-eyed accuracy the clothes and accoutrements of the courts of north-western Europe.

It has been possible to use several Hiberno-Norse artefacts in connection with the Lewis chessmen: the silk headcovering, the leather shoes, the carved animal head chair terminals and the ceremonial horn of the Kings of Leinster. They illustrate the cultural links between Scandinavia and both the Hiberno-Norse and Irish people that were still strong. However, as we have seen, the general elite culture of Christian courts made up of kings, queens, bishops, knights and soldiers was recognised in the 12th century in many parts of Europe and of course in Norway. It is somehow satisfying to know that the Kavanagh royal horn was given to Donall MacMorrough, son of Dermot MacMourrough, King of Leinster in AD 1175, probably no more than a few years later than the carving of the Lewis chessmen by the master craftsmen of Trondheim.

Acknowledgements

I would like to thank Lise Bender Jørgensen for very kindly sending me the information on the chessmen from Hitra, Norway, Krambugata, Sweden and Øm, Denmark.

Notes

1   Some single finds of chess pieces in Europe:
    1. King, southern Italy? late 11th century. Musée Cluny, Paris.
    2. King, Cistercian monastery at Øm, Denmark, before AD 1495. Skanderborg Museum, Denmark.
    3. King, Hitra, Bekkvika, Norway, walrus ivory, Vitenskapsmuseet, Trondheim, Norway (McLees 2001).
    4. King, Sweden, Krambugata, Folkbibliotekstomta, Sweden (McLees 1989).
    5. Queen, County Meath, Ireland, 12th century, National Museum of Ireland, Dublin.
    6. Queen, Spain? (place of manufacture based on headdress style) ivory? 12th century, Walters Art Museum, Baltimore, USA.
    7. Queen, (D 12367.278), Greenland, Qeqertaq Island, Sisimiut, height 84 mm, walrus ivory, in poor condition, AD 1200–1300s, National Museum of Denmark.
    8. Chess piece, Italy, now in British Museum.
    9. Chess piece now lost, Dunstaffnage Castle, Argyll, west coast of Scotland (Madden 1832).
    10. Chess piece, Island of Skye, Scotland, (M.A thesis, Dorothy Bosomworth, Edinburgh University).
2   These garments had an earlier history of being in common, secular use in the Late Roman Empire and northern Europe beyond the Empire (Wild 1963, 193–202; Wincott Heckett 2001, 91–97).

Bibliography

Eales, R. (2007) Changing Cultures: The Reception of Chess into Western Europe in the Middle Ages. In I. L. Finkel (ed.), *Ancient Board Games in perspective Papers from the 1990 British Museum colloquium, with additional contributions*, 162–168. London, The British Museum Press.

Fentz, M. (1998) En hørskjorte fra 1000-arene. In J. Hjermind, M. Iversen and H. Krongaard Kristensen (eds), *Viborg Søndersø 1000–1300: Byarkæologiske undersøgelser 1981 og 1984–85*, 249–66. Højbjerg and Aarhus, Jysk Arkæologisk Selskab/Aarhus Universitetsforlag.

Hägg, I. (1983) Viking Women's Dress at Birka: A Reconstruction by Archaeological Methods. In N. B. Harte and K. G. Ponting (eds), *Cloth and Clothing in Medieval Europe*, 316–350. London, Heinemann Educational Books/The Pasold Research Fund Ltd.

Hald, M. (1950) *Olddanske Textiler*. Nordiske Fortidsminder V, Copenhagen.

Hald, M. (1980) *Ancient Danish Textiles from Bogs and Burials. A Comparative Study of Costume and Iron Age Textiles*. Copenhagen, Publications of the National Museum. Archaeological-Historical Series Vol. XXI.

Lang, J (1988) *Viking-Age Decorated Wood*. Medieval Dublin Excavations 1962–81 Ser B vol. 1. Dublin, National Museum of Ireland/Royal Irish Academy.

McLees, C. (1989) Sjakk-Kongen fra Folkbibliotekstomta, *SPQR fortidsnytt fra midt-norge* 2, 1989, 51.

McLees, C. (2001) Nye funn: Kongen fra Bekkvika – om en gammel sjakkbrikke fra Hitra. *SPQR Nytt fra fortiden*, 1, 50–51.

Madden, F. (1832) Historical remarks on the introduction of the game of chess into Europe and of the ancient chessmen discovered in the Isle of Lewis. *Archaeologia* XXIV, 203–291.

Østergård, E. (2004). *Woven into the Earth – Textiles from Norse Greenland*. Aarhus, Aarhus University Press.

Ó Floinn, R. (2002) Later Medieval Ireland AD 1150–1550. In P. F. Wallace and R. Ó Floinn (eds), *Treasures of the National Museum of Ireland, Irish Antiquities*, 257–300. Dublin, Gill & Macmillan.

O'Rourke, D. (1997) Leather Artefacts. In M. F. Hurley, O. M. B. Scully and S. W. J. McCutcheon (eds), *Late Viking Age and Medieval Waterford Excavations 1986–1992*, 703–736. Waterford, Waterford Corporation.

Robinson, J. (2004) *The Lewis Chessmen* London, The British Museum Press.

Schmedding, B. (1978) *Mittelalterliche textilien in kirchen und klöstern der Schweiz*. Berne, Abegg-Stiftung.

Stratford, N. (1997) *The Lewis Chessmen and the enigma of the hoard*. London, British Museum Press.

Wallace, P. (2002) Viking Age Ireland AD 850–1150. In P. F. Wallace and R. Ó Floinn (eds), *Treasures of the National Museum of Ireland, Irish Antiquities*, 213–256. Dublin, Gill & Macmillan.

Wild, J. P. (1963) The *Byrrus Britannicus. Antiquity* 37, 193–202.

Wincott Heckett, E. (1985) The Textiles. In M. Hurley, Excavations of part of the medieval city wall at Grand Parade, Cork. *Journal of the Cork Historical and Archaeological Society* 90, 85–87.

Wincott Heckett, E. (2001) Beyond the Empire: an Irish mantle and cloak. In P. Walton Rogers, L. Bender Jørgensen and A. Rast-Eicher, (eds), *The Roman Textile Industry and its Influence*, 91–97. Oxford, Oxbow Books.

Wincott Heckett, E. (2003) *Viking Age Headcoverings from Dublin*. Dublin, Royal Irish Academy.

46 Headwear, Footwear and Belts in the Íslendingasögur and Íslendingaþættir

by Anna Zanchi

In my doctoral dissertation I analysed the portrayal and literary significance of dress in medieval Icelandic literature, specifically in the *Íslendingasögur* and *Íslendingaþættir* – the 13th-century Sagas and Tales of Icelanders (Zanchi 2007). The study is a critical assessment of the two corpuses through the lens of costume and textile history, aiming to shed light on the possible connections or discrepancies between dress and identity within the narrative fabric. Chapter 7 of the dissertation occupies a small but nonetheless crucial role in the study, focusing on the role of accessories as they are portrayed in the literature in question. References to headwear, footwear and belts are particularly frequent and are employed, beyond their descriptive nature, in the construction of specific episodes and literary motifs. I have noted how saga authors ingeniously make use of these additional items of dress to animate their characters and illuminate their narratives.

Not only is the analysis of accessories in the Old Norse literary corpus a valuable medium by which to assess the literary significance of specific items of clothing: the study also sheds light on the variegated nature of medieval Icelandic accessories, both from an archaeological and historical perspective, and it is aimed at providing an effective source of inspiration for future dress and textile research in the field. The study is thus multidisciplinary in approach, taking into account the evidence of written sources, which are analysed for their descriptive as well as symbolic qualities, but also that of textile archaeology, costume history, and history of art, so as to offer a wide-ranging and comprehensive view of the topic under discussion.

Headwear

The Old Norse literary corpus is exceptionally rich in episodes describing or dealing with headgear. All instances have been meticulously listed and catalogued by Falk in his comprehensive work on medieval Scandinavian fashions (Falk 1919, 90–116), while the archaeological data and the results of their scholarly interpretation have been summarised by Ewing in his recent *Viking Clothing* (Ewing 2006, 52–55 and 117–122).[1]

Male Headwear

Disproving the outdated notion of horned Viking helmets once and for all, Ewing correctly notes how men's hats seem to have been typically conical in shape throughout the Viking Age (Ewing 2006, 117–122). This interpretation is extensively corroborated by the contemporary pictorial evidence of the Gotlandic stones, particularly those from Lärbro Tängelgårda and Sanda, Sweden, but also the Anglo-Scandinavian stones in Kirklevington and Middleton, England (*cf.* Ewing 2006, 93, 110, 120 and 130). The bronze figurines from Eyrarland, near Akureyri, Iceland, and Rällinge, Sweden, probably depicting the gods Þórr and Freyr respectively, also point to the same conclusions (*cf.* Perkins 2001, 86, 87 and 100).[2] Interestingly, the pictorial rune stone from Lärbro Tängelgårda, as several others of its Gotlandic counterparts, unequivocally suggests a style of conical hat which implied its point to be hanging down the men's necks. Ewing rightly observes that, rather than representing a long-tailed liripipe hood, which came into fashion not earlier than in the 13th century; the headwear in question might correspond to the description found in the 10th-century Persian geographical treatise *Hudud al-'Alam* (The Regions of the World). Here, it is said of the Rus that they 'wear woollen bonnets with tails let down behind the necks' (Minorsky 1937, ch. 44; *cf.* also Birkebæk 1975, 51–52), which also seems to fit with the floppy silk-trimmed hats uncovered in the Birka graves Bj. 581 and Bj. 644 (*cf.* Ewing 2006, 117–121), the points of which were decorated with a silver tip apparently manufactured in Kiev.

The *gerzkr* or *girzkr hattr* (Russian hat) often mentioned in the sagas might be this type of headcovering. Its likely eastern provenance is confirmed by a passage in the *Laxdæla Saga* dealing with the wealthy merchant Gilli, from whom Hǫskuldr, an Icelandic chieftain, had purchased Melkorka, a slave. Sitting in his *skrautligt* (showy) tent, Gilli welcomes the chieftain dressed in a outfit of *guðvefr* and wearing a *gerzkr hattr*, and introduces himself as Gilli inn gerzki (Gilli the Russian) (*Laxdæla Saga*, 22–23). The same air of ostentation can be perceived in the description of another saga episode featuring a man in a Russian hat. The *Gísla Saga* relates Þorkell's impressive arrival at the spring assembly in

Þorskafjǫrðr, dressed in a grey cloak fastened at the shoulder by a gold brooch, a Russian hat and a fine sword. The sons of Vésteinn, disguised as vagabonds, approach him, and the elder one asks Þorkell: 'Hverr er sá inn gǫfugligi, er hér sitr? Eigi hefi ek sét vænna mann né tíguligra' (Who is the magnificent man sitting here? I have never seen anyone more handsome or more majestic). Oblivious of the motives underlying the boys' flattery, and implicitly enjoying the attention received, Þorkell allows the elder to have a look at his sword, with which the unsuspecting Þorkell is immediately slain (*Gísla Saga*, 90–91; *cf.* also Roscoe 1992, 141).

The implicit value of Russian hats is also corroborated by their bestowal as precious royal donations. During his campaign abroad, Gunnarr departs the court of King Haraldr Gormsson in Denmark with the amicable gift of the king's own *tignarklæði*, a pair of gloves embroidered with gold, a headscarf ornamented with golden knots and, *dulcis in fundo*, a Russian hat (*Njáls Saga*, 82–83). Similarly, Earl Hákon of Norway sends a *gerzkr hattr* and a Russian axe to the two Icelandic *goðar* Guðmundr inn ríki and Þorgeirr Þorkelsson 'til trausts' (for protection), in order to smooth the homecoming of the troublesome Sǫlmundr (*Ljósvetninga Saga*, 6).

A less elaborate style of headgear may have been represented by the '*húfa*'-caps often featured in the *Íslendingasögur*, which seem to have been a popular choice among Viking Age and medieval Icelanders. Their material and value could vary: Hǫskuldr is said to have worn a *línhúfa* (linen cap) on the day of his death (*Njáls Saga*, 319), while Ljótr inn bleiki, the showy Viking of the *Svarfdæla Saga*, makes an entrance dressed in a red scarlet tunic, a dark-blue hooded cloak and a '*hlaðbúna húfu á hǫfði*' (cap trimmed with tablet-weaving on his head) (*Svarfdæla Saga*, 136).[3] It is however difficult, in most cases, to identify the actual cut of this garment; it may have been a pointed hat, as frequently seen in Viking Age art, or perhaps a bonnet, as another pictorial rune stone from Lärbro Tangelgårda hints at (*cf.* Ewing 2006, 119). The intrinsic literary significance of the *húfa* is also ambiguous, and we might here be dealing with a general term used to describe headwear in a variety of situations (*cf.* Camilla Dahl in this volume).

The *hǫttr* or *hattr* and *hetta*, all referring to similar hood styles, are by far the most common form of male head-covering in the *Íslendingasögur* and *Íslendingaþættir*. Whether sewn to the neck of a cloak or, more frequently, worn on top of it, hoods are almost invariably portrayed, in the literature in question, as a means of concealment. Particularly in the case of the *hetta*, however, this type of garment is also mentioned with reference to its obvious protective function against unfavourable weather conditions. For instance, it is said of Grettir that he always travelled with his *hetta* down on his shoulders, 'hvárt sem var betra eða verra' (whether in good and in bad weather). Having journeyed to Glaumbær, Iceland, *hettulauss* (bare-headed) on a particularly cold winter night, the people at the farm are surprised by Grettir's apparent insensitivity to the weather, which further highlights the man's incredible strength and resilience (*Grettis Saga*, 224–225). Similarly, the farmer Sveinungr in the *Fljótsdæla Saga* reproaches his son for having wished to fetch his hood and gloves before bringing the sheep back to the farm on a stormy night: ' "Þá er vér vorum ungir, þurftum vér hvórki hött né vöttu" ' ("When we were young we did not need either hood or gloves") (*Fljótsdæla Saga*, 274).

The *kollhetta* featured in the *Kjalnesinga Saga* is, on the other hand, of more dubious interpretation. The coal-biter Kolfiðr's humble outfit includes one such garment, of which he has 'kneppt blöðum milli fóta sér' (tied its two laps between his legs) (*Kjalnesinga Saga*, 17). Falk interprets this very rare term, in my opinion correctly, as indicating a round, close-fitting hood, devoid of the long 'tail' often associated with medieval hoods (Falk 1919, 96). Similarly to the *skauthetta* or *skauthekla* (*cf.* Helgi Guðmundsson 1967, 13–14), the head-piece also comprised a front and a back skirt, which, as Jóhannes Halldórsson too points out (*Kjalnesinga Saga*, 18, n.1), reminds one of the *kjafal* described in *Eiríks saga rauða*: 'þat var svá gǫrt, at hǫttr var á upp ok opit at hliðunum ok engar ermar á ok kneppt saman milli fóta með knappi ok nezlu' (it was so fashioned that it included a hood at the top and it was open at the sides with no sleeves and tied together between the legs with a button and a loop) (*Eiríks saga rauða*, 223).

The function of hoods as concealment devices is portrayed at length in the *Íslendingasögur*. A saga character wearing a *hǫttr* is very likely to be involved in situations implying subterfuge and requiring the disguise of his identity, and the same can be said of men donning a *hattr*, which is most often described as *síðr* (low). Legal stratagems and hidden schemes are often carried out by or with the help of men donning low hoods. Vémundr kǫgurr manages to peacefully settle his case with Steingrímr by infiltrating himself in Steingrímr's farm, disguised as a wanderer dressed in a low hood (*Reykdæla Saga*, 196–197). Grettir's soon to be friend Hallmundr makes his first encounter with the outlaw wearing a 'síðan hatt á hǫfði ok sá ógløggt í andlit honum' ('low hood on his head, and his face could not be made out') and naming himself Loptr, while, in the same saga, Grettir himself inquires about Þórir's whereabouts disguised in 'annan búning ok hafði síðan hǫtt niðr fyrir andlitit ok hafði staf í hendi' (another outfit, with a low hood drooping down on his face and a staff in his hand) (*Grettis Saga*, 175 and 206–207). Njáll's plans for Gunnarr to spy on Hrútr also involve camouflage, and Gunnarr is advised to 'láta slota hatt þinn mjǫk' ('let your hood hang down very low') (*Njáls Saga*, 59).[4]

Fugitives are often described wearing a low hood over their face. The *Laxdæla Saga* and *Gunnars páttr Þiðrandabana* relate Gunnarr's arrival at Guðrún's farm to ask for support in the legal case concerning the murder of Þiðrandi. In both narratives, Gunnarr appears on the day of Guðrún and Þorkell's wedding with a 'hatt síðan á hǫfði' (low hood on his head), which induces Þorkell to inquire about the man's identity (*Laxdæla Saga*, 202 and *Gunnars páttr Þiðrandabana*, 210). Similarly, the *Fóstbræðra Saga* mentions one Gestr and his request to sail from Norway to Greenland aboard Skúfr's ship, on which Þormóðr Kolbrúnarskáld is also sailing, having asked permission to leave King Óláfr Haraldsson's court. Gestr makes an entrance dressed in a 'síðan hǫtt á hǫfði, ok máttu þeir ekki sjá í andlitit á honum' (low hood on his head, and they could not make out his face), and is reluctant to disclose much about his origins and intentions. However, the man's true identity is revealed by King Óláfr himself who, appearing to Þormóðr's host in a dream, warns him about Gestr being in fact an Icelander named Steinarr, who had journeyed to Greenland with the intent of avenging Þorgeirr Hávarsson's death (*Fóstbræðra Saga*, 221 and 255).

Hoods are frequently employed in situations where an armed attack is executed or suffered by the wearer of this garment. Bjǫrn Hítdælakappi grows uncomfortable after having dreamt a nightmarish dream the night before. Venturing into the woods for timber with his farmhands the same day, he bears a large sword and a shield, his head covered by a *hǫttr*. Soon enough, the company is approached by Kálfr illviti and his men, and battle ensues (*Bjarnar Saga*, 196–204). Seeking revenge against Berg-ǫnundr, Egill sails to the island of Herǫla, where the man and his followers are stationed. He disembarks fully armed one night, with a helmet on his head and hood on top of it. A little later, Egill employs the same outfit to secretly visit his friend Arinbjǫrn in York, where King Eiríkr is also stationed (*Egils Saga*, 167 and 178). A hood is also worn with explicitly murderous intent on one occasion. *Fóstbræðra Saga*'s Þormóðr seeks revenge for the attack against Þorgeirr, which Þorgrímr had directed, by entering Þorgrímr's booth at the assembly at Garðar, having turned his cloak to its *svartr* side and having pulled a hood over his head. His identity appears clear to Þorgrímr only too late, when Ótryggr Tortryggsson slashes his head open with an axe (*Fóstbræðra Saga*, 231–233).[5]

Female Headwear

Female headwear is also portrayed at length in the *Íslendingasögur*. At least five types are mentioned in the narratives under discussion, namely the *motr*, *faldr*, and *sveigr*, and the more generally named *hǫfuðdúkr* and *lín*. The headpiece constitutes part of a woman's bridal dress and is featured in wedding scenes. Even more conspicuously, all styles are featured in *Laxdæla Saga*, three of which – the *motr*, *sveigr* and *lín* – only appear in this narrative, which somehow fits with the saga's attention to women's affairs and the many marriages described in it.

While it is impossible to identify with precision the nature of the *lín*-headdress worn by Guðrún's maids-in-waiting (*Laxdæla Saga*, 202), the *hǫfuðdúkr* (headcloth), its use and design, can be understood by the situations in which it is described. The twenty-ells-long *hǫfuðdúkr* that Gísli wishes to give to Ásgerðr implies a garment that was repeatedly wrapped around a woman's head. As Ásgerðr is already married, there is no reason to believe that this particular headpiece was meant singularly for the purposes of a marriage ceremony (*Gísla Saga*, 42). Ástríðr's stratagem to save Helgi's life in *Njáls Saga* also includes a *hǫfuðdúkr*, which is this time draped around a man's head. What is interesting to note at this point of the analysis are Ástríðr's instructions on how to accomplish the camouflage: '"Gakk þú út með mér, ok mun ek kasta yfir þik kvenskikkju ok falda þér við hǫfuðdúki. […] Ástríðr vafði hǫfuðdúki at hǫfði honum"' ("Walk out with me and I will throw a woman's cloak over your shoulders and wrap a head-cloth around your head. […] Ástríðr wove the head-cloth around his head") (*Njáls Saga*, 329).[6] The verb *at falda* (to fold) is also indicative of another common style of headwear, namely the *faldr*. Aside from a group of women who are glimpsed in the process of *at falda sér* (wrap their heads) in their quarters in *Vatnsdæla Saga* (*Vatnsdæla Saga*, 116), all other instances in which a *faldr* is mentioned imply wedding ceremonies or married women.

The association between marriage and enveloping head-dresses emerges from *Rígsþula*, where the married Móðir, just like Edda, is once portrayed in a *faldr*, which is described as 'jutting out' – *at keista* – on the woman's head (*Rígsþula*, st. 2 and 29). Amma is also described wearing a *sveigr* in the same poem (st. 16), a term which, as its etymology suggests (*sveigr* > *at sveigja* (to sway)), may entail a headcloth twisted around the head.[7] As Ewing too correctly observes, the words *faldr* and *sveigr* do not seem to define different garments as such, but different ways of wearing the headscarf (Ewing 2006, 52–53). Whichever the case, a single *sveigr* features in the *Íslendingasögur*, namely the *sveigr mikill* (large head-cloth) worn by Guðrún on the occasion of her husband Bolli's murder (*Laxdæla Saga*, 168). The *faldr*, on the other hand, is more frequently mentioned. Þorgerðr Hǫlgabrúðr (Hǫlgi's bride) is depicted wearing a *faldr* in *Njáls Saga* (*Njáls Saga*, 214),[8] while *Svarfdæla Saga*'s Yngvildr is abducted dressed only in her *serkr* (shift) and *faldlaus* (without head-covering) by Karl at her husband's farm early one morning (*Svarfdæla Saga*, 197) – which might also suggest that a married woman was not considered to be fully clothed without her *faldr*. The *krókfaldr* (crooked head-cloth) appearing in *Laxdæla Saga* is more interestingly employed by its author as a means to an end. Guðrún dreams of one such item and asks her kinsman Gestr to interpret her vision:

Úti þóttumk ek vera stǫdd við lœk nǫkkurn, ok hafða ek krókfald á hǫfði ok þótti mér illa sama, ok var ek fúsari at breyta faldinum, en margir tǫldu um, at ek skylda þat eigi gera. En ek hlýdda ekki á þat, ok greip ek af hǫfði mér faldinn, ok kastaða ek út á lœkinn, – ok var þessi draumr eigi lengri.

(I seemed to be standing by a stream, with a crooked head-cloth woven around my head, which I felt did not suit me, and I was eager to change the head-dress but many told me that I should not do that. But I did not pay attention to that, and pulled the head-dress off my head and threw it into the stream, – and that was the end of this dream.) (*Laxdæla Saga*, 88).

Gestr interprets the dream as a clear allusion to her first marriage, which will turn out to be an unhappy one:

Bœndur mantu eiga fjóra, ok vændir mik, þá er þú ert inum fyrsta gipt, at þat sé þér ekki girndaráð. Þar er þú þóttisk hafa mikinn fald á hǫfði, ok þótti þér illa sama, þar muntu lítit unna honum, ok þar er þú tókt af hǫfði þér faldinn ok kastaðir á vatnit, þar muntu ganga frá honum. Því kalla menn á sæ kastat, er maðr lætr eigu sína ok tekr ekki í mót.

(You will have four husbands, and I expect that the first man to whom you will be married will not have been a love match. As you thought you wore a large head-dress on your head, which you felt did not suit you, you will love this man little, and since you took the head-dress off your head and threw it into the water, you will walk away from him. People say things have been thrown into the sea when one gets rid of his possessions and does not get anything in return.) (*Laxdæla Saga*, 89–90).

Not only does the dream suggest Guðrún's lack of affection for her first husband Þorvaldr; the *krókfaldr*, its large size and imposing weight, convey a feeling of claustrophobia and of a burdening sensation, and highlight the obviously manoeuvred nature of Guðrún's marriage. As Roscoe points out, Guðrún, wealthy and powerful, is given in marriage to a much lower-ranked man, her only comfort being the finery and jewellery which Þorvaldr has sworn to provide for her. The *krókfaldr* signifies precisely these material goods and boastful display of wealth, which are aimed at compensating for an evident lack of status and love (Roscoe 1992, 134–135).

The term *motr* appears only in the *Laxdæla Saga* and refers almost certainly to a mitre-like head-dress of foreign origin (*cf.* Ewing 2006, 169–171 and Falk 1919, 26, 103 and 105–106). The overly sumptuous white gold-embroidered garment is given Kjartan by Ingibjǫrg during his stay in Norway, and is intended as a wedding gift to his fiancée Guðrún. In Ingibjǫrg's words, 'vil ek, at þær Íslendinga konur sjái þat, at sú kona er eigi þrælaættar, er þú hefir tal átt við í Nóregi' ("I wish for Icelandic women to see that the woman with whom you have been in Norway is not of mean extraction") (*Laxdæla Saga*, 131). The events that immediately follow pivot around the headdress: once Kjartan has disembarked in Iceland, Hrefna accidentally comes across the precious item in the man's chest. She unfolds it – *rekr í sundr* – and wishes to *falda sér* with the *motr*, which she later also calls *faldr*. This not only points to the obvious similarities between the two styles of headwear, but also to their similar function as bridal headdresses. As soon as Kjartan notices Hrefna wearing the luxurious headscarf, he apparently forgets about Guðrún and marries Hrefna shortly afterwards, thus giving her the *motr* as a wedding gift (*Laxdæla Saga*, 133 and 138). Events come to a head on the day of the couple's wedding, to which Guðrún is also invited. Guðrún manages to persuade Hrefna to show her the garment, despite Kjartan's protests, and is little short of petrified at the sight of the magnificent *brúðargjǫf*: not only has Hrefna deprived her of the *motr* but, with it, also of Kjartan, or so Guðrún feels (*Laxdæla Saga*, 139–140). Guðrún avenges herself by way of the same garment, stealing it from Hrefna and, figuratively, stealing the woman's beauty and appeal with it, as well as the newly-weds' happiness (*Laxdæla Saga*, 143–144). As a seemingly guilty and ashamed Kjartan finally reproaches his wife for mocking Guðrún: '"myndi Guðrún ekki þurfa at falda sér motri til þess at sama betr en allar konur aðrar"' ("Guðrún would not need to wrap her head in a headcloth to look better than all other women") (*Laxdæla Saga*, 145).

Footwear

According to *Hávamál*, the art of shoemaking was not to be taken lightly:

[...] skósmiðr þú verir né skeptismiðr,
nema þú siálfom þér sér;
skór er skapaðr illa, eða skapt sé rangt,
þá er þér bǫls beðit.

(you shall not be a shoemaker nor a shaftmaker,
save for yourself;
a badly crafted shoe, or a crooked shaft
shall win you misfortune.) (*Hávamál*, st. 126)[9]

Extensive finds of early and late medieval footwear have been uncovered at two archaeological sites in Norway, Borgund and Gullskoen. The results of both excavations have been catalogued and analysed by Anne J. Larsen in 1970 and 1992 respectively. Larsen aimed at hightlighting the extensive and highly organised shoemaking industry which flourished in Norway at least from the end of the Viking Age, pointing to the essential role of shoemakers, as well as tanners, in the local economy (Larsen 1970 and Larsen 1992; *cf.* also Þorkelsson and Jonsson (eds.) 1943, 127–128).

Larsen's archaeological investigations also helped to shed light on the various styles and makes of shoes in vogue in the Viking Age and early medieval times. Interestingly, many of them find representation in the *Íslendingasögur* and *Íslendingaþættir*. Although there does not seem to have been a significant difference between male and female footwear, and most examples in the sagas simply refer to *skór* (shoes), worn as a rule by male characters, other types are also alluded to. Both the *kolbítr* Kolfiðr of the *Kjalnesinga Saga* and the prophetess Þorbjǫrg of *Eiríks saga rauða* are pictured wearing a pair of *loðnir kálfskinnskór* (furry calfskin shoes) on one occasion (*Kjalnesinga Saga*, 18 and *Eiríks saga rauða*, 206–207).[10] Boots, or *upphávir skór* (tall shoes) are also frequently mentioned, seemingly denoting men of wealth and status. Hrútr is said to have owned one such item (*Njáls Saga*, 50), while Skarpheðinn's outfit on the day of the settlement with Flosi includes a pair of 'uppháva svarta skúa' (black boots) (*Njáls Saga*, 304).[11] On the other hand, the *bótar* (boots) appearing in the same scene on top of Flosi's pile of silver may have constituted a rich donation, but also, if we are to judge from the *Norges gamle love*, a *feminine* one. The Bergen bylaws of 1282 mention this type of footwear as *kvenmanz botar* (woman's boots), and so do similar regulations from 1377 and 1384 (*Norges gamle love*, 13, 201 and 219; *cf.* Larsen 1992, 65). This would fit with Flosi's outraged reaction at the sight of the garments which accompanied his monetary compensation. On a more gruesome note, Valbrandr's sons are raking hay at their father's farm barefoot and undressed, having taken their *upphávir skór* and clothing off. When Hávarðr Ísfirðingr arrives and asks them to help him avenge his son's death, the two boys run to fetch their clothes, but when they try to pull their boots on their feet, they realise that they had 'skorpnat í skininu. Þeir stigu í ofan sem skjótast, svá at þegar gekk skinnit af hælunum, ok er þeir kómu heim, váru skórnir fullir af blóði' (shrunk in the sunlight. They stepped into them so quickly that the skin on their heels was immediately stripped off, and when they arrived home their shoes were full of blood) (*Hávarðar saga Ísfirðings*, 321–322).[12]

There seems to have been little glory in losing one's shoes in the thick of the action. Skarpheðinn's mishap at the battle of Markarfljót is exemplary in this respect. As related in *Njáls Saga*:

[þ]at varð Skarpheðni, þá er þeir hljópu ofan með fljótinu, at støkk í sundr skóþvengr hans, ok dvalðisk hann eptir. "Hví hvikask þér svá, Skarpheðinn?" segir Grímr. "Bind ek skó minn," segir Skarpheðinn. "Føru vér fyrir," segir Kári, "svá lízk mér sem eigi muni hann verða seinni en vér."

(It happened to Skarpheðinn that, while they were running down along the river, his shoe-lace snapped and he fell behind. "Why do you hesitate, Skarpheðinn?" says Grímr. "I am tying my shoe-lace," says Skarpheðinn. "Let us go ahead," says Kári, "since I believe that he will be no slower than us.")

It seems as if Skarpheðinn feels taunted and challenged by this inopportune stroke of bad luck and the ensuing anxiety of his companions. As soon as he has secured his shoe-thong, he immediately springs up and runs at full speed with his axe in his hand, to then glide on top of an ice-slab 'svá hart sem fogl flygi' (as fast as a flying bird). He slides towards Þrándr with such speed as to hit him hard enough to plant his axe into the enemy's skull and down to the jaw. "'Karlmannliga er at farit'" ("That was manly"), Kári tells him afterwards, and Skarpheðinn's initial hesitation is fully excused (*Njáls Saga*, 233).

The similar use of a *skóþvengr* as a literary pretext is frequent in the sagas and, as the author of *Stjörnu-Odda draumr* observed, even the occurrence of a shoelace coming undone, 'þó at lítils vægis þykki vera' (although it may seem of little importance), can dictate the unfolding of the events (*Stjörnu-Odda draumr*, 471). The ingenious *addendum* of a shoe-thong in an action scene, where tragicomedy is also involved, can be observed for instance in an episode of the *Eyrbyggja Saga*. Þorbrandr's slave Egill sterki is trying to win himself his freedom but, in order to be granted it, he first needs to spy on the men of Breiðavík. According to the narrative, 'Egill hafði skúfaða skóþvengi, sem þá var siðr til, ok hafði losnat annarr þvengrinn, ok dragnaði skúfrinn' (Egill wore shoes with tasseled thongs, as it was in fashion then; one thong had become untied so that the tassel trailed along). Egill is trying to enter the farm at Breiðavík 'hljóðliga' (silently), but walking across the threshold 'þá sté hann á þvengjarskúfinn, þann er dragnaði; ok er hann vildi hinum fœtinum fram stíga, þá var skúfrinn fastr, ok af því reiddi hann til falls, ok fell hann innar á gólfit; varð þat svá mikill dynkr, sem nautsbúk flegnum væri kastat niðr á gólfit' (he then stepped over the thong's tassel that was dragging along; and when he tried to step forward with the other foot, the tassel got stuck, which caused him to trip and fall down onto the floor; it was such a cracking noise as if a cow's skinned carcass had been thrown onto the floor). Egill's hope for freedom is now dissipated (*Eyrbyggja Saga*, 117). The same incompetence and clumsiness can also be detected in Hallgerðr's slave Melkólfr. He is sent to Otkell's farm at Kirkjubær to steal some food, but one of his shoelaces comes undone on the way. He takes off his knife to fix the thong but, stupidly, leaves the weapon behind. When this is recovered by Skamkell, Otkell's friend, the thief is identified and trouble ensues (*Njáls Saga*, 123–124). A slave's foolishness, but also his master's scornful ruthlessness, is highlighted in another *skóþvengr* episode. *Hávarðar saga Ísfirðings* relates how Steinþórr's slave Svartr is asked to join his master in a wrestling game, despite Svartr's reluctance on account of the work that he still needs to do at the farm that day. At each attack, Svartr's shoes come off, perhaps on account of their poor quality, and he spends a long time tying them back into place after each fall. Steinþórr's men make much fun of the slave and his clumsiness, although Hávarðr is not too impressed (*Hávarðar saga Ísfirðings*, 346).

Skóþvengir are often featured in more violent scenes, which usually end with the death of the unfortunate wearer of an untied shoe. This is certainly the case for *Egils Saga*'s Þrándr, Steinarr's slave, who declares himself so defiant of Þorsteinn's threats should he not abandon his pasture, that he kneels down in front of him to tie his shoelace, at which point Þorsteinn slays him without delay (*Egils Saga*, 281). Shoelaces inadvertently coming undone at duels are also frequent, and the vignette usually seems to pause on the apparent embarrassment which follows the kneeling or bending down of one opponent and the dignity of the other in not profiting from the advantage point. For instance, *Þorsteins þáttr stangarhǫggs* tells of Þorsteinn's lengthy duel with Víga-Bjarni. Eventually, Bjarni's shoe-thong becomes loose, and Þorsteinn invites his opponent to tie it back into place. 'Nú lýtr Bjarni niðr' (Now Bjarni bends down), while Þorsteinn, rather than exploiting Bjarni's momentary vulnerability, walks into the hall to fetch more effective weapons for the fight. The respect is mutual, and neither man fails to acknowledge his enemy's sense of honour: "'Orðit hafa mér svá fœri í dag á þér, at ek mætta svíkja þik, ef ógæfa mín gengi ríkara en lukka þín, ok mun ek eigi svíkja þik,' sagði Þorsteinn. 'Sé ek, at þú ert afbragðsmaðr,' sagði Bjarni' ("I had the chance to take advantage of you today, if my ill-luck had been richer than your good-luck, and I shall not cheat you," said Þorsteinn. "I see that you are an exceptional man," said Bjarni) (*Þorsteins þáttr stangarhǫggs*, 75–76). Grímr's behaviour towards Skúta in the *Reykdœla Saga* is, on the other hand, more treacherous. Skúta has offered Grímr hospitality, unaware of his ulterior motives. The two men are out retrieving some fishing nets one day when Skúta's *skóþvengr* comes undone, at which he stops and bends down to tie up the lace. At that moment Grímr strikes at Skúta, without realising that his host is wearing armour under his cowl. Skúta is unarmed, but wants to know the reason of the attack: Grímr reveals himself to have been sent by Þorgeirr with the intent of killing him, and a merciful Skúta lets him go (*Reykdœla Saga*, 217–218).[13] The same nobility is shown by Kerþjálfaðr, King Brjánn's foster-son, to Þorsteinn Siðu-Hallsson at the battle of Clontarf, as related towards the end of *Njáls Saga*. When King Sigtryggr's army is defeated, Þorsteinn 'nam staðar, þá er aðrir flýðu, ok batt skóþveng sinn' (stopped, whereas others fled, and tied his shoe-lace). Kerþjálfr asks him why he is not fleeing too, but Þorsteinn wittily replies: '"Því [...] at ek tek eigi heim í kveld, þar sem ek á heima út á Íslandi"' ("Because [...] I cannot reach home tonight, since my home is in Iceland"') – at which point Kerþjálfaðr spares his life (*Njáls Saga*, 451–452).[14]

Overall, it is interesting to note how the amount of sympathy given to men whose shoelaces come undone seems to vary according to rank: while the unfortunate occurrence seems to confirm a slave's clumsiness and stupidity, as in the case of Egill or Svartr, heroes like Bjarni and Kerþjálfr are treated with more forbearance, as if the incident represented, in their case, just a stroke of bad luck.

Belts

A close analysis of the *Íslendingasögur* also brings to light, I believe, a fundamental difference in style and function between Viking Age and medieval belts. On the one hand we have belts which were left visible to the eye, on the other those used to support a man's trousers and which were kept hidden under overgarments. To the first category belong the *reip* (rope), *taug* (cord) or *strengr* (string), and the more refined *belti* (belt). The first three terms are invariably mentioned in connection with lowly individuals, or characters disguised as such, dressed in an unimposing *kufl* (cowl). Gísli the outlaw is portrayed, on the day of his death, 'í kufli grám ok hafði gyrt at sér með reipi' (in a grey cowl and had a rope around his waist), a rope which he later uses to contain his entrails during the last fight (*Gísla Saga*, 112 and 114). Króka-Refr disguises his identity in Norway by way of a dark-blue cowl tightened at the waist with a *svarðreip* (rope of walrus hide) (*Króka-Refs Saga*, 151), while Bárðr Snæfellsáss uses an analogous camouflage on two occasions (*Bárðar Saga*, 127 and 129). Grettir also conceals himself with a cowl and *basttaug* (bast cord) in order to ambush Þórir's sons (*Grettis Saga*, 130).[15]

The term *belti*, in contrast, is consistently used in connection with characters of status and wealth or with reference to royal gifts, which clearly highlights the superior nature of this accessory. Interestingly, a *belti*, be it a gift or a mark of status, is almost always combined with a knife, and occasionally with a pouch, which seems to indicate both its use as an item of outerwear and the objects that would customarily be strapped to it. The term *silfrbelti*, referring to belts bejewelled with silver inserts, is particularly indicative of the ornamental function of exterior belts. Such a lavish accessory would hardly have been worn under a tunic, but most likely flaunted on top of it. *Laxdæla Saga*'s Þorgils Hǫlluson, who was earlier defined as *inn mesti ofláti* (most boastful), arrives at his meeting with Guðrún dressed in a russet tunic and a *breitt silfrbelti* (broad silver belt) around his waist (*Laxdæla Saga*, 170 and 194), just as *Finnboga Saga*'s Álfr, the Norwegian nobleman, makes an entrance in a red tunic of scarlet cloth encircled by a *digrt silfrbelti* (broad silver belt) (*Finnboga Saga*, 276). Some of *Njáls Saga*'s prominent characters are also said to have owned or won themselves one such item. Both Gunnarr and Skarpheðinn are once portrayed wearing a *silfrbelti* at public occasions, the first at a horse fight, the latter at the Alþingi (*Njáls Saga*, 150 and 304). In addition, Þorgeirr skorargeirr donates a *silfrbelti* to Guðmundr inn ríki in exchange for his hospitality, while Kári receives one from Mǫrðr as a generous – yet treacherous – parting gift (*Njáls Saga*, 414 and 276). Two women are also depicted sporting a silver belt, namely *Kjalnesinga Saga*'s Princess Fríðr upon Finnbogi's arrival at the cave of the giant king Dofri (*Kjalnesinga Saga*, 29) and *Njáls Saga*'s Hallgerðr, on the occasion of her engagement to Glúmr – about which it is said that she had tucked her long hair under the belt (*Njáls Saga*, 44).[16]

From a literary point of view, it is curious to consider how belts, another seemingly innocuous item of clothing just as headwear and footwear, are employed by saga authors as a means to an end. The term *bróklindi*, referring to a simple band or cord used to tie one's trousers at the waist (*cf.* Cleasby-Vigfússon, s.v. *lindi* and Ewing 2006, 102), is particularly interesting in this respect. The killing of an enemy, or an opponent in a fight, is perpetrated, in at least three saga episodes, by way of one such item of clothing. When Króka-Refr turns down Gellir's insistent requests to join him at a wrestling match, he is forced to fight *nauðigr* (unwilling) against him there and then. However, Refr proves to be hardier than expected and, when Gellir's attacks begin to lose their momentum, Refr grasps his adversary between the shoulders with one hand and 'undir bróklindahaldi' (under the belt buckle/fastening) with the other and throws him off violently, causing him a nasty fall. Gellir stands up at once and attacks Refr with a spear, to then escape with his companions. Refr retaliates shortly afterwards by murdering Gellir, for which he is then compelled to flee to Greenland (*Króka-Refs Saga*, 129–131).

On the other hand, *Bárðar Saga* reports an incident in which a man's death is brought about by his *bróklindi* coming undone. Einarr Sigmundarson gets even with Lón-Einarr for having accused his mother of witchcraft. After a lengthy fight, '[þ]á gekk í sundr bróklindi Lón-Einars, ok er hann tók þar til, hjó Einarr hann banahǫgg' (then the belt of Lón-Einarr's breeches came apart and as he grabbed his breeches Einarr dealt him his death blow') (*Bárðar Saga*, 120–121).[17] There seems to be something unmanly about the manner of Lón-Einarr's death. Lón-Einarr himself is aware of the embarrassment implied, should his trousers fall to his ankles during the duel, and his first reaction at the loosening of the *bróklindi* is to try to hold his breeches up. Einarr Sigmundarsson nevertheless takes advantage of his enemy's momentary weakness. The situation is mirrored in a further saga episode, namely the final duel between Þormóðr and Falgeirr in *Fóstbrœðra Saga* (*cf.* Zanchi 2004, 92–94). Þormóðr journeys to Einarsfjǫrðr with the intent to kill Bǫðvarr and Falgeirr, who had had him outlawed for the death of Þorgrímr. A farmhand tells him that Bǫðvarr is not at home, while Falgeirr and his brothers have gone out fishing. Þormóðr sees the men rowing back to the shore and, after having killed his brothers, he attacks Falgeirr. The fight is protracted and consuming and, in the end, both men fall weaponless and badly wounded from a cliff into the sea. Þormóðr's energy is fading but, as the author tells us, 'fyrir því at Þormóði varð eigi dauði ætlaðr, þá slitnaði bróklindi Falgeirs; rak Þormóðr þá ofan um hann brœkrnar' (since Þormóðr was not destined to die, the belt of Falgeirr's beeches snapped; Þormóðr then pulled down his breeches round him).

This way, Falgeirr is put out of action and cannot swim: he keeps going under, swallowing mouthfuls of water, and finally drowns. It is said that, in his final moments, Falgeirr 'skýtr þá upp þjónum ok herðunum, ok við andlátit skaut upp andlitinu; var þá opinn muðrinn ok augun, ok var þá því líkast at sjá í andlitit, sem þá er maðr glottir at nǫkkuru' (his buttocks and shoulders then rise up out of the water, and so did his face at the moment of his death; his mouth and eyes were open, and the look on his face was like when someone grins at something). Þormóðr later describes his enemy's death to his followers with a rather scornful verse:

> Skoptak enn, þás uppi
> undarligt á sundi
> – hrók dó heimskr við klæki –
> hans razaklof ganði;
> alla leitk á Ulli
> eggveðrs hugar gleggum
> – setti gaurr ok glotti –
> goðfjón – við mér sjónir.
>
> (I was still floating,
> when his anus glared at me
> strangely from the sea;
> – the fool died a shameful death –
> I saw all abomination
> on that mean warrior
> – the man looked at me
> and grinned.) (*Fóstbrœðra Saga*, 239–242)

The episode on Falgeirr's death is particularly revealing when analysed from the point of view of 'grotesque realism'. In her article 'Bróklindi Falgeirs', Helga Kress considers medieval comedy and the ironic undertone of the *Fóstbrœðra Saga*. According to Kress, the saga in question is not heroic in subject matter to the same extent as the *Laxdœla Saga* or *Njáls Saga*, and has never been considered as such. The saga is not told from the perspective of chieftains, as most heroic sagas are, but from that of common people. The *Fóstbrœðra Saga* is a comedy where heroic values are ridiculed, its humour almost carnivalesque. The narrative is alive with ironic and grotesque personages and descriptions, where a keen interest for bodily parts – buttocks and noses – and bodily functions – eating, drinking and digesting – is clearly discernible, as is that for bodily harm, injury and mistreatment (Kress 1987, 51–54 and 60–61).

Several elements distinctive of 'grotesque realism' are present in the episode in question. Falgeirr's belt splits and Þormóðr plainly drags his trousers down in order to drown him. He then mocks the 'fool', the 'mean warrior', who died a 'shameful death', underscoring the particular of Falgeirr's anus abominably glaring at him strangely from the sea. What is more, the author tells us that Falgeirr had died with a grin on his face: this final gesture could be interpreted as a grotesque reflex attributed to a drowning man, but it also seems as if Falgeirr himself were realising, perhaps with a hint of self-irony, the shameful manner of his own death. He may also have perceived the grotesque, if not tragicomic, charge of the event: the great warrior, *Ullr eggveðrs*, dies a dishonourable death and laughs at his own misery.

Conclusions

All in all, the analysis of accessories in the Old Norse literary corpus has proven to be a valuable medium by which to assess the literary significance of specific items of clothing. The study has also shed light on the multifaceted nature of medieval Icelandic accessories, both from an archaeological and historical point of view, and it is hoped that the article will constitute an effective source of inspiration for future dress and textile research in the field.

Whether mentioned for their descriptive or functional value, headwear, footwear and belts are frequently featured in the *Íslendingasögur* and *Íslendingaþættir* as an integral part of a man or woman's outfit. The seemingly innocuous allusion to accessories, be they ornamental or practical, incorporates, more often than not, ulterior motives, and saga authors skilfully make use of these supplementary items of dress to give life to their characters and illuminate their narratives. Wealthy or high-born male characters are glimpsed wearing silver-encrusted belts, tall boots and Russian hats, particulars which draw attention to their status or their boastful flamboyance. A woman's nobility and prosperity are equally signalled by lavish belts and elaborate headcoverings, employed to highlight her beauty and brilliance in the context of public occasions and wedding scenes. Yet, all that glisters is not gold: a man donning a low hood will, almost invariably, be involved in circumstances requiring subterfuge and concealment – an underhand scheme, a furtive mission, a flight, even murder. A woman's bridal headscarf may mark the burden of an unasked-for marriage, or may ignite jealousy and strife. Even more conspicuously, the unfastening of a shoe or the loosening of a belt are ingeniously employed time and again in the construction and rendering of salient scenes, be it for the sake of comedy, tragedy, or both. A *skópvengr* coming undone at inopportune moments can dictate the unfolding of the events, leading to a man's unfortunate death, and underscoring his clumsiness in dealing with the mishap or the unscrupulousness of those taking advantage of this stroke of bad luck. A loosened *bróklindi* foretells the shameful death of the wearer who, preoccupied with the keeping of his trousers, and his dignity, is mercilessly attacked in this flash of weakness – and even scorned for it. Once more, the artistry of saga authors does not fail to entertain, captivate and let us ponder.

Notes

1. *Cf.* also Geijer 1983, 80-99 and Hägg 1983, 316–350 and 1984, 177–188.
2. On a different interpretation of the Eyrarland image *cf.* Kristján Eldjárn 1981, 73–84.
3. *Cf.* Also Steingrímir ørnólfsson, a *skrautmenni it mesta* (the showiest dresser), dressed in a white shirt and a *hlaðbúin húfa* in the *Reykdæla Saga*, 182.
4. *Cf.* also Ófeigr karl's torn dark-brown outfit, comprising a

low hood, upon arrival to the assembly, where he intends to defend Oddr in his legal case (*Bandamanna saga*, 318–319) and Nollarr's similar disguise which he dons in order to spy on Helgi Ásbjarnarson (*Fljótsdæla Saga*, 249–250).

5 *Cf.* also Ásbjǫrn vegghamar in *Fljótsdæla Saga*, whose low hood and coloured outfit anticipate the imminent armed conflict, in which he loses his life (*Fljótsdæla Saga*, 261).

6 *Cf.* the outfit of the female ghost appearing to Herdís Bolladóttir in a dream in *Laxdæla Saga*, 223: 'sú var í vefjarskikkju ok faldin hǫfuðdúki' (she wore a woven mantle and her head was wrapped in a head-cloth). The character in question is however a *vǫlva* (prophetess): her dress might not represent typical female attire.

7 It should be noted that *Rígsþula* defines Móðir as the ancestress of nobility, Edda of slaves as a social class and Amma of the farming class. However, a connection between the usage of a *faldr* or a *sveigr* and class distinction seems to me debatable.

8 Þorgerðr is in fact not a real woman, but the ideal of a goddess. The exact form and meaning of her nickname are also debatable – *cf.* McKinnell 2002, 265–290.

9 The warning against making shoes for someone else may be related to the rather low status of tanners. *Cf.* the story of the quarrel between the smith and the tanner in *Sneglu-Halla þáttr*, 267–269, where the tanner is compared to the giant Geirrøðr and the dragon Fáfnir.

10 Þorbjǫrg's attire, which includes furry calfskin shoes with long laces decorated with tin buttons on their ends, is clearly ritualistic, and will not be discussed in the present article.

11 Another pair of (tanned) black shoes is featured in *Hænsna Þóris Saga*, 29.

12 For an accurate record of all shoe types featured in Old Norse literature, *cf.* Falk 1919, 129–139. *Cf.* also Ewing 2006, 58-59 and 123–124. For the manufacture and use of socks, hoses, leg-bindings and other *skóklæði*, which will not be discussed in this article, *cf.* Valtýr Guðmundsson 1914, 77–87.

13 *Cf.* also Önundr's spontaneous death while kneeling down to tie his shoelace in *Harðar Saga*, 62.

14 *Cf.* also Gunnarr Þiðrandabani's flight in *Fljótsdæla Saga*, 271, where his shoes are untied as are his breeches.

15 *Cf.* also the nickname of Sigurðr ullstrengr (Wool-String) in *Gísls þáttr Illugarsonar*, 336.

16 On the connection between belts and unmarried women *cf.* Ewing 2006, 46–49.

17 The episode finds its source in the *Sturlubók* version of *Landnámabók*, and has been employed by the author of *Bárðar Saga* almost word for word. *Cf. Landnámabók*, 108–109 and *Bárðar Saga*, lxxvi.

18 For those unacquainted with Icelandic literature, please note that Icelandic scholars are always referred to by first name then family name in works of reference.

Bibliography[18]

Primary Sources

Bjarni Vilhjálmsson and Þórhallur Vilmundarson, eds (1991) *Harðar saga. Bárðar saga Snæfellsáss, Flóamanna saga, Gull-Þóris saga, Harðar saga, Stjörnu-Odda Draumr*. Íslenzk fornrit 13. Reykjavík, Hið íslenska fornritafélag.

Björn Sigfússon, ed. (1940) *Ljósvetninga saga. Ljósvetninga saga, Reykdǿla saga ok Víga-Skutu*. Íslenzk fornrit 10. Reykjavík, Hið íslenska fornritafélag.

Björn K. Þórólfsson and Guðni Jónsson, eds (1943) *Vestfirðinga sögur. Fóstbrǿðra saga, Gísla saga Súrssonar, Hávarðar saga Ísfirðings*. Íslenzk fornrit 6. Reykjavík, Hið íslenska fornritafélag.

Einar Ól. Sveinsson, ed. (1934) *Laxdǿla saga. Laxdǿla saga*. Íslenzk fornrit 5. Reykjavík, Hið íslenska fornritafélag.

Einar Ól. Sveinsson, ed. (1939) *Vatnsdǿla saga. Vatnsdǿla saga*. Íslenzk fornrit 8. Reykjavík, Hið íslenska fornritafélag.

Einar Ól. Sveinsson, ed. (1954) *Brennu-Njáls saga*. Íslenzk fornrit 12. Reykjavík, Hið íslenska fornritafélag.

Einar Ól. Sveinsson and Matthías Þórðarson, eds (1935) *Eyrbyggja saga. Eiríks saga rauða, Eyrbyggja saga*. Íslensk fornrit 4. Reykjavík, Hið íslenska fornritafélag.

Guðni Jónsson, ed. (1936) *Grettis saga. Bandamanna saga, Grettis saga Ásmundarssonar*. Íslenzk fornrit 7. Reykjavík, Hið íslenska fornritafélag.

Guðni Jónsson and Sigurður Nordal, eds (1938) *Borgfirðinga sögur. Bjarnar saga Hítdǿlakappa, Gísls þáttr Illugasonar, Heiðarvíga saga, Hǿnsa-Þóris saga*. Íslenzk fornrit 3. Reykjavík, Hið íslenska fornritafélag.

Jakob Benediktsson, ed. (1968) *Landnámabók. Landnámabók*. Íslenzk fornrit 1. Reykjavík, Hið íslenska fornritafélag.

Jóhannes Halldórsson, ed. (1959) *Kjalnesinga saga. Finnboga saga, Kjalnesinga saga, Króka-Refs saga*. Íslenzk fornrit 14. Reykjavík, Hið íslenska fornritafélag.

Jón Jóhannesson, ed. (1950) *Austfirðinga sögur. Fljótsdǿla saga, Gunnars þáttr Þiðrandabana, Þorsteins þáttr stangarhǫggs*. Íslenzk fornrit 11. Reykjavík, Hið íslenska fornritafélag.

Jónas Kristjánsson, ed. (1956) *Eyfirðinga sögur. Sneglu-Halla þáttr, Svarfdǿla saga*. Íslenzk fornrit 9. Reykjavík, Hið íslenska fornritafélag.

Keiser, R., and Munch, P. A., eds (1849) *Norges gamle Love indtil 1387*. Christiania.

Minorsky, V., ed. and trans. (1937) *Hudud al-'Alam*. London.

Neckel, G., and Kuhn, H., eds (1983) *Edda: die Lieder des Codex Regius nebst verwandten Denkmälern*. 5th edn. Heidelberg.

Sigurður Nordal, ed. (1933) *Egils saga Skalla-Grímssonar*. Íslenzk fornrit 2. Reykjavík, Hið íslenska fornritafélag.

Secondary Sources

Birkebæk, F. (1975) *Norden i vikingetiden*. Copenhagen.

Cleasby, R., and Guðbrandur Vígfússon (1975) *Icelandic-English Dictionary*. Oxford, Oxford University Press.

Kristján Eldjárn (1981) The Bronze Image from Eyrarland. In H. Bekker-Nielsen, U. Dronke, Guðrún P. Helgadóttir and G. W. Weber (eds), *Speculum Norroenum. Norse Studies in Memory of Gabriel Turville-Petre*, 73–84. Odense.

Ewing, T. (2006) *Viking Clothing*. London, Tempus.

Falk, H. (1919) *Altwestnordische Kleiderkunde*. Kristiania, Oslo.

Valtýr Guðmundsson (1914) 'Úr sögu íslenskra búninga', in *Afmælisrit til dr. phil. Kr. Kålunds bókavarðar við safn Árna Magnússonar 19. ágúst 1914*, 66–87. Copenhagen.

Helgi Guðmundsson (1967) Um Kjalnesinga sögu – Nokkrar athugarnir. In S. J. Þorsteinsson (ed.), *Studia Islandica 26*, Reykjavík.

Helga Kress (1996) Bróklindi Falgeirs – *Fóstbræðra saga* og hláturmenning miðalda. In Helga Kress (ed.), *Fyrir dyrum fóstru – Greinar um konur og kynferði í íslenskum fornbókmenntum*, 45–65. Reykjavík. First published in *Skírnir* 161 (1987), 271–286.

Larsen, A. J. (1970) *Skomaterialet fra utgravningene i Borgund på Sunnmøre 1954–1962*. Årbok for Universitetet i Bergen, Humanistisk Serie 1. Bergen.

Larsen, A. J. (1992) *Footwear from the Gullskoen Area of Bryggen*. The Bryggen Papers, Main Series 4. Bergen.

McKinnell, J. (2002) Þorgerðr Hölgabrúðr and *Hynduljóð*. In W. Heizmann and R. Simek, (eds), *Mythological Women. Studies in Memory of Lotte Motz*. Studia Medievalia Septentrionalia 7, 265–290. Vienna.

Perkins, R. (2001) *Thor the Wind-Raiser and the Eyrarland Image*. Viking Society for Northern Research Series. London.

Zanchi, A. (2004) Klæðskipti í *Íslendingasögum* – Norræna fornbókmenntir undir sjónarhorni kynjafræða. Unpublished MA thesis, University of Iceland, Reykjavík.

Zanchi, A. (2007) Dress in the *Íslendingasögur* and *Íslendingaþættir*. Unpublished PhD thesis, University College London, London.

47 The Use of Horsehair in Female Headdresses of the 12th–13th Century AD Latvia

by Irita Žeiere

Alongside the traditional materials used for textiles such as wool and linen, sometimes different materials of animal or plant origin were also used. One of these materials was horsehair, the evidence for which can also be found in Latvian archaeological textile materials. Horsehair was used as the auxiliary material for threading various ornaments and creating decorative braiding, as well as for tablet-woven bands or specially woven ribbons. There are not many of these finds in Latvia, and thus far they have not been studied in detail but only mentioned in several articles (Zariņa 1960, 79–95; 2006, 139–143).

The oldest horsehair find in Latvia is used as auxiliary material for the closing of a purse dated to the 9th century AD. As material for parts of clothing, it is used only from the 12th century onwards and was still in use in the costumes of the 18th–19th centuries. Most of the horsehair finds uncovered in archaeological excavations are from the eastern part of Latvia and are dated to the 12th–13th century AD (Fig. 47.1). The samples were found only in female burials and, in all cases, horsehair was used in connection with the preparation and decoration of these headdresses.

In the 12th–13th century, horsehair in female headdresses was mainly used in three ways:

– First, the most simple and common use was to secure various ornaments such as glass beads, tin rosettes or small bronze spirals to *vainags* (fabric headdresses), arranging them in various geometrical ornaments. In some cases, horsehair was braided prior to use.

On rare occasions, horsehair and bronze spirals were used in appendages of fabric headdresses. At the back of the headdresses, the ends were joined with metal appendages, woven, or braided together with bronze ornaments. The ends of the appendage cords were always decorated – most commonly with rhomboid braiding which was often made of horsehair and covered with bronze spirals (Fig. 47.2). This type of bronze spiral braiding with horsehair as its auxiliary material is typical for the territories of the Baltic Finns.

– Second, sometimes, horsehair was used in very narrow *vainags* headdresses, which have been found together with other kinds of headdresses. Thus far, four headdresses have been discovered in which horsehair is used together with bast

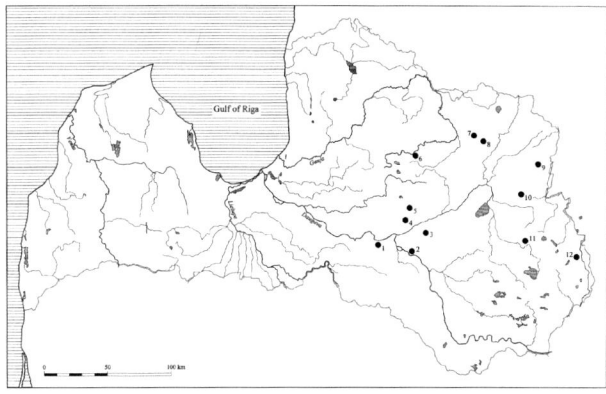

Fig. 47.1. Find places of horsehair ribbons and headdress fragments: 1. Sēlpils Lejasdopeles; 2. Krustpils Oglenieki; 3. Sāviena; 4. Kalsnavas Daktiņi;. Bērzaunes Pakalnieši; 6. Jaunpiebalga; 7. Stāmerienes Annasmuiža; 8. Litenes; Leški; 9. Šķilbēnu Daņilovka; 10. Bērzpils Bonifaceva; 11. Salenieku Makašēni; 12. Nirza.

Fig. 47.2. The end of the appendage cords braided from horsehair and bronze spirals. Annasmuiža (Photo: R. Kaniņš).

Fig. 47.3. Fragment of headdress made of horsehair, bast and wool threads from Jaunpiebalga (Photo: R. Kaniņš).

Fig. 47.4. Fragments of leather strap braiding in a horsehair headdress from Daņilovka (Photo: I. Tuņa).

Fig. 47.5. The back of a headdress with the fragments of different horsehair ribbons on each side from Pakalnieši (Photo: R. Kaniņš).

Fig. 47.6. Fragment of a horsehair ribbon made in a tapestry-like technique from Sāviena (Photo: R. Kaniņš).

and wool threads of different colours (Fig. 47.3). The width of these headdresses is only 1–1.3 cm, and it is possible that they were tied at the back. The bases of these headdresses were made of bast plaits with horsehair twisted around them. In order to join both ends together, ornaments made of leather straps were braided through the horsehair on both sides. However, only small fragments of these ornaments are extant (Fig. 47.4).

Because the exact find places have not been noted for any of these finds, it is impossible to determine whether the fragments belong to the headdress itself, or to the cord of its appendage. They are 1 cm wide and have four bronze spirals threaded on the warp. The fragments demonstrate that horsehair of different thickness and colours was used.

– Third, the commonest and most decorative finds are the separately woven ribbons made of horsehair of various colours. Fragments of such ribbons have been discovered in 10 burials. The horsehair ribbons were used separately or sewn on the ends of the fabric headdresses. The ribbons were sewn on vertically on each of the ends, and the ribbon itself was different on each of those ends (Fig. 47.5). The finds demonstrate that, often, instead of using one ribbon all around the headdress, two separate ribbons of up to 8 cm long were sewn on its ends. The end of the ribbon was folded inwards and sewn onto the fabric with a thick plied wool thread.

The 0.8–2.2 cm wide ribbons were made of horsehair of different thickness and colour in a tapestry-like technique (Fig. 47.6). Nine to 24 bundles of horsehair were used as warp; each of these bundles was approximately 0.75 mm thick and made of 4–8 separate horsehairs. The weft folded around the static warp created various patterns. Tabby weave is mostly used, although in some cases 2/2 weft-faced twill can also be seen. Depending on the size and the number of the materials used, more or less dense horsehair weft-wrapping of various colours is created. The density in ribbons is up to 24 wrappings per cm.

The horsehair was used both in its natural colour and dyed. The information about the use of horsehair in female headdresses and their dyeing can be found in the 17th

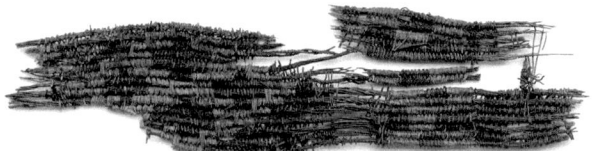

Fig. 47.7. The fragments of horsehair ribbons and their pattern: Jaunpiebalga (Photo: R. Kaniņš; Drawing: A. Alksne-Alksnīte).

Fig. 47.8. The fragments of horsehair ribbons and their pattern: Lejasdopeles (Photo: R. Kaniņš; Drawing: A. Alksne-Alksnīte).

century *Livonian Chronicles* written by the dean of Vīlande, Dionisius Fabricius. He writes: "The servants and unmarried women wear headdresses made of colourful horsehair and they know how to dye the horsehair by themselves" (Spekke 1969, 29).

The horsehair that was woven into the ribbons formed geometrical ornaments formed as crosses and rhombi in different combinations, which were repeated throughout the entire length of the ribbon separated by plain vertical areas of different width and colour (Figs 47.7–47.8). Even though the main elements of the ornaments are repeated, they are in fact never the same.

There is no doubt about the local origin of the ribbons, which have the typical headdress ornaments such as beads and bronze spirals threaded to the warp horsehair. The rest of the horsehair materials are also assumed to have been created in the territory of Latvia. However, it is possible that the idea of using horsehair for weaving comes from Scandinavia, where horsehair was commonly used in tablet-woven bands already at the end of the Migration Period (Nockert 1991, 70–71, 83–92). The horsehair finds of Latvia demonstrate that the use of this material was quite different from that in Scandinavia and had been adapted to local traditions. It maintained its form up to the 19th century.

The questions that still need to be answered are why such delicate horsehair ribbons were only used for the decoration of headdresses, and whether horsehair as a material had any special or symbolic meaning. Hopefully, this and other interesting issues concerning the use of horsehair will be further investigated in the future.

Bibliography

Nockert, M. (1991) *The Högom find and other Migration Period textiles and costumes in Scandinavia.* Umeå.

Spekke, A. (1969) *Vecākie latvju tautas apģērba zīmējumi.* Čikāga.

Zariņa, A. (1960) Latgaļu vainagi laikā no 6. līdz 13. gadsimtam. *Arheoloģija un etnogrāfija* II, Rīga.

Zariņa, A. (2006) Apģērba fragmenti Lejasdopeļu kapulaukā. *Latvijas Nacionālā vēstures muzeja raksti* Nr. 11. Rīga.

48 Two Early Medieval Caps from the Dwelling Mounds Rasquert and Leens in Groningen Province, the Netherlands

by Hanna Zimmerman

The Groninger Museum has in its collection two preserved early medieval woollen caps, woven in diamond twill. They were found in Rasquert and Leens (Fig. 48.1).

Rasquert *(GM 1928/VIII:1)*

During the archaeological excavations on the dwelling mound (*wierde*) of Rasquert in 1928, Professor A. E. van Giffen recovered a headcovering in the shape of a peaked cap (Fig. 48.2). This cap is dated to AD 800–1200. All parts are of the same fabric, woven in a very regular diamond twill of 17z/9s threads per cm (Fig. 48.3). The cap is beautifully made. The crown consists of two parts. At the front, a peak is attached. In its current state, the peak is folded inwards several times, so that it appears smaller than its actual size. The sides are somewhat bias cut. At the back of the cap is an area of wear, where the surviving weft threads are darned with thick, coarsely spun wool yarn. This repair is probably contemporary. Other repairs were carried out in the course of conservation, using a double thread of cotton machine-spun yarn in the dark grey colour of the fabric.

Along the entire bottom edge, at about 2 mm from the seam allowance, backstitching, which also continues along the edge of the peak, is visible. The 8 mm wide hem along the bottom is finished on the inside with overcast stitches, which are barely visible on the outside. The crown and the sides are joined with an unusual, decorative stitch. The seam across the crown and the back seam are executed in the same stitch. The actual join is made with simple stitches, for which a still somewhat reddish, S-twined thread was used. The decorative effect of the seam was obtained by drawing two threads through these stitches (Fig. 48.4). On the inside, an allowance about 5 mm wide is visible on all seams and was secured with overcast stitching.

Leens *(GM 1939/IV:13)*

In the dwelling mound of Leens, a headcovering was brought to light by commercial quarrying in 1939. This woollen cap (Fig. 48.5) was badly damaged but, thanks to conservation, various details have remained clearly observable. This cap is dated to AD 700–1000.

The two parts of the cap's sides are in diamond twill; one part has 12z/9s threads per cm, the other 8z/7s. The crown of the cap consists of three strips, the central one of which is a herringbone twill of 10z/10s. The strips on either side of it are in such poor condition that the type of twill can no longer be determined.

The sides consist of two parts, with seams both at the front and the back. Since only 6 cm remain of the sole surviving seam, and the other seam has completely disappeared, it is no longer possible to distinguish the front and the back. The seam allowances of the two parts of the crown are secured on the inside with 5 mm wide stitches. The crown is attached to the sides in the same way. The bottom edge of the sides is folded twice over and secured with a double thread of Z-twisted wool yarn in very coarse, oblique overcast stitches at least 1 cm apart. In the conservation of this cap too, machine-spun cotton was used.

Conclusions and Perspectives

An examination of the weaving as well as the sewing of seams and the way it is finished on the inside, of the cap GM 1928/VIII:1 excavated from the Rasquert mound indicates that it must have been meticulously made. The fact that the cap GM1939/IV:13 from Leens is put together from at least three, perhaps more weaves, may be an indication that it is put together from remnants. Furthermore, the Rasquert cap with the peak had possibly been made as protection against the sun, while the Leens cap with its sou'wester form protected the wearer against the rain. Although the caps are well conserved, further analysis has never been done.

Apart from the aforementioned caps, the Leeuwarden Museum also has two caps, and there is a partially preserved cap in the museum Het Admiraliteishuis in Dokkum. All of them are excavated from comparable mounds, and I intend to publish an article about these five caps.

48 Two Early Medieval Caps from the Dwelling Mounds Rasquert and Leens in Groningen Province

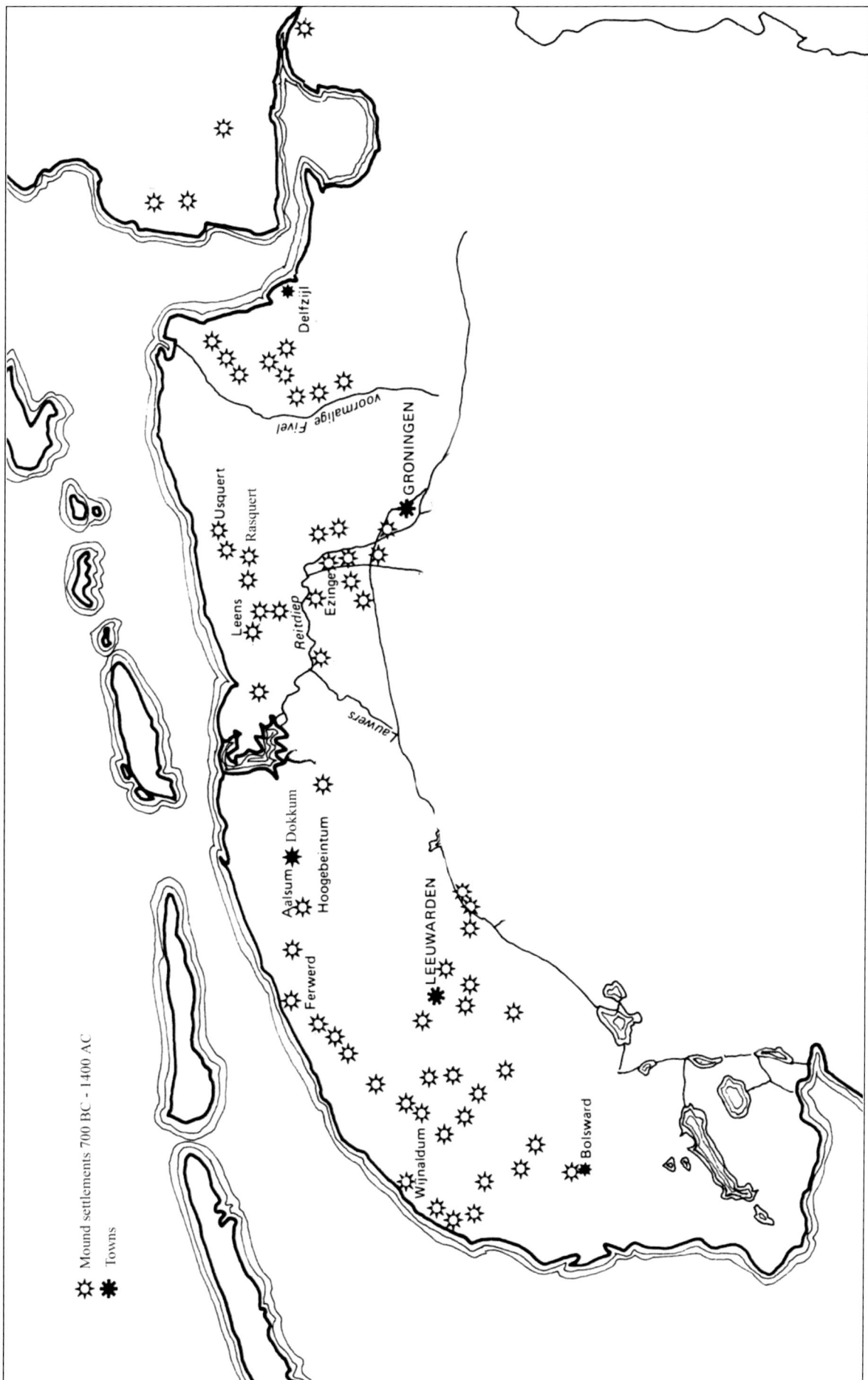

Fig. 48.1. Map of mounds in the provinces of Groningen and Friesland inhabited in AD 700–1400 (Adapted from Handwerken zonder Grenzen 1981/4, 19).

Fig. 48.2. Cap found in the mound of Rasquert, GM 1928/ VIII, 1 (AD 800–1200) (Photo: M. de Leeuw, Groninger Museum).

Fig. 48.4. Decorated seam on the Rasquert cap (drawing by H. Zimmerman).

Fig. 48.3. Diamond twill (Drawing by H. Zimmerman).

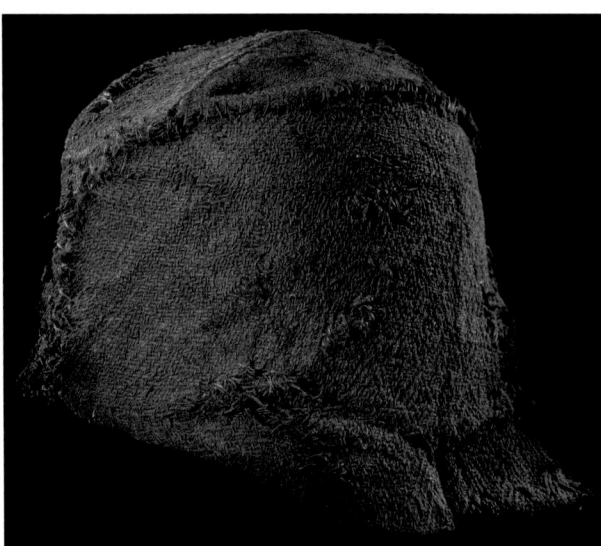

Fig. 48.5. Cap, model sou'wester, found in the mound of Leens, GM 1939/ IV: 13 (AD 700–1000) (Photo: M. de Leeuw, Groninger Museum).

49 Studies of the Textiles from the 2006 Excavation in Pskov

by Elena S. Zubkova, Olga V. Orfinskaya, and Kirill A. Mikhailov

Fig. 49.1. General view of the block of soil prior to its disassembling (Photo: © The Authors).

In the summer of 2006 in North-West Russia, during a rescue excavation in the southern section of the *Okolnyi Gorod* (Roundabout City) of Pskov, directed by Elena A. Yakovleva, a third wooden chamber burial of the Scandinavian type was uncovered (Burial no. 1 had previously been excavated by E. A. Yakovleva; Burial 2 by A.V. Mikhailov). The chamber had been looted in antiquity and contained no human remains. Nevertheless, the nature of the set of goods preserved enabled the identification of the grave as a female burial and to date it by analogy with Burial no. 1 to the mid-10th–early 11th century AD (Yakovleva 2006, 70).

A block of soil with traces of textiles recovered from under the remains of the floor of the chamber is the focus of this study (Fig. 49.1). The cleaning and disassembling of the block of soil (0.3 × 0.26 m) was carried out under laboratory conditions. This process included several phases. In order to examine the features of the textiles, microscopic methods were applied with the use of MBS-10 and POLARM R-211 microscopes; the nature of the fibres was identified using histochemistry and spectrometry, and microscopic analyses in reflected and transmitted light with a magnification of 16–400×.

The aim of the first stage of the investigation was to distinguish the outlines and dimensions of the textile remains, as well as to identify other objects present in the soil block. This study has enabled us to state firstly that the textiles were placed in a birch-bark container, possibly with a leather cover. Secondly, we were able to clean and unravel the bundle of textiles and extricate an oval bronze 'tortoise' brooch from it. Thirdly, a number of textile types were identified and divided into two main groups: fabrics from plant fibres interwoven with linen threads and silk fabrics. Fourthly, the plant fibres were identified as linen. And finally, the technological features of each group of textiles were defined.

It became evident already at the first stage of the studies that the tabby linen cloth was present in practically all layers of the investigated soil block (Table 49.1). The plant fibres of the tabby were destructurized and had lost their morphological parameters completely. These remains presented just a black layer of decomposed organics, which covered the relatively well-preserved silk cloth. Therefore, it was decided to concentrate on the silk textile. All layers of the tabby textile were removed from the silk, although some of the separate fragments that still preserved their structure were strengthened and stabilized.

The aim of the second stage of the work was the final cleaning and unfolding of the main mass of textiles of silk. The process resulted in separating two fragments from it. From the first, three elements of silk fabric were sorted out. Two of these were identified as sleeve cuffs from a garment

1	Layer of sand with pieces of wood and birch bark
2	Layer of destructurized linen mixed with sand and fragments of silk fabric
3	Layer of sand
4	Layer of silk fabric
5	Layer of silk and tabby fabric
6	Layer of tabby fabric
7	Layer of sand with pieces of tabby fabric
8	Layer of sand with pieces of wood and decomposed wood
9	Layer of wood
10	Layer of birch bark
11	Layer of sand

← Fragment 2 o/м
} Fragment 1 o/м

} (Bottom of a cylindrical birch bark container)

Table 49.1. Layers 1–7, main bulk of textiles.

Fig. 49.2. Cuff, detail no. 5 (Photo: © The Authors).

Fig. 49.3. Remains of straps and collar on the pin of fibula no. 2 (Photo: © The Authors).

Fig. 49.4. General view of detail no. 6 (Photo: © The Authors).

(Fig. 49.2). The third element separated from fragment 1 of the main part of the textiles was a scrap of narrow (4.5 cm wide) silk band. In the process of unfolding fragment 2, a second bronze oval brooch was discovered. On its pin, straps of linen and a fragment of a collar from a garment made from a similar linen textile were preserved (Fig. 49.3). Fragment 2 was a large single element (Fig. 49.4). Its total length was 1.5 m, and width was over 0.3 m.

After the complete dismantling of the block of soil, the following objects were found: the remains of the base of a birch-bark container reinforced by wood, 11 elements of clothing made of two kinds of fabric (linen and silk), and two oval brooches, which once held the items of costume together.

The various analyses conducted give us grounds to state that the main bulk of the textiles placed in the container consisted of linen tabby cloth folded in several layers. In addition, it was ascertained that all the elements made of silk were covered on the reverse side by a thin layer of degraded linen tabby, with the exception of the tucked-in edges and connecting seams. These facts, as well as traces of the sewing threads, enable us to state that all the silk parts were sewn onto a linen garment. A meticulous examination of all the elements identified led us to the conclusion that the birch bark container contained a set of female clothes, which consisted of two articles sewn from thin blue linen cloth and faced with silk. The date and type of the burial chamber, as well as the presence of two oval bronze brooches undoubtedly belonging to the Viking Age, define the search for parallels of the Pskov find among Scandinavian materials.

From the preliminary studies of the Pskov grave, we have already achieved some interesting results. After the final disassembling of the block of soil and the identification of considerable remains of textiles, of particular interest was detail no. 6 (Fig. 49.4). As mentioned above, this item was sewn together from a number of strips of silk cloth. A meticulous examination of each constituent of detail no. 6 showed that they were cut from different fabrics of varying quality. Although in terms of the weave they all belong to a type of samite, each has its own peculiarities, and thus it has proved possible to distinguish three kinds of silk (Table 49.2).

Silk of type I was used for parts Ia, Ib, IIIa, IIIb, IV, V of detail no. 6 (Fig. 49.5); from silk of type II parts IIa and IIb were cut. Cloth of type III was used as trimming in area Ib.

In the type I textile, elements of a woven pattern, which included zoomorphic and anthropomorphic motifs, as well as a floral design, were discernible. It has been possible in the drawings of the single pieces of parts I, III, IV and V in detail no. 6 to identify the fragments of a woven design representing a well-known hunting scene involving the Sassanian prince, Bahram Gur, who ruled Persia in the 5th century AD (Figs 49.6–49.7). Fabrics with a similar motif are known from finds throughout Europe (Muthesius 1997, 68–72). One example is housed in Milan, Italy (Muthesius 1997, 174, pl. 25a, cat. M 31). It was used for upholstering a part of the

Type I

Colour	golden-pink (visual evaluation) pattern against blue background. In some areas, green bands are discernible instead of blue (interchangeable wefts)
Pattern	zoomorphic and anthropomorphic representations with elements of floral design; the repetition of the cloth pattern was impossible to restore
Type of textile	samite (direction of twill – S)
Warps	two yarns of the inner warp (in some areas up to three yarns together) and one yarn of the binding warp
Yarns of the inner warp	1 order, X-twist (step 1.00 mm), thickness 0.20 – 0.30 mm
Yarns of the binding warp	1 order, Z-twist (step 0.60 – 0.80 mm), thickness 0.25 – 0.35 mm
Yarns of the weft	four: blue, vague yellow, vague pink, green
Blue weft	without twist, thickness 0.25 – 0.5 mm
Green weft	without twist, thickness 0.25 mm
Vague pink weft	without twist, thickness 0.25 mm
Vague yellow weft	without twist, thickness 0.50 – 0.60 mm
Density	16 yarns of the inner warp, 8 yarns of the binding warp and 18 – 40 yarns of the weft per 1 cm

Type II

Colour	reddish-violet (visual evaluation)
Pattern	fine treatment of details; the repeat of the cloth pattern impossible to identify
Type of textile	samite (direction of twill – S)
Warps	two yarns of the inner warp and one yarn of the binding warp; up to four yarns of warp together are distinguishable in some areas
Yarns of the inner warp	1 order, Z-twist (step – 0,75 mm), thickness – 0.10 – 0.15 mm
Yarns of the binding warp	1 order, Z-twist (step – 1.00 – 0,80 mm), thickness – 0.20 mm
Yarns of the weft	three – red-violet, yellow, green
Red-violet weft	without twist, thickness 0.35 mm
Green weft	without twist, thickness – 0.20–0.25 mm
Yellow weft	without twist, thickness – 0.50 mm
Density	40 yarns of the inner warp per 20 yarns of the binding warp and 20–45 yarns of the weft per 1 cm

Type III

Colour	unidentifiable
Pattern	unidentifiable
Type of textile	samite (direction of twill – S)
Warps	one yarn of the inner warp and one yarn of the binding warp
Yarns of the inner warp	1 order, Z-twist (step – 1.00 mm), thickness – 0.25 mm
Yarns of the binding warp	1 order, Z-twist (step – 1.00 mm), thickness 0.20 – 0.25 mm
Yarns of the weft	three (colour unidentifiable)
Weft 1	without twist, thickness 20 mm
Weft 2	without twist, thickness 0.50 mm
Weft 3	without twist, thickness 0.35 mm
Density	20 yarns of the inner warp per 20 yarns of the binding warp and 20–50 yarns of the weft per 1 cm

Table 49.2. Characteristics of silk types.

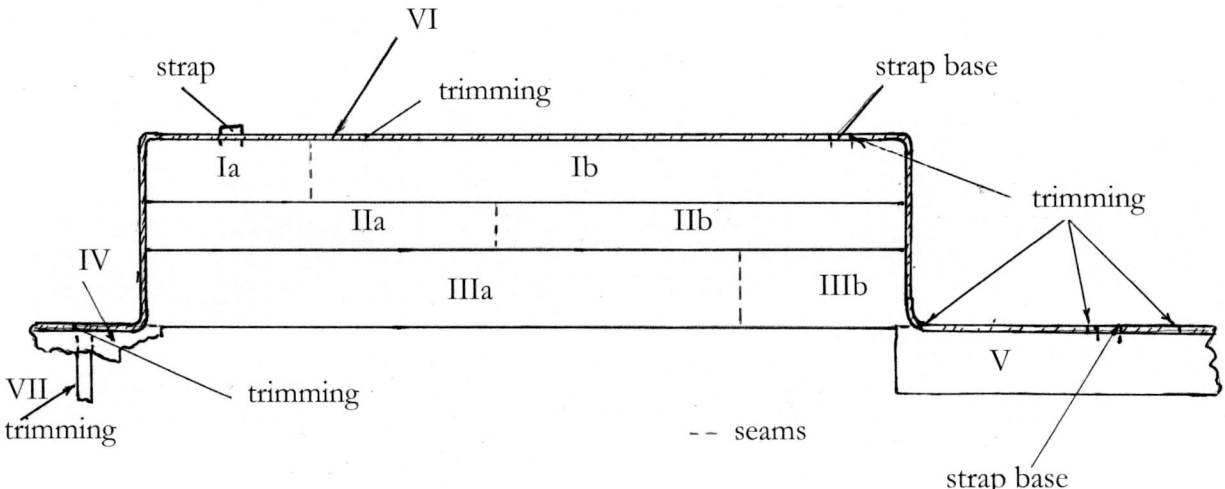

Fig. 49.5. Schematic drawing of the pattern of detail no. 6 (Drawing © The Authors).

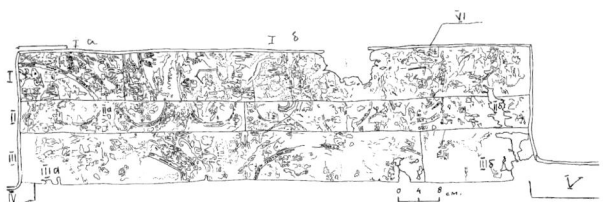

Fig. 49.6. Schematic drawing of the woven pattern of detail no. 6 (Drawing © The Authors).

Fig. 49.7. Reconstruction of the pattern of the type I textile (© The Authors).

golden throne of St. Ambrose, manufactured in *c.* AD 835. Another example of fabric with a depiction of Bahram Gur was found in the tomb of St. Kunibert in Cologne, Germany (Muthesius 1997, pl. 25b, cat. M 347). It is believed by scholars that, the cloth was placed there during the opening of the tomb in AD 1168. The third example of a woven scene of the Sassanian prince hunting was found on the binding of the large parchment Gospel from the Library of the Royal Capital of Prague, Czech Republic (Muthesius 1997, cat. M 350a, pl. 79a). The Gospel was written in the 9th century, but the fabric was used as a binding, presumably in the 15th century. The fourth example may be found at the Church of Saint-Calais, Sarthe, France (Muthesius 1997, cat. M 350c, pl. 79b). According to N. P. Toll, the manufacture of these fabrics begins in the mid-6th century AD (Toll 1924, 192). A. A. Ierusalimskaja dates their manufacture to the 1st half of the 7th–9th century AD (Ierusalimskaja 1961, 49; 1996, 239, abb 222). In Russia, silk textiles with similar designs are so far known only from two archaeological excavations of rock tombs in the cemeteries of Moshchevaja Balka and Nizhnij Archyz in the North Caucasus (Ierusalimskaja 1996, 239 abb. 222). These burial grounds were used only during the 8th–9th centuries AD. That was the period when a branch of the Great Silk Road from China to Byzantium went via the North Caucasus.

At present, it is extremely difficult to judge the manufacture of the Pskov silk textiles. In terms of their technological features, they belong to the kind of Byzantine textiles, which are copies of earlier Sassanian cloths. The design of the pattern undoubtedly depicts a hunting scene with Bahram Gur, but the execution of certain elements and details of the pattern display numerous differences from the hitherto known parallels. This is, however, a subject for further study. Here, we will limit ourselves to descriptions of a number of observations recorded during the careful investigation of detail no. 6.

As mentioned earlier, the middle band in the central part of detail no. 6 was composed of red-violet silk of type II (IIa, IIb in Fig. 49.5). Microscopic analysis of details no. 1, 2, 3, 4 and 5 (scraps of trimming bands and cuffs) showed that they were cut from a similar cloth. This discovery has allowed us to interpret the scraps of silk bands (with an identical width of 4.5 cm) as remains of the decoration of the lower hem of the underdress, especially as they are identical in colour to the trimming of the sleeves. A further fact that supports this argument is that during the examination of parts of detail no. 6 from the silk of type I (Ia, Ib, IIIa, IIIb, IV, V), it was found that the design with the depiction of Bahram Gur is repeated in them with varying sets of elements taken from the complete scene. Moreover, the orientation of the pattern is not followed in these pieces. First, it looks as if detail no. 6 was sewn together from different scraps of silk. Meanwhile, an attempt to unite all parts of detail no. 6 made from cloth of Type I has demonstrated that the fragments all fit together, supplementing each other and composing a complete repetitive design of the hunting scene. The scene is repeated twice throughout the length of the cloth (Fig. 49.8). Thus, knowing the width of each piece of cloth, it was possible to define the size of the entire woven silk textile from which they were cut. The width did not exceed 46 cm, and the length was at least 1.05 m as defined from the length of the central bands (I, II, III).

The colours of the middle (red-violet) band and of the outer ones (blue-green with a yellow tint) setting the former off, which comprise detail no. 6, as well as the differing directions of the type I fabrics sewn together, give grounds to suggest that the person who trimmed the garment, was mainly interested in the colour combinations of the various parts, and neglected the integrity and direction of the fabric designs. It is probable that the red band (coupled with the same colour of the trimming of the sleeves) decorated the hem of the underdress. Moreover, the narrow strip (VII in Fig. 49.5) sewn onto detail no. 6 (so that in all probability it was positioned on the line of the lateral seam of the blue

Fig. 49.8. Schematic drawing of the cutting of the woven piece of type I textile for detail no. 6 (Drawing © The Authors).

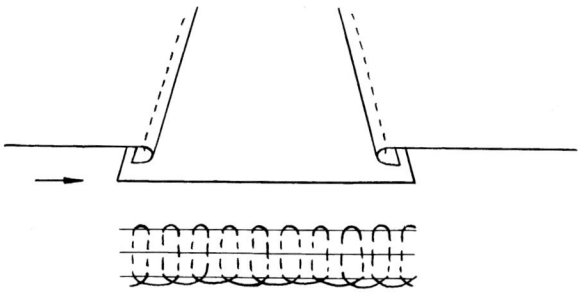

Fig. 49.10. Drawing of the seam between parts I, II and III of detail no. 6 (Drawing © The Authors).

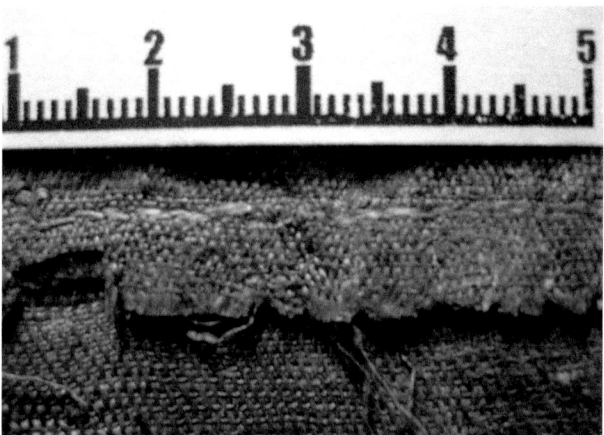

Fig. 49.9. Area of the connecting seam between parts I and II of detail no. 6: reverse side (Photo: © The Authors).

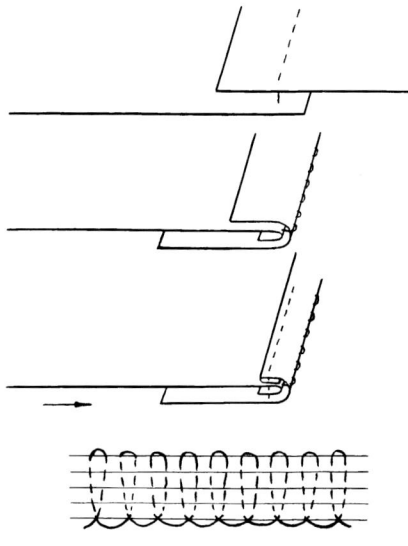

Fig. 49.11. Drawing of the hemming of silk details (Drawing © The Authors).

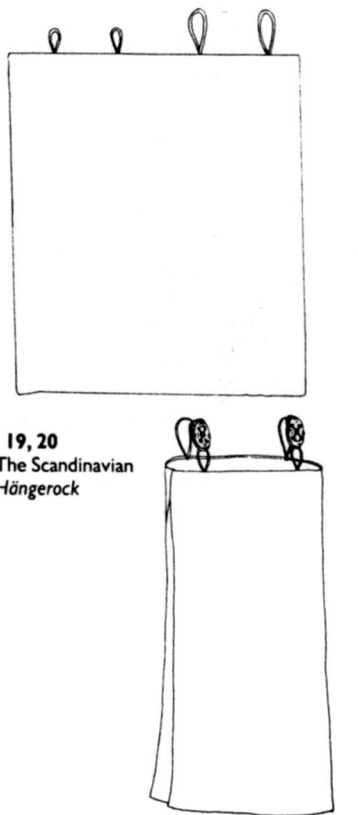

Fig. 49.12. Reconstruction of the apron according to A. Geijer (After Geijer 1938, 154, Abb. 49).

Fig. 49.13. Fragment of the collar of the underdress – detail no. 7 (Photo: © The Authors).

sarafan), was perhaps red-violet too (it was impossible to identify its colour during the investigation).

On detail no. 6, sewn together from seven parts, two types of seams are recognizable. The first is known as an open seam (Khanus 1991). This type of seam is used to join parts Ia and Ib, IIa and IIb, IIIa and IIIb, III and IV, and III and V of detail no. 6 and to sew details no. 3 and 5 (cuffs). The elements sewn to each other are folded with the right side in and stitched together at a distance of 0.5 cm from the edge. The second type of seam, is termed a patching seam (Khanus 1991) and was used to join parts I, II and III of detail no. 6. This seam both connects and is decorative at the same time. One of the edges of parts I and III was turned inside out, superimposed onto the right side of part II and sewn to it (Figs 49.9–49.10). The two types of seam were both sewn with backstitch (Khanus 1991, 21, 37, 43, 62). Stitches of that kind form very strong seams. On the reverse side, their length is two-three times greater that on the front side. In our case, it was 1 mm on the front and 2–3 mm on the reverse.

The top edge of detail no. 6, the edges of the cuffs at the wrist and one of the edges of details no. 1, 2 and 4 were trimmed. A strip of cloth about 3 cm wide was overlaid on the right side of the detail, overlapping it 1 cm from the edge, and was sewn with running stitch. The free hem of the band was then turned round the edge of the detail, folded and sewn onto the reverse side. The opposite edge of the trimming was folded and fixed on the front side of the detail (width 0.5 cm) with back stitch (Fig. 49.11).

Scandinavian Comparisons

At present, there are a number of interpretations of the Viking Age female costume, suggested by researchers such as A. Geijer (1938), M. Hald (1950), I. Hägg (1974), and F. Bau (1982). These reconstructions are based on contemporary archaeological textile finds and iconographic evidence, particularly the textiles from the large burial ground in Birka (Sweden) and burials in Denmark and Norway. Despite certain variations in the details, the female costume is in general regarded to be a set of four garments: the underdress with a long-sleeved overdress or gown over it; over the gown, a garment which looked like a pinafore or apron was worn. It was held in place by a pair of oval 'tortoise' brooches that fastened the straps sewn on the front and back of the apron. A mantle or cloak was used to complete the costume. Its edges were drawn together and fastened by means of an equal-armed or round brooch (Geijer 1938, 149–156; Hägg 1974; Bau 1982). Regrettably, neither archaeological textiles (in view of their fragmentary state), nor iconographic evidence (by virtue of its relatively subjective character) can give us a complete idea of the costume type or cut. The so-called apron especially aroused great interest among textile scholars. Its identification and reconstruction was first made by A. Geijer (1938) and later revised by I. Hägg (1974). This part of the female costume seemed to be made of a single piece of cloth folded in two. The longitudinal edges were sewn together by a short seam running from the top down, and placed at the side. On the front short straps and in the back long straps passing over the shoulders were sewn on. The straps were held

Fig. 49.14. Fragment of detail no. 8, apron strap (Photo: © The Authors).

However, as already mentioned above, this evidence does not allow us to determine the construction of these garments with certainty. In that respect, the Pskov find considerably extends the scope of the reconstruction of the costume in general, and such parts as the apron, in particular.

The Costume from Pskov

After the complete disassembling of the block of soil with the remains of textiles, 11 elements of clothing made of linen and silk textiles were identified. Their careful examination allowed us to suggest the presence of an ensemble of a female garment consisting only of two articles. These two were both sewn from thin blue linen. The first is an underdress to which details nos. 3, 5, 7 and 10 may be related. The two latter are the remains of a collar (Fig. 49.13), which was turned inside out along the edge of its neck, cut out and gathered in fine pleats. In front, there was an upright slit in the middle. The folds were fixed with a thin band, which pulled together the edges of the collar. Details no. 3 and 5 made of silk were cuffs – evidently sewn onto the long tapering sleeves of the underdress. The length of the complete cuff (Fig. 49.2) was 12.5 cm, the width near the wrist 10 cm when folded. The lower hems of the cuffs were trimmed with a narrow band of cloth; the upper ones were rolled inside out.

The second article, in Slavic areas traditionally called a *sarafan*, may be akin to the Scandinavian apron. This is connected to details no. 6 (Fig. 49.4) and no. 8 – an apron strap (Fig. 49.14). Detail no. 6 is of particular interest in terms of the reconstruction of the type of garment which it decorated. This detail, sewn from several strips of silk cloth, probably served as the trim of the top edge of the *sarafan* (as mentioned above, over the entire surface of the reverse side, a destroyed layer of blue linen and the remains of sewing threads have been observed). On the broad central area of detail no. 6, at equal distances from the centre, the base of a strap of blue linen on one side, and the traces of needle holes and the remains of sewing threads on the other, have been identified. The distance between the holes was equal to the width of the preserved fragment of the strap. On one of the narrow lateral strips of the detail described, at a distance of 20 and 25 cm from the place to which it is attached on the broad central part, the remains of threads and traces of sewn on straps have also been recognized (Fig. 49.5). The general symmetry of detail no. 6, as well as the symmetrical position of the remains of one of the straps, and the traces of a similar one on the central part, all suggest the presence of the identical straps on its second narrow lateral strip. Furthermore, since any trace of sewing on the straps is absent on the narrow strip, we may suppose that these were long straps. Such straps are also mentioned by I. Hägg (1974) and F. Bau (1982) in their descriptions of the apron. These consisted of a narrow long strip of cloth turned into several layers and folded in two in the middle. The ends of the strips (each separately but close to each other) were sewn to the top edge of the clothing from the back, thus forming a long loop.

Thus, all in all this suggests that detail no. 6 constitutes the entire top section of the *sarafan,* this being the most

together and fastened to the underlying garment by means of fibulae (Fig. 49.12).

Continuing these studies, F. Bau proposed several other suggestions on how to reconstruct the apron. These are based on the comparative analysis of the positions of the remains of the straps corroded onto the pins of brooches, and representations of female figurines (Bau 1982, 25, fig. 9):

A. An apron was a piece of cloth wrapped around the body covering the back and sides and leaving the front open. At the top corners straps were sewn on and two similar straps were placed in the middle of the top edge. The back straps then passed from the back – one over each shoulder to the front – and fastened to the corners by the brooches.

B. To the open front of *variant A,* a plastron (like a long bib) was added, fastened to the fibulae through additional straps. An apron of exactly this kind is believed to be discernible on a figurine from Tissø in Zealand, Denmark (Jørgensen 1999, 62; 2000, 4.19, p. 85, fig. 4.19).

C. Finally, on the figurines from Tuna in Alsike parish, Uppland and Sibble in Grödinge parish, Södermanland, Sweden (Holmqvist 1960, 113, fig. 23–24), a third arrangement adding a train (again fastened with straps to the same brooches) to the pinafore and bib may be observed.

There is no doubt that the costume of Scandinavian Viking Age women was composed of several articles. Their number possibly varied but the presence of an underdress and (or) overdress over which an apronlike item of clothing was worn is confirmed by several sources dated to the Viking Age.

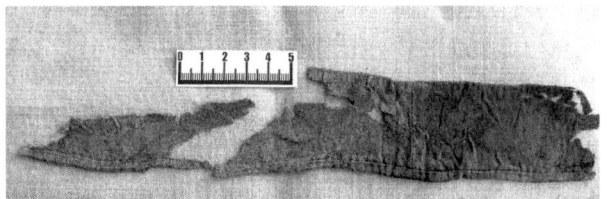

Fig. 49.15. One of the scraps of trimmings of the shirt hem– detail no. 4 (Photo: © The Authors).

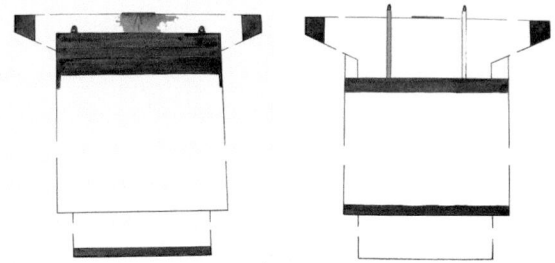

Fig. 49.16. Reconstruction of the costume (© The Authors).

informative for a reconstruction of this type of clothing. Its broad central part was sewn to the front, passing to the back by narrow strips which were evidently joined at the centre. On the chest, short straps were sewn to the top edge of the *sarafan* and the long ones were passed from the back over each shoulder (fastened probably to the middle of the back cloth). The front and back straps were fastened by means of brooches, thus fixing the *sarafan* to the underlying garment. A possible reconstruction of the second article from the Pskov burial most closely resembles the apron reconstructed by A. Geijer. The completely decayed lower part of this garment unfortunately cannot give us any idea of its length nor its design below the silk trimming. Nevertheless, a small detail in the form of a 2 cm wide strip (measured without the folds) sewn onto the narrow lateral parts of detail no. 6 (VII in Fig. 49.5) apparently served either as the trim of the cut in the lateral seam or, on the contrary, covered the lateral seam. None of these suggestions contradicts the reconstruction of the apron as proposed by A. Geijer (Geijer 1938, 153–155, abb. 49; Owen-Crocker 1986, 29, fig.19, 20). Moreover, detailed examination of the inner parts of the brooches with the traces of a pair of straps on the pins have led us to question the presence of either a pinafore or a train as proposed in F. Bau's reconstructions (Bau 1982).

The other details (nos. 1, 2 and 4), which are fragments of strips of various lengths (Fig. 49.15), cannot be reliably connected with either of the two above described costumes. The 4.5 cm wide strips turned inside out on one side and trimmed by an additional detail on the other edge probably were sewn onto the hem of either the underdress or the *sarafan*. On the reverse side of details no. 1, 2 and 4, similar to the cuffs and detail no. 6, a very thin layer of almost completely degraded linen and the remains of sewing threads were recognizable. The general view of the costume from the birch-bark container is shown in Figure 49.16. This is a reconstruction based on the evidence and the results of various analyses.

Conclusions

The unique character of the Pskov find not only allows us to expand our knowledge of the female Viking costume, but also yields rich evidence for understanding the techniques of needlework, principles of tailoring and finally the aesthetic tastes of medieval women. Hopefully, the further examination of the material presented here will shed more light on these issues.

Finds of clothing remains are not uncommon during archaeological excavations in Russia. However, these are mostly small fragments in which the seams are rarely preserved. From this perspective, the Pskov find promises significant information as to the character of the seams and type of the stitches. Summarizing the results of the studies described above, it can be concluded that the Pskov find is of undoubted interest for researchers of ancient textiles, the technology of sewing and the history of costume.

Bibliography

Bau, F. (1982) Seler og slæb i vikingetid. Birkas kvindedragt i nyt lys. *Kuml 1981. Årbog for Jysk Arkæologisk Selskab.* Nordisk Forlag, København.

Geijer, A. (1938) *Birka III. Die Textilfunde aus den Gräbern.* Uppsala, Almkvist and Wiksells B.A., Kungl. Vitterhets Antikvitets Akadamien.

Hald, M. (1950) *Olddanske Tekstiler.* Nordiske Fortidsminder 5. København.

Holmqvist, W. (1960) The Dancing Gods. *Acta Archaeologica* 31, 101–127.

Hägg, I. (1974) *Kvinnodräkten i Birka.* Uppsala.

Ierusalimskaja A. A. (1961) Tkan's Bakhramom Gurom iz mogil'nika Moshchevaya Balka // *Kul'tura i iskusstvo narodov Vostoka.* Trudy Gosudarstvennogo Ermitazha Vol. V. Leningrad.

Ierusalimskaja A. A. (1996) *Die Gräber der Moščevaja Balka: frühmittelalterliche Funde an der nordkaukasischen Seidenstrasse.* München.

Jørgensen L. (1999) *Der Herrenhof am Tissø.* Archäologie in Deutschland 2.

Jørgensen, L. (2000) Political Organization and social life. In W. W. Fitzhugh and E. I. Ward (eds), *Vikings: The North Atlantic Saga.* Washington and London.

Khanus S. (1991) *Kak shit'?* (Translated from the original Polish). Moscow.

Muthesius, A. (1997) *Byzantine Silk Weaving AD 400 to AD 1200.* Wien, Verlag Fassbaender.

Owen-Crocker, Gale R. (1986) *Dress in Anglo-Saxon England.* Wolfeboro, NH, Manchester University Press.

Toll, N.P. (1924) *Sasanidskie tkani s izobrazheniem Bakhrama Gura.* "Seminarium". Part 3. Praga.

Yakovleva, E.A. (2006) Kamernoe pogrebenie X veka iz Starovoznesenskogo raskopa. In *Arkheologiya i istoriya Pskova i Pskovskoy zemli. Materialy nauchnogo seminara 2005 g.* Pskov.

North European Symposium for Archaeological Textiles X

Colour Plates

Plate 1: Figure 2.3 (left)

Plate 2: Figure 2.4 (right)

Plate 3: Figure 2.5 (below)

Mitterberg, Fn 4

Hallstatt-Textil 124
(Inv.Nr. 89833)

Hallstatt-Textil 179
(Inv.Nr. 90180)

Hallstatt-Textil 134
(Inv.Nr. 89842)

Hallstatt-Textil 123
(Inv.Nr. 89832)

Hallstatt-Textil 20 (Inv.Nr. 73345)

Hallstatt-Textil 58 (Inv.Nr. 75904)

Plate 4: Figure 3.6

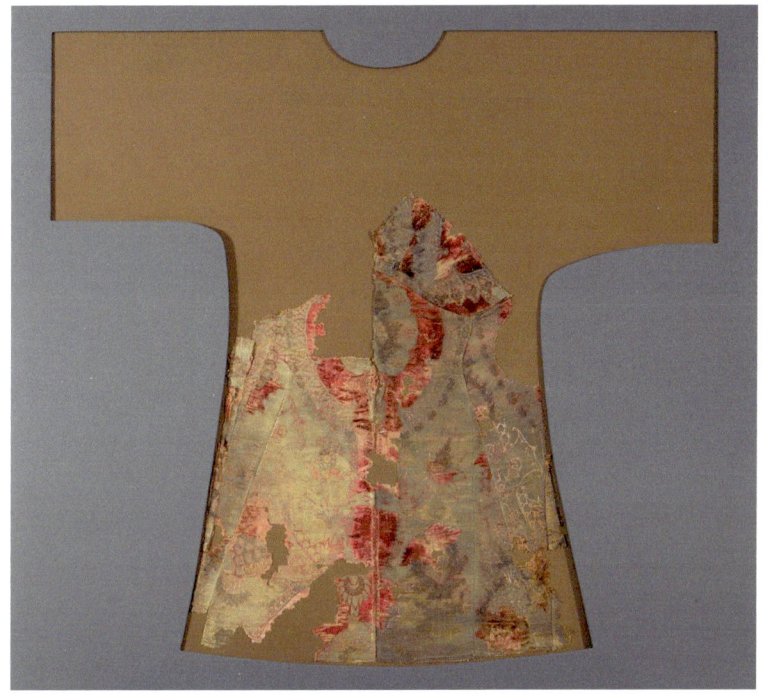

Plate 5: Figure 6.6

Plate 6: Figure 6.7

Plate 7: Figure 7.3

Plate 8: Figure 7.6

Colour Plates 303

Plate 9: Figure 16.3

Plate 10: Figure 20.9

304   Colour Plates

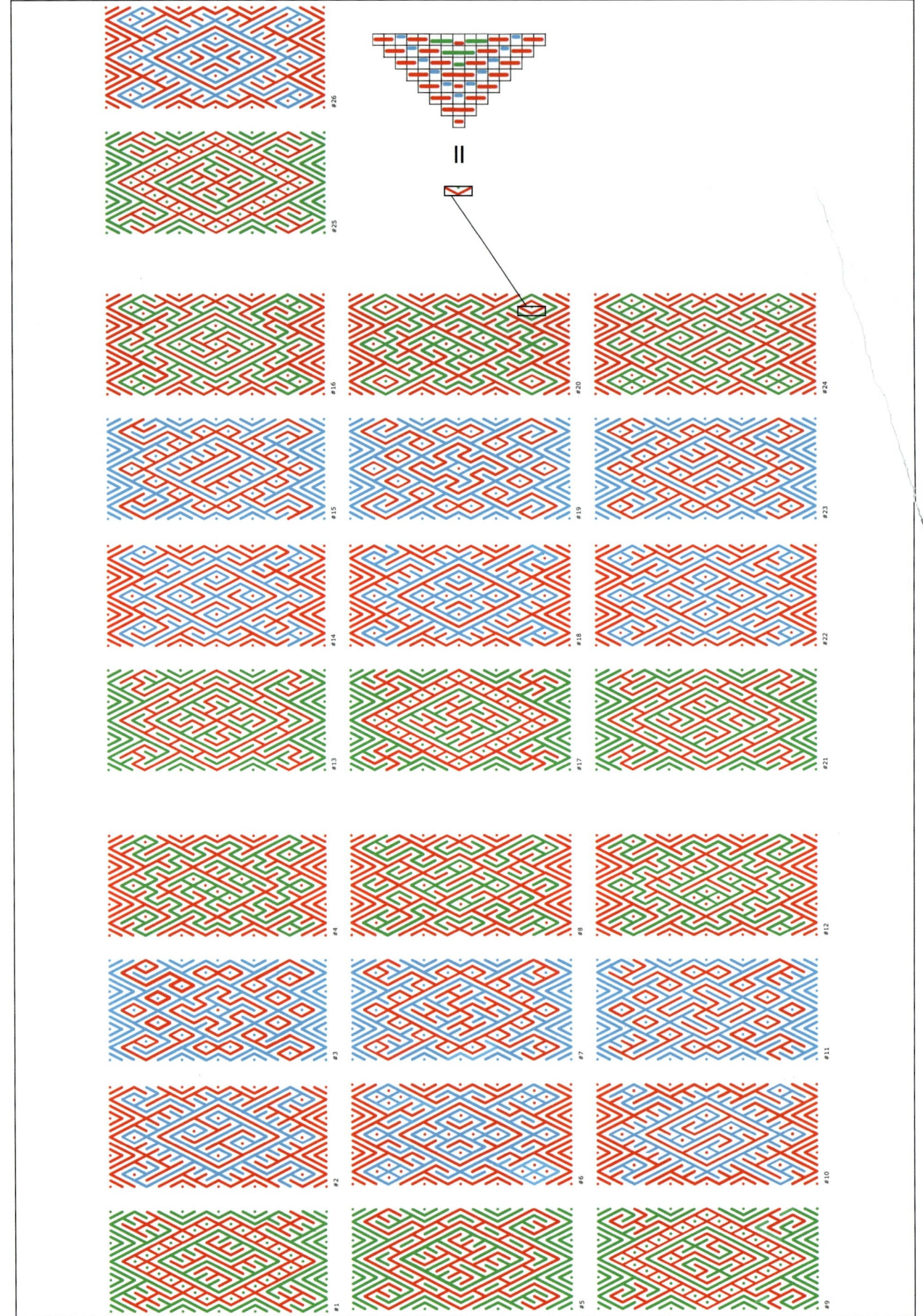

Plate 11: Chapter 20 Appendix b

Plate 12: Figure 36.3

Plate 13: Figure 36.4

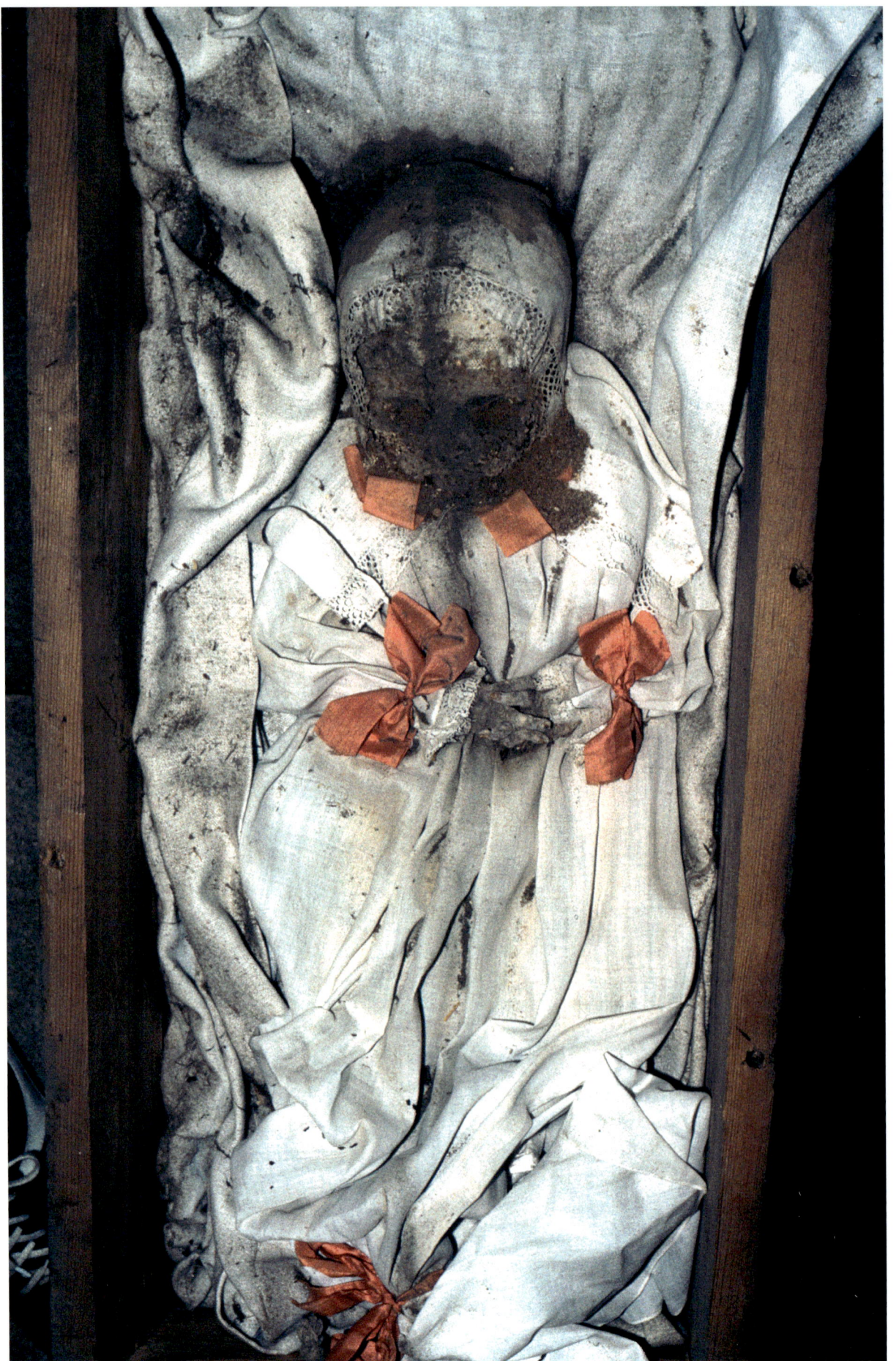

Plate 14: Figure 43.3